Absorption of Ionizing Radiation

Absorption of Ionizing Radiation

David W. Anderson, Ph.D.

written while
Professor of Radiological Sciences (Radiation Physics)
University of Oklahoma Health Sciences Center
and
Professor of Physics
University of Oklahoma

Illustrations
by
Carolyn Martel, M.S.

University Park Press • Baltimore

University Park Press
International Publishers in Medicine and Human Services
300 North Charles Street
Baltimore, Maryland 21201

Sponsoring editor: Ruby Richardson
Production editor: Michael Treadway
Cover and text design by: Caliber Design Planning, Inc.
Typeset by Kingsport Press
Manufactured in the United States of America by Halliday Lithograph

Cover illustration is adapted from Figure 7.12.

Library of Congress Cataloging in Publication Data
Anderson, David W.
Absorption of ionizing radiation.

Includes bibliographies and index.
1. Ionizing radiation. I. Title.
QC795.3.A5 1983 539.7′2 83–16727
ISBN 0–8391–1821–X

Contents

Preface

The purpose of this book is to provide an introduction to the basic interactions of ionizing radiations in matter and to some of the more immediate physical effects of the energy absorbed. The material was assembled from notes for lectures given to radiological physics students at the University of Oklahoma Health Sciences Center. The mathematical level of the presentation is appropriate for upper-level undergraduate students and graduate students. The scope of the material is sufficiently broad to be of interest in total or in part to students studying in the disciplines of medical physics, radiation biophysics, radiation physics, health physics, and nuclear engineering. A good background in calculus and in atomic or modern physics is assumed and familiar concepts included in most texts on these subject areas are not explicitly introduced. Familiarity with topics in advanced electricity and magnetism and advanced quantum mechanics is not necessary.

Physicists contribute uniquely to advances in radiation science through their ability to apply mathematical techniques to the solution of concrete problems. Since this ability must be developed, the number of mathematical descriptions used in the text is substantial. However, the differential equations included are elementary, and the developments are usually classical or semiclassical rather than quantum mechanical. Approximations in mathematical developments are numerous of necessity, since the overall development is intended to reinforce conceptual foundations rather than to obtain results with complete generality. Review articles and publications at more advanced levels are often referenced.

Topics covered in this work have been included in a variety of books that are already available. Although many of these source works are excellent, no single book was found with the particular scope that was desired. Many texts available on the subject of radiological physics have been written for radiology or radiation therapy residents, at least in part. In this case, basic formulae are often given without development and with minimal use of differential equations and statistical principles. Other texts that deal with dosimetric concepts may include a high level of mathematical physics. Many are excellent reference works for professionals, but often they are not written with students in mind. Furthermore, SI units have not been included in many of the books in print and introduction of these units is important in a basic text.

The scope of the included subjects is broad. Interactions of photons, neutrons, and charged particles are all discussed in some detail. Energy transfer is central to our description of all ionizing radiations, and some of the descriptive and mathematical techniques can be applied to several different types of radiations. We intend to include all ionizing radiations of substantial practical significance in these discussions.

After the introduction to ionizing radiation and definitions of some basic terms are given in Chapter 1, the subject headings can be divided into three basic categories. Chapters 2 through 4 deal with charged particle interactions and energy absorption patterns from charged particle beams. Chapters 5 through 9 deal with uncharged radiations, specifically photons and neutrons, their interactions, and relevant energy absorption patterns.

Chapters 11 through 13 describe excitation, ionization, and de-excitation processes for gaseous, liquid, and solid crystalline absorbers. In Chapter 14, the effects of radiation on macromolecules present in biological systems are introduced. Subjects related to the absorption of ionizing radiation, which would usually be studied in some detail in a complete curriculum built around ionizing radiation, include measurement and detection methods, radiation chemistry, and radiation biology. Chapters 10 through 14 are intended to serve as connecting links between the primary physical interactions and detailed studies in each of these related areas.

I wish to especially acknowledge the work of Carolyn Martel, M.S., for the clear illustrations that emerged from my rough sketches. I am indebted to Glenda Sims and Phyllis Joseph for expert typing of drafts and help with grammar and punctuation. Much of my understanding of the part physics has to play in an interdisciplinary understanding of the effects of radiation on biological systems came during professional association with Gail D. Adams, Ph.D. I am very grateful for these contributions and for encouragement from many other sources.

David W. Anderson, Ph.D.
Director of Radiological Physics
City of Faith Medical and Research Center
Tulsa, Oklahoma

Absorption of Ionizing Radiation

1

Fundamental Radiation Concepts

1.1 Introduction

The major purpose of this first chapter is to review fundamental definitions and discuss basic concepts that will be used extensively in succeeding chapters. The material presented here should help bring readers with diverse backgrounds to a more common level of understanding before they are thrust into the more substantial concepts. Furthermore, it is generally helpful to introduce some symbols used for physical parameters in early chapters so the more detailed equations presented later are not populated with entirely unfamiliar symbols.

We define radiation and attempt to delimit ionizing radiation in the next section. In Section 1.3 we define the fluence, exposure, kerma, and dose. In Section 1.4 we identify radiation quality and discuss energy distributions.

A secondary purpose of this chapter is to set the perspective for the succeeding material. The presentations given will certainly be from the point of view of the physicist, but they are intended to be set in the context of biological effects and medical applications of ionizing radiation. Examples and problems often will be taken from this context. Section 1.5, on radiation action on living systems, is included to solidify this point of view in the early going.

1.2 Directly and Indirectly Ionizing Radiation

Radiations are electromagnetic waves or atomic and subatomic particles that are agents of energy transfer. When radiation is absorbed, energy is transferred from a radiation beam to a medium. The energy transfer processes and the resulting alterations in the medium are both of interest. Probabilities for total absorption, scatter, and transmission of the energy are important. Usually, consideration of these topics leads to delimitation into several types of radiations. Ionizing and nonionizing radiations are two major categories.

Ionizing radiations are capable of producing a separation of charge in matter by transfer of sufficient energy to overcome the electron-binding energy in the atom or molecule. The separation occurs directly or through the production of secondary radiations. Thus, the known group of ionizing radiations can be subdivided further into two types:

1. Directly ionizing radiations are charged particles, such as electrons, protons, alpha particles, and charged mesons, which ionize by means of the particle-to-particle coulomb force.
2. Indirectly ionizing radiations are uncharged species, such as electromagnetic quanta (photons), neutrons, and uncharged mesons, which undergo interactions releasing energetic charged particles.

Photons and neutrons release electrons and protons, respectively, in aqueous media. Of course, each photon or neutron can produce an ionization event directly, during the initial interaction. Nevertheless, most of the ionization is accomplished subsequent to that event by the charged particle that is released. Thus, the name "indirectly ionizing radiation" is derived.

Ionizing radiations differ from nonionizing radiations because they have an increased amount of energy available for a single transfer event. Charge separation in a molecular system often requires more than 10 eV of transferred energy, although the lower limit depends on the particular medium. For particles with mass, the kinetic energy must exceed this lower limit. For photons the quantum energy must exceed the lower limit for ionization.

The energy cutoff between ionizing and nonionizing radiations is not precise, since low-energy radiations may be capable of producing ions in one medium and not in another. For some evidence of this fact, refer to Table 1.1, where ionization potentials (1) for several molecules are given. The reader will recall that the ionization potential is numerically equal to the minimum energy in electron volts required to remove an electron from the atom or molecule. Note that ionization in H_2O requires at least 12.6 eV of energy, whereas the energy required for N_2 is 15.6 eV and for C_6H_6 is 9.3 eV. Thus, radiation that can ionize in benzene might not be able to ionize in nitrogen.

Table 1.1 Average Experimental Values of Ionization Potentials

Molecule	*Ionization Potential*[a]	*Molecule*	*Ionization Potential*[a]
H_2	15.4	CO_2	13.8
N_2	15.6	CH_4	13.0
O_2	12.1	C_2H_4	10.5
H_2O	12.6	C_6H_6	9.3

Data from Ref. 1.
[a] Numerically equal to minimum energy to ionize (in eV).

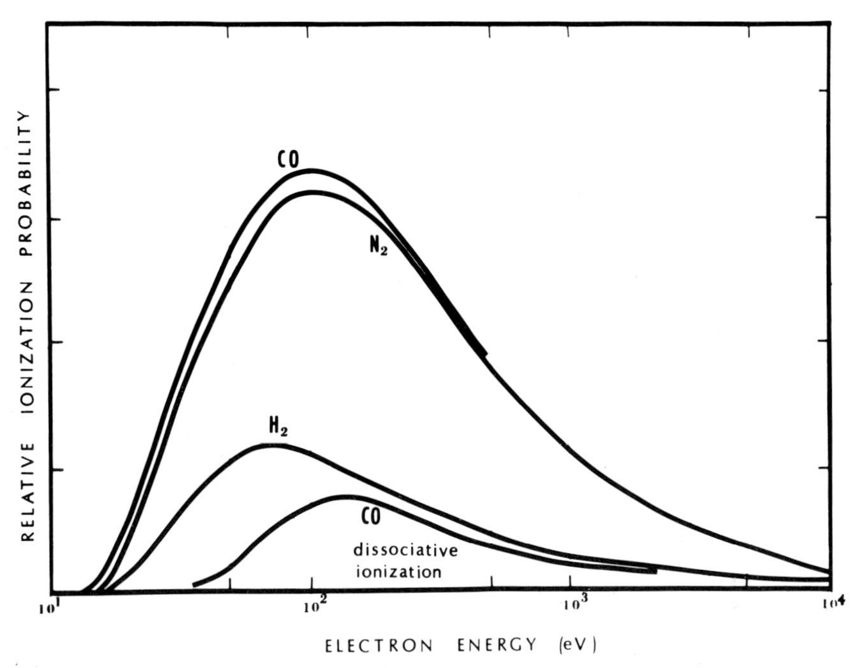

Figure 1.1 The ionization probability for electrons increases from zero at the ionization potential to a maximum at about 100 eV for common gas targets (2). Dissociation is often concurrent with ionization.

A large ionization probability is not necessarily indicated whenever the available energy exceeds the ionization potential. Consider an example with electrons. As Figure 1.1 shows, the ionization probability is very small for electrons with kinetic energy slightly above the ionization potential of the gases considered (2). Ionization events occur frequently only if the electron energy is substantially greater than the ionization potential. For the gases shown, the ionization probability increases with energy until around 100 eV kinetic energy.

A process with energy transfer involving only bound absorber states is called an excitation. Excitations occur for both ionizing and nonionizing radiations. When electron removal does not occur but the configuration or state of motion of the system is changed because of the additional energy, the molecule is said to be excited. Excitation and ionization can occur in the same system during the same event if sufficient energy can be transferred.

For our purposes it is useful to separate ionizing radiations from nonionizing radiations on the basis of their effects on living systems. Chemical bonding energies for the molecular constituents of living systems are commonly between 1 and 8 eV (3), as is shown in Table 1.2. Since energy requirements for ionization are generally larger than this, bond cleavage and dissociation often accompany the absorption of ionizing radiation in molecules in living systems. Thus, ionizing radiations can be the initiators of chemical changes through the production

Table 1.2 Bond Energies for Dissociation

Bond	*Energy (eV)*	*Bond*	*Energy (eV)*
C—C	3.5	O—O	1.5
C=C	6.3	O—H	4.7
C≡C	8.6	H—H	4.5
C—H	4.3	N—N	1.6
C—Cl	3.4	N—H	4.0
C—O	3.4	H—Cl	4.4
C=O	7.5	Cl—Cl	2.5

Data from Ref. 3.

of molecular fragments, especially those with unpaired electrons, called free radicals (4).

Nonionizing radiations, especially optical radiation and microwaves, can cause damage because of their thermal effects (5). Generally, the minimum amount of energy required for possible lethal damage to a living cell is substantially more for a thermal process than for a dissociative free radical process (6). It follows that living systems are more sensitive (per unit energy absorbed) to the ionizing radiations than to thermally active nonionizing radiations. In some measure this sensitivity difference can be attributed to the relatively uniform distribution of thermal energy, whereas energy deposited by ionizing radiation may be distributed very nonuniformly and localized for maximum effect.

1.3 Fundamental Quantities

In the following discussion, suppose that radiation from a generator of some kind or a radioactive source passes through a region in space. To fully specify the radiation, it is necessary to utilize algebraic concepts relating to quantity and quality that can be evaluated in that region. The primary goal of the succeeding paragraphs is to introduce concepts that can be used to specify radiation quantity. Quality will be discussed in Section 1.4.

The quantities defined below are macroscopic since they are valid for regions containing uncountable numbers of molecules and in principle involve averages over large numbers of individual processes. Because of this property, statistical fluctuations in the quantities are ignored, and the analytical representations are usually considered continuous and differentiable in space and time coordinates (7). In Chapter 14 microscopic quantities and fluctuation levels will be considered.

The definitions given below generally follow the recommendations of the International Commission on Radiation Units and Measurements [ICRU (7)]. The International System of Units (SI) will be utilized (8, 9), with length, mass, and time specified by meters (m), kilograms (kg), and seconds (s), respectively. Charge is given in coulombs (C) and energy in joules (J). SI units will be used in all algebraic relationships given in the following chapters, although graphs and tables may show other units if they are more convenient. A resumé of the SI units is given in Appendix 2, in the back of the book. Some traditional radiological quantities in common usage, which are not included in this system, will nonetheless be defined and used in discussion.

It is necessary to answer the question "How much?" to specify the quantity of radiation. How much radiation is emitted from an x-ray generator? How much radiation is incident on a detector? Since numerical values are often needed, carefully defined units are necessary.

In the following definitions, the term "particles" is meant to include photons as well as entities with mass. Furthermore, the definitions of fluence below are given in terms of total numbers entering a geometrical volume rather than differences between numbers entering and numbers leaving.

The particle fluence Φ is the quotient of dN by da, where dN is the number of particles that enter a sphere with circular cross-sectional area da:

$$\Phi = \frac{dN}{da} \tag{1.1}$$

The energy fluence Ψ is the quotient of dE by da, where dE is the sum of the energies (exclusive of rest mass energies) of all particles entering a sphere with circular cross-sectional area da:

$$\Psi = \frac{dE}{da} \tag{1.2}$$

Notice that for these quantities multidirectional incidence of the radiation on a spherical surface is considered. Unidirectional beams can easily be treated as a special case of multidirectional incidence. The situation is illustrated in Figure 1.2. The spherical surface with center at point P is defined as the locus of points swept out by the circumference of the circular area da when rotated about an axis in the plane of the circle that passes through the center. The

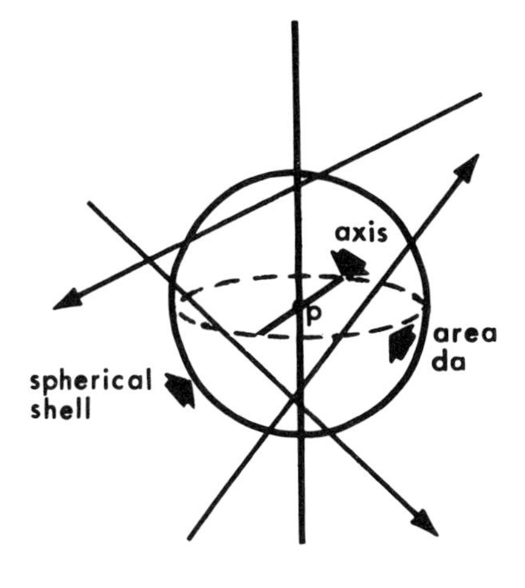

Figure 1.2 The radiation fluence is defined for multidirectional incidence through a spherical shell formed by rotation of area *da* about a bisecting axis. Adapted from W. C. Roesch and F. H. Attix, "Basic Concepts of Dosimetry," in *Radiation Dosimetry*, Vol. I, edited by F. H. Attix and W. C. Roesch (Academic Press, New York, 1968).

arrows on the figure indicate that many directions of travel are possible for the radiation.

Because of the historical significance of the quantity and its common usage, exposure will be defined next. However, it must be noted that exposure is used for photons only and must be measured in air. Because of this, it is less fundamental than the concept of fluence.

The exposure X is the quotient of dQ by dm, where dQ is the sum of the electrical charges of one sign on all the ions produced in air when all the electrons liberated by photons in a volume of air of mass dm are completely stopped in air:

$$X = \frac{dQ}{dm} \tag{1.3}$$

Exposure in SI units is given in coulombs per kilogram. The roentgen (R) is the special unit of exposure and is defined such that

$$1\ \mathrm{R} = (2.58)10^{-4}\ \mathrm{C/kg} \tag{1.4}$$

Several comments on the definition of exposure are worthwhile since application of the concept is not as simple as might first be assumed. The phrase "when all the electrons liberated . . . are completely stopped in air" is particularly important. This means that if exposure is to be measured directly, the lateral dimensions of the ion collection chamber must exceed the photon beam dimensions at least by the maximum length of the path of any liberated electron. Furthermore, the air thickness where photons enter the active volume of the chamber must be equivalent to this path length to ensure that ionization produced by electrons escaping from the exit side of the volume is balanced by the ionization produced by electrons entering at the front.

For high-energy photon beams, the path lengths of the liberated electrons become so large in air that huge air ionization chambers would be required for direct measurement. Table 1.3 shows that the maximum electron range is several meters when the photon energy approaches 1 MeV (10). When the ions produced in the chamber are collected at plates, the charge collection efficiencies for such large chambers are substantially less than one, unless very large plate voltages are used.

Of course, ion chambers of convenient size can be applied when plastic walls are employed (6). Unfortunately, attenuation of the photon beam in the

Table 1.3 Path Length of Electrons Produced by Photons

Photon Energy (MeV)	*Maximum Compton Electron Energy (MeV)*	*Maximum Path Length in Air (m)*
0.3	0.16	0.3
1.0	0.80	3.0
3.0	2.76	12.2
10.0	9.75	40.9

Data from Ref. 10.

plastic entrance wall may introduce significant errors in the response of chambers designed for equilibrium at very high energies. Because of the problems outlined, the concept of exposure is rarely applied when photon energies are greater than $h\nu = 3$ MeV.

The three quantities, particle fluence, energy fluence, and exposure, are sufficient for our purposes to specify the incident quantity of radiation. The time rate of change of each of these quantities also may be used. Thus, the exposure rate is often specified in roentgens per second or roentgens per minute. The particle fluence rate, often called the particle flux density, and the energy fluence rate, called the intensity, also may be used.

In a change of emphasis, energy absorbed from a beam of radiation by a material medium will now be discussed. The distinction between incident energy fluence (or exposure) and energy absorbed in a medium is important because radiation effects depend on net quantities absorbed rather than on total amounts of radiation incident on a surface. For example, in describing the characteristics of the output of a radiation generator, the emitted fluence rate, the exposure rate, or quantities derived from these are appropriate. But in a discussion of the effects of the radiation on an object that intercepts the beam, the amount of energy absorbed per unit volume is more fundamental than the fluence or exposure.

The energy absorbed per unit volume is the physical variable most widely used to correlate the amount of radiation with the effects of radiation. This does not mean that all other variables can be disregarded. The energy deposited per unit length along the track of the radiation is important. Biological damage is more substantial when the deposited energy is concentrated in a small region instead of being dispersed uniformly. The duration of time involved in delivery of the energy (11) is important for living systems, which are capable of repairing radiation damage. Other variables such as the overall condition of a living system and the size and position of the irradiated region must also be considered.

Figure 1.3 can be used in the derivation of the expression for absorbed energy density. Notice the element of mass m located in a large volume V of uniform composition. The element is small and the radiation field can be considered homogeneous within it. Particles A and B (as well as many others) are assumed to pass through this volume and deposit energy.

Consider that:

E_D is the energy imparted to the volume;
ΣE_{in} is the sum of energies (excluding rest energy) of all ionizing radiations that have entered the volume;
ΣE_{out} is the sum of energies (excluding rest energy) of all ionizing radiations that have left the volume;
ΣQ is the sum of all the net energy released by mass conversion in nuclear and elementary-particle reactions in the volume.

Then,

$$E_D = \Sigma E_{in} + \Sigma Q - \Sigma E_{out} \tag{1.5}$$

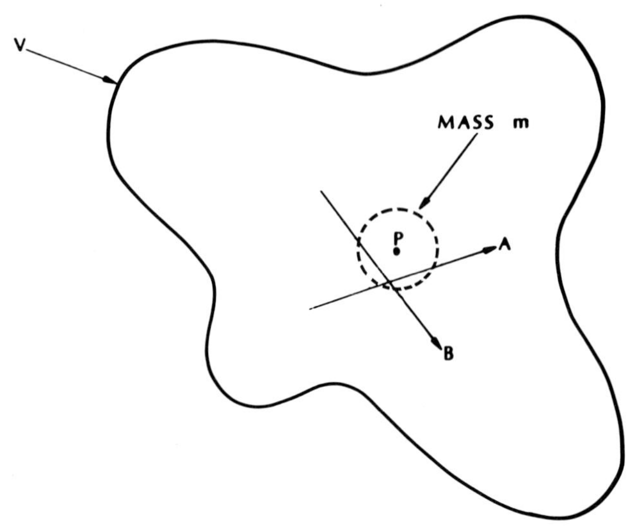

Figure 1.3 Energy deposited in the element *m* by particles *A* and *B* contributes to the dose. Adapted from W. C. Roesch and F. H. Attix, "Basic Concepts of Dosimetry," in *Radiation Dosimetry,* Vol. I, edited by F. H. Attix and W. C. Roesch (Academic Press, New York, 1968).

Suppose m and E_D shrink to the infinitesimals dm and dE_D. The absorbed dose D is the quotient of dE_D by dm, where dE_D is the mean energy imparted by ionizing radiation to a volume element, and dm is the mass of that volume element:

$$D = dE_D/dm \tag{1.6}$$

The SI unit of dose is the gray (Gy); another unit used for absorbed dose is the rad (r). The definitions are

$$1\ \text{Gy} = 1\ \text{J/kg}, \quad 1\ \text{rad} = (1.0)10^{-2}\ \text{J/kg} \tag{1.7}$$

Thus, one rad is equivalent to one centigray.

For indirectly ionizing radiation, another quantity called the kerma is often of interest. Suppose that indirectly ionizing radiations produce some number of charged particles in the volume element with mass dm. The kerma K is the quotient of dE_K by dm, where dE_K is the sum of the initial kinetic energies of all the charged particles liberated by indirectly ionizing radiations in a volume element, and dm is the mass of that volume element:

$$K = \frac{dE_K}{dm} \tag{1.8}$$

The kerma is useful in mathematical calculations because ionization events that follow the primary interactions need not be considered. It is more easily calculated than dose in many situations because spatial distributions of charged secondary particles can be ignored. It is virtually equal to the dose if liberated charged particles have path lengths that are very small and if emission of electromagnetic energy can be neglected.

1.4 Radiation Quality

In the succeeding paragraphs, the term "quality" will be used in reference to charged-particle beams and neutron beams as well as photon beams. This usage is consistent with a comprehensive viewpoint of radiation science. For our purposes the radiation quality is specified when the type of radiation and the energy characteristics are specified.

Radiation quality is significant because it relates most directly to the fractional penetration of radiation through an absorbing object. Thus, the transmission through a barrier of lead put around a radioactive source depends on the energy of the gamma rays emitted. The fraction of the incident x rays that penetrate to a deep human tumor during radiation therapy depends on the energies of the photons involved.

If the particles (quanta) within a beam have a single energy value, the quality is completely specified by identifying the type of particles and their energy; such radiations are termed monoenergetic. If optical electromagnetic radiation is involved, the word "monochromatic" is often used. Strictly speaking, truly monoenergetic beams do not exist. The uncertainty principle indicates that there is always a finite uncertainty in the energy and that this uncertainty is inversely related to the emission time for the radiation source. Within the resolution limits of common detection systems, the gamma-ray fluence from a radioactive source and the characteristic x-ray fluence from a target can be used as though they were composed of discrete monoenergetic photon components.

If the radiation beam contains many discrete energies, the average energy can be computed as an aid in specifying the quality. If a continuous distribution of energies is involved, an average or an effective energy can be obtained. In any case, the thickness of an absorber that reduces the beam fluence rate to 50% of its incident value is called the half-value thickness (HVT). In some circumstances the HVT is considered sufficient to specify quality. When needed, the effective energy can be estimated from the HVT. These ideas are discussed further in Chapter 5.

Complete specification of radiation quality for complex radiation beams requires a specification of the differential distribution of the particle fluence as a function of energy. A plot of a differential fluence distribution versus energy often is called a spectrum. An example of a spectrum is shown in Figure 1.4 for an x-ray beam (12). The notation $\phi(E)$, or $d\Phi(E)/dE$, indicates the differential distribution in particle fluence as a function of available energy E. This quantity is defined so that the particle fluence with energy between E and $E + dE$ is given by $\phi(E)dE$. In many instances, the lowercase letter is used to

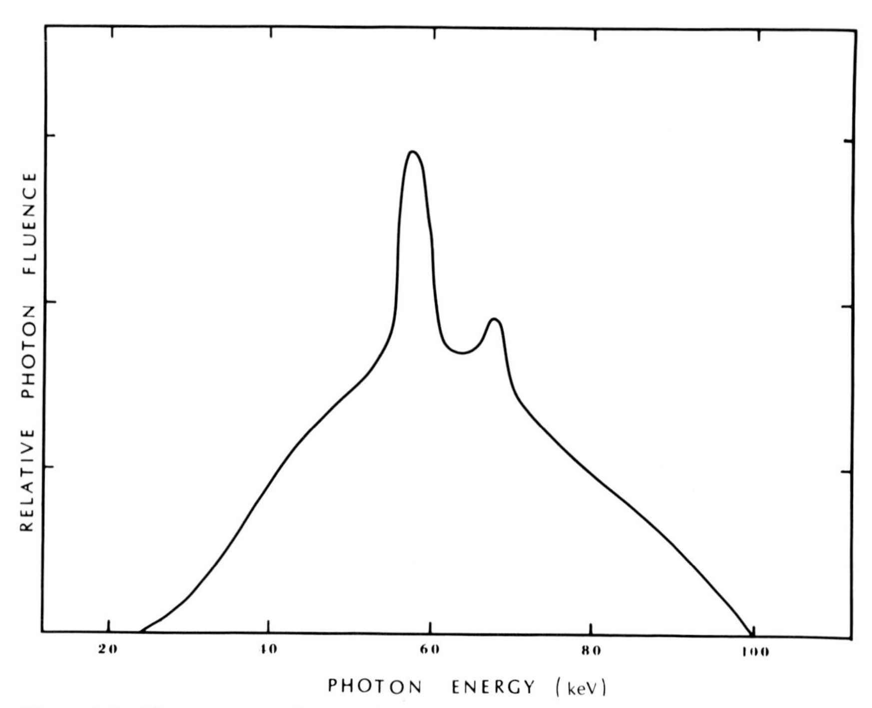

Figure 1.4 The spectrum of x rays from a generator contains a continuous distribution of photon energies (12).

indicate the differential distribution in some variable and the capital letter indicates the integrated value. For example,

$$\Phi(T)=\int_0^T \phi(E)dE=\int_0^T \frac{d\Phi(E)}{dE}dE \tag{1.9}$$

The total fluence $\Delta\Phi$ between some energy T_1 and a limiting energy T_2 is given by

$$\Delta\Phi_{T_1,T_2}=\int_{T_1}^{T_2} \phi(E)dE \tag{1.10}$$

Distributions for photon beams often are given in terms of the quantum energy $h\nu$ rather than E as the variable.

The other quantities mentioned, such as the HVT and the average energy, can be calculated (in principle) if the radiation spectrum is known. Generally, the average values are obtained from the differential distribution by integration. For example, the average energy $\langle E\rangle$ between T_1 and T_2 is given by

$$\langle E\rangle=\frac{\int_{T_1}^{T_2} E\phi(E)dE}{\int_{T_1}^{T_2} \phi(E)dE} \tag{1.11}$$

1.5 Radiation Action on Living Systems

In succeeding chapters only the physical processes involved in the absorption of radiation will be discussed in detail. Nevertheless, the intent is to present the material in the context of the possible biological effects of ionizing radiations. In this context, pertinent physical events are viewed as a part of a complete picture of larger proportions rather than in isolation. Some of the chemical and biological ramifications of physical absorption processes are indicated briefly in the following paragraphs to facilitate such a perspective.

A schematic representation of this larger picture is given in Figure 1.5. Suppose that the beam of radiation, in this case x rays or neutrons, is incident on a living system. The boxes indicate stages in a time sequence; the arrows indicate particles or molecular species present during the sequence. As an example, suppose x-ray photons undergo scattering or absorption reactions in the first stage, and fast electrons are produced. The electrons dissipate their kinetic energy in a variety of ways. Most of this energy results only in increased thermal agitation, which raises the temperature of the system slightly. Less numerous but more important processes produce molecular dissociation and thereby may cause permanent changes in the chemical structure of the molecular constituents. These chemical changes may prove harmless, or they may manifest themselves in biological damage.

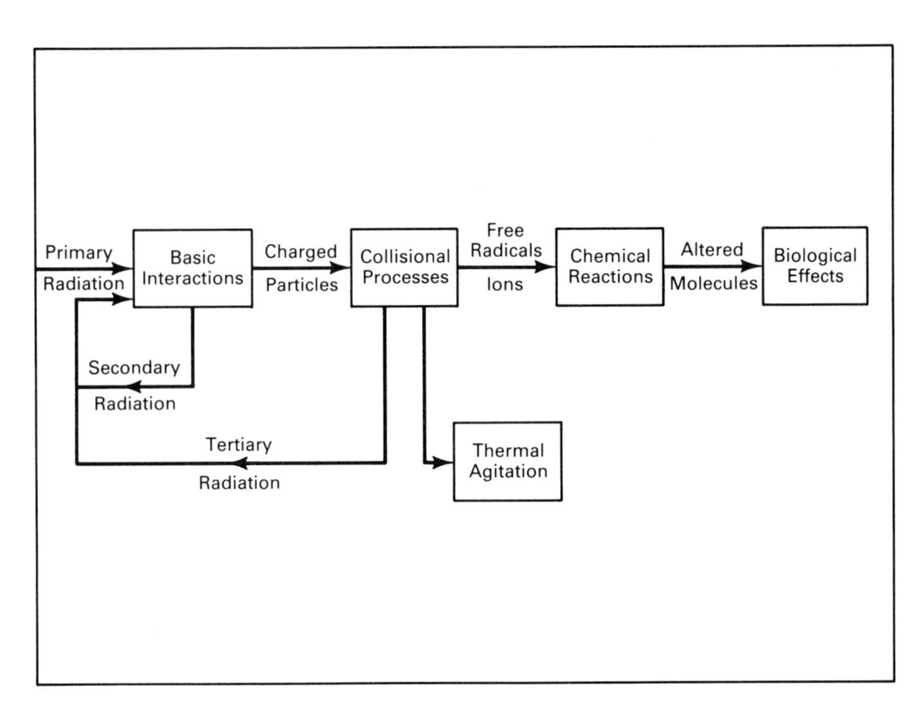

Figure 1.5 The sequence of the action of radiation on living systems is represented schematically, with boxes indicating important stages.

The physical processes are more complex when scattering produces secondary indirectly ionizing radiation or when some of the kinetic energy of the charged particles is reconverted into electromagnetic energy as bremsstrahlung or characteristic radiation. This radiation is then cycled through the sequence again. Thus, many individual interactions at many sites may be required to degrade completely the energy of the initial photon or neutron.

Platzman (13) has discussed the action of radiation on living systems in terms of specific stages that are related in some ways to the sequence shown in Figure 1.5. In the physical stage, the primary processes are initiated and the energy carried by the radiation is distributed among the molecular constituents. Excited molecules are formed in a nonuniform spatial distribution. In the physiochemical stage, some of the unstable excited molecules undergo de-excitation processes, bond dissociation occurs, and electrons are thermalized. The chemical stage occurs after thermal equilibrium has been attained, and during this stage most free radical reactions occur. During the biological stage, the organism as a whole must respond to the chemical changes that have occurred.

Boag (14) has estimated the time intervals for the sequence of radiation processes. Figure 1.6 is an illustration of similar events for millimolar concentra-

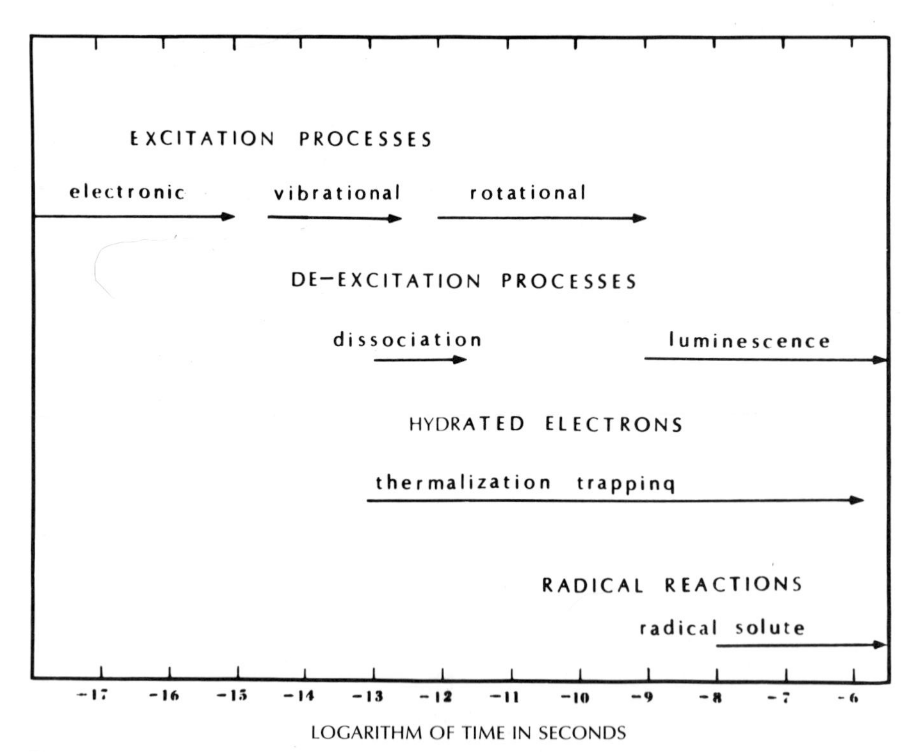

Figure 1.6 The passage of a 1-MeV electron through a solute molecule in a water solution can initiate a complex sequence of events. A typical time scale is shown. Adapted from Ref. 14.

tions of solute molecules in a water solution at room temperature. Assume that a 1-MeV electron passes through the molecule of interest from time zero to about 10^{-17} s; the arrows extend through the appropriate elapsed times for the molecular processes. Included are electronic, vibrational, and rotational processes. The electronic excitation interval is very short, whereas rotational motions are the slowest. Notice that molecular dissociation occurs simultaneously with vibrational excitation for the most part and that luminescence can occur as early as 10^{-9} s after the initial stimulus (fluorescence) or be much delayed (phosphorescence). These processes are basically physical in nature.

At the lower part of Figure 1.6, the processes more closely associated with the chemical stage are shown. Note that the electrons that were set free in the solution may be trapped and form a species called the hydrated electron, e^-_{aq}. The hydrated electron is a combination of several water molecules loosely held to the electron because of the strong electric dipole moment of the water molecules. As shown in the figure, reactions of the free radicals occur after most of the physical processes have run their course.

1.6 Problems

1. Convert the energies given in Table 1.2 into units of kilocalories per mole, which are often used in chemistry texts.
2. Estimate the lower limit imposed by the uncertainty principle for the time involved for an ionization from an outer shell of a water molecule. Use the ionization potential for a gross estimate of the energy uncertainty.
3. Estimate the time involved for ionization in hydrogen by calculating the time for a 100-eV incident electron [with maximum ionization probability (Figure 1.1)] to cross the atomic diameter.
4. Refer to Figure 1.6. Why is the dissociation time similar to the vibrational excitation times? Why is the trapping time for the hydrated electron similar to rotational excitation times for the molecules?
5. Prove that the definition $1\ R = 2.58 \times 10^{-4}$ C/kg is equivalent to the statement $1\ R = 1$ esu/cm^3 at standard conditions of temperature and pressure (esu stands for electrostatic units).
6. a) How many ion pairs are produced in 1 cm^3 of air at standard conditions when the exposure is 1 R?
   b) What is the minimum fractional standard deviation expected in ions collected for a 1-cm^3 chamber for a 1-R exposure if the collection efficiency is 1.00?
7. a) An absorbed dose of 4.0 Gy to the whole body often would be lethal for humans. Suppose that the energy equivalent of this dose is distributed uniformly in a 70-kg man and that no heat is lost. What temperature rise would be expected for the man? The energy equivalent of heat is 4.18 J/cal. Assume the specific heat is similar to that for water.
   b) Comment on the importance of the thermal effects of ionizing radiation.
8. A 2-g piece of plastic receives a dose of 2 Gy. The piece is cut exactly in

half. How much dose did each half get? How much energy did each half get?

9. A photon-beam pulse is accurately described as a continuous distribution in energy $h\nu$ (in megaelectron volts) with $\phi(h\nu)dh\nu = (10 - 2E)dh\nu$. Calculate the total fluence with $h\nu \leq 5$ MeV, $\Phi(5)$, and the average photon energy in the interval $0 \leq h\nu \leq 5$ MeV.

1.7 References

1. L. G. Christophorou, *Atomic and Molecular Radiation Physics* (Wiley-Interscience, New York 1971), p. 618.
2. L. J. Kieffer, Joint Institute for Laboratory Astrophysics, Report No. JILA 30, Boulder, CO (1965).
3. K. S. Pitzer, *Quantum Chemistry* (Prentice Hall, Englewood Cliffs, NJ, 1953), p. 170.
4. D. J. Pizzarello and R. L. Witcofski, *Basic Radiation Biology* (Lea and Febiger, Philadelphia, 1967), p. 86.
5. G. V. Dalrymple, "Microwaves," in *Medical Radiation Biology,* edited by G. V. Dalrymple, M. E. Gaulden, G. M. Kollmorgen, and H. H. Vogel, Jr. (W. B. Saunders, Philadelphia, 1973), p. 262.
6. H. E. Johns and J. R. Cunningham, *The Physics of Radiology,* 3rd Ed. (Charles Thomas, Springfield, IL 1969), pp. 197, 675.
7. *Radiation Quantities and Units, ICRU, Report 33* (International Commission on Radiation Units and Measurements, Washington, 1980).
8. *Le Système International d'Unités,* 3rd ed. (OFFLIB, Paris, 1977).
9. Editors, Phys. Med. Biol. **24**, 685 (1979).
10. M. J. Berger and S. M. Seltzer, "Tables of energy losses and ranges of electrons and positrons," in *Studies in Penetration of Charged Particles in Matter,* Publication 1133, (National Academy of Sciences–National Research Council, Washington, 1964), p. 267.
11. F. Ellis, Clin. Radiol. **20**, 1 (1969).
12. G. Drexler and M. Gossrau, *Spektren Gefilterter Röntgenstrahlungen für Kalibrierzwecke, Ein Catalog* (Gesellschaft Für Strahlen-forschung, München, 1968), p. 36.
13. R. L. Platzman, "Energy spectrum of primary activations in the action of ionizing radiation," in *Radiation Research,* edited by G. Silini (North-Holland, Amsterdam, 1967), p. 20.
14. J. W. Boag, "The events following primary activations," in *Radiation Research,* edited by G. Silini (North-Holland, Amsterdam, 1967), p. 43.

2

Energy Loss by Heavy Charged Particles

2.1 Introduction

In succeeding chapters the transfer of energy from fast charged particles to a material medium will be considered in some detail. This topic is of obvious importance when beams of electrons or protons from accelerators or alpha or beta emissions from radioactive sources are to be utilized. It is also of general importance in radiation science since a complete description of the distribution of energy from beams of photons or neutrons must include a discussion of local energy transfer by the liberated secondary charged particles. Thus, concepts developed in this and the several succeeding chapters will be widely used later on.

Table 2.1 shows the important properties (1, 2) of the more common particles considered in this chapter. For our purposes the most important heavy charged particles are protons and alpha particles, although all other ionized nuclear species must be included. Charged mesons are also included, although they are not generally considered to be heavy ions. The developments utilized in this chapter are generally applicable to both charged mesons and heavy ions but are not always applicable to electrons or positrons.

The list of the interactions of fast charged particles in a material medium can be divided into several categories, including both elastic and inelastic processes. In an elastic process, the total kinetic energy remains unchanged. Some kinetic energy may be transferred from an incident particle to a target particle, but changes in the state of the target system that require energy or are precursors to photon emission are not allowed. For an inelastic process, the total kinetic energy is changed either because of excitations or emission of electromagnetic energy or both.

Table 2.1 Properties of Charged Mesons and Heavy Particles

Name	*Charge*	*Mass (u)*	*Mc² (MeV)*
Muon	$\pm e$	0.1134	105.7
Pion	$\pm e$	0.1498	139.6
Proton	$+e$	1.0073	938.3
Deuteron	$+e$	2.0136	1875.6
Triton	$+e$	3.0155	2808.9
Alpha	$+2e$	4.0015	3727.3

Data from Refs. 1 and 2.

Four categories for consideration are:

1. Elastic processes with electrons or groups of electrons;
2. Inelastic processes with electrons or groups of electrons;
3. Elastic processes with nuclei or groups of nuclei;
4. Inelastic processes with nuclei or groups of nuclei.

For the first two categories, the incident particles interact via the coulomb force with atomic electrons. In the last two categories, the incident particles interact with atomic nuclei either through the coulomb force or through the strong interaction (nuclear force).

Fast charged particles lose most of their energy by processes belonging in category 2; ionizations and excitations are particularly important. Elastic collisions of charged particles with atomic electrons are dominant when the particle energy is so small that ionization and excitation are not common (3). In the following discussions the term "collisional energy loss" will refer specifically to transfer of energy from a charged particle to an atomic electron by means of the coulomb force. In this sort of collision, particle-to-particle contact is not necessary since the force acts at a distance.

Several important interactions of charged particles with nuclei will be discussed in succeeding chapters. Multiple scattering, due mainly to processes in category 3, will be covered in Chapter 4. Bremsstrahlung emission, which belongs to category 4, is discussed in the next chapter with electrons. Nuclear reactions of heavy charged particles are of some practical importance since they are utilized in the production of radioactive isotopes for research, medicine, and industry (4–6) and in the production of neutron beams for radiation therapy (7, 8). Nevertheless, because the reaction probabilities are relatively small, they will not be specifically considered in these chapters.

2.2 Collisional Energy Loss Cross Section

Collisional energy loss has been discussed by several authors (9–11). The most notable early contributions were made by Bohr in 1913 and 1915 (12, 13). Later fundamental work was published by Bethe (14–16), Möller (17) and Bloch

(18, 19). The expression relating the energy loss per unit length to constants of the medium is often called the Bethe-Bloch formula because of the contributions by these physicists. The simplified development given in this section and the next will be classical and nonrelativistic. More advanced results will be discussed briefly in Section 2.5.

A digression on the concept of the reaction cross section is appropriate at this point. Figure 2.1 shows some particles that are incident on the surface of a slab of area A and infinitesimal thickness dx. The cylindrical regions represent target particles of cross-sectional area σ imbedded in the slab. The number of these targets per unit volume is n_v. A reaction is said to occur when by chance the trajectory of an incident particle falls within the cross-sectional area of one of the targets. Suppose ϕ is the incident fluence and $d\phi$ represents the incremental fluence initiating reactions in the slab. Then the fraction of particles causing reactions is seen to be equal to the fraction of the total area filled by the targets:

$$\frac{d\phi}{\phi} = \frac{dA}{A} = \frac{(n_v A dx)\sigma}{A} = n_v \sigma dx, \qquad \sigma = \frac{d\phi/\phi}{n_v dx} = \frac{d\phi/n_v dx}{\phi} \tag{2.1}$$

In Equations 2.1 the cross section is shown to be equal to the quotient of the reaction probability per incident particle divided by the number of targets per unit area ($n_v dx$). However, the concept is sufficiently flexible to allow one to focus attention on the targets and define the cross section as the reaction probability per target divided by the number of incident particles per unit area (ϕ). Either definition is dimensionally equivalent (m^2) and equally valid.

One should realize that the concept of the reaction cross section can be applied to situations in which target dimensions are not easily defined. Figure 2.1 is only schematic. For subatomic targets, the geometrical cross-sectional area may have little meaning. The concept of the cross section is useful through the abstract definition.

Reference is made to Figure 2.2 to derive an expression for the cross section for collisional energy loss by heavy charged particles. In this figure

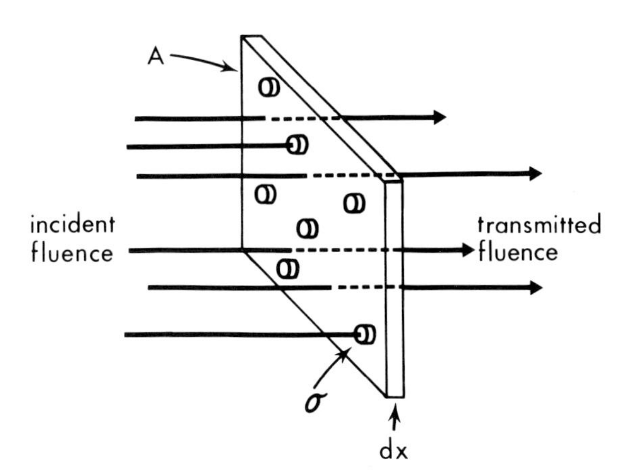

Figure 2.1 The concept of cross section can be understood by considering particles incident on a slab with area A and thickness dx. The target particles are presumed to have projected cross sectional area σ.

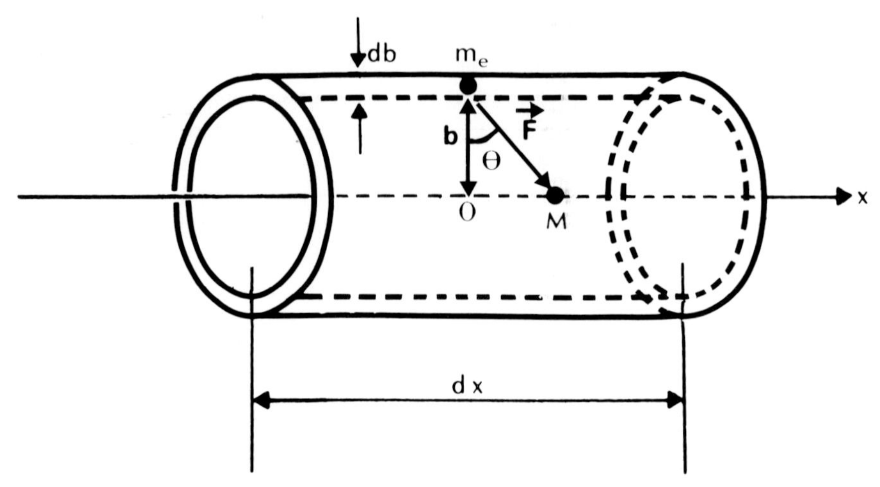

Figure 2.2 The geometrical parameters necessary to define a collision are illustrated. Note that **F** is the force on the electron, which has mass m_e. The incident ion with mass M is assumed to be positively charged. Closest approach occurs at the origin ($x = 0$).

the incident ion is traveling in the positive x direction with speed v and kinetic energy T. The net charge on the ion is $+ze$, where e is the magnitude of the electronic charge, and the value of z is often less than the ion atomic number. The incident ion (of mass M) is traveling in a medium characterized by mass density ρ and atomic density n_v atoms per cubic meter. Z is the atomic number of the medium.

The incident ion interacts with an atomic electron of charge $-e$ and mass m_e which is assumed to be located in a cylindrical shell of radius b, length dx, and thickness db. The radius b is often called the impact parameter since it represents the smallest perpendicular distance between the incident-particle path and the electron. Time $t = 0$ corresponds to closest approach with $x = 0$. The interaction is due to the coulomb attractive force given by

$$\mathbf{F} = -e\mathbf{E} = eE\hat{r} = \frac{1}{4\pi\epsilon_0}\left[\frac{ze^2}{r^2}\right]\hat{r} \tag{2.2}$$

where $\mathbf{E}$ is the electric field surrounding the incident ion and ϵ_0 is the permittivity of free space. Since the force acts on the atomic electron, $\hat{r}$ is the unit vector pointing from that electron toward the positive ion.

The development we utilize is semiclassical and approximate rather than rigorous. However, the resultant expressions show the same basic dependence on the parameters as advanced derivations. The following assumptions are necessary for this development:

1. Movement of the atomic electron during the collision can be neglected.
2. The initial energy of the electron in its atomic orbit can be neglected.
3. Changes in the incident-particle trajectory and speed during a single collision are small.

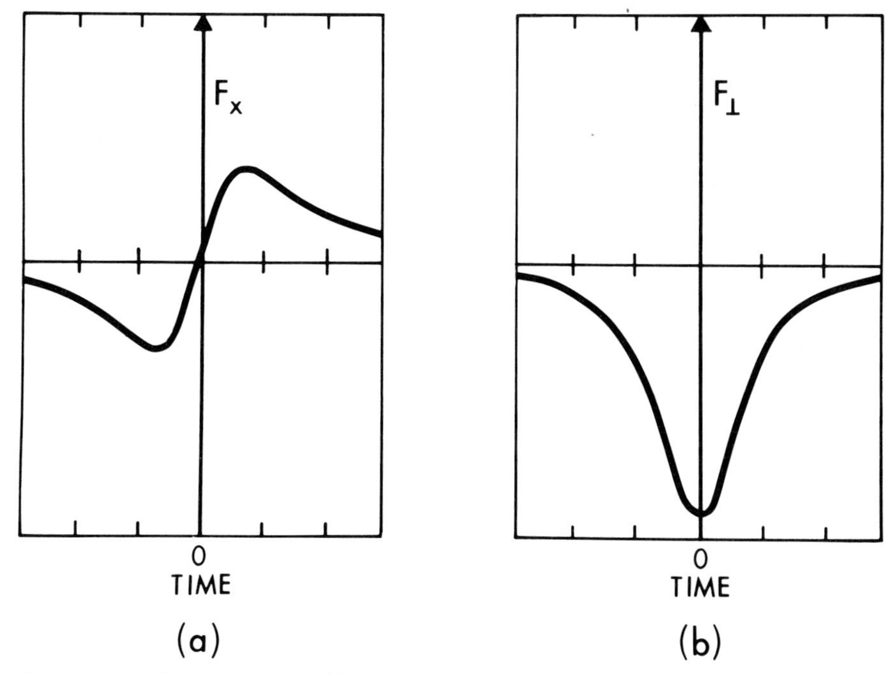

Figure 2.3 The components of force on an atomic electron during a collision are shown as a function of time. Side a shows the force in the x direction; side b shows the force in the direction perpendicular to the path and radially outward through the electron.

Under these assumptions simple impulse expressions can be used. In effect, number 1 above requires that collision times be short and that an average electron position be utilized. Assumption 3 implies that the incident particle loses its kinetic energy in many low-energy loss collisions rather than in a few high-energy loss collisions. Under this condition the deviation in the ion trajectory is likely to be small during a single collision, since it has a large mass when compared to the atomic electron mass. When the incident particle is itself an electron, assumption 3 is more troublesome.

In Figure 2.3 the force F on the electron is shown as a function of time. Suppose that the direction perpendicular to the ion path is taken to be positive when pointing radially outward from the path and through the electron. Then $F_\perp$ is negative with a magnitude that is initially small. The magnitude increases to a maximum at closest approach; it then diminishes as the ion moves on. The force parallel to the ion path in the x direction is negative at first ($-x$ direction) and then positive ($+x$ direction). Under the assumptions made, the impulse in the x direction for $t < 0$ cancels the impulse when $t > 0$. Thus, we need only consider the perpendicular component of the force on the electron.

The increase in electron momentum during the collision, $\Delta p_\perp$, equals the integral of the impulse over the collision time:

$$\int_{t_1}^{t_2} F_\perp dt = \int_{p_1}^{p_2} dp_\perp = \Delta p_\perp \tag{2.3}$$

But from Figure 2.2,

$$\cos\theta = b/r, \quad r = b/\cos\theta, \tag{2.4}$$

$$\Delta p_\perp = -\int_{t_1}^{t_2} F\cos\theta\, dt = -\int_{t_1}^{t_2}\left[\frac{1}{4\pi\epsilon_0}\right]\frac{ze^2}{r^2}\cos\theta\, dt = -\frac{ze^2}{4\pi\epsilon_0}\int_{t_1}^{t_2}\frac{\cos^3\theta}{b^2}\, dt \tag{2.5}$$

Of course, θ changes with time during the collision:

$$\tan\theta = \frac{x}{b}, \quad \frac{d(\tan\theta)}{dt} = \frac{d(\tan\theta)}{d\theta}\frac{d\theta}{dt} = \sec^2\theta\,\frac{d\theta}{dt} = \frac{1}{b}\frac{dx}{dt} = \frac{v}{b} \tag{2.6}$$

After solving for dt in Equation 2.6,

$$\Delta p_\perp = -\frac{ze^2}{4\pi\epsilon_0}\int_{-\pi/2}^{\pi/2}\frac{\cos^3\theta}{b^2}\left[\frac{b\,d\theta}{v\cos^2\theta}\right]$$

$$\simeq -\frac{ze^2}{4\pi\epsilon_0 bv}\int_{-\pi/2}^{\pi/2}\cos\theta\, d\theta = -\frac{2ze^2}{4\pi\epsilon_0 bv} \tag{2.7}$$

In the last step, the collision was assumed to start with the particle at $x \to -\infty$ ($\theta = -\pi/2$) and end at $x \to +\infty$ ($\theta = +\pi/2$). Of course, v must change during the collision since energy is transferred to the electron. Factoring v out of the integrand in Equation 2.7 can be justified only when the change in ion speed during a single collision is a very small fraction of the incident speed.

The energy E transferred from the incident particle to the atomic electron is given by

$$E = \frac{(\Delta p_\perp)^2}{2m_e} = \left[\frac{1}{4\pi\epsilon_0}\right]^2\frac{4z^2e^4}{2m_e v^2 b^2} = \frac{2z^2e^4}{(4\pi\epsilon_0)^2 m_e v^2 b^2} \tag{2.8}$$

Equation 2.8 is written under the assumption that the initial electron momentum is negligible. Further manipulation yields

$$b^2 = \frac{2z^2e^4}{(4\pi\epsilon_0)^2 m_e v^2 E}, \quad b\,db = \frac{z^2e^4}{(4\pi\epsilon_0)^2 m_e v^2}\left[-\frac{dE}{E^2}\right] \tag{2.9}$$

The minus sign means that the energy increases when the impact parameter b decreases. For a given incident particle speed and charge, the energy transferred is determined by the impact parameter.

The differential cross section per electron, $\frac{d\sigma(E)}{dE}\,dE$, for transferred energy between E and $E + dE$ is easily obtained from Equation 2.9 since

$$d\sigma(E) = \frac{d\sigma(E)}{dE}\,dE = |2\pi b\,db| = \frac{2\pi z^2e^4}{(4\pi\epsilon_0)^2 m_e v^2}\left[\frac{dE}{E^2}\right] = 2z^2\pi r_0^2\frac{m_e c^2}{\beta^2}\left[\frac{dE}{E^2}\right]$$

$$= \frac{2z^2\pi r_0^2 m_e c^2}{T(T+2Mc^2)/(T+Mc^2)^2}\left[\frac{dE}{E^2}\right] \tag{2.10}$$

In this expression, the classical electron radius r_0 and the incident particle kinetic energy T have been used:

$$r_0 = \frac{e^2}{4\pi\epsilon_0 m_e c^2}, \quad \beta^2 = \left[\frac{v}{c}\right]^2 = \frac{T(T+2Mc^2)}{(T+Mc^2)^2} \tag{2.11}$$

The classical electron cross-sectional area πr_0^2 is included in Equation 2.10 as should be expected.

In Equation 2.10 the cross section is shown to depend on the square of the net charge on the incident ion. Actually, an incident positive ion may pick up an electron during a collision and at least partially neutralize its charge. When the ion speed is high, the net effect of this tendency is not significant, since the newly attached electron is usually lost in a succeeding encounter. However, when the speed of the ion is comparable to the orbital velocity of the atomic electrons, the probability of capture increases concomitantly with a decrease in the electron-loss probability (20). Thus, on the average, the net positive charge on an incident ion decreases when the speed decreases. Figure 2.4 shows the dependence of the beam average or effective charge on speed for alpha particles and for protons (11).

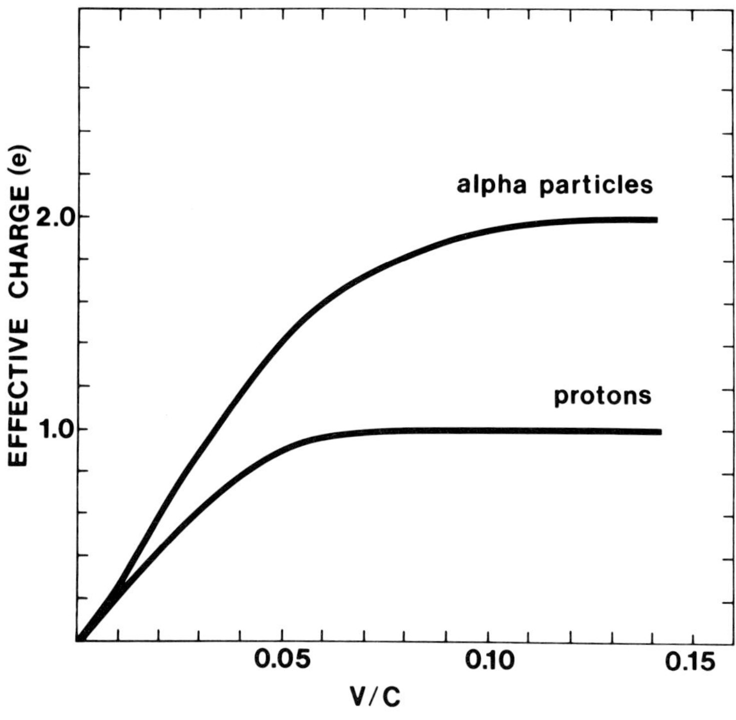

Figure 2.4 The effective charge for protons and alpha particles is reduced at low speed, because of electron attachment (20).

Figure 2.5 shows the ionization cross section for atomic hydrogen by protons (3, 21). For larger values of the proton kinetic energy, the cross section is approximately proportional to $1/v^2$ as indicated by Equation 2.10. However, when the incident proton speed is comparable to the electron orbital velocity, the dependence changes. At these lower energies, the cross section is greatly reduced, as is shown in the figure. A decrease in effective charge contributes to this reduction. A second cause is the binding energy of the atomic electron, which was ignored in the preceding development. Of course, the cross section for ionization during a collision must decrease to zero at the ionization potential.

Figure 2.6 shows the product of the differential ionization cross section and the ejected electron energy as a function of energy. The curve is representative of data taken with 1.4-MeV protons incident on molecular nitrogen (22). The quantity shown on the abscissa depends inversely on the ejected electron energy in a limited intermediate energy region. At higher energies the cross section falls off very rapidly because the amount of energy transferred in a single collision is limited. The maximum energy transferred will be discussed in Section 2.3. At lower energies experimental data may be enhanced by secondary electrons. One source of secondaries is the Auger effect, a result of ionization from inner atomic shells (23). The peak in Figure 2.6 at about 0.5 keV is an Auger peak. The Auger effect is discussed in Chapter 11.

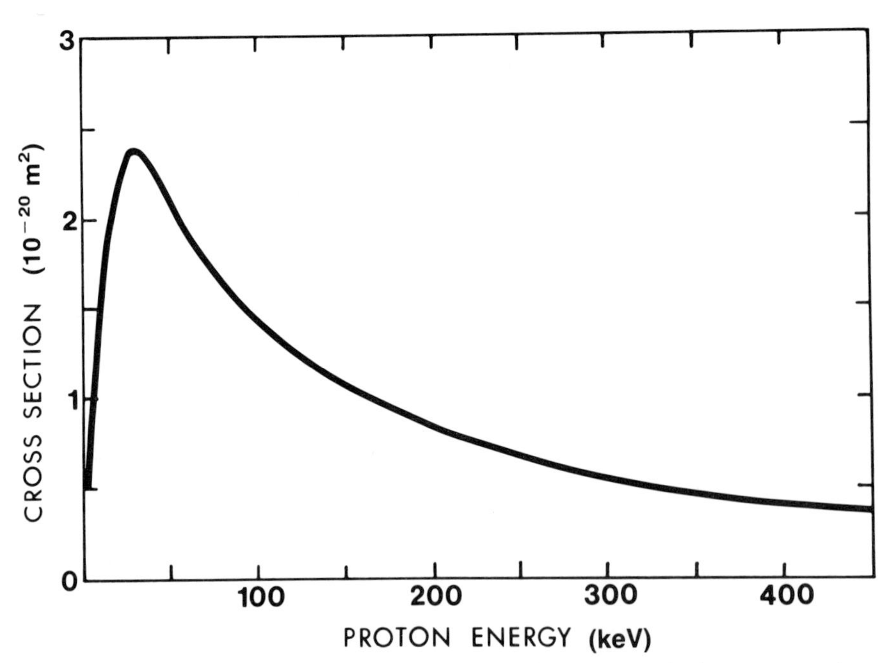

Figure 2.5 The ionization cross section for atomic hydrogen by incident protons is found to rise to a maximum at a proton energy of about 40 keV (3).

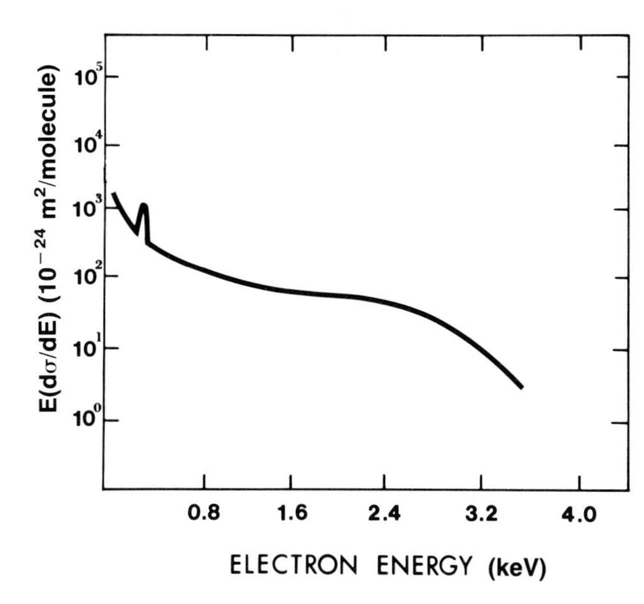

Figure 2.6 The product of the ionization cross section multiplied by the ejected electron energy is shown as a function of the ejected electron kinetic energy for molecular nitrogen. The incident proton energy is 1.4 MeV (22).

2.3 Stopping Power Formulation

Charged-particle energy loss per unit length is called stopping power. The stopping power S_c for collisional losses for charged particles is defined to be the quotient of $-dT$ by dx, where $-dT$ is the average energy lost in traversing a distance dx (24). Thus,

$$S_c = (-dT/dx)_c \tag{2.12}$$

The subscript c indicates that only collisional energy losses are included.

Equation 2.10 can be used to obtain a stopping-power expression. Consider again the cylindrical shell shown on Figure 2.2. The volume d^2V of the shell is a second-order differential:

$$d^2V = |2\pi b db| dx = \frac{d\sigma(E)}{dE} dE\, dx \tag{2.13}$$

The total number of electrons in the shell is equal to the electron density $n_v Z$ times the shell volume:

$$n_v Z(d^2V) = n_v Z \frac{d\sigma(E)}{dE} dE\, dx \tag{2.14}$$

From Equation 2.10,

$$n_v Z(d^2V) = 2z^2 \pi r_0^2 \frac{m_e c^2}{\beta^2} n_v Z \left[\frac{dE}{E^2}\right] dx \tag{2.15}$$

The product of the total number of electrons in length dx and the energy transferred per electron, when integrated over all possible values of transferred energy, gives the total energy loss in length dx. Thus,

$$\frac{-dT}{dx} = 2z^2\pi r_0^2 \frac{m_e c^2}{\beta^2} n_v Z \int_{E_{min}}^{E_{max}} \frac{dE}{E} = 2z^2\pi r_0^2 \frac{m_e c^2}{\beta^2} n_v Z \ln\left[\frac{E_{max}}{E_{min}}\right] \tag{2.16}$$

The energy transferred to an atomic electron per collision was specifically limited in the integration in Equation 2.16 by a maximum value E_{max} and a minimum value E_{min}. Of course, the maximum value of the energy transferred in Equation 2.16 cannot exceed the incident-particle energy. In fact, it must be much smaller than this. The minimum value of the energy loss must be greater than zero if the energy loss per unit distance is to remain finite. Indeed, infinitesimal energy transfers are not possible, since bound electrons that exist in discrete energy levels are involved.

The maximum energy transfer possible in a single head-on elastic collision can be obtained by applying momentum and energy conservation. Classical energy expressions will be used, since the incident particles considered in this chapter are assumed to be nonrelativistic ($v \ll c$) and the energy transferred to the atomic electron will be much less than $m_e c^2$. Since v is the incident ion speed before the collision, we use v' to designate the speed after the collision. The electron speed is negligible before the collision but has the value v_e' afterwards. Then, energy and momentum conservation for a head-on collision yields

$$\frac{1}{2}Mv^2 = \frac{1}{2}M(v')^2 + \frac{1}{2}m_e(v_e')^2, \quad Mv = Mv' + m_e v_e' \tag{2.17}$$

Eliminating v' by substitution means that

$$\frac{1}{2}Mv^2 = \frac{(Mv - m_e v_e')^2}{2M} + \frac{1}{2}m_e(v_e')^2,$$

$$m_e v\, v_e' = \frac{m_e(m_e + M)}{2M}(v_e')^2, \quad v_e' = \frac{2Mv}{m_e + M} \tag{2.18}$$

The maximum electron kinetic energy is

$$E_{max} = \frac{m_e(v_e')^2}{2} = \frac{2m_e M^2 v^2}{(m_e + M)^2} \tag{2.19}$$

When the heavy-particle limit is taken with $M \gg m_e$,

$$E_{max} = 2m_e v^2 = 2m_e c^2 \beta^2 \tag{2.20}$$

Note that v^2 is the square of the *ion* speed and m_e is the atomic *electron* mass.

The discrete energy differences between bound electron states in atomic or molecular systems must be considered in an estimate of the minimum energy transferred, E_{min}. Since electron binding energies are quite different for different shells and subshells, a variety of energy transfer events are possible for any given target. Possible electron transitions for such an atomic system are represented on Figure 2.7 by the arrows. The average excitation energy I is the average of all such transition energies for a given atomic target when weighted by the probability of the transition.

The minimum energy transferred in a collision can be related to I^2 when the momentum transferred to the entire atomic system is considered (10, 14–16):

$$E_{\min} = I^2/2m_e v^2 \tag{2.21}$$

Using Equations 2.20 and 2.21 for $E_{\max}$ and $E_{\min}$ and using Equation 2.16, the result for the stopping power can be written

$$S_c = 2z^2\pi r_0^2 \frac{m_e c^2}{\beta^2} n_v Z \ln\left[\frac{2m_e v^2}{I}\right]^2 = 4z^2\pi r_0^2 \frac{m_e c^2}{\beta^2} N_A(Z/M_m)\rho \ln\left[\frac{2m_e c^2\beta^2}{I}\right]$$

$$= 4z^2\pi r_0^2 m_e c^2 N_A(Z/M_m)\rho \frac{(T+Mc^2)^2}{T(T+2Mc^2)} \ln\left[\frac{2m_e c^2 T(T+2Mc^2)}{I(T+Mc^2)^2}\right] \tag{2.22}$$

In this expression, the atomic density n_v in atoms per cubic meter has been replaced by an expression involving Avogadro's number of atoms in a mole N_A, the mass per mole M_m, and the mass density ρ:

$$n_v = \frac{N_A}{M_m}\rho \tag{2.23}$$

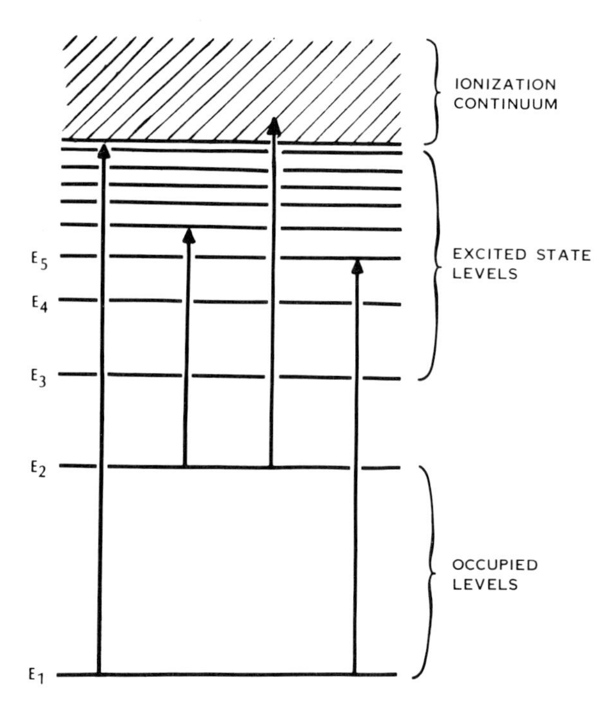

Figure 2.7 Some possible transitions in an atomic or molecular system are indicated by arrows originating at levels for filled electron states.

The functional form of Equation 2.22 is appropriate for the variables of interest in the nonrelativistic limit. In particular the electron mass m_e has been incorporated into the electron rest energy expression for simplicity in calculation.

The principle features of the stopping power expression can be divided into two parts:

1. **Ion Properties**
   The collisional stopping power depends directly on the square of the incident-particle net charge. The stopping power depends inversely on the square of the speed. In addition, a logarithmic term containing the square of the particle speed in the numerator is involved.

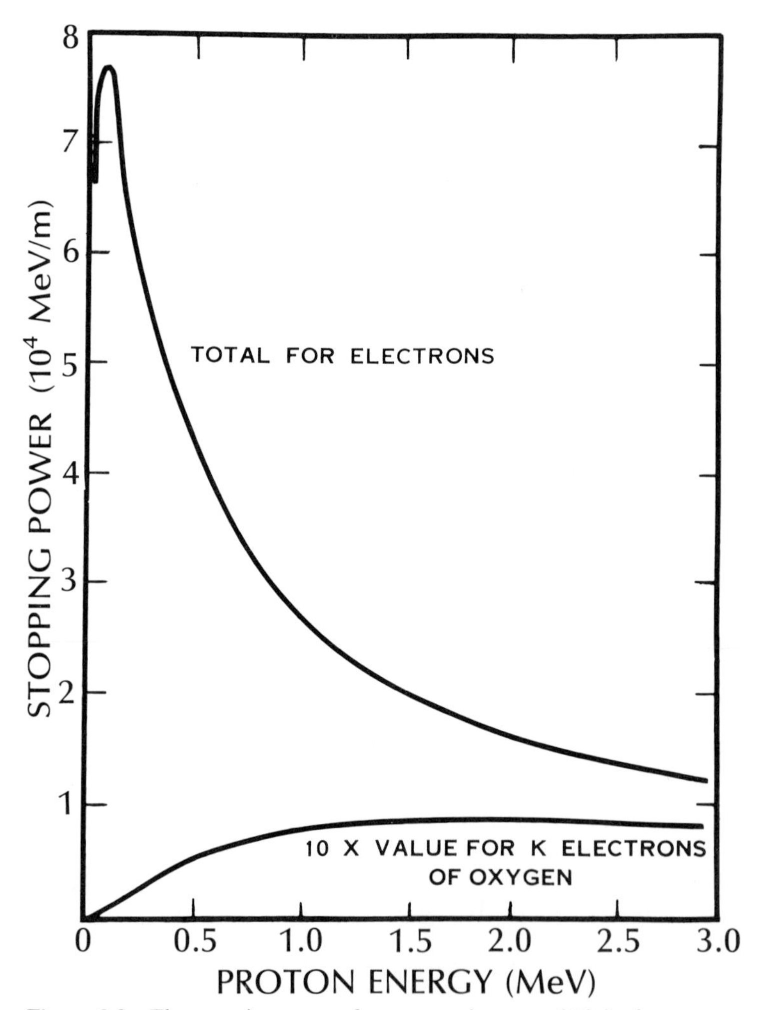

Figure 2.8 The stopping power for protons in water (25) is shown. The stopping power for the *K* electrons of oxygen contributes very little to the total.

2. **Medium Properties**
   The collisional stopping power depends directly on the electron density of the absorbing medium. The stopping power depends directly on a logarithmic term containing the average excitation energy in the denominator.

The dependence of the collisional energy loss cross section on particle charge and speed has been discussed in Section 2.2. These comments apply to the stopping power also. Figure 2.8 illustrates the dependence of the stopping power of a proton in water on the inverse of the square of the speed (25). The cross section for only the oxygen K electrons is also shown in Figure 2.8. It is evident that outer-shell electrons contribute greatly to the stopping of incident ions. The stopping power due to nuclear collisions is less than 0.1% of the value for atomic electrons at most energies (25).

2.4 Linear Energy Transfer

The linear energy transfer is a concept closely related to the collisional stopping power. It is especially important in applications of radiation to biological systems because concentration of the deposited energy is significant in those cases.

The linear energy transfer may be characterized as a restricted collisional stopping power (24). The linear energy transfer L_Δ of charged particles in a medium is defined as the quotient of $-dT$ by dx, where dx is the distance traversed by the particle and $-dT$ is the mean energy loss due to collisions with energy transfers less than some specified value Δ. Thus,

$$L_\Delta = (-dT/dx)_c, \quad E \leq \Delta \tag{2.24}$$

Although the symbol L is often used in algebraic relationships without specifying Δ, note that the definition allows for specification of a cutoff parameter Δ, which is to be evaluated as the largest energy included. Nevertheless, it is intended that the energy losses included will be local to the incident ion track, because the distance traveled by a liberated electron as it moves away from the track is limited by its energy. High-energy liberated electrons, which deposit most of their energy far away from the path of the incident particle, make a contribution to the stopping power. They make no contribution to the linear energy transfer if Δ is chosen to exclude them. Microscopic biological damage is most clearly related to energy density deposited locally around the path of an incident particle. Thus, the distinction between linear energy transfer and stopping power is a useful one.

Specific ionization is another concept often mentioned in context with linear energy transfer. The specific ionization i is defined as the number of ions produced per unit path length for a given particle. The average specific ionization $I = \langle i \rangle$ for a beam of identical particles is related to the stopping power for collisional loss by

$$S_c = WI \tag{2.25}$$

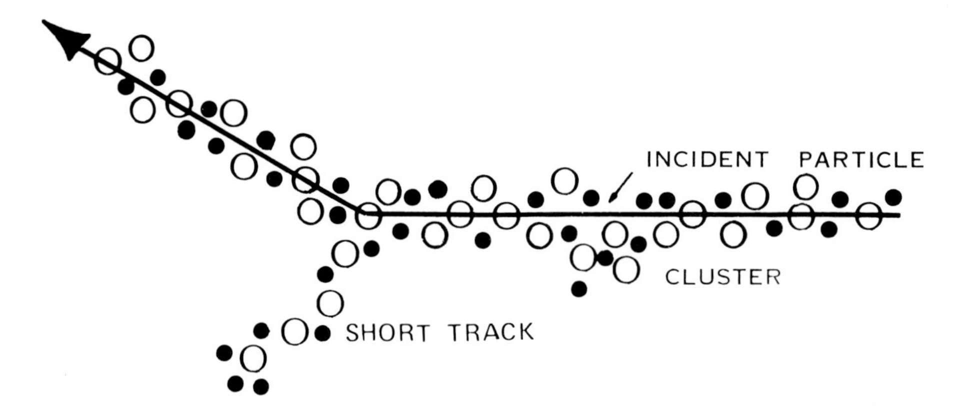

Figure 2.9 The path of an ion is shown, as well as examples of a cluster and a secondary short track. Adapted from H. E. Johns and J. R. Cunningham, *The Physics of Radiology,* 3rd Ed. (Charles C Thomas, Springfield, Ill., 1969).

where W stands for the average energy loss per ion pair. W is calculated by dividing the total energy loss that occurred in a given path length by the total number of ion pairs produced along that path segment for a number of identical ions. The value therefore includes energy lost in excitations as well as ionizations. Since W is nearly constant at large ion speed (26), Equation 2.25 can be applied directly except near the end of the path of an ion. The average energy per ion pair is discussed in detail in Chapter 11.

In some circumstances the path of a charged particle can be delineated by the ionization events occurring nearby. Thus, emulsions or bubble chambers can be used to detect fast charged particles. Figure 2.9 is an illustration of a charged-particle track. Several secondary groupings are shown on the figure in addition to the main track.

High-energy electrons that have been liberated from atoms are often called delta rays. Delta rays have sufficient energy to produce a path of ionization

Table 2.2 Cluster Size and Energy

Cluster Energy (eV)	*18-keV Electrons (% Energy)*	*440-keV Electrons (% Energy)*
6–100	38	64
100–500	12	11
500–5000 (short track)	50	25

Data from Ref. 27.

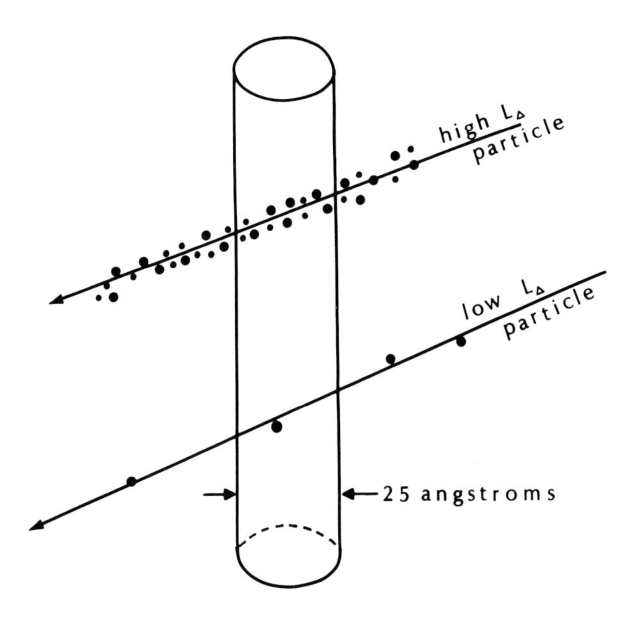

Figure 2.10 The ionization density in and around a segment of a DNA molecule is illustrated for the cases of a low linear energy transfer particle and a high linear energy transfer particle. Adapted from H. E. Johns and J. R. Cunningham, *The Physics of Radiology,* 3rd Ed. (Charles C Thomas, Springfield, Ill., 1969).

and excitation of their own, a short track. Such a track can be distinguished from clusters of ion pairs produced near the path of the primary particle, as shown in Figure 2.9. Depending on the choice of cutoff energy Δ, energies for liberated electrons producing short tracks may be excluded from the linear energy transfer. Table 2.2 shows some data (27) on relative amounts of energy deposited by short tracks in a water medium. It is evident that a substantial part of the total energy is excluded if Δ is set equal to 500 eV or less.

Figure 2.10 is intended to illustrate why the linear energy transfer is important for biological effects. Suppose the cylinder shown represents the volume of a DNA double strand in the nucleus of a cell. A low linear energy transfer ion may deposit some energy in the region of the DNA in a single collisional loss event, but the probability of depositing a great deal of energy in a single pass is very small. A single energy deposition event of this sort probably will not cause much permanent damage. However, the high linear energy transfer particle is quite likely to severely damage the DNA because so much energy is deposited in a single pass.

2.5 Restricted Energy Transfer Expressions

A short form of the complete expression (11) for the restricted collisional stopping power for heavy ions can be written as

$$L_\Delta = 2z^2\pi r_0{}^2 m_e c^2 N_A (Z/M_m)\rho \frac{(T+Mc^2)^2}{T(T+2Mc^2)} C(T,Z,M,\Delta) \tag{2.26}$$

The function $C(T,Z,M,\Delta)$ is given by

$$C(T,Z,M,\Delta)=\left\{\ln\left[\frac{2T(T+2Mc^2)\Delta/m_ec^2}{(Mc^2)^2(I/m_ec^2)^2}\right]-\frac{\Delta/m_ec^2}{2(T+Mc^2)^2/(Mc^2)^2}-\frac{T(T+2Mc^2)}{(T+Mc^2)^2}-\delta-2\lambda\right\} \quad (2.27)$$

In these equations, the incident particle kinetic energy T is used as the variable in place of β for convenience, I is the average excitation energy, and Δ is the cutoff parameter. When

$$\Delta=\Delta_{\text{max}}=2m_ec^2\frac{T(T+2Mc^2)}{(Mc^2)^2} \quad (2.28)$$

the expressions give the stopping power.

There are four categories for additional factors and correcting terms inside the braces in Equation 2.27 that are not given in Equation 2.22. They are:

1. Factors involving the energy cutoff parameter Δ;
2. Relativistic corrections involving T and Mc^2;
3. Relativistic density effect corrections involving δ;
4. Electronic shell corrections involving λ.

When the corrections are applied, Equations 2.26 and 2.27 are typically accurate to about 1% over a variety of experimental conditions. They are not this accurate (9, 10, 11) in the very low velocity regions ($T/Mc^2 < 0.001$).

The relativistic corrections inside the braces in Equation 2.27 generally increase the energy loss per unit length at high velocity. These corrections are due to well known results from special relativity, such as mass increase of the incident particle, as well as considerations involving momentum transfer by virtual photon processes (10). Another relevant relativistic effect is the Lorentz contraction of the electric field of the particle, which leads to the density effect. The relativistic density effect correction term δ in Equation 2.27 is included for completeness, but it is not important for heavy particles unless kinetic energies that are a sizable fraction of the rest mass energy are attained (28–30). In brief, the correction occurs because polarization of the medium around the incident particle tends to shield its charge. At relativistic speeds the Lorentz contraction changes the electric field pattern of the incident particle in a way that enhances the effect of the polarization and reduces the energy loss. Thus, the density effect term reduces the stopping power at relativistic speeds. The relativistic density effect is discussed in more detail in Chapter 3 since it is more often significant for electrons.

The shell correction term λ is necessary since electrons are in constant motion in the atomic system, a fact not included in the development in Section 2.2. The probability of energy transfer is reduced and that of orbital deformation increased when the electron orbital speeds are comparable to the incident particle speed. Since the electrons exist in atomic shells, the orbital speeds in the K,

L, M shells, and so forth, are separately considered. Thus, the shell correction term is often written as

$$\lambda = \sum_i C_i/Z, \tag{2.29}$$

where the sum is over individual values C_i for the various shells ($i = 1$, *K* shell; $i = 2$, *L* shell; and so forth). Electron shell effects produce a reduction in the calculated stopping power. Values for the adjusted shell correction term λ_a are shown in Figure 2.11 for two elements (11). The adjusted shell correction values (with subscript *a*) are for use with the adjusted average excitation energy denoted I_a, which will be discussed later in this section.

The average excitation energy *I* was described briefly in Section 2.3. When the average excitation energy is calculated (11, 31), it is usually obtained from

$$Z \ln I = \sum_n f_{n0} \ln(E_n - E_0) \tag{2.30}$$

The subscripts 0 and *n* denote the ground state and final states, respectively, of the target atom with atomic number *Z*.
This expression indicates that ln *I* is a weighted average of the natural logarithm of the energy difference. The weighting function f_{n0} is the summed oscillator

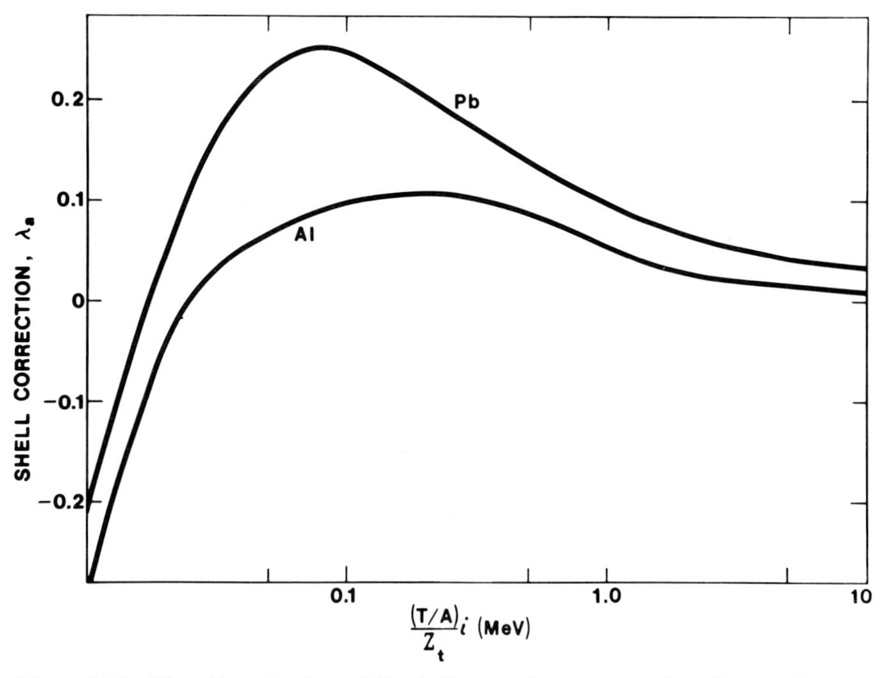

Figure 2.11 The adjusted values of the shell correction term are given for two elements. The ordinate is the kinetic energy per nucleon (*T/A*) of the incident ion *i* divided by the atomic number *Z* of the target *t*. Adapted from Ref. 11.

strength for all electrons in the atom. It is presumed to give the probability of occurrence of the particular transition from state 0 to state n.

The summed oscillator strength is given by (3)

$$f_{n0} = \frac{2m_e}{\hbar^2}(E_n - E_0)\left|\int u_n^* \sum_i x_i u_0 \, d\tau\right|^2 = \frac{2m_e}{\hbar^2}(E_n - E_0)\left|(\Sigma x_i)_{0n}\right|^2 \tag{2.31}$$

The stationary state wave functions represent the ground state u_0 and the final state u_n of the target atom. The position coordinate of the ith electron is labeled x_i.

Values of ln I can be calculated from Equations 2.30 and 2.31. Accurate determination of the average excitation energy by this route is difficult since the wave functions are not well known except for very light atoms.

In most cases I is determined from measurements related to the stopping power. To accomplish this, all the quantities in Equations 2.26 and 2.27 except I, λ, and δ must be evaluated. For heavy-particle stopping powers with $\Delta = \Delta_{max}$ and $T/Mc^2 \ll 1$, the density effect term δ is negligible and a term ln $I + \lambda$ will be left. A common practice is to define an adjusted value of the average excitation energy I_a and use this quantity with the adjusted value of the shell correction term such that

$$\ln I_a + \lambda_a = \ln I + \lambda, \quad \lambda_a \to 0, \quad T/Mc^2 \to \infty \tag{2.32}$$

Under these conditions a recommended set of values of I_a has been given (32). They are shown on Table 2.3 for values of $Z \leq 13$. For elements with $Z > 13$, the adjusted average excitation energy in electron volts can be calculated from (28)

$$I_a = 9.73Z + 58.8Z^{-0.19} \tag{2.33}$$

Figure 2.12 is a plot of I_a/Z versus Z. Notice that at large Z values, the adjusted average excitation energy is proportional to Z since I_a/Z is constant. This proportionality was first demonstrated by Bloch (18, 19) and is sometimes called the Bloch relationship.

Table 2.3 Values of The Adjusted Average Excitation Energy

Element	I_a*(eV)*
Hydrogen	18.9
Helium	42
Lithium	38
Beryllium	60
Carbon	78
Nitrogen	85
Oxygen	89
Neon	131
Aluminum	163

Data from Ref. 32.

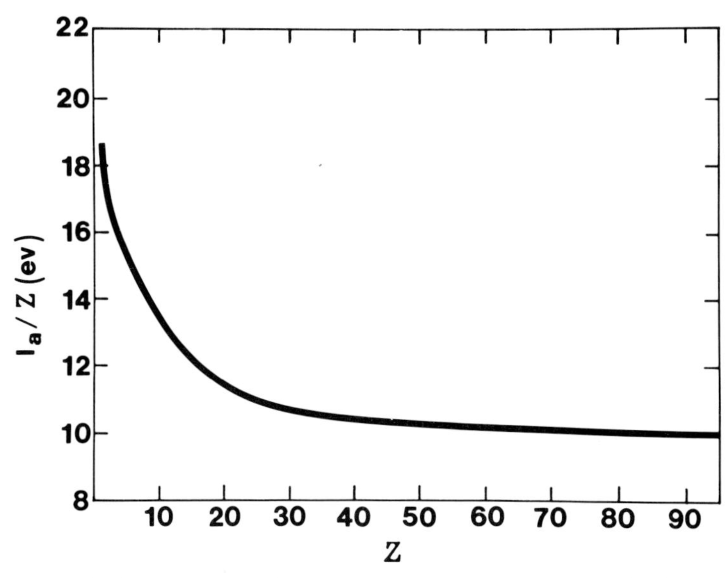

Figure 2.12 The adjusted average excitation energy divided by Z is shown as a function of the atomic number of the target.

2.6 Systematics of Mass Stopping Power

The mass stopping power S/ρ is the quotient of the stopping power divided by the mass density of the medium:

$$S/\rho = \frac{1}{\rho}(-dT/dx) \tag{2.34}$$

Mass stopping power values for collisional losses by incident protons for several elements are given in Appendix 6. Interpolation using this table to obtain mass stopping power values for elements that are not included and conversion of values for atomic materials to those for molecular substances will be discussed in succeeding paragraphs.

As can be seen from Equations 2.22 and 2.26, the stopping power is proportional to ρ, the mass density. Thus, stopping power values for the different physical states of the same element can be radically different. Furthermore, the mass density varies from element to element in the solid state in a manner difficult to predict. Because of this, any interpolation for stopping power between listed elements will not be very accurate. On the other hand, the mass stopping power varies in a more easily predictable manner because the density differences have been divided out. Thus, a limited table of mass stopping power values can have wide usefulness, and interpolation with reasonable accuracy is possible. Because of this, the tables in the Appendixes 6 and 7 give mass stopping power values.

Consider the problem of interpolation between elements when using a mass stopping power table. If the mass stopping powers for two different absorbers are to be compared, different values of Z/M_m, I, λ, and δ must be considered. However, the value of I enters through a logarithmic term that changes slowly. Thus, effects of small differences in the argument of this term can usually be neglected. Since λ and δ are small quantities, Z/M_m appears to be the most significant parameter. For monoisotopic elements the molar mass in atomic mass units is approximately equal to the mass number A of the stable isotope. Thus, the ratio Z/A can be considered.

The isotopic proton-to-neutron ratio is approximately equal to 0.5 except for the case of hydrogen. In heavy nuclei a neutron excess persists, and as a consequence, the Z/A value is somewhat less than 0.5. Generally Z/A variation from element to element is minimal. Table 2.4 shows some values of Z/A for various elements and average proton-to-neutron ratios for important substances. Note that the value for Z/A for hydrogen is unity, and thus substances containing hydrogen have relatively large average Z/A values.

It is evident from Table 2.4 and Equation 2.26 that Z/M_m or Z/A ratios can be utilized when interpolating for mass stopping power. For example, to find the collisional mass stopping power for element 2 from the known value for the neighboring element 1, a good approximation is obtained from

$$\left(\frac{S}{\rho}\right)_2 \simeq \frac{(Z/M_m)_2}{(Z/M_m)_1}\left(\frac{S}{\rho}\right)_1 \tag{2.35}$$

If one must calculate the mass stopping power for a substance containing several elements, Bragg's additivity rule can be used. Under this rule the mass stopping power for the substance is taken to be equal to the weighted sum of the mass stopping powers of the constituent atoms. Thus,

$$S/\rho = \sum_i \epsilon_i (S/\rho)_i \tag{2.36}$$

Table 2.4 Atomic-Number-To-Mass-Number Ratio

Material	*Z/A*
Hydrogen	1.00
Helium	0.50
Carbon	0.50
Oxygen	0.50
Aluminum	0.48
Copper	0.46
Iodine	0.42
Gold	0.40
Lead	0.40
Water	0.55
Muscle	0.56
Lithium fluoride	0.46
Air	0.50

In this expression ϵ_i is the fraction by weight of the ith atom in the molecule. Equation 2.36 yields very accurate results for high Z constituents. However, for light elements a substantial proportion of the total number of electrons are valence electrons, and molecular binding will affect their average excitation energy and thus the stopping power. Nevertheless, the stopping powers for common organic molecules can be calculated to an accuracy of about 1% if the Bragg rule is followed (33). If greater accuracy is desired, one can assign an average excitation energy that is specific for the chemical binding configuration of the constituent atom in the molecule (34). Values shown in Table 2.5 give average excitation energies for several binding configurations for hydrogen, carbon, and oxygen atoms.

Table 2.5 Average Excitation Energies for Atoms Bound in Molecules

Substance	*I(eV)*
Hydrogen (gas)	20.7
Hydrogen (saturated compounds)	17.6
Hydrogen (unsaturated compounds)	14.8
Carbon (graphite)	78.4
Carbon (saturated)	77.3
Oxygen (—O—)	98.5
Oxygen (O═)	88.9

Data from Ref. 34.

2.7. Problems

1. Evaluate the total cross section per electron given in Equation 2.10 by integrating over all energies between the limits for E_{min} and E_{max} given in Section 2.3. Show that in the limit when the ion kinetic energy is very much larger than the energy of the atomic electrons, the result is inversely proportional to I^2. Furthermore, show that the result decreases to zero as the ion kinetic energy decreases.
2. Using an equation in the chapter, justify the relationship used in heavy particle emulsion studies:

$$\frac{dN}{dx} = \frac{Cz^2}{\beta^2}\left[\frac{1}{E_1} - \frac{1}{E_2}\right]$$

   where N is the number of delta rays with kinetic energy between E_1 and E_2 per unit length of track, C is a constant, and $\beta = v/c$.
3. For incident protons with energy 1.4 MeV, calculate the maximum energy given to an atomic electron under classical assumptions. Relate this value to the dependence on Figure 2.6.
4. Calculate the rate of energy loss of a 2.5-MeV proton in aluminum. Use Equations 2.26 and 2.27 with no shell corrections or density corrections. Use the I_a value from Table 2.3.

5. Amend the calculations of problem 4 by adding the shell correction and the effective charge correction.
6. Calculate the rate of energy loss per unit length for a 5-MeV alpha particle (^{239}Pu) in water. Use Equations 2.26 and 2.27 and the I_a values from Table 2.3. Assume Bragg's rule.
7. Amend the calculation of problem 6 by using values of I_a from Table 2.5.
8. Compute the number of ion pairs per millimeter of path generated by a 2.5-MeV proton in water. Assume that the total energy dissipated for each ion pair produced is 33.8 eV.
9. Show that $\beta^2 = T(T + 2Mc^2)/(T + Mc^2)^2$.
10. If the rate of energy loss for a 2.5-MeV proton in graphite of density $(2.25)10^3$ kg/m^3 is 32.3 GeV/m, estimate the rate of energy loss in gaseous nitrogen of density 1.29 kg/m^3. Ignore differences in logarithmic terms.
11. If the rate of energy loss for a 2.5-MeV proton in water is 13.8 GeV/m, estimate the same quantity for a 10-MeV alpha particle.

2.8. References

1. L. A. Koenig, J. E. Mattauch, and A. H. Wapstra, Nucl. Phys. **31**, 18 (1962).
2. A. H. Rosenfeld, A. Barbaro-Goltieri, W. H. Barkas, P. L. Bastien, J. Kirz, and M. Roos, Rev. Mod. Phys. **32**, 633 (1965).
3. L. G. Christophorou, *Atomic and Molecular Radiation Physics* (Wiley-Interscience, New York, 1971), pp. 189–320, 589.
4. M. M. Ter-Pogossian, J. O. Eichling, D. O. Davis, and M. J. Welch, J. Clin. Invest. **49**, 381 (1970).
5. R. S. Tilbury, J. R. Dahl, and J. P. Mamacos, Int. J. Appl. Radiat. Isot. **21**, 277 (1970).
6. E. Lebowitz, M. W. Greene, R. Fairchild, P. R. Bradley-Moore, H. L. Atkins, A. N. Ansari, P. Richards, and E. Belgrave, J. Nucl. Med. **16**, 151 (1975).
7. C. J. Parnell, G. D. Oliver, Jr., P. R. Almond, and J. B. Smathers, Phys. Med. Biol. **17**, 429 (1972).
8. C. A. Kelsey, Med. Phys. **2**, 185 (1975).
9. F. K. Richtmyer, E. H. Kennard, and J. N. Cooper, *Introduction to Modern Physics,* 6th ed. (McGraw-Hill, New York, 1969), pp. 646–650.
10. J. E. Turner, Health Phys. **13**, 1255 (1967).
11. H. Bischel, "Charged particle interactions," in *Radiation Dosimetry,* vol. 1, edited by F. H. Attix and W. C. Roesch (Academic, New York, 1968), p. 157–228.
12. N. Bohr, Philos Mag. **25**, 10 (1913).
13. N. Bohr, Philos Mag. **30**, 581 (1915).
14. H. A. Bethe, Ann. Phys. (Leipzig) **5**, 325 (1930).
15. H. A. Bethe, Z. Phys. **76**, 293 (1932).
16. H. A. Bethe, "Quanten Mechanik der ein- und zwei-Elektronen Probleme," in *Handbuch der Physik,* vol. 24, 2nd ed. edited by H. Geiger and K. Scheel (Springer, Berlin, 1933), p. 273.
17. C. Möller, Ann. Phys. **14**, 531 (1932).
18. F. Bloch, Ann. Phys. **16**, 285 (1933).

19. F. Bloch, Z. Phys. **81**, 363 (1933).
20. R. D. Evans, *The Atomic Nucleus* (McGraw-Hill, New York, 1955), p. 636.
21. H. B. Gilbody and J. V. Ireland, Proc. R. Soc. London Ser A **277**, 137 (1964).
22. L. H. Toburen, Phys. Rev. A **3**, 216 (1971).
23. G. N. Ogurtsov, Rev. Mod. Phys. **44**, 1 (1972).
24. *Radiation Quantities and Units, ICRU Report 33* (International Commission on Radiation Units and Measurements, Washington, 1980).
25. R. L. Platzman, in *Symposium on Radiobiology,* edited by J. J. Nickson (John Wiley and Sons, New York, 1952), p. 97.
26. I. T. Myers, "Ionization," in *Radiation Dosimetry,* vol. 1, edited by F. H. Attix and W. C. Roesch (Academic, New York, 1968), pp. 317–330.
27. A. Mozumber and J. L. Magee, Radiat. Res. **28**, 203 (1966).
28. R. M. Sternheimer, Phys. Rev. **103** 511 (1956).
29. R. M. Sternheimer, Phys. Rev. **145**, 249 (1966).
30. E. Fermi, Phys. Rev. **57**, 485 (1940).
31. J. E. Turner, "Values of I and Iadj suggested by the subcommittee," in *Studies in Penetration of Charged Particles in Matter, Publication 1133* (National Academy of Sciences–National Research Council, Washington, 1964), p. 99.
32. L. Pages, E. Bertel, H. Joffre, and L. Sklavenitis, At. Data **4**, 3 (1972).
33. T. J. Thompson, University of California Laboratory Report UCRL-1910 (1952).
34. T. E. Burlin, "Cavity chamber theory," in *Radiation Dosimetry,* vol. 1, edited by F. H. Attix and W. C. Roesch (Academic, New York, 1968), p. 337.

3

Energy Loss by Electrons and Positrons

3.1 Electron and Positron Collisional Energy Loss

In this chapter both collisional and radiative energy loss processes for electrons and positrons will be discussed. We will use the term "electrons" to indicate the negatively charged particles with mass $(9.1)10^{-31}$ kg; the term "negatron" will not be utilized. The positively charged species with identical mass will always be referred to as a positron.

The description of collisional energy loss for electrons and positrons is similar in many respects to the description of energy loss for heavy ions. The parameters of the collision and the basic mechanism for energy transfer with resulting ionization and excitation of the target atoms is similar for all incident charged particles. The terminology used to describe the mathematical expressions is common to all charged particles; stopping power, linear energy transfer, and specific ionization are valid concepts for electrons and positrons as well as heavy ions.

Unfortunately, the assumptions that were necessary in the development of the energy loss cross-section expression in Chapter 2 are often invalid when electrons or positrons are considered. For example, Equation 2.19 can be used to show that it is possible for an incident electron to transmit *all* of its kinetic energy in a single head-on collision with an atomic electron. In fact, collisions causing the transfer of a substantial fraction of the available kinetic energy are relatively frequent for electrons, and they may involve a substantial change in the direction of motion of the incident particle.

In this section we will show expressions from quantum mechanical fomulations for the collisional loss cross section and stopping power for electrons and positrons. Arguments will be made by analogy to the heavy-ion discussions in Chapter 2, but the electron-electron collisional results will be quoted directly from references without any development. It will be apparent that the expressions describing electron and positron losses are similar in many ways to those describing heavy-particle losses.

The Möller cross section (1, 2) for energy between E and $E + dE$ imparted to an atomic electron by an incident electron of kinetic energy T can be written as

$$d\sigma(T,E) = \frac{d\sigma(T,E)}{dE}\, dE$$
$$= \frac{2\pi r_0^2 m_e c^2}{T(T+2m_e c^2)/(T+m_e c^2)^2}\left[\frac{1}{E^2} + \frac{1}{(T-E)^2} - \frac{(2T+m_e c^2)m_e c^2}{E(T+m_e c^2)^2(T-E)} + \frac{1}{(T+m_e c^2)^2}\right] dE \tag{3.1}$$

When $E \leq T/2 \ll m_e c^2$, the expression reduces to the nonrelativistic form first given by Mott (3). It is similar to Equation 2.10 for heavy ions except that the expression in the brackets has more terms. However, when $E \ll T/2 \ll m_e c^2$, the first term in the brackets dominates and Equation 3.1 reduces to the form of Equation 2.10.

The reasons behind the appearance of the extra terms in the brackets are at least twofold. One is that during an electron-electron collision, the two emerging particles cannot be distinguished and the electron that leaves with the most kinetic energy is usually defined to be the incident electron. Under this definition, the maximum energy transferred in a single collision is taken to be $T/2$, half of the initial electron kinetic energy.

Relativistic effects also give rise to complications in Equation 3.1. Relativistic formulations often are necessary when properties of electron beams must be described, but they are often unnecessary in descriptions of energy loss for incident heavy charged particles. For example, the rest mass equivalent of a proton is about 931 MeV. Protons liberated as secondaries during neutron irradiation usually have kinetic energy $T \ll 931$ MeV. However, the electron or positron rest mass is only 511 keV, and electron beams used in many applications have energies comparable to or in excess of this value. Even electrons liberated as secondary particles from ^{60}Co gamma-ray beams can have kinetic energy greater than 511 keV. Thus, accurate descriptions for these electrons must include relativistic corrections.

Figure 3.1 shows some theoretical ionization cross sections for argon expressed as a function of the incident electron energy (4). They have been integrated over all possible energy transfer values. Notice that the cross sections for several of the electron shells are shown. The appropriate ionization cross section is zero at the binding energy characteristic of each electron subshell and shows a maximum at about 100 eV for the $3s$ and $3p$ shells at least. Of course, binding energy effects at low energy are not incorporated into Equation

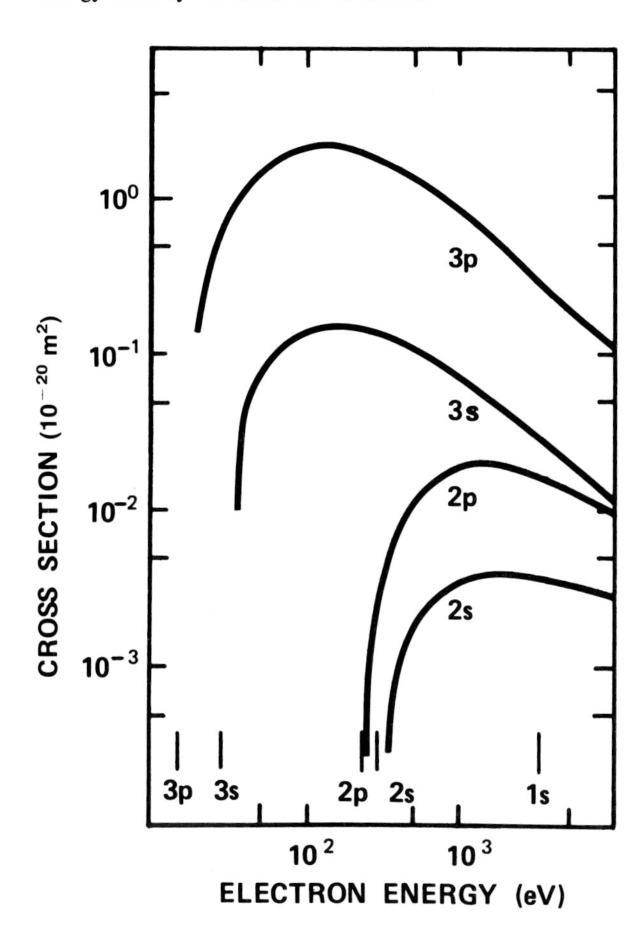

Figure 3.1 The partial ionization cross sections for electrons incident on argon atoms are shown as functions of the incident electron energy. The labels $2s$, $2p$, $3s$, and $3p$ indicate the subshells involved. Adapted from Ref. 4.

3.1. It is evident from the figure that less tightly bound outer-shell electrons contribute more to the energy transfer, since their cross sections peak at higher values. In most cases the inner-shell electrons contribute only about 10% to the total ionization (5).

The differential ionization cross section for 2-keV electrons incident on nitrogen targets (6) is shown in Figure 3.2 as a function of the energy of the liberated electrons. The cross section depends inversely on the square of the ejected electron energy, at least in the intermediate energy region. At smaller energies the binding energy of the atomic electrons in their parent atoms causes a reduction in the cross section.

The complete expression for the restricted stopping power for the incident electrons and positrons of kinetic energy T is given by Rohrlich and Carlson (7) as

$$L_{\Delta}^{\pm} = 2\pi r_0^2 m_e c^2 N_A (Z/M_m)\rho \frac{(T + m_e c^2)^2}{T(T + 2m_e c^2)} C^{\pm}(T, Z, \Delta) \tag{3.2}$$

$$C^{\pm}(T, Z, \Delta) = \left\{ \ln\left[\frac{2(T + 2m_e c^2)}{m_e c^2 (I/m_e c^2)^2}\right] + F^{\pm} - \delta - \lambda \right\} \tag{3.3}$$

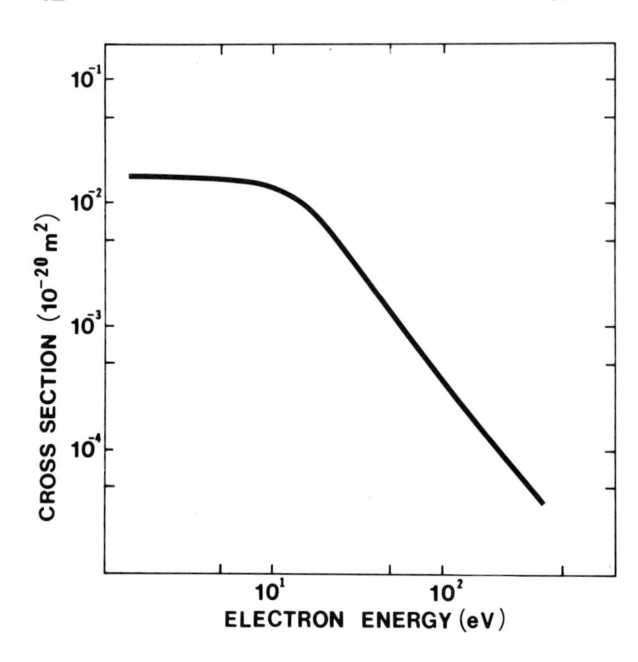

Figure 3.2 The ionization cross section for a nitrogen target by 2-keV incident electrons is shown as a function of the energy of the emitted electron. Adapted from Ref. 6.

For negative electrons the quantity F^- in Equation 3.3 is

$$F^- = -1 - \frac{T(T+2m_ec^2)}{(T+m_ec^2)^2} + \ln\left[\frac{(T-\Delta)\Delta}{(m_ec^2)^2}\right] + \left[\frac{T}{T-\Delta}\right] + \frac{\Delta^2/2 + (2T+m_ec^2)m_ec^2\ln(1-\Delta/T)}{(T+m_ec^2)^2} \tag{3.4}$$

For positrons the quantity F^+ in Equation 3.3 is

$$F^+ = \ln\left[\frac{T\Delta}{(m_ec^2)^2}\right] - \frac{[(T+2m_ec^2)/T]^{1/2}}{(T+m_ec^2)m_ec^2}\left\{T+\Delta-\frac{5\Delta^2/4}{(T+2m_ec^2)} + \frac{(T+m_ec^2)(T+3m_ec^2)\Delta-\Delta^3/3}{(T+2m_ec^2)^2} - \frac{(T+m_ec^2)(T+3m_ec^2)(\Delta^4/4)(m_ec^2)^2 - T\Delta^3/3+\Delta^4/4}{(T+2m_ec^2)^3}\right\} \tag{3.5}$$

The variable I is the average excitation energy and Δ is the cutoff energy, as was the case for heavy ions. Values of I_a shown in Table 2.3 can be used in Equation 3.3.

For electrons the stopping power (not restricted) can be obtained by setting

$$\Delta = T/2 \tag{3.6}$$

For positrons the stopping power (not restricted) can be obtained by setting

$$\Delta = T \tag{3.7}$$

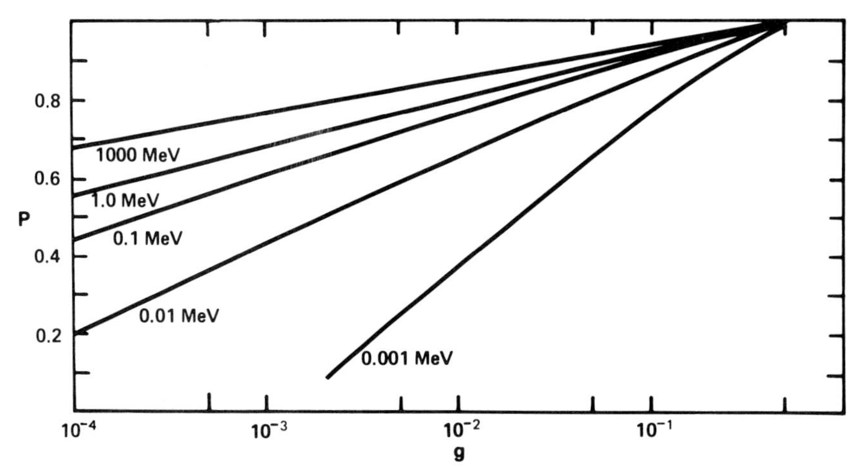

Figure 3.3 The ratio of the linear energy transfer to the collisional loss stopping power is shown as a function of electron energy (8). The abscissa is the parameter $g = \Delta/T$. Adapted from Ref. 8.

Figure 3.3 illustrates the effect of Δ on the energy loss expressions. P, the ratio of the restricted collisional stopping power to the total collisional stopping power, is shown as a function of $g = \Delta/T$ for several values of electron kinetic energy (8). The reduction factor is especially important when the incident-electron kinetic energy is small.

Figure 3.4 shows the electron stopping power in aluminum metal (9). Notice that the maximum stopping power occurs a few electron volts above

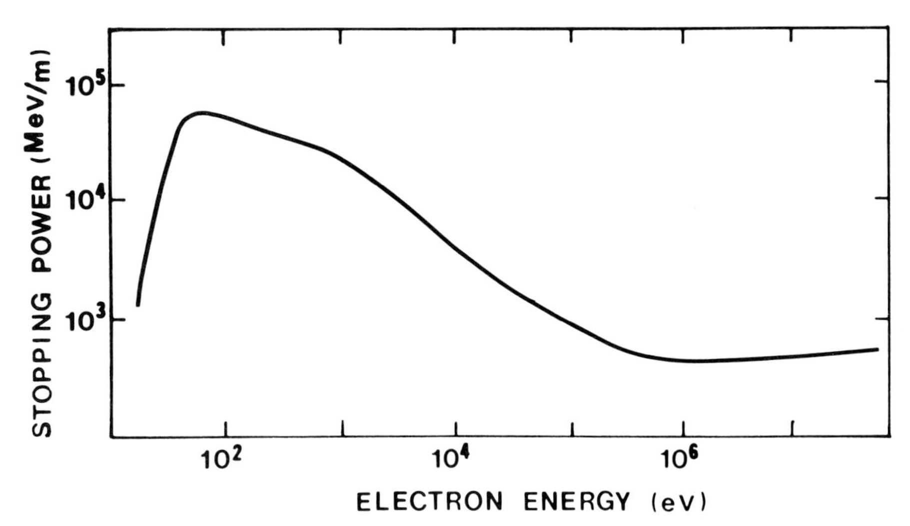

Figure 3.4 The stopping power for electrons in aluminum is shown as a function of the electron energy above the bottom of the conduction band (9).

the conduction band edge energy and the stopping power decreases thereafter with increasing energy until the MeV region. Of course, details in the functional dependence of the stopping power in metallic aluminum at low energies cannot be included in Equations 3.2 through 3.5. Nevertheless, the gross features at higher energies are accurately represented by these expressions. Figure 3.4 also shows that the collisional stopping power for aluminum increases above 1.5 MeV. This increase is a relativistic effect since it is contrary to the basic $1/v^2$ dependent.

The variable δ in Equation 3.3 is called the relativistic density effect correction term. Sternheimer has evaluated the term (10, 11). As was mentioned in Chapter 2, the electric field surrounding an incident charged particle induces a rearrangement of charge in the absorbing medium. The effect is to displace the bound atomic electrons from their parent nuclei along the line of the impressed field. The net result is the formation of aligned dipoles in the medium. This is termed medium polarization. Since the polarization is produced by the electric field of the incident particle, it decays away after the charge has passed by. The polarization partially neutralizes the effect of the charged particle field by screening the charge and thus reducing the stopping powers.

Medium polarization is most noticeable when target atoms are close to the incident-particle trajectory. This means that the effect of the phenomenon is more noticeable in solids than in low-density gases. Since the screening is dependent on the density of the medium, the stopping power reduction due to polarization is often called the density effect. Reductions due to the low-energy density effect are included when experimental values of I, the average excitation energy, are used. However, one must use values measured for the appropriate state of matter.

At high energies and relativistic speeds, the Lorentz contraction changes

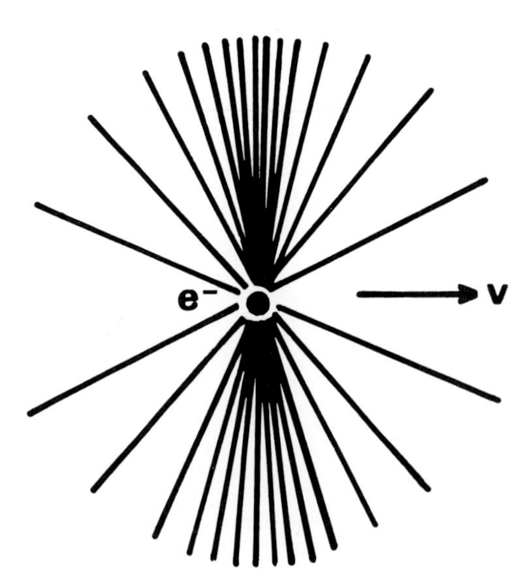

Figure 3.5 The pattern of the electric field lines around an electron e^- with relativistic speed v is shown.

Table 3.1 Percent Reduction of Collision Energy Loss Due to Density Effect

Electron T (*MeV*)	H_2[a]	*C*	*Al*	*Cu*	*Au*
0.1	0.0	0.0	0.0	0.0	0.0
0.2	0.0	0.4	0.0	0.0	0.0
0.5	0.0	1.2	0.5	0.5	0.0
1.0	0.0	2.7	1.5	1.5	0.7
2.0	0.0	4.8	3.4	3.4	2.0
5.0	0.0	8.5	6.8	6.8	4.9
10	0.0	11.8	9.8	9.9	7.6
20	0.0	15.2	13.1	13.3	10.7
50	0.7	19.5	17.3	17.6	14.9
100	3.3	22.5	20.3	20.7	18.1
200	6.6	25.1	23.1	26.6	21.1
500	10.6	28.1	26.4	27.0	24.8
1000	13.4	30.1	28.6	29.2	27.3

Data from Ref. 8.
[a] At normal pressure.

the electric field pattern of the incident particle as seen in the medium. As Figure 3.5 shows, the effective electric field for the incident electron tends to concentrate in the plane perpendicular to the velocity rather than maintaining the usual isotropic distribution. This change in pattern increases the effect of screening due to polarization. The correction term δ present in Equation 3.3 is due to this relativistic density effect. Table 3.1 gives some data for percentage stopping power reduction due to the relativistic density effect (8).

The term λ in Equation 3.3 is due to shell corrections. Shell corrections were discussed in Chapter 2 for incident heavy ions. The shell corrections occur because of a reduction in the energy transfer probability when incident particle speeds are comparable with electron speeds in their atomic orbitals. They are usually of less practical importance for electrons than they are for heavy charged particles. The speed of an incident electron, in most situations of interest, is considerably larger than the speed of the orbital electrons involved.

3.2 Bremsstrahlung Cross Section

The bremsstrahlung process is the emission of a photon by a free charged particle in a coulomb force field. The name "bremsstrahlung" means "braking radiation." Energy loss by bremsstrahlung emission is often termed radiative loss. We will devote considerable space to a discussion of bremsstrahlung emission during electron collisions with the nucleus. The probability for the process is quite large for high-energy electrons and positrons. In fact, at electron beam energies produced by many present-day accelerators, the electron radiative energy loss exceeds the collisional energy loss in heavy elements. In the following sections only emissions from electrons are discussed, but the basic results are applicable to positrons also.

Classical principles (12) can be used to derive the relationship between the power radiated as electromagnetic energy, dW/dt, and the acceleration a of a nonrelativistic particle with net charge ze:

$$\frac{dW}{dt} = \frac{(ze)^2 a^2}{6\pi\epsilon_0 c^3} \tag{3.8}$$

Since the acceleration is inversely proportional to the mass of the particle, the power radiated is inversely proportional to the square of the mass of the particle. Thus, if m_e is the electron mass and M is the mass of a heavy particle with net charge $\pm e$, and both experience the same force, then

$$\frac{(dW/dt)_M}{(dW/dt)_{m_e}} = \left(\frac{m_e}{M}\right)^2 \tag{3.9}$$

According to this equation, the radiative losses for electrons or positrons may be considerable, even when the losses for heavy ions with the same kinetic energy are entirely negligible.

The bremsstrahlung cross section has been discussed by many authors (13, 14). Fundamental contributions were made by Sommerfeld (15), Bethe (16), Heitler (17), and Schiff (18). The development to follow is semiclassical and much simplified (19, 20); some results from advanced formulations will be discussed later. Much of the functional dependence resulting from our discussions can be related easily to equations developed using more advanced theories.

The schematic diagram appropriate for the bremsstrahlung cross section is shown in Figure 3.6. It is similar to Figure 2.2 except that the particles

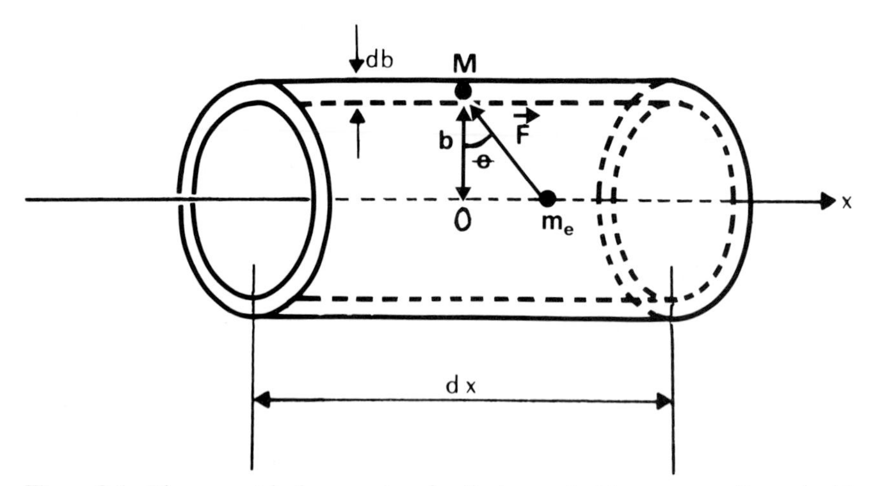

Figure 3.6 The geometrical parameters for the bremsstrahlung process for an incident electron of mass m_e and a target nucleus of mass M are illustrated. **F** is the force on the incident electron.

with mass M_e and M have been exchanged and the force is operating on the incident particle in this case. The incident electron is traveling in the positive x direction with speed v and kinetic energy T. The medium is characterized by mass density ρ and atomic density n_v atoms per cubic meter. The electron interacts with the nucleus of an atom of mass M and nuclear charge $+Ze$. The nucleus is assumed to be located in a cylindrical shell of radius b, length dx, and thickness db. At time $t = 0$, the incident particle is at closest approach with $x = 0$. The coulomb force is responsible for the acceleration ultimately resulting in emission of electromagnetic energy.

The electromagnetic energy is emitted as a photon with total energy content $h\nu$. The electron is slowed by the emission process since total energy is conserved. Neglecting any nuclear recoil,

$$T = h\nu + T', \quad h\nu_{\max} = T \tag{3.10}$$

where T' is the kinetic energy of the electron after the emission process. The maximum possible energy content of the photon equals the incoming kinetic energy. Other than this, there are no explicit limitations on the bremsstrahlung energy spectrum. In fact, the spectrum is continuous in frequency ν, since the incoming electron has a continuum of final energy states available. Assumptions necessary for our development include:

1. Electrons are not relativistic.
2. Movement of the nuclear target during the collision is negligible; effects of atomic electrons can be ignored.
3. Changes in the incident electron trajectory and speed during the effective collision times are small.

The assumptions are similar in some ways to those in Section 2.2 and are necessary to allow a simple expression for the acceleration of the electron. The restriction on nuclear motion causes little difficulty since nuclear recoil is usually not significant. Atomic electrons screen the nuclear charge quite effectively until the incident electron is near the nucleus. Thus, these electrons decrease the effect of long-distance collisions and decrease the effective collision time. The restriction on change in electron speed is a severe limitation. It means that the expressions we obtain will be valid only when a small fraction of the electron energy is radiated in a single collision. The high-frequency region of the bremsstrahlung spectrum will not be accurately represented by our functional results.

Figure 3.7 shows the force components on the incident electron attributable to the nuclear charge. The solid line follows the time dependence of the force if there were no atomic electrons to screen the nucleus; the dashed line indicates the time dependence of the force as the incident electron penetrates through the electron cloud into the nuclear region. Since the power radiated is related to the square of the acceleration, both the x component and the perpendicular component of the force are important. Both can induce emission of electromagnetic energy. For simplicity we will consider only the force component $F_\perp$ in the direction perpendicular to the incident-electron path and passing through the nuclear center.

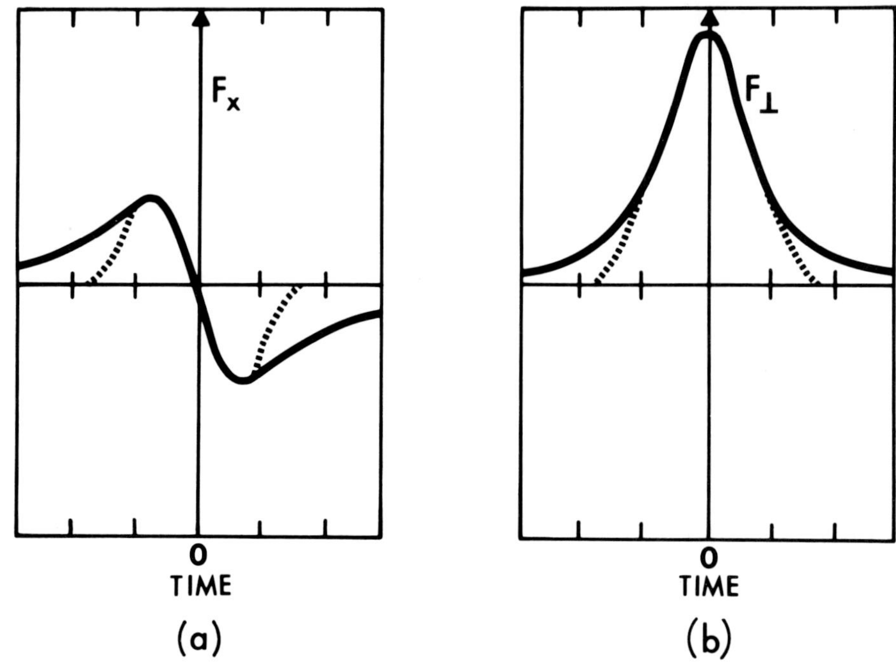

Figure 3.7 The components of the force on the incident electron are shown as a function of time when closest approach occurs at $t = 0$. The dashed curve shows how screening of the nucleus shortens the effective collision time.

The coulomb force on the incoming electron produces an acceleration in the perpendicular direction given by

$$a_\perp(t) = \frac{F}{m_e}\cos\theta = \frac{Ze^2}{4\pi\epsilon_0 m_e r^2}(b/r) = \frac{Ze^2}{4\pi\epsilon_0 m_e}\left[\frac{b}{(b^2+v^2t^2)^{3/2}}\right] \tag{3.11}$$

where r is the distance between the electron and the nucleus. The relationships

$$\cos\theta = b/r, \qquad r^2 = b^2 + v^2t^2 \tag{3.12}$$

were utilized in Equation 3.11. They can be verified with reference to Figure 3.6.

At this point a brief digression on the properties of a Fourier transform pair is desirable. A more complete discussion of Fourier transforms can be found in Ref. 21. Recall that if $f(t)$ and $g(\omega)$ are a transform pair of functions, they can be related by

$$g(\omega) = \frac{1}{\sqrt{2\pi}}\int_{-\infty}^{\infty} f(t)\exp(-i\omega t)dt, \quad f(t) = \frac{1}{\sqrt{2\pi}}\int_{-\infty}^{\infty} g(\omega)\exp(i\omega t)d\omega \tag{3.13}$$

where t and ω are the conjugate variables. An important property of the Fourier transformation is that the squares of the absolute values of the transform functions yield equal results when integrated over their appropriate conjugate variable spaces. Thus,

$$\int_{-\infty}^{\infty} |f(t)|^2 dt = \int_{-\infty}^{\infty} |g(\omega)|^2 d\omega \tag{3.14}$$

Suppose $f(t)$ is real. Then from Equation 3.13,

$$g(-\omega) = \frac{1}{\sqrt{2\pi}} \int_{-\infty}^{\infty} f(t) \exp(+i\omega t) dt = g^*(\omega), \quad |g(-\omega)|^2 = |g(\omega)|^2,$$

$$\int_{-\infty}^{\infty} [f(t)]^2 dt = 2\int_{0}^{\infty} |g(\omega)|^2 d\omega \tag{3.15}$$

The Fourier transformation is often used in physical problems because it enables one to obtain distributions in frequency (or momentum) from functions describing the time (or spatial) dependence.

Since we are interested in obtaining the bremsstrahlung cross section as a function of frequency, Equation 3.13 can be utilized to transform the acceleration of Equation 3.11:

$$a_\perp(\omega) = \frac{1}{\sqrt{2\pi}} \int_{-\infty}^{\infty} a_\perp(t) \exp(-i\omega t) dt =$$

$$\frac{Ze^2 b}{\sqrt{2\pi}\, 4\pi\epsilon_0 m_e} \int_{-\infty}^{\infty} \frac{(\cos \omega t + i \sin \omega t) dt}{(b^2 + v^2 t^2)^{3/2}} \tag{3.16}$$

Since $\sin \omega t$ is an odd function, the imaginary part of Equation 3.16 will not contribute to the integrated result. Further simplification is possible utilizing the properties of the even function remaining:

$$a_\perp(\omega) = \frac{2Ze^2 b}{\sqrt{2\pi}\, 4\pi\epsilon_0 m_e} \int_0^{\infty} \frac{\cos \omega t\, dt}{(b^2 + v^2 t^2)^{3/2}} = \frac{2Ze^2}{\sqrt{2\pi}\, 4\pi\epsilon_0 m_e v^2}\, \omega K_1\!\left(\frac{\omega b}{v}\right) \tag{3.17}$$

$K_1(\omega b/v)$, the modified Bessel function of order 1, has an integral form (22) similar to that shown in Equation 3.17. In order to obtain this Bessel function form, v^3 was factored out of the integral. This step will introduce anomalies in the result unless the energy radiated $\hbar\omega \ll T$. $K_1(\omega b/v)$ is real, and, as Figure 3.8 shows, it is very large at small values of the argument but rapidly decreases as the argument increases.

From Equations 3.8, 3.14, and 3.15, it follows that the total energy radiated for a single collision at impact parameter b can be obtained from

$$W_s = \int \left[\frac{dW_s}{dt}\right] dt = \frac{e^2}{6\pi\epsilon_0 c^3} \int_{-\infty}^{\infty} [a_\perp(t)]^2 dt$$

$$= \frac{e^2}{6\pi\epsilon_0 c^3} \int_{-\infty}^{\infty} [a_\perp(\omega)]^2 d\omega = \frac{e^2}{3\pi\epsilon_0 c^3} \int_0^{\infty} [a_1(\omega)]^2 d\omega \tag{3.18}$$

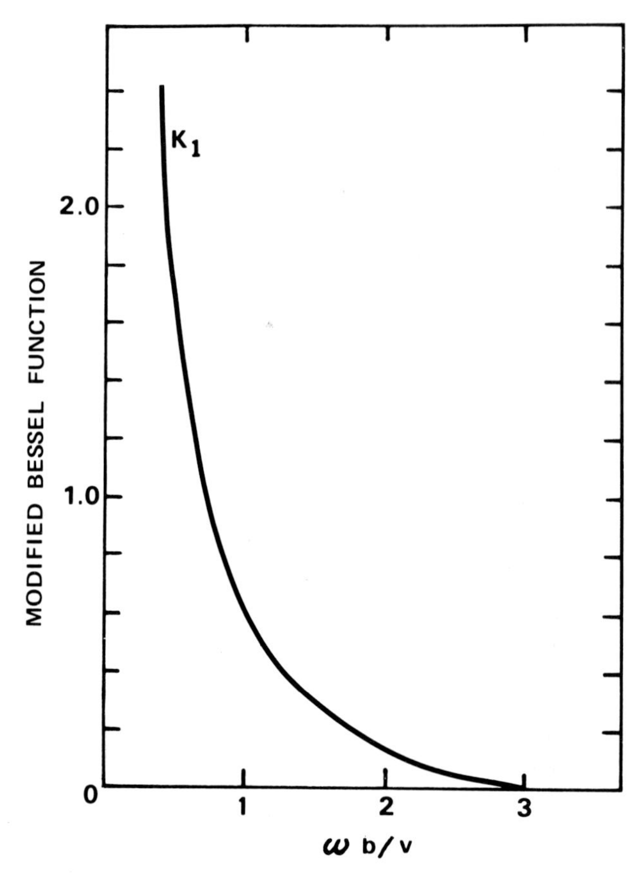

Figure 3.8 The dependence of the modified Bessel function K_1 on its argument is shown (22).

The subscript s indicates that only a single collision is involved. Equation 3.14 was utilized because $a_{\perp}(t)$ and $a_{\perp}(\omega)$ are identified as a real transform pair. By definition,

$$W_s = \int_0^{\omega_{\max}} \left[\frac{dW_s}{d\omega}\right] d\omega = \frac{e^2}{3\pi\epsilon_0 c^3} \int_0^{\omega_{\max}} [a_{\perp}(\omega)]^2 d\omega \tag{3.19}$$

The maximum possible value of the angular frequency $\omega_{\max}$ was used instead of the infinite upper limit in Equation 3.18. Identification of integrands in Equation 3.19 leads to

$$\frac{dW_s}{d\omega} = \frac{e^2}{3\pi\epsilon_0 c^3} [a_{\perp}(\omega)]^2 = \frac{e^2}{3\pi\epsilon_0 c^3} \left[\frac{2Ze^2\omega K_1(\omega b/v)}{\sqrt{2\pi}\, 4\pi\epsilon_0 m_e v^2}\right]^2$$

$$= \frac{8}{3\pi} Z^2 r_0^2 \alpha \left[\frac{c}{v}\right]^2 \left[\frac{\omega}{v}\right]^2 \hbar \left[K_1\left(\frac{\omega b}{v}\right)\right]^2 \tag{3.20}$$

In this expression, the classical electron radius r_0 and the fine structure constant α have been used:

$$r_0 = \frac{e^2}{4\pi\epsilon_0 m_e c^2}, \qquad \alpha = \frac{e^2}{4\pi\epsilon_0 \hbar c} \tag{3.21}$$

The total energy loss for all values of the impact parameter is equal to the energy loss per collision (Equation 3.20) multiplied by the number of collisions and integrated over the impact parameter. If W_T indicates total energy radiated,

$$d^3 W_T = \left[\frac{8}{3\pi} Z^2 r_0{}^2 \alpha \left[\frac{c}{v}\right]^2 \left[\frac{\omega}{v}\right]^2 \left[K_1\left(\frac{\omega b}{v}\right)\right]^2 \hbar d\omega\right] [n_v(2\pi b db)dx] \tag{3.22}$$

Then

$$\frac{d^2 W_T}{d\omega dx} = Z^2 r_0{}^2 \alpha n_v \hbar \left\{\frac{16}{3}\left[\frac{c}{v}\right]^2 \int \left[K_1\left(\frac{\omega b}{v}\right)\right]^2 \left[\frac{\omega}{v}\right]^2 b db\right\}$$

$$= Z^2 r_0{}^2 \alpha n_v \hbar \Phi(T, Z, h\nu) \tag{3.23}$$

The integration indicated in the braces in Equation 3.23 must be taken over all allowed values of the impact parameter. The intensity function $\Phi(T, Z, h\nu)$ is introduced in Equation 3.23 for simplicity in the notation. In this case Φ does not refer to particle fluence. The arguments T and $h\nu$ follow from the dependence of the integrand in Equation 3.23 on v and ω. The Z dependence is not so evident but occurs because of nuclear screening by atomic electrons.

The electromagnetic energy is radiated in the form of photons with energy $\hbar\omega = h\nu$. The number of photons emitted, d^2N, is equal to the total energy emitted divided by the energy of each photon:

$$d^2 N = \frac{d^2 W_T}{d\omega dx}\left[\frac{d\omega dx}{h\nu}\right] = Z^2 r_0{}^2 \alpha n_v \Phi(T, Z, h\nu) \frac{dh\nu}{h\nu} dx \tag{3.24}$$

The differential cross section for emission of a photon with energy between $h\nu$ and $h\nu + dh\nu$ per unit energy is equal to the photon emission probability per incident electron divided by the number of nuclear targets per square meter:

$$\frac{d\sigma(h\nu)}{dh\nu} = \frac{d^2 N}{dh\nu(n_v dx)} = Z^2 r_0{}^2 \alpha \frac{\Phi(T, Z, h\nu)}{h\nu} \tag{3.25}$$

Figure 3.9 shows the results of measurements of the bremsstrahlung cross section by Motz (23). Notice that the measured values have been multiplied by $h\nu$ and divided by Z^2 so the value plotted is proportional to the intensity function $\Phi(T, Z, h\nu)$. The function must equal zero when $h\nu = T$, as indicated by energy conservation. The functional form given in Equation 3.25 is compatible with that obtained from realistic calculations (17, 18). The result for $\Phi(T, Z, h\nu)$, which is obtained from the integration in Equation 3.23 and involves K_1, deviates from experimental results when $h\nu \simeq T$. However, when $h\nu \ll T \ll m_e c^2$, Equation 3.23 is adequate (19). In relativistic formulations $\Phi(T, Z, h\nu)$ is a rather complicated function of the parameters.

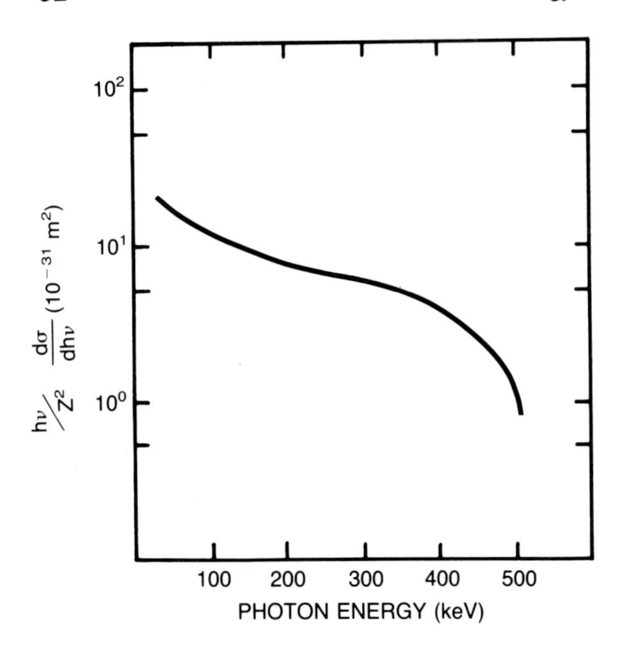

Figure 3.9 The experimental results for the bremsstrahlung spectrum from a 500-keV electron are shown (22).

3.3 Radiative Stopping Power and Related Concepts

Since a charged particle loses energy in an absorber by radiative as well as collisional processes, the stopping power must include both effects. The total stopping power S_t is the sum of the collisional stopping power S_c and the radiative stopping power S_r:

$$S_t = S_c + S_r \tag{3.26}$$

S_r is defined with analogy to Section 2.3 (24). The stopping power for electromagnetic energy radiated is the quotient of $-dT$ by dx, where $-dT$ is the average energy radiated by a charged particle in distance dx. Thus,

$$S_r = (-dT/dx)_r \tag{3.27}$$

An analytical expression for the average energy radiated by one electron can be obtained by multiplying the number of target nuclei per unit area, $n_v dx$, by the energy radiated per electron, per target per unit area. Equation 3.25 can be used for the differential bremsstrahlung cross section per unit energy. The total energy radiated can be written as

$$-dT = (n_v dx)\int_0^T h\nu \frac{d\sigma(h\nu)}{dh\nu}\, dh\nu = (n_v dx)Z^2 r_0{}^2\alpha\int_0^T \Phi(T,Z,h\nu)dh\nu \tag{3.28}$$

Then

$$S_r = (-dT/dx)_r = Z^2 r_0{}^2 \alpha n_v \int_0^T \Phi(T,Z,h\nu)dh\nu$$

$$= Z^2 r_0{}^2 \alpha n_v R(T,Z)\int_0^T dh\nu = Z^2 r_0{}^2\alpha(N_A/M_m)\rho R(T,Z)T \tag{3.29}$$

The substitutions utilized in Equation 3.29 are

$$n_v = \frac{N_A}{M_m}\rho, \quad R(T,Z) = \frac{\int_0^T \Phi(T,Z,h\nu)dh\nu}{\int_0^T dh\nu} \tag{3.30}$$

where N_A is Avogadro's number, M_m is the molar mass, ρ is the mass density, and $R(T,Z)$ is the spectrum average of the intensity function.

The principal feature of the stopping power expression can be divided into two parts.

1. **Electron Properties**
   The radiative stopping power depends directly on the kinetic energy of the incident electron. It also depends on the average of the intensity function, which includes a slow dependence on kinetic energy.
2. **Medium Properties**
   The radiative stopping power depends directly on the square of the nuclear atomic number and also depends directly on the nuclear density of the medium. The average of the intensity function produces an additional slow dependence on atomic number.

The radiation length is often used to compare the efficiency of various materials as bremsstrahlung emitters. The radiation length l_r is defined to be the path length that will reduce the energy of an incident electron to 0.368 of its initial value due to radiation loss. From Equation 3.29, for an incident electron of energy T_0,

$$\int_{T_0}^{T} \frac{dT'}{T'} = -Z^2 r_0^2 \alpha \left(\frac{N_A}{M_m}\right)\rho R(T,Z) \int_0^x dx' = -\frac{1}{l_r}\int_0^x dx',$$

$$\ln T - \ln T_0 = -\frac{x}{l_r}, \quad T = T_0 \exp(-x/l_r) \tag{3.31}$$

Therefore,

$$l_r = [Z^2 r_0^2 \alpha (N_A/M_m)\rho R(T,Z)]^{-1} \tag{3.32}$$

An approximation for $R(T,Z)$ that is often used in calculations of l_r for high-energy electrons is (13)

$$R(T,Z) = 4 \ln(183 Z^{-1/3}) \tag{3.33}$$

This function includes the effects of screening of the nuclear charge by atomic electrons. Partial corrections for electron-electron bremsstrahlung can be included in Equations 3.29 and 3.32 by replacing Z^2 with $Z(Z + 1)$. More exact expressions for the average value of the intensity function can be quite complicated.

Recommended values for the radiation length, including the energy dependence, are given in Table 3.2. Notice that the radiation length is only a

Table 3.2 Radiation Lengths

Material (*Z*)	ρ (*kg/m*^{3})	l_r (*m*)	$l_r\rho$ (*kg/m*^{2})	$l_r^{40}\rho$ (*kg/m*^{2})	$l_r^{120}\rho$ (*kg/m*^{2})
Water	$(1.0)10^3$	0.506	506		
Carbon (6)	$(2.5)10^3$	0.206	517	546	485
Aluminum (13)	$(2.7)10^3$	0.0974	263	296	266
Copper (20)	$(8.9)10^3$	0.0148	132	154	140
Tantalum (73)	$(16.6)10^3$	0.0039	64	80	73
Tungsten (74)	$(19.4)10^3$	0.0033	64		
Platinum (78)	$(21.5)10^3$	0.0028	61		
Lead (82)	$(11.3)10^3$	0.0052	59		

$l_r\rho$ computed from Equations 3.32 and 3.33; $l_r^{40}\rho$, at 40 MeV (14); $l_r^{120}\rho$, at 120 MeV (14).

few millimeters for high Z materials like tantalum [$Z = 73$, $\rho = (16.6)10^3$ kg/m^3], whereas the radiation length for low Z materials like graphite [$Z = 6$, $\rho = (2.5)10^3$ kg/m^3] is quite large.

3.4 Bremsstrahlung Yields from Thick Absorbers

Although some bremsstrahlung accompanies the degradation of electrons in tissue and aqueous solutions, the total amount of energy radiated is usually quite small when compared to the losses by ionization and excitation. Since the stopping power for radiative loss is proportional to Z^2, bremsstrahlung is much more significant when electrons impinge on high atomic number materials. The energy radiated from metallic absorbers, particularly the targets in x-ray machines and linear electron accelerators, is of special interest.

Before discussing the amount of electromagnetic energy emitted in traversing an absorber, it is useful to introduce two categories: thin absorbers and thick absorbers. In a thin absorber (or thin target), the probability of appreciable energy loss prior to emission of a photon is negligible. In a thick absorber (or thick target), emission of several photons as well as considerable collisional loss by a single electron is quite probable in the absorber.

Figure 3.10 shows the energy fluence spectrum that might be emitted from a thick absorber as compared with the spectrum for a thin absorber. Of course, the total energy radiated by the thick absorber would substantially exceed that of the thin absorber in most cases. However, in this figure the spectra are normalized for equal total energy to compare the shapes more easily. Notice that the spectrum for the thick target contains relatively more low-energy photons. This is because many electrons penetrate well into a thick target before emitting a photon. Since the electron energy has been degraded due to collisional losses before emission, the emitted photon spectrum must be degraded. In the next few paragraphs the amount of energy radiated in thick absorbers will be discussed with due concern for energy degradation effects.

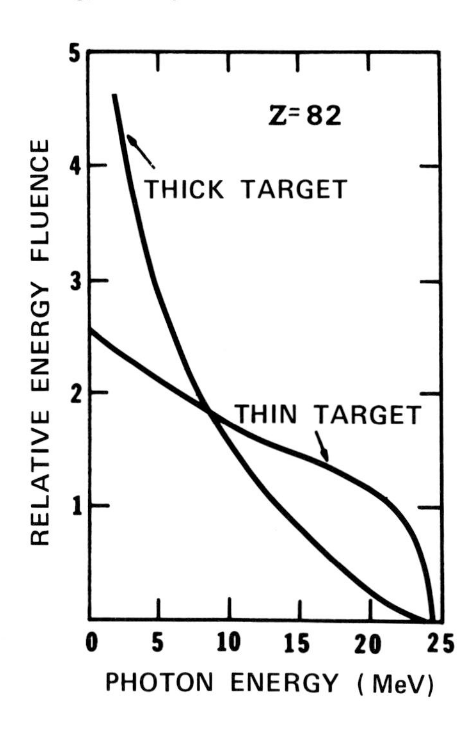

Figure 3.10 The bremsstrahlung energy fluence emitted in a thick target is compared to the thin target fluence curve. The total energy from the thick target is usually much greater than the thin-target radiated energy. In this case the two curves have been normalized for equal total energy.

Suppose that E_r is the total energy radiated in a thick absorber. If the electron is completely stopped in the absorber,

$$E_r = \int_0^s S_r\,dx = \int_T^0 \frac{S_r dT'}{(dT'/dx)} = \int_0^T \frac{S_r dT'}{(-dT'/dx)} = \int_0^T \frac{S_r dT'}{S_t} \tag{3.34}$$

In this equation, s is the total path length of the electron in the absorber, x is the path length variable, and T' is an integration variable for kinetic energy. The limits of integration in Equation 3.34 are used because $x = 0$ at entrance into the absorber when the energy is T. When the distance variable equals the total path length s, the kinetic energy is completely dissipated and $T = 0$.

The radiation yield $Y(T,Z)$ is the fraction of the kinetic energy radiated in the course of slowing down. When nonrelativistic electrons are considered, $S_r \approx S_c \gg S_r$. The radiation yield can be expressed by

$$Y(T,Z) = \frac{E_r}{T} \simeq \frac{1}{T}\int_0^T \left[\frac{S_r}{S_c}\right] dT'$$

$$= \frac{Z^2 r_0^2 \alpha (N_A/M_m)\rho}{\pi r_0^2 (m_e c^2)^2 N_A (Z/M_m)\rho}\left(\frac{1}{T}\right)\int_0^T \frac{R(T',Z)T'dT'}{C(T',Z)/T'} \tag{3.35}$$

Equations 3.2, 3.29, and 3.34 were used to obtain this expression with ($T \ll m_e c^2$). Since $R(T',Z)/C(T',Z)$ varies slowly with kinetic energy, $Y(Y,Z)$ depends approximately on ZT^2.

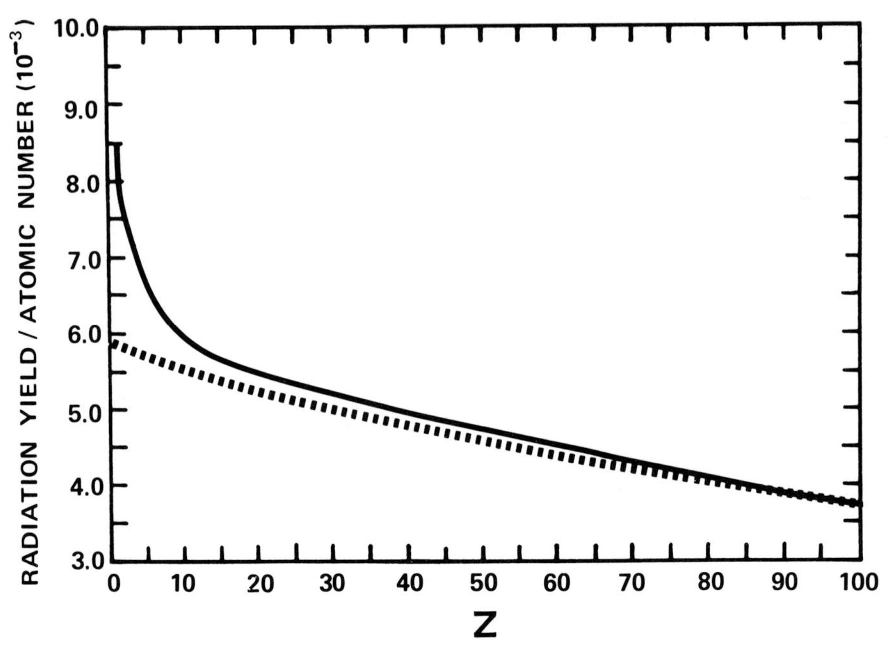

Figure 3.11 The ratio of the radiation yield to the atomic number Z for 10-MeV electrons incident on a target is shown as the solid line; the dashed line gives the result of Equation 3.36. All ordinate values must be multiplied by 10^{-3}.

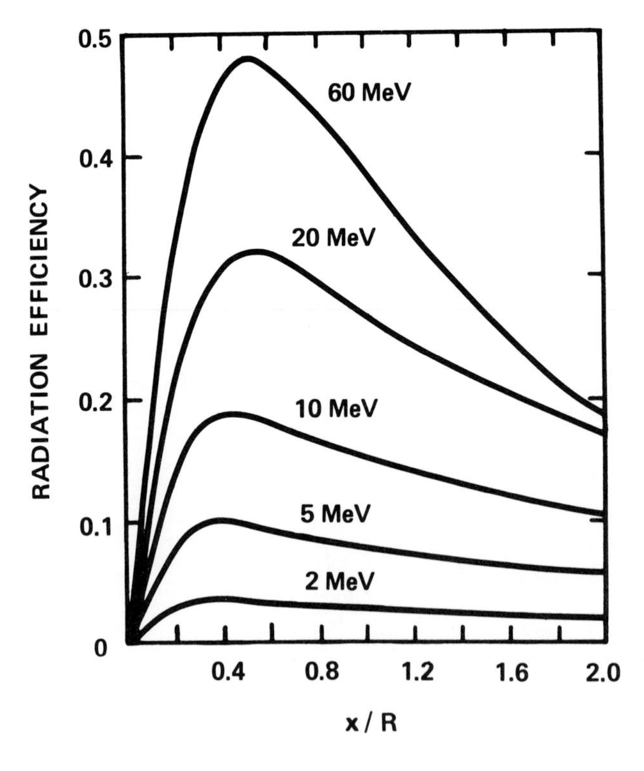

Figure 3.12 The radiative efficiency for electrons incident on a tungsten target is shown as a function of relative target thickness. R is the incident electron range; x is the actual thickness. Adapted from Ref. 26.

A function that fits experimental data when $T \gg m_e c^2$ for high atomic number absorbers (9, 25) is

$$Y(T,Z) = \frac{(6.0)10^{-4}ZT}{1 + (6.0)10^{-4}ZT} \tag{3.36}$$

where T is expressed in megaelectron volts. Figure 3.11 shows that this functional form is appropriate for absorbers with atomic number greater than about 10.

Of course, the effects of self-absorption of the radiation in the radiating material are not included in Equations 3.35 or 3.36. Self-absorption has been included in the function shown in Figure 3.12, which represents emitted efficiency for tungsten targets (26). The abscissa is the target thickness divided by the path length of the electron in the target. Notice that for all the energies considered, the maximum radiation efficiency occurs when the abscissa is 0.5 or less.

3.5 Systematics of Electron and Positron Mass Stopping Power

The mass stopping power values for electrons in several absorbers (8) are given in Appendix 7. Values are included for many energies and both collisional and radiative loss rates are tabulated. Many of the comments given in Section 2.6 for ions apply for electrons also. Stopping power definitions and the reasons for tabulating mass stopping powers are similar for electrons and ions.

Equation 3.2 in this chapter shows that the collisional energy loss stopping power for electrons is proportional to Z/M_m for the medium. Thus, one can estimate the collisional loss mass stopping power for element 2 from the tabulated value for a neighboring element 1 by

$$\left(\frac{S_c}{\rho}\right)_2 \simeq \frac{(Z/M_m)_2}{(Z/M_m)_1}\left(\frac{S_c}{\rho}\right)_1 \tag{3.37}$$

Of course, this relationship is approximate and is valid only for neighboring elements. The logarithmic dependence of the stopping power on the average ionization potential has been ignored (as well as density effect and shell correction terms). In most cases the mass number A could be substituted for M_m in this relationship.

It is evident from Equation 3.29 that the radiative stopping power for electrons is proportional to Z^2/M_m for the medium. In this case one can estimate the radiative mass stopping power for element 2 from the value for neighboring element 1 using

$$\left(\frac{S_r}{\rho}\right)_2 \simeq \frac{(Z^2/M_m)_2}{(Z^2/M_m)_1}\left(\frac{S_r}{\rho}\right)_1 \tag{3.38}$$

For better accuracy, Z^2 can be replaced by $Z(Z+1)$ to include atomic electron effects. Of course, the ratio of the logarithmic dependence of Equation 3.33 can be included if it is desired.

Bragg's additivity rule can be used to calculate the electron mass stopping power for a molecule containing several elements. This has already been discussed in Chapter 2. Equation 2.36 from Section 2.6 can be used for stopping powers of incident electrons as well as those for ions.

As was mentioned in Section 3.3, the total energy loss is the sum of the loss due to collisional processes and that due to the emission of bremsstrahlung. For electrons at low energies, the collisional loss dominates. At high energies the radiative loss dominates. Figure 3.13 shows that the collisional stopping power for electrons in water is greater than the radiative stopping power unless T is greater than 50 MeV. Notice that the radiative loss in tungsten equals the collisional energy loss at about 9 MeV.

From Equations 3.2 and 3.29,

$$\frac{S_r}{S_c} = \frac{Z^2 r_0^2 \alpha (N_A/M_m)\rho R(T,Z)T}{2\pi r_0^2 m_e c^2 N_A (Z/M_m)\rho[(T + m_e c^2)^2/T(T + 2m_e c^2)]C(T,Z)}$$

$$= \frac{ZT/m_e c^2}{M(T,Z)} \tag{3.39}$$

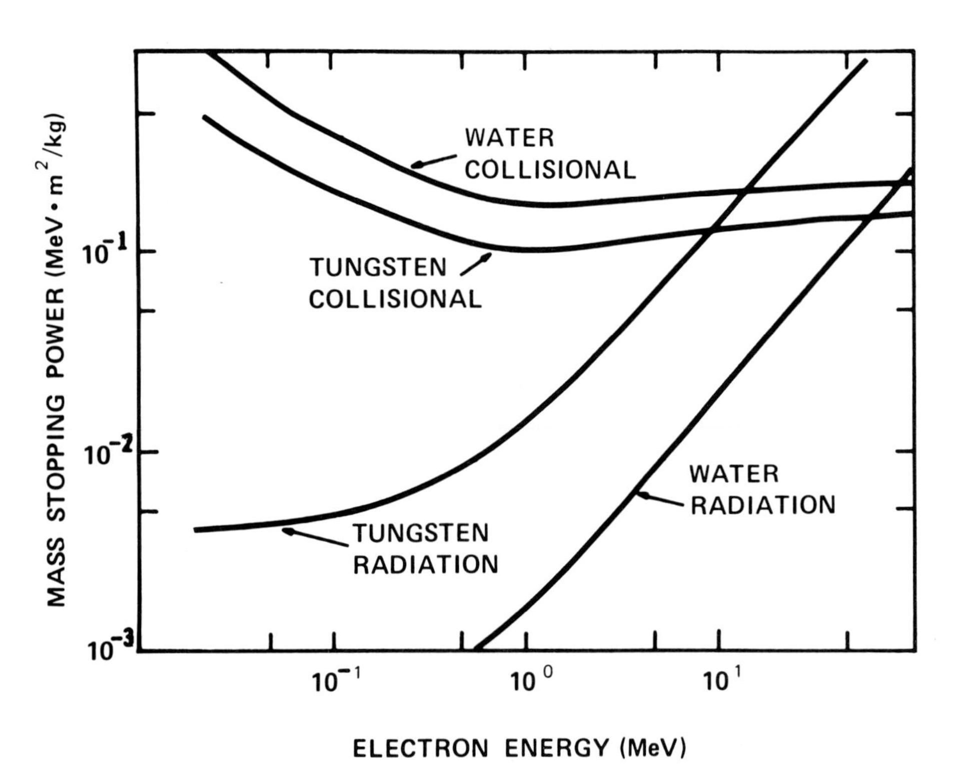

Figure 3.13 The mass stopping power functions for collisional loss and for radiative loss are shown (8) for tungsten and water absorbers.

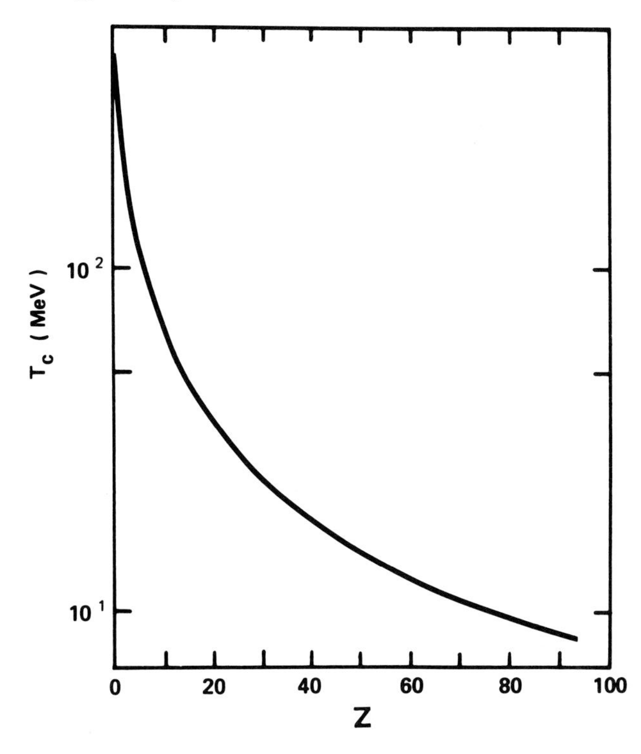

Figure 3.14 The critical energy T_c is shown as a function of target atomic number Z. Adapted from Ref. 8.

We have defined the collisional to radiative loss function $M(T,Z)$ by this equation. $M(T,Z)$ is slowly varying in T and Z when the energy $T \gg m_e{}^2$.

The critical energy T_c is defined as the energy such that the radiative stopping power equals the collisional stopping power. When $T = T_c$,

$$S_r/S_c = \frac{ZT_c/m_ec^2}{M(T_c,Z)} \qquad T_c = \frac{M(T_c,Z)m_ec^2}{Z} \tag{3.40}$$

A good approximation to the stopping power ratio and to the critical energy is given by (8)

$$S_r/S_c = (Z + 1.2)T/800, \qquad T_c = 800/(Z + 1.2) \tag{3.41}$$

where T is in megaelectron volts. Figure 3.14 shows the critical energy as a function of Z.

With regard to the critical energy, it may be of some interest that materials used for targets in x-ray machines (with $Z > 70$) have critical energies in the neighborhood of 10 MeV. Because diagnostic x-ray generators operate well below this energy, collisional loss dominates in the target, and the bulk of the electron energy incident on target is converted to heat. Thus, x-ray generators are relatively inefficient bremsstrahlung producers. When high-intensity beams are needed for long durations, targets must be externally cooled.

3.6 Positron Annihilation

Preceding sections of this chapter give information on radiative stopping powers, radiation yields, and related concepts for negative electrons with no direct reference to positrons. In most cases the expressions obtained are applicable to positrons as well as electrons because the mass and charge magnitude are the same for either particle. Of course, electrons are of primary concern because they are intermediaries in most absorption processes, and electron beams are produced in many devices. Positron beams are less common, but the particles occur in medical and industrial applications or of radiations, either as a product of radioactive decay (β^+ decay) or after absorption of high-energy photons (pair production). After degradation to thermal energies, electrons tend to diffuse through an absorbing medium until charge equilibrium is established. No unique phenomenon accompanies their thermalization. We include this section because positrons disappear by annihilation during thermalization and thereafter.

Positron annihilation is the process wherein a positron and an electron disappear simultaneously and electromagnetic energy is radiated away in their place. The process is illustrated on Figure 3.15 for the most probable case of emission of two monoenergetic annihilation photons. Three-photon emission occurs only 1 time in about 370, and the possible photon energies show a continuous spectrum of values (27). Single-quantum emission in the electric field of a nucleus is even more unlikely. The result of positron annihilation is the production of secondary photons, as is the case for the bremsstrahlung process. For the most part, annihilation photons have a single characteristic energy value, easily distinguished from the continuous energy spectrum of the bremsstrahlung.

The equations for energy and momentum conservation for two-photon annihilation shown in Figure 3.15 are

$$T_+ + m_e c^2 + T_- + m_e c^2 = h\nu_1 + h\nu_2,$$

$$\mathbf{p}_+ + \mathbf{p}_- = \mathbf{p}_1 + \mathbf{p}_2 \tag{3.42}$$

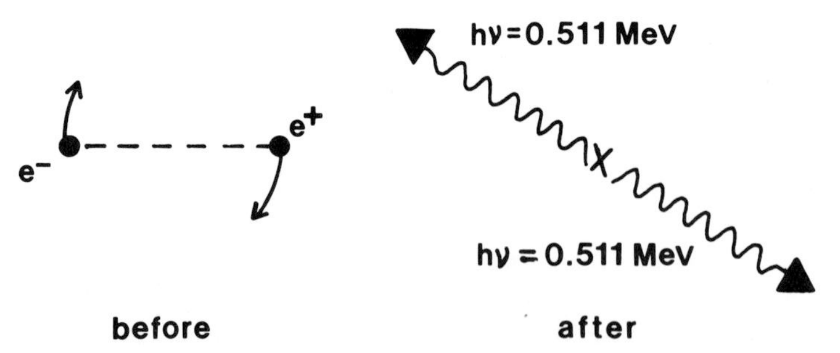

Figure 3.15 The annihilation of a positron with an electron is shown for the case when two photons are emitted.

In these equations the subscripts + and − indicate positrons and electrons, respectively, while the subscripts 1 and 2 are for the photons. T indicates kinetic energy and $\mathbf{p}$ stands for momentum as usual.

Dirac derived an expression for the cross section for positron annihilation in 1930 (28). In the limit with $v_+ \ll c$,

$$\sigma = \pi r_0^2 \frac{c}{v_+} \tag{3.43}$$

Low-energy positron annihilation is clearly favored. In fact, degrading collisional and radiative losses usually dominate for fast positrons until they are thermalized. The positrons then are annihilated, usually with valence electrons (27, 29).

If the momentum and kinetic energy of the positron and electron are negligible upon annihilation, Equations 3.42 simplify:

$$h\nu_1 + h\nu_2 = 2m_ec^2, \quad \mathbf{p}_1 = -\mathbf{p}_2, \quad h\nu_1/c = h\nu_2/c,$$

$$h\nu_1 = h\nu_2 = m_ec^2 \tag{3.44}$$

The 0.511-MeV photons are characteristic of the annihilation process. Under the assumption of negligible input momentum, the two photons must emerge at directions separated by 180°.

In fact, the momentum of the positron and electron before annihilation is not completely negligible. Because of residual momentum, the photons depart at relative angles distributed around 180° with an angular width of about 1° or less. The photon energy spread is often about 10 eV for the same reasons, as long as $T_+ \ll m_ec^2$, $T_- \ll m_ec^2$ (27). Relativistic positrons can annihilate in flight prior to thermalization. In this case the photon emitted in the forward direction gets the bulk of the available energy.

In noncrystalline materials, a positron and an electron can form a bound system, called positronium, prior to annihilation. The two particles move about their center of mass in this quantized system (28). If the spins of the electron and positron are antiparallel (singlet, $S = 0$), the system is parapositronium. If the spins are parallel (triplet, $S = 1$), the system is called orthopositronium. Since the ground state of the positronium system has no orbital angular momentum, parapositronium ($S = 0$) is annihilated with the emission of two photons. The ground state annihilation of orthopositronium ($S = 1$) produces at least three photons in free space. The mean lifetime of parapositronium is about 10^{-10} s. Orthopositronium lives considerably longer, with mean life about 10^{-7} s.

3.7 Problems

1. Calculate the collisional loss stopping power for a 1-MeV electron in aluminum. Ignore shell and density effect corrections. Obtain the linear energy transfer for a cutoff energy of 1 keV using the appropriate value from Figure 3.3.
2. Calculate the linear energy transfer for a 1-MeV electron in aluminum

using Equation 3.2. Use an energy cutoff of 1 keV in the calculation. Ignore shell and density effect corrections.

3. From Figure 3.1 evaluate the ratio of the ionization cross section for $2p$ and $2s$ electrons to the ionization cross section for $3s$ and $3p$ electrons in argon at 1 keV.
4. Evaluate the integral for $\Phi(T, Z, h\nu)$ in Equation 3.23 using the low frequency limit for the modified Bessel function, $K_1(\omega b/v) = v/\omega b$. Estimate the maximum impact parameter allowed by screening as v/ω and the minimum impact parameter by the reduced de Broglie wavelength of the electron. Show the resulting expression.
5. Calculate the radiated energy in kiloelectron volts per nucleus per square meter per electron when $T = 2$ MeV and $Z = 74$.
6. Calculate the radiation length of a tungsten target in SI units using Equations 3.32 and 3.33.
7. How much energy is radiated by a 4-MeV electron in a tungsten target thick enough to stop the electron? Use the radiational yield expression of Equation 3.36.
8. For a 4-MeV incident electron, estimate the energy actually emitted from a tungsten target thick enough to stop the electron when self-absorption is considered. Use Figure 3.12. Suppose the target was 0.2 radiation lengths thick. How much would be emitted?
9. Estimate the radiative loss stopping power at 5 MeV from a gold target from the value for a tungsten target in Appendix 7.
10. Evaluate the ratio of the radiative energy loss rate to the collisional loss rate at 10 MeV and 1 MeV for a tungsten target.
11. Evaluate the ratio of the electron radiative energy loss rate to the collisional loss rate at 10 MeV in a water absorber using Equation 3.41.
12. Show that the Dirac (28) expression for the positron annihilation cross-section

$$\sigma = r_0^2\pi \frac{1}{\beta^2\gamma^2(\gamma+1)}\{(\gamma^2+4\gamma+1)\ln[\gamma+(\gamma^2-1)^{12}]-\beta(\gamma+3)\}$$

reduces to Equation 3.43 when $v \ll c$, $\beta \to 0$.

3.8 References

1. C. Möller, Ann. Phys. **14**, 531 (1932).
2. H. A. Bethe and J. Ashkin, "Passage of radiation through matter," in *Experimental Nuclear Physics VI*, edited by E. Segre (John Wiley and Sons, New York, 1953), pp. 166–379.
3. N. F. Mott, Proc, R. Soc, London Ser. A **126**, 259 (1930).
4. S. J. Wallace, R. A. Berg, and A. E. S. Green, Phys. Rev. Sect. A **7**, 1616 (1973).
5. D. H. Madison and E. Merzbacher, "Theory of charged-particle excitations," in *Atomic Inner Shell Processes*, edited by B. Crasemann (Academic, New York, 1975), p. 18.

6. E. C. Beatty, Radiat. Res. **64**, 70 (1975).
7. F. Rohrlich and B. C. Carlson, Phys. Rev. **93**, 38 (1954).
8. M. J. Berger and S. M. Seltzer, "Tables of energy losses and ranges of electrons and positrons," *Studies in Penetration of Charged Particles in Matter, Publication 1133* (National Academy of Sciences–National Research Council, Washington, 1964), pp. 205–227.
9. J. W. McConnell, R. D. Birkhoff, R. N. Hammond, and R. H. Ritchie, Radiat. Res. **33**, 216 (1968).
10. R. M. Sternheimer, Phys. Rev. **103**, 511 (1956).
11. R. M. Sternheimer, Phys. Rev. **145**, 247 (1966).
12. W. K. Panofsky and M. Phillips, *Classical Electricity and Magnetism* (Addison-Wesley, Reading, MA, 1955), pp. 297–309.
13. E. Segre, *Nuclei and Particles* (W. A. Benjamin, New York, 1965), pp. 57–67.
14. J. L. Matthews and R. O. Owens, Nucl. Instrum. Methods **3**, 157 (1973).
15. A Sommerfield, Ann. Phys. **11**, 257 (1931).
16. H. A. Bethe, Proc. Cambridge Philos. Soc. **30**, 524 (1934).
17. W. Heitler, *Quantum Theory of Radiation,* 2nd ed. (Oxford University, London, 1944).
18. L. I. Schiff, Phys. Rev. **83**, 252 (1951).
19. W. T. Grandy, Jr., *Introduction to Electrodynamics and Radiation* (Academic, New York, 1970).
20. G. B. Rybicki and A. P. Lightman, *Radiative Processes in Astrophysics* (John Wiley and Sons, New York, 1979), pp. 58–59, 134.
21. R. V. Churchill, *Fourier Series and Boundary Value Problems* (McGraw-Hill, New York, 1941), p. 92.
22. F. W. J. Olver, "Bessel functions of integral orders," in *Handbook of Mathematical Functions,* edited by M. Abramowitz and I. A. Stegum, (National Bureau of Standards, Washington, 1968), pp. 375–376.
23. J. W. Motz, Phys. Rev. **100**, 1560 (1955).
24. *Radiation Quantities and Units, ICRU Report 33* (Internal Commission on Radiation Units and Measurements, Washington, 1980).
25. H. W. Koch and J. W. Motz, Rev. Mod. Phys. **31**, 29 (1959).
26. M. J. Berger and S. M. Seltzer, Phys. Rev. Sect. C **2**, 621 (1970).
27. S. De Benedetti, "Annihilation of positrons," in *Beta and Gamma Ray Spectroscopy,* edited by K. Sieghahn (North-Holland, Amsterdam, 1955), p. 673.
28. P. A. M. Dirac, Proc. Cambridge Philos. Soc. **26**, 361 (1930).
29. J. D. McGervey, *Introduction to Modern Physics* (Academic, New York, 1971), p. 446.

4

Charged-Particle Energy Deposition in Depth

4.1 Path Length, Range, and Related Concepts

In the last two chapters, expressions for the energy loss per unit length along the path of a charged particle were developed. In this section these expressions will be related to the penetration of such particles into an absorber. We will introduce the concepts of path length and projected path length for individual charged particles, and range and projected range for beams of charged particles. In most cases, these names follow conventions used previously (1). Since the path length and the projected path length refer to a given particle, they are subject to a high level of fluctuation. Because the range and the projected range are averages over large numbers of particles, fluctuations are usually ignored. The range and projected range will be related analytically to the stopping power expression in the next section. In succeeding sections they will be used to describe some properties of the pattern of the energy deposited in depth in an absorber.

The path length s is the total distance traversed by a charged particle without relation to direction. By reference to Figure 4.1,

$$s = \sum_i \Delta s_i \tag{4.1}$$

where Δs_i represents the distance traveled between collision $i-1$ and collision i.

The range R is defined as the average path length for many identical monoenergetic charged particles:

$$R = \langle s \rangle = \sum_{j=1}^{N} s_j / N \tag{4.2}$$

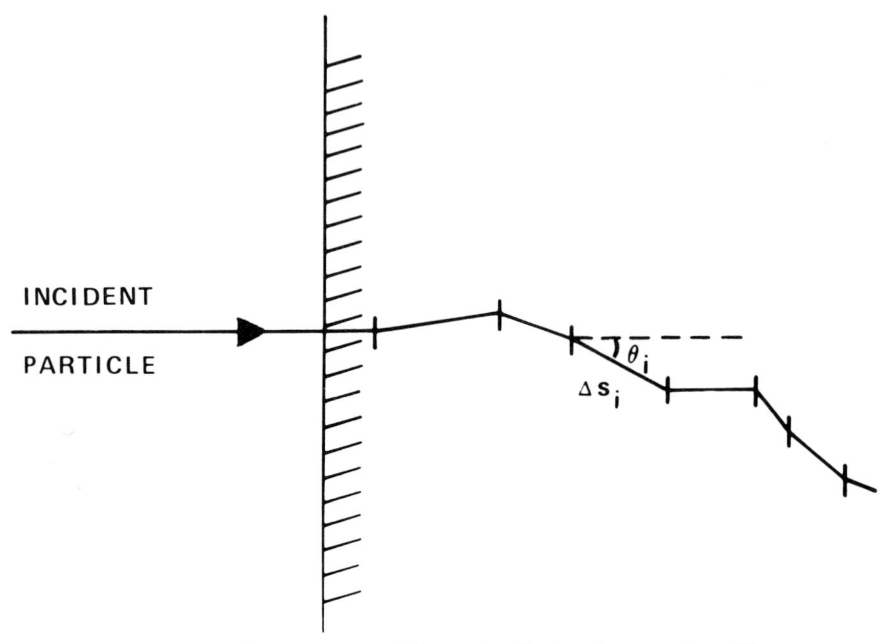

Figure 4.1 The path of a charged particle can be displayed as a series of line segments between collision points. The path length is the sum of the segment lengths.

where $\langle\,\rangle$ indicates that an average has been made for the N incident particles and j is the summation index for individual particles. Note that the distance specified in equation 4.1 includes that covered from incidence of a particle on an absorber surface to thermalization within the absorber. Of course, after reaching thermal energies, a particle may diffuse some distance. Charged-particle diffusion will not be included in our discussions.

The projected path length t is defined as the sum of the projections of individual path-length increments along the incident direction. If θ_i is the deflection angle shown on Figure 4.1, then

$$t = \sum_i \Delta s_i \cos \theta_i \tag{4.3}$$

Note that t is equivalent to the depth of penetration of an individual particle.

The projected range, R_t, is the average projected path length for many identical monoenergetic particles. It relates most closely to average depth of penetration. Thus, for N such particles,

$$R_t = \langle t \rangle = \sum_{j=1}^{N} t_j / N \tag{4.4}$$

From these definitions, it is evident that the projected path length cannot be greater than the path length, $t \leq s$. The difference between s and t is especially noticeable for incident electrons, which can be deflected appreciably in a single collision. The two quantities are more nearly equal for heavy ions since they are not deflected through large angles except in infrequent nuclear collisions at small-impact parameter. The cumulative effects of multiple scattering collisions are discussed in Section 4.4.

The relationship between the projected path length and the projected range can be clarified by reference to a transmission curve. Figure 4.2 shows such a curve for monoenergetic electrons. The transmission is defined as the number of particles that penetrate through an absorbing sheet divided by the total number of particles incident on its surface. Because of deviations in direction during scattering collisions (multiple scattering) and fluctuations in energy loss rate (energy straggling), the curve is not rectangular, as one might suppose (2). Rather, a distribution in projected path length is indicated, with some particles penetrating further than others. Thus, after an initial plateau, the transmission falls gradually to zero. It is apparent that the projected range R_t is equal to the thickness corresponding to a transmission of 0.5 if the fall-off is symmetrical. The extrapolated range R_e, which is the intercept obtained by extrapolating the linear part of the transmission curve to the axis, is also shown on the figure.

For large numbers of incident monoenergetic particles, the projected range can be related to the projected path length of an individual particle by the relationships:

$$R_t = \langle t \rangle = \int_0^\infty tP(t)dt, \quad P(t)dt = \frac{1}{\sqrt{2\pi V_t}} \exp[-(t - R_t)^2/2V_t]dt \tag{4.5}$$

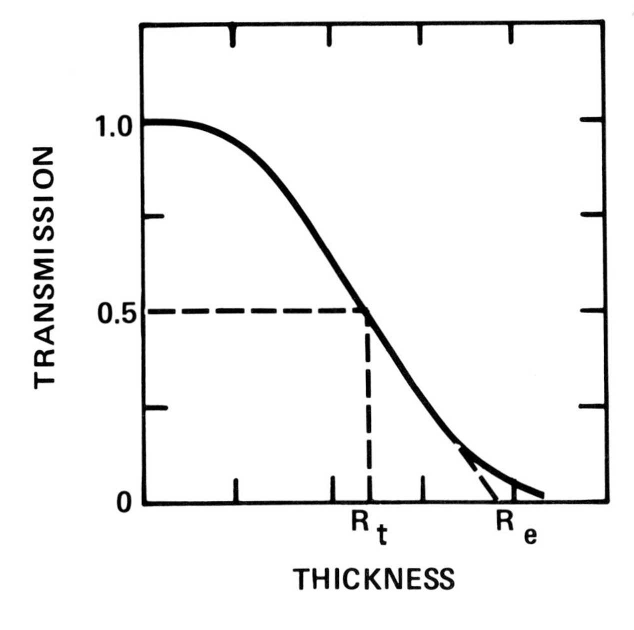

Figure 4.2 The transmission of electrons through an absorber is shown. The projected range R_t and the extrapolated range R_e are indicated.

In these equations, $P(t)dt$ is taken to be the probability an individual particle will have a projected path length between t and $t + dt$. We have assumed as an approximation that this distribution is similar to a normal curve with projected path-length variance V_t. Since the normal distribution is symmetrical, the number of particles transmitted through an absorber of thickness R_t is 0.5 of the incident particle number. If asymmetries in the distribution are considered, a different analysis must be used.

From Figure 4.2 the extrapolated range R_e can be related to R_t through the variance V_t of the normal distribution. Note that the slope on the transmission curve is related to $P(t)$ through

$$\frac{1}{N_0}\frac{dN}{dt} = \frac{dN/N_0}{dt} = \frac{P(t)dt}{dt} = \frac{1}{\sqrt{2\pi V_t}} \exp[-(t - R_t)^2/2V_t] \tag{4.6}$$

dN is the number of particles stopped at a thickness between t and $t + dt$ and N_0 is the total number of particles incident on the absorber suface. The slope from Figure 4.2 can be evaluated from the point $N/N_0 = 0.5$ at $t = R_t$:

$$\frac{0.5}{R_e - R_t} = \frac{1}{N_0}\frac{dN}{dt} = \frac{1}{\sqrt{2\pi V_t}}, \qquad R_e = R_t + \sqrt{\pi V_t/2} \tag{4.7}$$

Thus, R_e is larger than R_t because of the fluctuations in projected path length.

A special case of an electron transmission curve occurs for beta particles from the radioactive decay of a nucleus. The spectrum of electrons from beta decay (3) is continuous and decreases to zero at the end-point energy (see Figure 4.3). Because of this spectrum, the beta-particle transmission curve is approxi-

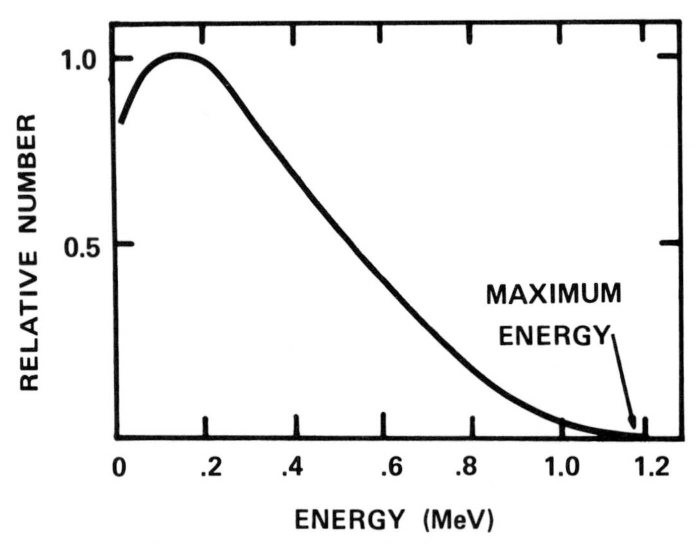

Figure 4.3 The beta spectrum from ^{210}Bi is shown (3).

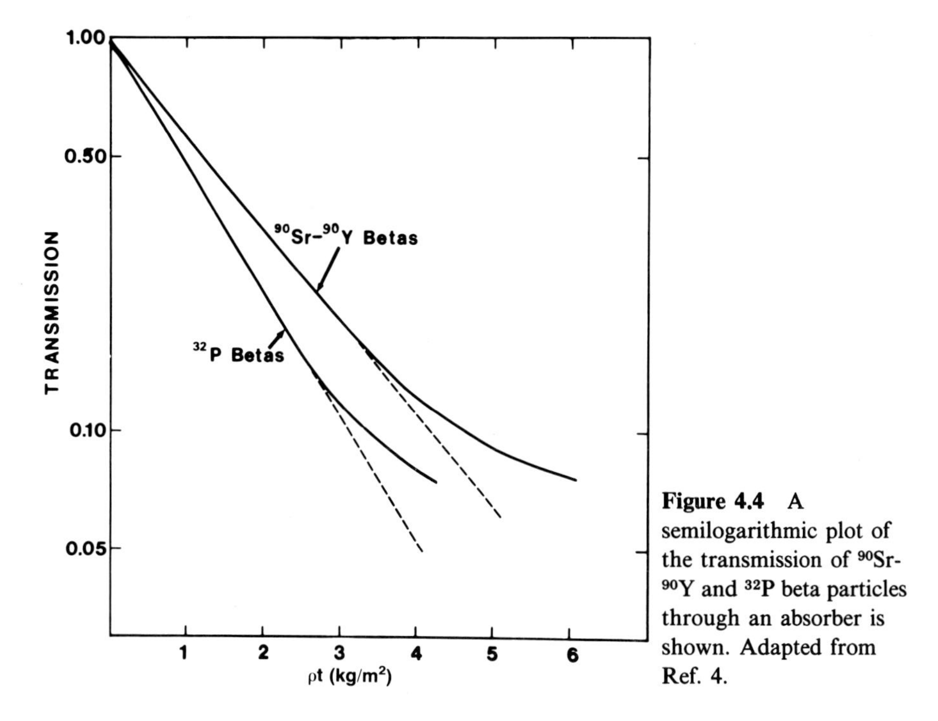

Figure 4.4 A semilogarithmic plot of the transmission of ^{90}Sr-^{90}Y and ^{32}P beta particles through an absorber is shown. Adapted from Ref. 4.

mately exponential in an intermediate region (4), as shown in Figure 4.4. Thus, at intermediate transmission values,

$$N/N_0 = \exp[-\beta \rho x], \quad \beta = \frac{1.7}{(T_{max})^{1.14}} \tag{4.8}$$

where β is the effective beta mass attenuation coefficient. The numerical value of β in square meters per kilogram can be calculated from the second equation (4–7) if T_{max} is the spectrum end-point energy in megaelectron volts.

4.2 Expressions for Range Versus Energy

A theoretical approach to the determination of charged-particle range utilizes stopping power expressions similar to those discussed in Chapters 2 and 3. The value found is often called the CSDA range, an acronym for Continuous Slowing Down Approximation (8, 9). Since range is an average value, fluctuations can be ignored and losses are assumed to be continuous. The CSDA range can be calculated from

$$R = \int_0^R dx = \int_{T_0}^0 \frac{-dT}{(-dT/dx)} = \int_0^{T_0} dT/S = \int_{T_1}^{T_0} dT/S + R_1(T_1) \tag{4.9}$$

where dx is the path-length variable of integration and S is the stopping power. In this expression, T_0 is the initial kinetic energy of the charged particle. The last part of the path length is usually not accurately calculated when an analytical expression is used for the stopping power. Most such expressions are inaccurate at low energies. Because of this, a finite lower limit can be utilized, and $R_1(T_1)$ can be estimated from experimental results. Extensive CSDA range calculations for electrons have been published (10, 11).

A simple nonrelativistic CSDA evaluation of the range (5) can be made for heavy ions. Using Equation 2.22,

$$S_c = -\frac{dT}{dx} = \frac{2z^2\pi r_0^2 m_e c^2 N_A (Z/M_m)\rho \ln(2m_e c^2\beta^2/I)^2}{\beta^2} \tag{4.10}$$

But

$$dT = d\left[\frac{Mv^2}{2}\right] = \frac{Mc^2}{2}\, d(\beta^2) = Mc^2\beta\, d\beta,$$

$$\frac{dT}{S_c} = \frac{Mc^2\beta^3\, d\beta}{2z^2\pi r_0^2 m_e c^2 N_A (Z/M_m)\rho \ln(2m_e c^2\beta^2/I)^2} \tag{4.11}$$

Changing to the variable u simplifies the integration to follow:

$$u = (2m_e c^2\beta^2/I)^2 = \left[\frac{4m_e T}{MI}\right]^2, \quad du = \frac{16(m_e c^2)^2\beta^3}{I^2}\, d\beta \tag{4.12}$$

From Equations 4.9, 4.11, and 4.12,

$$R = \frac{Mc^2 I^2}{32z^2\pi r_0^2 (m_e c^2)^3 N_A (Z/M_m)\rho} \int_{u_1}^{u_0} \frac{du}{\ln u} + R_1(T_1)$$

$$= \frac{Mc^2 I^2}{32z^2\pi r_0^2 (m_e c^2)^3 N_A (Z/M_m)\rho} [\mathrm{Ei}(\ln u_0) - \mathrm{Ei}(\ln u_1)] + R_1(T_1) \tag{4.13}$$

The notation $\mathrm{Ei}(\ln u)$ is for the exponential integral with argument $\ln u$. Values of the exponential integral have been tabulated (12), and the functional form is shown in Figure 4.5.

The unit density range $R\rho$ is the product of the range and the density of the medium. From Equation 4.13 it is evident that the range is inversely proportional to the mass density. If the unit density range is used, the dependence on mass density is multiplied out and the result can be used for accurate interpolation from element to element. Unit density ranges for protons and electrons are tabulated in Appendixes 6 and 7.

Formulas for approximate values of $R\rho$ have been developed for restricted energy intervals. For protons in air in the energy range $10 \le T \le 200$ MeV, one can use (13)

$$(R\rho) = (2.2)10^{-2}T^n, \quad n = 1.8 \tag{4.14}$$

where T is in megaelectron volts. For electrons in aluminum in the energy range $0.01 \le T \le 3$ MeV, one can use (14)

$$(R\rho) = (4.1)T^n, \quad n = 1.27 - 0.1 \ln T \tag{4.15}$$

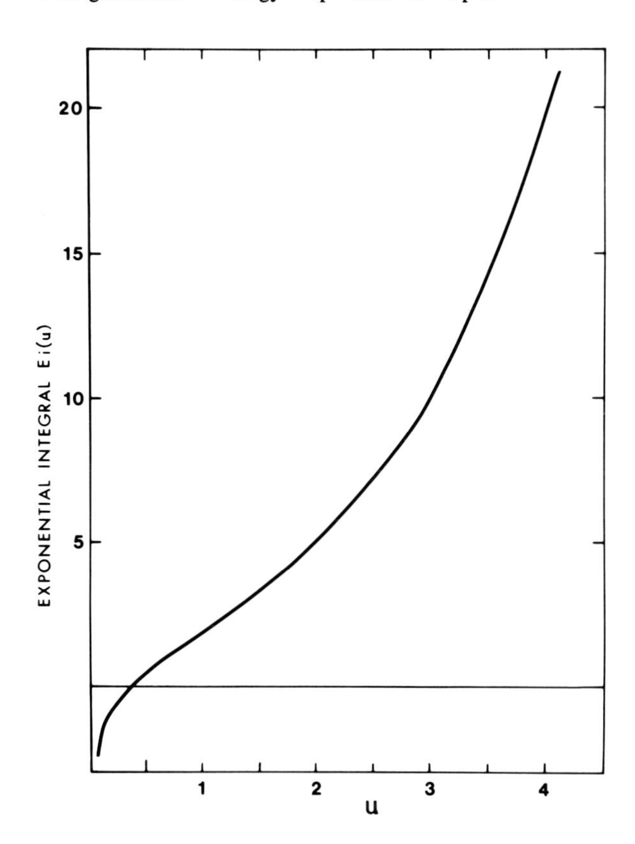

Figure 4.5 The exponential integral function increases with increasing argument from the zero at 0.372.

where T is in megaelectron volts. In both Equations 4.14 and 4.15 the $R\rho$ values are in kilograms per square meter.

The form of Equation 4.13 can be used to obtain approximate relationships between unit density range values for neighboring elements or for different nonrelativistic charged particles. The parameter of particular interest for these purposes is $\tau = T/Mc^2$ for the incident ion. When the variable $u = (4m_ec^2\tau/I)^2 \gg 1$, an asymptotic form of the exponential integral can be used (12):

$$\mathrm{Ei}(\ln u) \rightarrow \frac{u}{\ln u} = \frac{[(4m_ec^2/I)\tau]^2}{\ln[(4m_ec^2/I)\tau]^2} \tag{4.16}$$

For heavy ions when $u \gg 1$ and $R_1 \ll R$, one can estimate the range for a particle in an absorbing element 2 from the range in a neighboring element 1 using

$$(R\rho)_2 \simeq \frac{(Z/M_m)_1}{(Z/M_m)_2}(R\rho)_1 \tag{4.17}$$

In this relationship the mass numbers can be substituted for molar masses in most cases in the ratios. The direct dependence on I, the average excitation energy, cancels out of Equation 4.13 in the asymptotic limit and does not appear in 4.17. The remaining logarithmic dependence on I can often be ignored in a first approximation of the ratio for neighboring elements, since I varies slowly from element to element.

When $u \gg 1$, $R_1 \ll R$, and $\tau_a = \tau_b$, one can estimate the unit density range for a nonrelativistic particle b from the unit density range for nonrelativistic particle a in the same absorber:

$$(R\rho)_b \simeq \frac{(M/z^2)_b}{(M/z^2)_a} (R\rho)_a \tag{4.18}$$

When ions have the same kinetic-energy-to-mass-energy ratio, the masses and net charges determine the relative range.

4.3 Energy and Path-Length Straggling

Since radiative and collisional processes for charged particles occur randomly, the cumulative energy loss in an absorber must fluctuate from one particle to the next in a beam. Energy straggling is the fluctuation in energy loss for a given partial path length for identical monoenergetic particles. Because the energy loss for any partial path fluctuates, the total path length must also fluctuate. Path-length straggling is the fluctuation in total path length for identical monoenergetic particles.

Energy straggling of electrons is more pronounced than energy straggling for heavy ions. This is because the fraction of energy transferred in a single collisional loss can be much larger for electrons than heavy particles. Equation 2.20 shows that for nonrelativistic heavy particles, the maximum energy transferred is

$$E_{\max} = 2m_e v^2 = 4\frac{m_e}{M} T \ll T \tag{4.19}$$

For fast incident electrons the maximum energy transferred to atomic electrons is usually taken as half of the incident kinetic energy. High-energy transfer events lead to a high level of fluctuations in energy loss and path length. Furthermore, the probability of photon emission is much larger for an electron than for a heavy ion. Photon emission will contribute to energy straggling in high Z absorbers since the energy emitted is variable over a wide range.

In the following paragraphs, a description of energy straggling for heavy ions will be given (5, 15). Heavy particles are considered rather than electrons because simplified distributions can be used when single events with large loss are ignored. As before, the development given will be approximate. Some generalization to the case of electron straggling will be given at the close of the section.

Equation 2.15 shows that the total number of collisions, ΔN, an incident ion will sustain in path length Δs with energy loss between E and $E + dE$ is

$$\Delta N = 2z^2\pi r_0{}^2 \frac{m_e c^2}{\beta^2} n_v Z \left[\frac{dE}{E^2}\right] \Delta s \tag{4.20}$$

In this equation ze is the incident particle net charge, and β is the ratio of the incident particle speed to the speed of light. The number of electrons per unit volume is $n_v Z$ for the medium.

The number of collisions given by Equation 4.20 is random and subject to fluctuations. We will assume that the Poisson distribution can be applied (5); then the standard deviation in collision number is given by $\sigma = \sqrt{\Delta N}$. The fluctuation level in the total energy loss is of prime importance in energy straggling. Thus, the standard deviation in energy loss, $E\sqrt{\Delta N}$, and the energy loss variance, $E^2 \Delta N$, must be considered.

For independent processes, the total variance in energy loss, V_T, is obtained by summing the individual energy loss variance values (16). In this case, since continuous variables are involved, we integrate $E^2 \Delta N$:

$$V_T = 2z^2 \pi r_0^2 \frac{m_e c^2}{\beta^2} n_v Z \left[\int_{E_{\min}}^{E_{\max}} dE \right] \Delta s \tag{4.21}$$

Using Equation 2.20 when $E_{\max} \gg E_{\min}$,

$$E_{\max} - E_{\min} \simeq 2m_e v^2, \quad V_T = 2z^2 \pi r_0^2 \frac{m_e c^2}{\beta^2} n_v Z (2m_e v^2) \Delta s$$

$$= 4z^2 \pi r_0^2 (m_e c^2)^2 N_A (Z/M_m) \rho \Delta s \tag{4.22}$$

If a large number of monoenergetic particles are involved, and excessively large, single energy loss processes are ignored, the normal distribution in particle energy T can be used. Under these circumstances the probability $P(T)dT$ for residual kinetic energy between T and $T + dT$ after traversing a partial path length Δs in an absorber is

$$P(T)dT = \frac{1}{\sqrt{2\pi V_T}} \exp\left[-\frac{(T - \langle T \rangle)^2}{2V_T} \right] dT, \quad \langle T \rangle = T_0 - \int_0^{\Delta s} S dx \tag{4.23}$$

The variance V_T as given by Equation 4.22 was used in the distribution in *residual energy* as well as in the distribution in *energy loss,* since the two energies differ by a constant T_0, the incident energy. Notice that the width of the distribution increases as $(\Delta s)^{1/2}$ since V_T is proportional to Δs and $\sigma_T = \sqrt{V_T}$. An example will be shown later.

The residual path length of a charged particle is given by the product of residual energy and the reciprocal of the average stopping power. Thus, fluctuations in path length result from energy loss fluctuations. If the standard deviation associated with path length Δs is given by $\sqrt{V_T}/S$, the total path-length variance V_s is given by the sum of the segmental variance values V_T/S^2 (5, 15). Since continuous variables are involved, we substitute dx for Δs and integrate using Equation 4.22.

$$V_s = 4z^2 \pi r_0^2 (m_e c^2)^2 n_v Z \int_0^s \frac{dx}{S^2} = 4z^2 \pi r_0^2 (m_e c^2)^2 N_A (Z/M_m) \rho \int_0^{T_0} \frac{dT}{S^3} \tag{4.24}$$

If a large number of monoenergetic particles are considered and large energy loss collisions are ignored, the normal distribution in path length is useful. If $P(s)ds$ is defined as the probability for total path length between s and $s + ds$, then

$$P(s)ds = \frac{1}{\sqrt{2\pi V_s}} \exp\left[-\frac{(s-R)^2}{2V_s}\right]ds \tag{4.25}$$

From Equation 4.24 it is evident that path-length variance for heavy ions is inversely related to the stopping power. The dependence of the integrand on S^{-3} in that equation ensures the inverse relationship. As an example, this means that proton path-length straggling is more noticeable than alpha-particle straggling for the same incident energy, since the alpha-particle stopping power is much larger.

The basic conclusions from the foregoing descriptions using normal distributions are correct, but the numerical results are only approximate since large energy transfer events are not included. In fact, an occasional collision of a heavy particle with an atomic nucleus does involve transfer of a large amount of energy. Thus, the symmetrical distributions given in Equations 4.5, 4.23, and 4.25 are not entirely appropriate, and skewing towards smaller residual energy and smaller path length occurs.

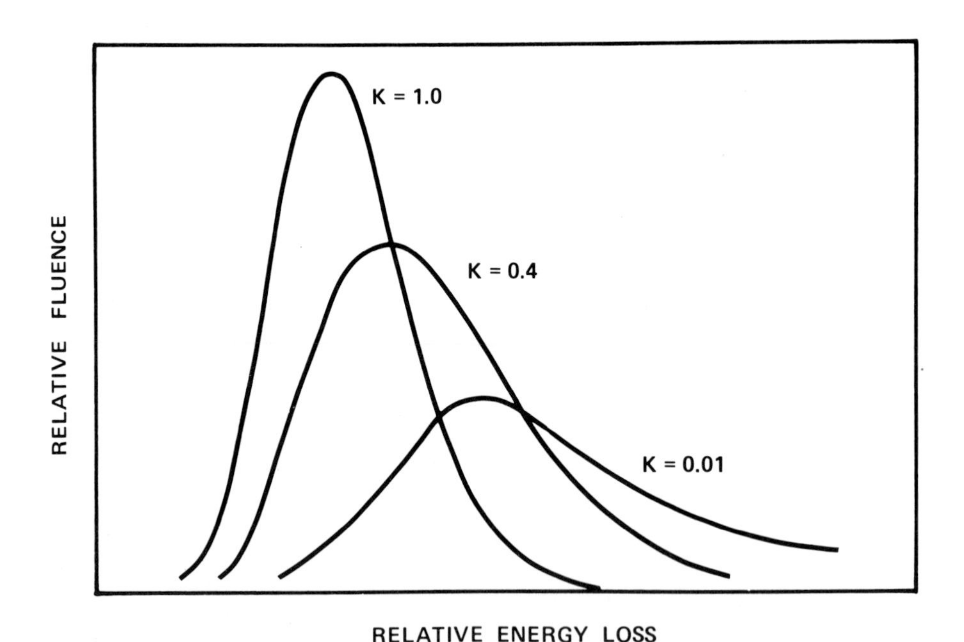

Figure 4.6 The Vavilov distribution (18) with $\beta^2 = 0.9$ is shown for several values of the parameter K. When $K = 1$, the distribution is nearly symmetrical; when $K = 0.01$ there is a pronounced tail toward large energy loss.

Experimental energy loss distributions for charged particles traversing a thin absorber are described by a theory developed by Vavilov (17). The normal distribution used in the foregoing descriptions can be considered a limiting case of the Vavilov distribution. The Vavilov parameter K is proportional to the mean energy loss in the path length divided by the largest possible energy loss in a single collision. Figure 4.6 shows the Vavilov distribution for various values of K (18). When K is small, the distribution has a very pronounced tail toward large energy losses. As K increases the distribution becomes more nearly symmetrical, and when $K \gg 1$ it approaches the normal distribution.

Figure 4.7 illustrates energy straggling for electrons in water (19). Note that the peak width increases as the residual energy decreases and the path length increases. This effect is expected by analogy to Equation 4.22. Large energy loss collisions occur with some regularity for electrons, so the distributions are skewed toward lower energy. The average residual energy is considerably less than the most probable residual energy.

Figure 4.8 shows some distributions in projected path length for electron beams of two energies in water (20). This figure is included to illustrate path length straggling. Notice that the distributions are skewed toward low depths. Single event, large energy loss collisions contribute to the skewing. In addition, fluctuations in the scattering angles of the electrons produce straggling in the projected path length, which broadens the peaks. This effect will be discussed in more detail in the next section.

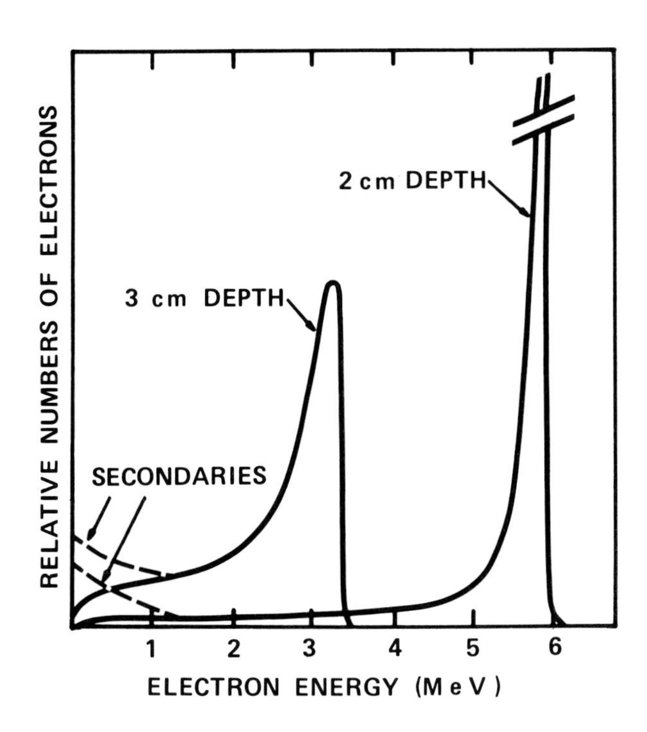

Figure 4.7 The energy distributions for 10 MeV incident electrons are shown at two depths in water; about 2 cm and about 3 cm. Adapted from Ref. 19.

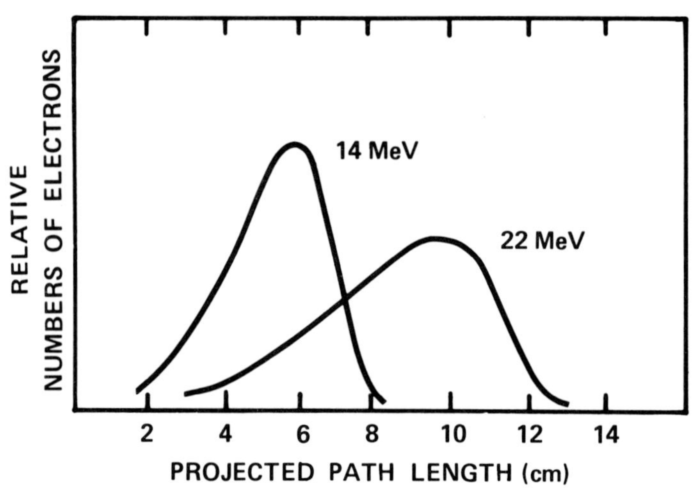

Figure 4.8 The distribution in projected path length is shown for a 14- and a 22-MeV electron beam in water (20).

4.4 Multiple Scattering

Multiple scattering is a series of collisions involving small individual deflections, producing a cumulative substantial deviation from the incident direction of the fast charged particle. The individual events that contribute most to multiple scattering are elastic coulomb collisions between the incident particle and an atomic nucleus. There is little energy transferred by these processes; they are significant because of the deflection caused. Contributions from incident-particle collisions with atomic electrons are often neglected when compared to the contributions from the nuclear collisions. The average deflection angle occurring in a collision with an atomic electron is less than the angle of deflection produced by a nuclear collision at an equivalent impact parameter.

Multiple scattering is the primary cause of the divergence of charged particles emerging from an absorber when the incident beam was parallel. Figure 4.9 is an illustration of this effect. Individual incident particles suffer deflecting collisions that tend to spread the beam into a cone of radiation. A plot of relative numbers of electrons emerging from a scattering foil per unit solid angle has the form shown in Figure 4.10 (21). Large-angle single collisions contribute to these distributions but have minor significance in the central region of the distribution. Large-angle heavy-ion collisions with nuclei have been widely studied. The Rutherford alpha-scattering experiments (22) were central in establishing the nuclear model of the atom.

We will develop an approximate expression for the angular deflection due to multiple scattering in the following paragraphs (15). Suppose that an incident particle with momentum p, energy T, and net charge $+ze$ collides with an atomic nucleus of charge $+Ze$ and mass M. Fortunately, the setup is familiar (Fig. 4.11). Because the target nucleus is so massive, there will be little energy

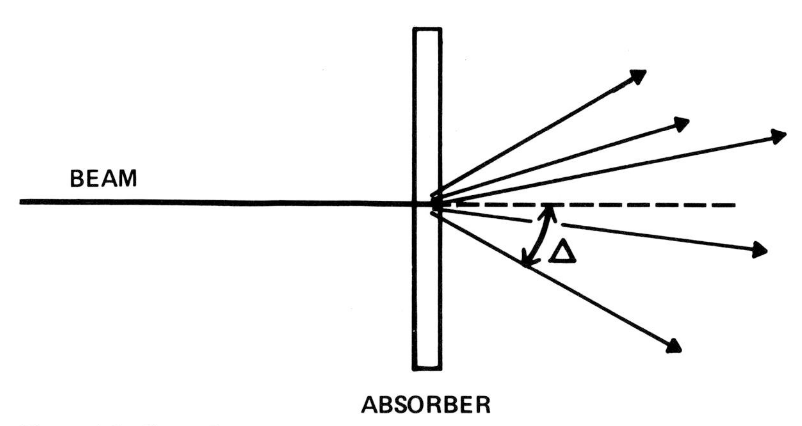

Figure 4.9 Scattering processes cause divergence in a beam emerging from an absorber.

transferred during the collision; rather, the less massive incident particle is deflected by the repulsive coulomb force **F**. The angle of deflection of the incident particle, δ, is our focus of interest in the next few paragraphs. The assumptions in the discussion to follow are similar to those in Sections 2.2 and 3.2. They include:

1. Movement of the nuclear scattering center during the collision is negligible.
2. The change in the incident-particle speed is negligible.
3. The angle of deflection δ is small.

Under these circumstances the net momentum transfer in the x direction is not significant since the force on the incident particle is nearly symmetrical

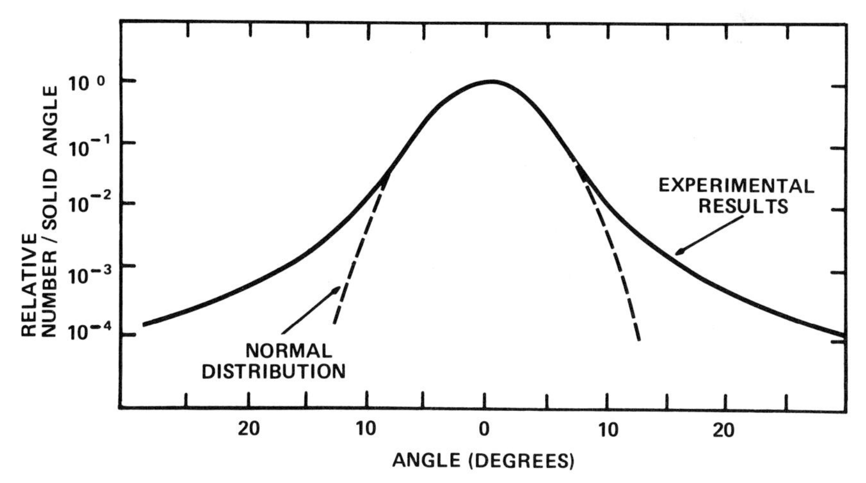

Figure 4.10 The angular distribution of 16-MeV electrons scattered from a gold foil is shown (21). Note that the normal distribution is not adequate to account for large angle scattering.

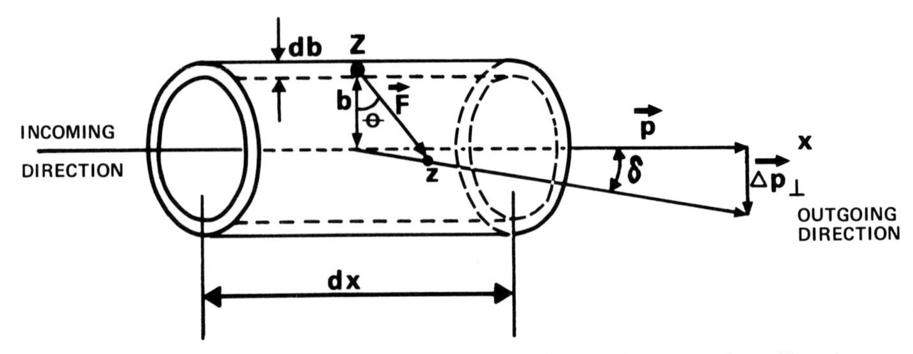

Figure 4.11 The geometrical parameters necessary for development of small-angle scattering equations are shown. **F** is the coulomb force.

in time. The expressions for the change in incident-particle momentum in the direction perpendicular to the x axis and passing outward through the nucleus is analogous to Equations 2.5 and 2.7. Then, for small δ, $\cos \theta = b/r$, $\tan \theta = x/b$, and

$$\Delta p_{\perp} = -\int_{t_1}^{t_2} F \cos \theta \, dt = \frac{-zZe^2}{4\pi\epsilon_0} \int_{t_1}^{t_2} \frac{\cos^3 \theta}{b^2} \, dt \tag{4.26}$$

From Equation 2.6, $dt = bd\theta/v \cos^2 \theta$; thus,

$$\Delta p_{\perp} = \frac{-zZe^2}{4\pi\epsilon_0} \int_{-\pi/2}^{\pi/2} \frac{\cos^3 \theta}{b^2} \left[\frac{bd\theta}{v \cos^2 \theta} \right] = \frac{-2zZe^2}{4\pi\epsilon_0 bv} \tag{4.27}$$

The last equality holds if the change in speed during the collision can be ignored. Using the momentum triangle on Figure 4.11,

$$\left| \frac{\Delta p_{\perp}}{p} \right| = \frac{2zZe^2}{4\pi\epsilon_0 pvb} = \tan \delta \simeq \delta \tag{4.28}$$

for small δ. Equation 4.28 is called the single-scatter relationship.

Next consider the effect of many scattering events on a single incident particle. Of course, the scattering angle can lie in any plane passing through the x axis, since the relative orientation of the incident particle with respect to the nucleus is random. Suppose the displacement due to the deflection is projected onto the yz (perpendicular) plane located at unit distance from the scattering center, as is shown in Figure 4.12. Since the angle is small and the plane is at unit distance, the projected displacement is assumed to be numerically equal to the value of the angle in radians:

$$\delta_y = \delta \sin \phi, \quad \delta_z = \delta \cos \phi \tag{4.29}$$

For several consecutive collisions the projections can be added. The total projections are given by

$$\Delta_y = \sum_i \delta_i \sin \phi_i, \quad \Delta_z = \sum_i \delta_i \cos \phi_i \tag{4.30}$$

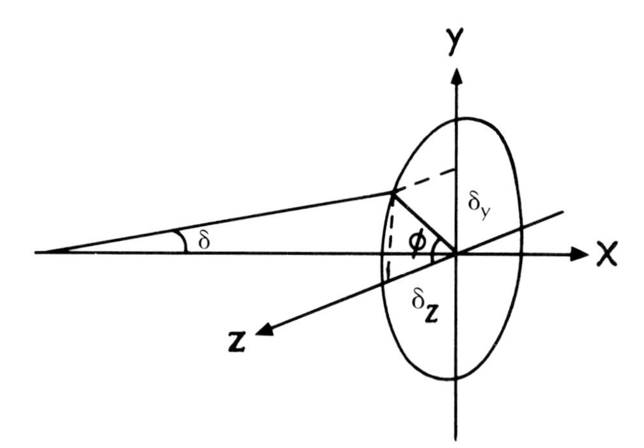

Figure 4.12 The scattering displacement distances are projected on the *yz* plane at unit distance from the scattering center.

where i is the summation index. Note that Δ_y and Δ_z are not usually zero, since they relate to a single particle which has been scattered many times. However, values of Δ_y or Δ_z for different particles in a beam will be distributed around zero.

If deflection angles given by Equations 4.30 are averaged over many incident particles, the result will approach zero. All angles are equally probable, and negative values should occur as often as positive values. However, the squares of Δ_y and Δ_z will be nonzero on the average. Thus, a mean square deflection angle can be obtained from

$$\langle\Delta^2\rangle = \langle\Delta_y{}^2\rangle + \langle\Delta_z{}^2\rangle = \left\langle\left[\sum_i \delta_i \sin\phi_i\right]^2\right\rangle + \left\langle\left[\sum_i \delta_i \cos\delta_i\right]^2\right\rangle \geq 0 \quad (4.31)$$

The $\langle\,\rangle$ indicate that an average over many incident particles has been made.

The individual terms in the sums from Equation 4.31 can be grouped as

$$\langle\Delta_y{}^2\rangle = \left\langle\left[\sum_i \delta_i \sin\phi_i\right]\left[\sum_j \delta_j \sin\phi_j\right]\right\rangle;$$

$$\langle\Delta_z{}^2\rangle = \left\langle\left[\sum_i \delta_i \cos\phi_i\right]\left[\sum_j \delta_j \cos\phi_j\right]\right\rangle \quad (4.32)$$

The sum of these products contains only pairs of terms like,

$$\langle\sin\phi_i \sin\phi_j\rangle + \langle\cos\phi_i \cos\phi_j\rangle \quad (4.33)$$

The sum of each pair of terms equals one if $i = j$. The values for $i \neq j$ average to zero since both positive and negative terms occur with equal probability (incoherent scatter). Equation 4.31 reduces to

$$\langle\Delta^2\rangle = \langle\Delta_y{}^2\rangle + \langle\Delta_z{}^2\rangle = \sum_i \delta_i{}^2(\sin^2\phi_i + \cos^2\phi_i) = \sum_i \delta_i{}^2 \quad (4.34)$$

The summation in Equation 4.34 is adequate if the number of collisions is relatively small. If the collision number is very large, integration is convenient. In the latter case, the number of nuclei in the infinitesimal cylindrical shell of radius b must be multiplied by the square of the appropriate deflection angle before the product is integrated. Since the number of nuclei is given by $n_v(2\pi b db)dx$, then

$$\langle \Delta^2 \rangle = \iint \delta^2 n_v (2\pi b db) dx = \iint \left[\frac{2zZe^2}{4\pi\epsilon_0 pvb} \right]^2 n_v (2\pi b db) dx$$

$$= 8\pi r_0^2 (m_e c^2)^2 N_A \int_{b_{\min}}^{b_{\max}} \frac{db}{b} \int_0^t \frac{dx}{(pv)^2} z^2 (Z^2/M_m)\rho \tag{4.35}$$

Here t is the thickness of the absorber and $n_v = N_A \rho / M_m$, where N_A is Avogadro's number, and $r_0 = e^2/4\pi\epsilon_0 m_e c^2$ is the classical electron radius. The values $b_{\max}$ and $b_{\min}$ are the maximum and minimum values of the impact parameter, respectively.

Rather than attempt to evaluate the integrals in Equation 4.35, we will define a function $B(T, Z, \rho t)$ with the units [(rad · MeV)^2 · kg · m^2/kg] for convenience in the discussion to follow. In the nonrelativistic case,

$$\langle \Delta^2 \rangle = 0.157\ B(T, Z, \rho t) \frac{z^2 (Z^2/M_m) \rho t}{p_0^2 v_0^2} = 0.157\ B(T, Z, \rho t) \frac{z^2 (Z^2/M_m) \rho t}{4T^2} \tag{4.36}$$

In the relativistic limit for electrons with $\beta \to 1$,

$$\langle \Delta^2 \rangle = 0.157\ B(T, Z, \rho t) \frac{z^2 (Z^2/M_m) \rho t}{p_0^2 v_0^2} = 0.157\ B(T, Z, \rho t) \frac{z^2 (Z^2/M_m) \rho t}{T(T + 2m_e c^2)} \tag{4.37}$$

The values p_0 and v_0 are for the incident momentum and speed, respectively. If the units of ρt are kilograms per square meter and T or pv is given in megaelectron volts, the resulting angle will be in radians. This functional form is useful because it is often sufficient to use $B(T, Z, \rho t) \simeq 1.0$ [(rad · MeV)^2 · kg · m^2/kg] for a first approximation for thin absorbers (1, 15). Some other values of $B(T, Z, \rho t)$ are given in Table 4.1 (1, 23, 24).

If single collisions producing large-angle deflections are ignored, a normal distribution can be used to describe the deflection probability for thin targets. For deflection angle Δ we can write (15),

$$P(\Delta) d\Delta = \frac{1}{\sqrt{2\pi V_\Delta}} \exp\left[-\frac{(\Delta - \langle \Delta \rangle)^2}{2V_\Delta} \right] d\Delta = \frac{1}{\sqrt{2\pi V_\Delta}} \exp\left[-\frac{\Delta^2}{2V_\Delta} \right] d\Delta$$

$$V_\Delta = \langle \Delta^2 \rangle - \langle \Delta \rangle^2 = \langle \Delta^2 \rangle, \quad \langle \Delta \rangle = 0 \tag{4.38}$$

The fraction of the total number of particles emerging with deflection angles less than or equal to some value θ can be obtained by integrating Equation 4.38 from $\Delta = 0$ to $\Delta = \theta$.

In summary, the major parameters determining the root mean square deflection angle $\sqrt{\langle \Delta \rangle^2}$ are:

1. Particle Properties

The root mean square deflection angle depends directly on the incident particle net charge and inversely on the product of the incident particle momentum and speed.

Table 4.1 Multiple Scattering Constant $B(T, Z, \rho t)$ for Incident Protons in (rad · MeV)^2 · kg · m^2/kg

		Z = 10			Z = 50	
v/c	ρt: 0.1	1.0	10	0.1	1.0	10
0.0	1.1	1.3	1.6	0.8	1.0	1.3
0.1	1.0	1.3	1.5	0.7	1.0	1.2
1.0	0.6	0.9	1.1	0.6	0.9	1.1

Data from Refs. 1 and 23.
Values of ρt are in kg/m^2.

2. Medium Properties
The root mean square deflection angle depends directly on $\sqrt{Z^2/M_m}$ and on the square root of the reduced thickness, $\sqrt{\rho t}$.

When comparing the scatter-induced divergence of a beam of electrons with that of a beam of heavy particles, one must remember that the electrons have very little forward momentum when compared to heavy particles with the same speed. Thus, they are more easily deflected. The dependence on $p_0{}^2 v_0{}^2$ in the denominator of Equations 4.36 and 4.37 ensures this tendency. When large-angle single scattering events are included for electron beams, the mean square deflection angle is further accentuated, especially near the end of the path. At low speeds, collisions with atomic electrons and delta rays released at large angles complicate the distribution. In fact, deflection angles are sufficiently large for electrons so that backscattering is significant (25, 26) at low energies.

The particle and medium properties outlined in the summary above carry over for more sophisticated descriptions of the multiple scattering problem. The functional form for the mean square scattering angle given by the Moliére theory (23, 24) is quite similar to Equation 4.36 except that Z^2 is replaced by $Z(Z + 1)$ to account for the scattering due to atomic electrons. However, the Moliére distribution function is not a normal distribution. Figure 4.10 shows that the normal distribution fails at large angles for electron scattering. The Moliére distribution can be used to describe the electron data much more closely, even at large angles, as long as the absorber thickness and electron energy are within appropriate limits.

4.5 Depth-Dose Curves for Charged-Particle Beams

The subject of charged-particle depth-dose curves provides a useful example for application of many of the concepts and distributions introduced in the last several sections. In the succeeding paragraphs, we will discuss Bragg ionization curves before depth-dose curves. Bragg curves for heavy particles are historically important, since many of the fundamental ideas about range and specific ionization were formulated from air chamber data. Furthermore, the shape of the depth-versus-dose curve is easily related to the shape of the Bragg curve.

A Bragg curve is a plot of the average specific ionization versus the projected distance along the central axis of a charged-particle beam in an absorbing medium. The name stems from the work of W. H. Bragg, who first examined the ionization along the path of alpha particles in air (27). Figure 4.13 shows a Bragg curve as compared with an ionization curve for a single particle (28). It is immediately noticeable that both curves start out with a plateau at small depth and rise to a peak before they drop to zero. Although Bragg curves are measured in air, the shape is largely independent of the medium (low Z).

The general appearance of the heavy-particle Bragg curve can be explained from principles developed in Chapter 2. The increase in ionization near the end of the path occurs because the stopping power is inversely related to the square of the particle speed. Thus, ionization density rises as the particle slows. Of course, the ionization must drop to zero when the energy of the incident particles has been dissipated. In fact, the ionization cross section falls to zero at the energy corresponding to the ionization potential of the medium. The two effects in sequence produce the Bragg peak.

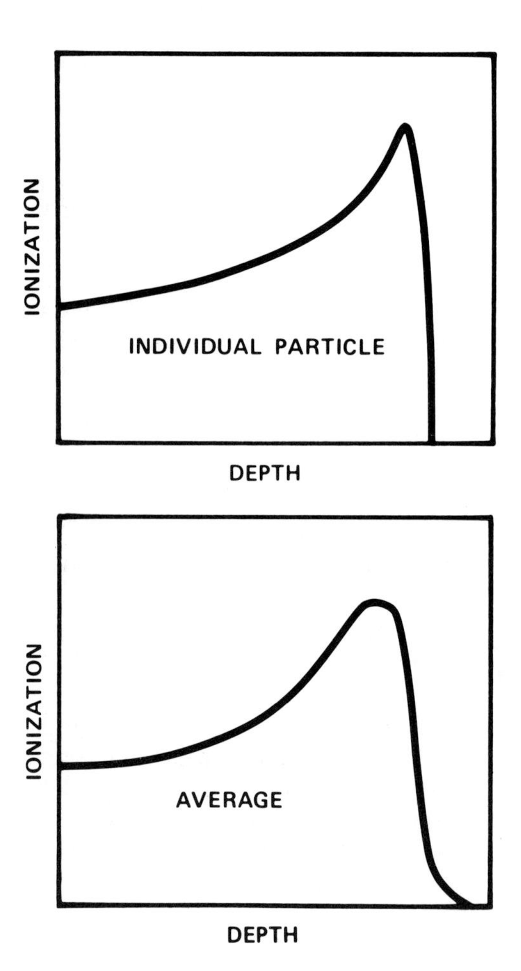

Figure 4.13 The specific ionization curve for a single particle and the Bragg curve for a beam of particles in air are shown (28).

From Figure 4.13 it is evident that the average specific ionization does not drop abruptly after the peak is reached but diminishes over a noticeable distance. Effects of the ionization cross-section curve and the effective charge have been discussed in Chapter 2. Major contributions to the width of the Bragg peak come from straggling and multiple scattering. As was discussed in Section 4.3, one particle may not ionize at exactly the same rate as another, even when they both have the same energy mass and charge. This contributes to the level of fluctuations in path length for particles in a beam. Furthermore, multiple scattering produces fluctuations in the net angle of deflection for identical particles in a beam. These angular deviations also contribute to fluctuations in penetration depth. Both effects tend to broaden the Bragg peak.

An analytical function relating the specific ionization of an individual particle and the average specific ionization for a beam of identical particles can be developed using the distribution in projected path length for the particles. Suppose $i(t - d)$ stands for the specific ionization for an individual particle given as a function of residual penetration distance, and $I(d)$ is the average specific ionization for monoenergetic incident particles in a beam. In this case t is the projected path length for the individual particle, and d is the central axis depth. Then, the differential contribution to the Bragg ionization function (5) is given by

$$dI(d) = i(t - d)\frac{d\phi}{\phi_0} \tag{4.39}$$

In Equation 4.39, $d\phi$ is the number of particles per unit area with projected path length between t and $t + dt$, and ϕ_0 is the total number of particles per unit area incident on the surface. If the projected path-length distribution is normal with variance V_t and projected range R_t, Equation 4.5 can be used:

$$I(d) = \left[\frac{F}{F + d}\right]^2 \frac{1}{\sqrt{2\pi V_t}} \int_d^\infty i(t - d) \exp\left[-\frac{(t - R_t)^2}{2V_t}\right] dt \tag{4.40}$$

The variable F is the distance from the point source of particles to the surface of the ionizing medium. The quantity in brackets indicates that divergence of the beam affects the average ionization along the central axis. It is assumed that $F \gg d$. The parameter V_t is related to the Bragg peak width in this equation. Notice that when d is larger than $R_t + \sqrt{V_t}$, the individual specific ionization is integrated over only a small fraction of the area of the normal distribution; thus, $I(d)$ decreases to zero when $d \gg R_t$.

When beams of charged particles are used in practical situations, it is usually important to know the deposited energy density as a function of depth. In particular, knowledge of the percent depth dose for the radiation beam at all relevant depths may be necessary. The percent depth dose $P(d)$ is defined as the quotient when $D(d)$, the dose at depth d, is divided by the maximum value of the dose, $D(m)$, and multipled by 100. Thus,

$$P(d) = 100\, D(d)/D(m) \tag{4.41}$$

$P(d)$ is often displayed as a function of depth along the central axis of the beam, although values off the axis may also be required.

It is evident that the shape of the central axis percent depth dose curve for charged particles is closely related to the shape of the Bragg ionization curve. Recall from Equation 2.25 that the product of the individual specific ionization i and the average energy per ion pair W, gives the rate of collisional energy loss. Suppose that a beam of particles is incident normally at central axis on area A. For a single particle, the energy per unit mass absorbed in an infinitesimal mass element of volume Adx centered on the particle path is

$$dE/dm = \frac{dE/dx}{\rho A} = \frac{(-dT/dx)_c}{\rho A} = \frac{Wi}{\rho A} \tag{4.42}$$

Equation 4.42 holds if the particle passes straight through the element dm and if the energy absorbed in element dm is the same as the energy lost by the particle in dx. For a diverging beam of N particles at normal incidence along the central axis on an area A at depth d:

$$D(d) = \frac{W}{\rho A} N\langle i(d)\rangle = \frac{W\phi_0}{\rho} I(d), \quad P(d) = 100\frac{I(d)}{I(m)} \tag{4.43}$$

In Equation 4.43 the dose is in grays at depth d along the central axis if ϕ_0 is the particle fluence at the surface of the absorber. $I(d)$ has the shape of the Bragg ionization curve described by Equation 4.40, although the absorber is usually not air. The relationships in Equation 4.43 are adequate if the effects of delta-ray energy losses, bremsstrahlung, and the scattering of primary particles out of the beam are not severe.

Several illustrative central axis percent depth-dose curves are shown in Figure 4.14 (29). The curve for the proton beam has the traditional sharp Bragg peak expected when straggling and multiple scattering are minimal. Both of the electron beam curves show decided effects of straggling and multiple scattering. Low-energy electron contamination in the incident beam may also complicate the situation. Indeed, the Bragg peak is not discernible at higher electron energies. The tail at large depth for the electrons is due to bremsstrahlung produced in the absorber. Some tables of electron percent depth dose in water are given in Appendix 8.

Figure 4.15 shows an isodose curve for an electron beam in water. An isodose curve is a display of contours of constant percent depth dose in some plane, usually through the central axis. Because of scattering effects at large depth, the peak area is reduced in lateral dimensions and the dose pattern extends out well beyond the limits indicated by field edge divergence.

A final example of a percent depth-dose curve is shown on Figure 4.16 for a high-energy negative pion beam (29, 30). Notice that the curve has a shape similar to a Bragg curve with a peak clearly defined. For the pion beam, the Bragg peak is greatly enhanced by the formation of stars near the end of the pion path. The stars are produced when nuclei in the medium absorb the pions and disintegrate into heavily ionizing fragments. Notice that the region

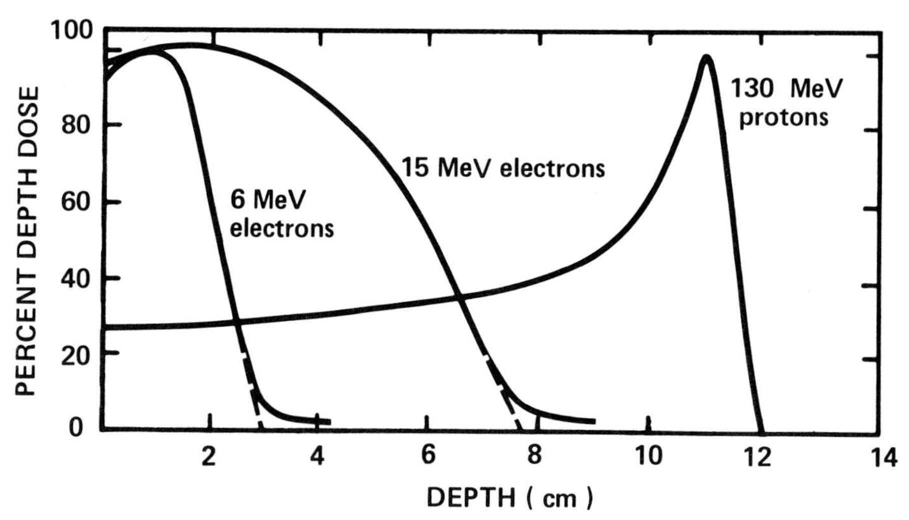

Figure 4.14 Percent depth dose curves for proton and electron beams are shown. Dashed lines indicate projections in the absence of bremsstrahlung. Adapted from H. E. Johns and J. R. Cunningham, *The Physics of Radiology,* 3rd Ed. (Charles C Thomas, Springfield, Ill., 1969).

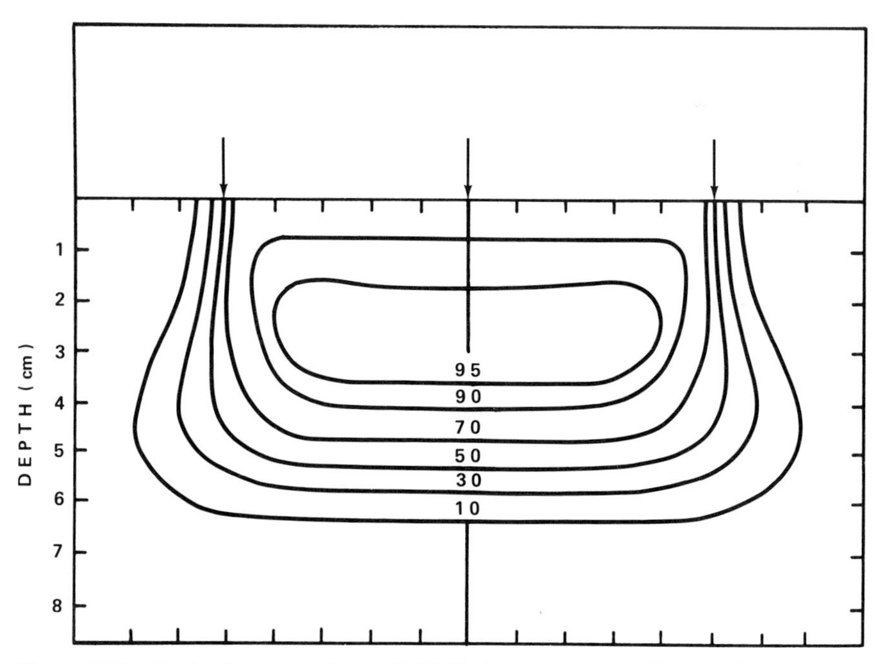

Figure 4.15 An isodose curve for a 13-MeV electron beam is shown. The numbers on the central axis are percent depth dose values.

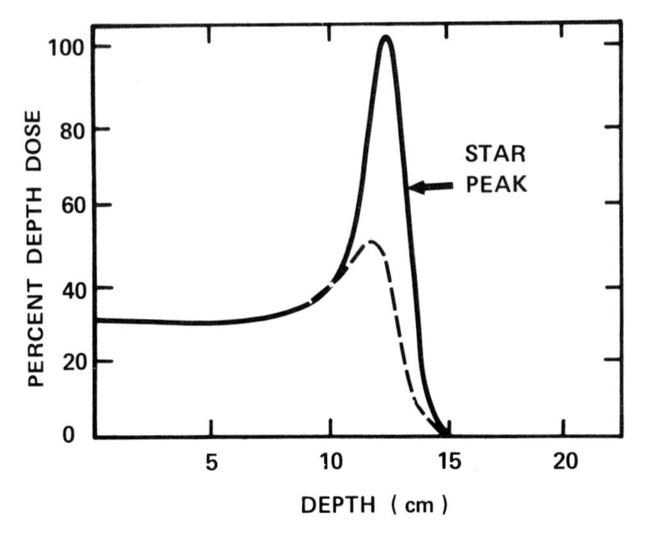

Figure 4.16 The percent depth dose curve for high-energy pions in water exhibits a large peak near the end of the path (29, 30). The dashed curve shows the relative energy density without star formation. Adapted from W. R. Hendee, *Medical Radiation Physics* (Yearbook, Chicago, 1970).

of the high dose is limited in extent since pion absorption by nuclei occurs only at low pion energy (degraded) and the range of the nuclear fragments is limited.

4.6 Problems

1. Using Equations 4.8, calculate the thickness of water that will transmit 50% of the beta rays emitted from a thin ^{90}Sr source.
2. Use Equation 4.13 to calculate the range of a 5-MeV alpha particle in N_2 gas at 760 mm Hg pressure. Assume that the exponential integral at T_1 (low-energy limit) and $R_1(T_1)$ can be neglected.
3. Calculate the range of 100-keV, 500-keV and 1-Mev electrons in aluminum from Equations 4.15.
4. If the range of an He^{++} ion was measured as 1.5 cm in aluminum, estimate the range in graphite.
5. Calculate the ratio of the range of a 14-MeV $^{14}N^{+++}$ ion to the range of a 1-MeV proton. Use Equation 4.18.
6. Calculate the ratio of the range of a 14-MeV $^{14}N^{+++++}$ ion to the range of a 1-MeV proton. Use Equation 4.18.
7. Calculate the standard deviation in the residual energy of 10-MeV protons after they have passed through 0.03 cm of aluminum.
8. Calculate the full width at half maximum (FWHM) of the distribution in residual energy of a 20-MeV proton beam after it has passed through 0.02 cm of aluminum and 0.04 cm of aluminum. FWHM $= 2.36\sigma$ for a normal distribution.
9. Multiple scattering determines the divergence of the bremsstrahlung beam in a betatron with 0.2-mm-thick tungsten target since the photons emerge

nearly tangent to the direction of travel of the electrons prior to emission. Suppose that the full width at half maximum (FWHM) of an 18-MeV beam is 5°. What would the FWHM be for a 25-MeV beam?

10. Obtain the algebraic expression for $B(T,Z,\rho t)$ in terms of the double integral using Equations 4.35 and 4.36.
11. What is the probability that a particle deflection will fall within a cone of half angle $\sqrt{\langle\Delta^2\rangle}$ for the normal distribution of Equations 4.38?
12. Estimate the ratio between the path length and the projected path length for 1-MeV electrons in a 0.2-mm-thick piece of tungsten.
13. What incident electron fluence is required to deposit 0.01 Gy at T = 10 MeV in water? Use Equation 4.43 and assume $W/e = 30$ eV.

4.7 References

1. H. Bischel, "Charged particle interactions," in *Radiation Dosimetry,* vol. 1, edited by F. H. Attix and W. C. Roesch (Academic, New York, 1968), pp. 157–228.
2. J. F. Perkins, Phys. Rev. **126**, 1781 (1962).
3. G. J. Neary, Proc. R. Soc. London Ser A **175**, 71 (1940).
4. B. R. Paliwal and P. R. Almond, Phys. Med. Biol. **20**, 547 (1975).
5. R. D. Evans, *The Atomic Nucleus* (McGraw-Hill, New York, 1955), pp. 621–671.
6. M. Curie, A. Debierne, A. S. Eve, H. Geiger, O. Hahn, S. C. Lind, S. Meyer, E. Rutherford, and E. Schweilder, Rev. Mod. Phys. **3**, 427 (1931).
7. L. Katz and A. S. Penfold, Rev. Mod. Phys. **24**, 28 (1952).
8. H. W. Lewis, Phys. Rev. **85**, 20 (1952).
9. U. Fano, Phys. Rev. **92**, 328 (1953).
10. M. J. Berger and S. M. Seltzer, "Tables of energy losses and ranges of electrons and positrons," in *Studies in Penetration of Charged Particles in Matter, Publication 1133* (National Academy of Sciences–National Research Council, Washington, 1964), pp. 205–227.
11. L. Pages, E. Bertel, H. Joffre, and L. Sklavenitis, At. Data **4**, 1 (1972).
12. W. Gautschi and W. F. Cahill, "Exponential integral and related functions," in *Handbook of Mathematical Fucntions,* edited by M. Abramowitz and I. A. Stegum (National Bureau of Standards, Washington, 1968), pp. 227–251.
13. R. R. Wilson, Phys. Rev. **71**, 385L (1947).
14. L. Katz and A. S. Penfold, Rev. Mod. Phys. **24**, 28 (1952).
15. E. Sergre, *Nuclei and Particles* (W. A. Benjamin, New York, 1965), pp. 39–47.
16. B. R. Martin, *Statistics for Physicists* (Academic, New York, 1971), pp. 39–41, 51–54.
17. P. V. Vavilov, JETP **5**, 749 (1957).
18. S. M. Seltzer and M. J. Berger, "Energy-loss straggling of protons and mesons: tabulation of the Vavilov distribution," in *Studies in Penetration of Charged Particles in Matter, Publication 1133* (National Academy of Sciences–National Research Council, Washington, 1964), pp. 187–203.
19. N. D. Kessaris, Radiat. Res. **43**, 281 (1976).
20. J. Van Dyck and J. C. F. McDonald, Phys. Med. Biol. **17**, 52 (1972).
21. R. D. Birkhoff, "The passage of fast electrons through water," in *Handbuch der Physik,* vol. 34, edited by S. Flügge (Springer, Berlin, 1958), pp. 53–138.
22. E. Rutherford, Philos. Mag. **37**, 537 (1919).

23. H. A. Bethe, Phys. Rev. **89**, 1256 (1953).
24. W. T. Scott, Rev. Mod. Phys. **35**, 231 (1963).
25. *Radiation Protection Design Guidelines for 0.1–100 MeV Particle Accelerator Facilities, NCRP Report 51* (National Council on Radiation Protection and Measurements, Washington, 1977).
26. M. Mladjenovic, *Radioisotope and Radiation Physics* (Academic, New York, 1973).
27. W. H. Bragg, *Studies in Radioactivity* (Macmillan, London, 1912).
28. M. G. Holloway and M. S. Livingston, Phys. Rev. **54**, 18 (1938).
29. M. R. Raju, J. T. Lyman, T. Brustad, and C. A. Tobias, "Heavy charged particle beams," in *Radiation Dosimetry,* vol. 3, edited by F. H. Attix and E. Tochilin (Academic, New York, 1969), pp. 151–193.
30. M. R. Raju, E. Lampo, S. B. Curtis, and C. Richman, Phys. Med. Biol. **16**, 599 (1971).

5

Systematic Relationships for Indirectly Ionizing Radiation

5.1 Introduction

In Chapter 1 indirectly ionizing radiations were described as agents of energy transfer that ionize primarily through liberated, charged secondaries. Their initial interactions in matter occur through some means other than the coulomb force, since they have zero net charge. Indirectly ionizing radiations often cause charge separation during an initial absorbing event. However, the bulk of the ionization occurs thereafter, during the degradation of the fast-electron or positive-ion secondaries.

Because of the nature of the process, it is helpful to divide energy transfer for indirectly ionizing radiations into two stages. A discussion of the first stage must include an outline of the reactions of primaries. For the second stage one should deal with degradation of secondaries. In succeeding chapters, we will consider the initial primary reactions; the general mechanism for energy transfer by fast charged particles has already been discussed in the preceding chapters. Some attention will be given to specific details about charged secondaries and their effects in later chapters.

The properties of several types of indirectly ionizing radiations (1) are given in Table 5.1. The species listed have few common characteristics other than the absence of a net charge. They react with other particles by three different basic interactions (2). Of the radiations listed in the table, only photons and the neutrons will be thoroughly discussed since they are the indirectly ionizing radiations with most practical significance. Photon beams are widely used in medical and industrial applications. Neutron reactions will be considered because of the applications to reactors and because accelerator-generated neutron beams may be more common in the near future in clinical centers.

Table 5.1 Properties of Several Indirectly Ionizing Radiations

Name	*Charge*	*Interaction*	*Atomic Mass (u)*	*Mc^2 (MeV)*
Photon	0	Electromagnetic		$<(2)10^{-21}$
Neutron	0	Nuclear	1.0087	939.6
Neutrino	0	Weak		$<(6)10^{-5}$
Pion (neutral)	0	Nuclear	0.1449	134.0

Data from Ref 1.

The properties of the photon have been fully discussed in other texts (3, 4). A brief review will be given here to contrast with the description of the neutron, which will follow. As the reader may recall, photons are pulses of electromagnetic radiation with specific energy content, which are emitted and absorbed as indivisible units. Their energy is directly proportional to the frequency of the associated oscillatory electric and magnetic fields. Photons are uncharged and have an intrinsic spin of 1 $\hbar$ and no magnetic moment. They are massless and travel with the speed of light in vacuum after emission. Photons can be diffracted and polarized because of the accompanying oscillatory fields. They interact by virtue of these fields with charged particles or collections of charged particles, which can sustain an electric or magnetic moment. The electric dipole moment of the absorber is important in common absorption reactions (4). Photons that are emitted from the nucleus are called gamma rays; photons emitted in extranuclear atomic processes are called x rays. In spite of the different names, a gamma ray cannot be distinguished per se from an x ray of identical energy.

The neutron symbolized by ${}^1_0n$ is a massive particle that decays in free space with a half-life of 10.6 min, according to the reaction (5):

$${}^1_0n \longrightarrow {}^1_1p + \beta^- + \bar{\nu} \tag{5.1}$$

where ${}^1_1p$ is a proton, β^- stands for a negative beta particle and $\bar{\nu}$ indicates an antineutrino. In spite of this reaction, the neutron is not a particular bound state of an electron and a proton. Rather, the internal structure must involve bound particles more massive than electrons, including mesons (6). Since the neutron has spin ½ $\hbar$ and a magnetic dipole moment of -1.9 nuclear magnetons (7), at least some of these particles are charged. As is shown in Figure 5.1, the charge distribution contains both negative and positive contributions (8).

The neutron reacts with other particles primarily via the nuclear force, which has very short range, about 10^{-14} m (2). At low energies, interactions involving the magnetic dipole moment of the neutron and the magnetic dipole moment of some other particle can occur, but these are usually of little practical consequence.

Since the photon and the neutron have such different properties, a discussion of the two of them in a common analytical framework may not seem plausible at first. Actually, the use of a common notation is helpful when the reactions involved are summarized mathematically in terms of reaction probabili-

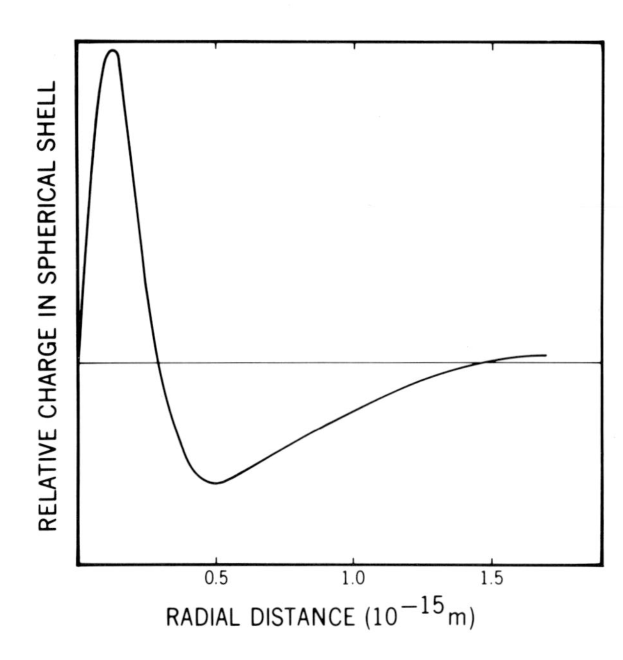

Figure 5.1 The charge distribution for the neutron has both positive and negative parts, although the net charge is zero. Since the negative charges occur at large radius, the magnetic moment is negative.

ties. Consideration of the two in a common framework is appropriate in our context, since absorption and scatter effects play an analogous role in energy transfer for both photons and neutrons. A systematic set of relationships involving reaction cross sections, the attenuation of beams, and deposition of energy will be given in Sections 5.2, 5.3, and 5.4. They can be applied to both neutron and photon beams.

In succeeding chapters the many reactions resulting in energy transfer and scatter will be considered individually for photons and then for neutrons. These processes can be summarized in a list of reaction categories similar to that given in Chapter 2 for charged particles. Indirectly ionizing radiations are involved in:

1. Elastic reactions with electrons or collections of electrons;
2. Inelastic reactions with electrons or collections of electrons;
3. Elastic reactions with nuclei or collections of nuclei;
4. Inelastic reactions with nuclei or collections of nuclei.

In Chapters 6 and 7, photon reactions that fall into categories 1, 2, and 4 will be considered. In Chapters 8 and 9, neutron-induced reactions will be considered; they fall into categories 3 and 4, since the strong interaction dominates.

5.2 Cross Sections and Coefficients

Individual particles in a beam of indirectly ionizing radiation often participate in reactions resulting in their complete removal from a beam in a single event.

These reactions are in contrast to the processes attributable to the coulomb force, which dominate for directly ionizing radiations. In most cases, directly ionizing radiations undergo an extended series of energy transfer events before they are thermalized.

The distinction between multiple energy transfer events for charged particles and single-event removal for uncharged radiations is not absolute. Single-event absorption occurs for charged particles in nuclear reactions (9), although these nuclear events have relatively low probability of occurrence. Furthermore, scatter reactions in broad beams of indirectly ionizing radiation can cause a gradual spectrum energy degradation with penetration depth, rather than a net loss in particles in the beam. Nevertheless, because of the importance of single-event removal from a narrow beam, the reaction cross sections for indirectly ionizing radiations are commonly related to removal coefficients and not to stopping powers. The coefficients in turn can be used to generate exponential beam transmission relationships for narrow monoenergetic beams. Furthermore, the concepts of path length and range that are helpful in discussing absorption of directly ionizing radiations are not so useful when photon and neutron beams are discussed. The mean free path and half value thickness (HVT) are much more important.

The reduction in the number of particles in a radiation beam as it passes through an absorber is called attenuation. A description of the attenuation of either photon or neutron beams usually begins with the reaction cross sections. Recall that the cross section σ can be defined as the probability per target per unit area for a given reaction. Using the notation discussed in Chapter 2,

$$\sigma = \frac{-d\phi/\phi}{n_v dx} \tag{5.2}$$

where the incident fluence is ϕ and $-d\phi$ is the fluence removed in a slab of absorber of infinitesimal thickness dx. Then $n_v dx$ is the number of targets per unit area for the absorber. In any case

$$-d\phi/\phi = n_v \sigma dx = \mu dx, \qquad \mu = n_v \sigma = n_m \rho \sigma \tag{5.3}$$

Equation 5.3 can be taken to define the linear attenuation coefficient μ as the product of the number of targets per unit mass n_m, the mass density ρ, and the cross section for the reaction. As such, it represents a reaction probability per unit thickness for an absorber of infinitesimal thickness and has units of reciprocal distance.

The ideas and symbols used to define the attenuation coefficient are familiar to students previously introduced to photon reactions. When neutron beam attenuation is discussed, the Greek letter Σ is often used instead of μ to symbolize the reaction probability per unit thickness and is called the macroscopic cross section in that context (10). In the following sections and chapters, the symbol μ and the phrase "linear attenuation coefficient," will be used for all indirectly ionizing radiations, neutrons as well as photons. This choice allows a unified discussion.

When absorption of ultraviolet light in solutions is discussed, the molar extinction coefficient ϵ is often utilized (11). One should realize that the extinction coefficient is directly proportional to the linear attenuation coefficient although it is expressed in units that are convenient in absorption chemistry. Then

$$\epsilon = (\mu/C_l)\ \ln 10 = 0.434\mu/C_l \tag{5.4}$$

The symbol C_l stands for the solute concentration in moles per liter. The extinction coefficient is expressed in liters per mole meter.

Probabilities for independent processes can be added to obtain the total probability. Since the cross section for a specific reaction can be viewed as a probability expressed in particular units, the total cross section for all relevant reactions can be given as a sum of cross sections for individual reactions. A similar statement can be made for the linear attenuation coefficient. Thus, for photon or neutron beams, it is sensible to write

$$\mu = \mu_s + \mu_a = n_m \rho(\sigma_s + \sigma_a) \tag{5.5}$$

where subscripts s and a stand for scatter reactions and absorption reactions, respectively. The cross section values σ_s or σ_a may each refer to the sum of several specific processes. For example, for neutrons σ_s would include the cross section for elastic scattering, whereas σ_a would be given by the sum of cross sections for all neutron-induced absorption reactions, including radiative capture reactions and neutron-induced fission. Similarly, for photons the cross section σ_s would be given by the sum of cross sections for coherent and incoherent scattering. The cross section σ_a for photons would include the cross sections for the atomic photoelectric effect, pair production, and photonuclear reactions.

There are several other coefficients, defined for indirectly ionizing radiations, that are closely related to the linear attenuation coefficient and have great utility in specific circumstances. Among these quantities are the mass attenuation coefficient, the mass energy transfer coefficient, and the mass energy absorption coefficient. In the next three paragraphs, these coefficients will be defined in a manner similar to that used by the ICRU (12).

The mass attenuation coefficient μ/ρ of a medium for indirectly ionizing particles is the quotient of dN/N by ρdx, where dN/N is the fraction of particles that experience reactions in distance dx in a medium of density ρ:

$$\mu/\rho = \frac{dN/N}{\rho dx} \tag{5.6}$$

Note that Equation 5.6 is consistent with Equations 5.3, since dN/N is equivalent to $d\phi/\phi$.

As was shown in Equations 5.3 and 5.5, the linear attenuation coefficient μ is a direct function of the density of the absorber ρ. Because of this, it has a different value for each different physical state of an atomic absorber: solid, liquid, or gaseous. Furthermore, the linear attenuation coefficient will fluctuate between elements with similar atomic number in the same physical state because of fluctuation of the mass density of neighboring elements. When the density is divided out, the effects of physical state and of density fluctuations are removed for the most part. Mass attenuation coefficients tabulated for a limited number of elements have wide utility because one can interpolate between elements on the table to obtain the coefficient for an unlisted element. The situation is analogous to that of the stopping power and the mass stopping power discussed in Chapter 2. A tabulation of photon mass attenuation coefficients for some absorbers is given in Appendix 9.

The mass energy transfer coefficient μ_{tr}/ρ of a medium for indirectly ionizing particles is the quotient of dE_{tr}/EN by ρdx, where E is the energy of each particle (excluding rest energy), N is the number of particles, and dE_{tr}/EN is the fraction of the incident particle energy that is transferred to kinetic energy of liberated charged particles in traversing a distance dx in a medium of density ρ:

$$\mu_{tr}/\rho = \frac{dE_{tr}/EN}{\rho dx} \tag{5.7}$$

μ_{tr}/ρ involves only energy losses that result in charged-particle kinetic energy. For photon beams the energy is transferred to electrons almost exclusively; for neutron beams in aqueous media, the energy is often transferred to protons.

The mass energy absorption coefficient μ_{en}/ρ of a medium for indirectly ionizing particles is the product of the mass energy transfer coefficient, μ_{tr}/ρ, and $(1 - g)$, where the symbol g stands for the fraction of the energy of the liberated charged particles that is lost to bremsstrahlung in the material:

$$\mu_{en}/\rho = (1-g)\mu_{tr}/\rho = [(1-g)dE_{tr}/EN]/\rho dx \tag{5.8}$$

μ_{en}/ρ and μ_{tr}/ρ values do not differ appreciably for neutron beams, since the ion secondaries from neutron absorption produce very little bremsstrahlung. The factor g cannot always be ignored when electron secondaries are liberated by high-energy photon beams. If the electron secondaries were monoenergetic, g would be equal to the radiation yield discussed in Chapter 3. For many processes g will involve an average of yield values for a spectrum of secondary electron energies. Note that the energy included in g is only due to bremsstrahlung and does not involve characteristic photons produced after ionization events in K and L shells.

If any of the foregoing mass coefficients must be found for a molecular substance and values $(\mu/\rho)_i$ are known for the constituent atoms, one can use

$$\mu/\rho = \sum_i \epsilon_i(\mu/\rho)_i \tag{5.9}$$

where ϵ_i is the fraction by weight of the ith element in the molecule. Generally, this equation is suitable because binding energy effects can be neglected. Equation 5.9 is an analog of Bragg's additivity rule for stopping power given in Chapter 2.

In the preceding paragraphs, reaction probabilities and energy transfer coefficients have been defined. The remaining topics relate to the penetration of photons or neutrons. There are two quantities commonly used: 1) the mean free path and 2) the HVT. Because there is no way to know in advance the actual distance a given photon or neutron will travel before being involved in a reaction, these quantities both involve averages for a beam of identical uncharged particles.

The mean free path λ for an indirectly ionizing particle is the average distance $\langle x \rangle$ between successive interactions:

$$\lambda = \langle x \rangle = \frac{\int xP(x)dx}{\int P(x)dx} = 1/\mu \qquad (5.10)$$

$P(x)dx$ is the probability that the particle will participate in a reaction after traversing a distance between x and $x + dx$. Since the linear attenuation coefficient is the reaction probability per unit distance, it must be equal to the reciprocal of the average distance traveled prior to a reaction.

The half value thickness, HVT, is the absorber thickness that will reduce the beam fluence to half of its original value. Since the attenuating properties of various absorbers are quite different, HVT values for different absorbers and different beam energies are often quite different. This subject is discussed in more detail in Section 5.3.

5.3 Attenuation of Beams

Exponential attenuation is the result of single-event removal of particles from a monoenergetic beam of indirectly ionizing radiation in an absorber. From Equation 5.3,

$$\int \frac{d\phi}{\phi} = -\mu \int dx', \quad \ln \phi = -\mu x + C,$$

$$\ln \phi_0 = 0 + C, \quad \ln(\phi/\phi_0) = -\mu x, \quad \phi = \phi_0 \exp(-\mu x) \qquad (5.11)$$

where C is the constant of integration, and the boundary condition is $\phi = \phi_0$ when $x = 0$.

Notice that an expression for the HVT of the radiation beam can be obtained from Equation 5.11. If

$$\tfrac{1}{2} = \phi/\phi_0 = \exp[-\mu(\text{HVT})], \quad \ln 2 = \mu(\text{HVT}), \quad \text{HVT} = 0.693/\mu \qquad (5.12)$$

The HVT is inversely proportional to the linear attenuation coefficient μ when attenuation follows the exponential law. The relationship between the HVT and the exponential form is illustrated in Figure 5.2.

It is important to realize that the relationships given in Equation 5.11 and 5.12 involve the effect of superposition of an absorber between the source of radiation and the point where the beam will be used. They do not include any effect on the fluence because of geometrical divergence of the beam. Thus, ϕ and ϕ_0 are values obtained at the same relative distance from the radiation source.

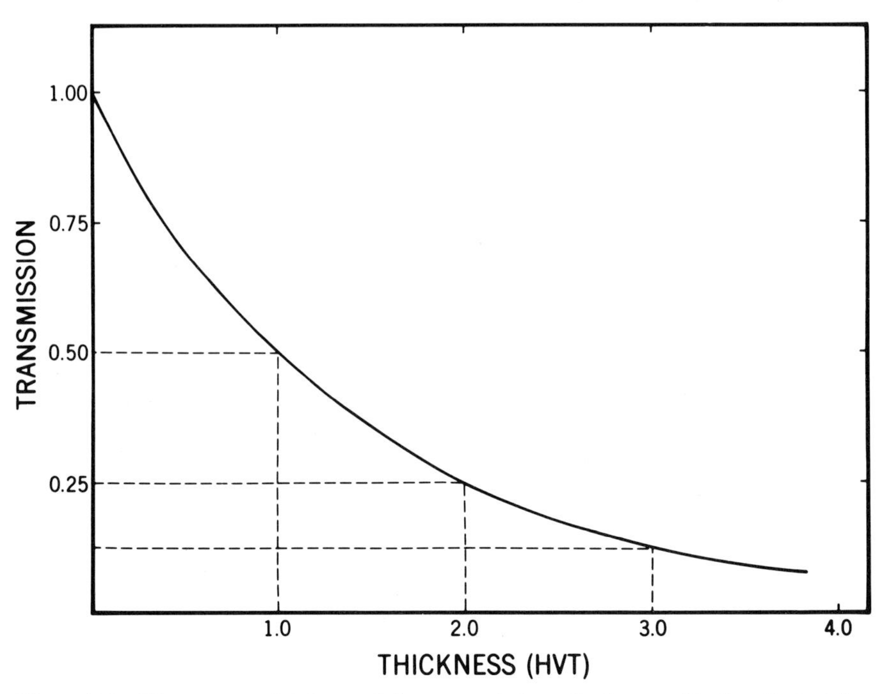

Figure 5.2 The exponential form of the transmission of a beam of indirectly ionizing radiation through an absorber is shown. Each added HVT of absorber reduces the transmission to half of its previous value.

The fractional reduction in fluence with increasing distance from a point source can be calculated using the inverse square law. Since emissions from a point source have radial directions, they intercept a spherical shell centered on the source at perpendicular incidence. The area of such a shell is $4\pi r^2$, where r is the radial distance from the center. Thus, the fluence at distance r is inversely proportional to r^2. This proportionality does not hold in practical situations unless source dimensions are much smaller than the distance r. For relatively large sources, the effective distance is more difficult to estimate.

In an experimental situation, a truly exponential relationship for attenuation is found for monoenergetic beams only if the measurements are made using narrow-beam geometry. This situation is illustrated in Figure 5.3. The condition for narrow-beam geometry is that the largest dimension of the beam spot on the absorbing material must be much less than the distance between the absorber and the position where the beam is to be used or measured. In this case the greatest part of the radiation scattered in the absorber will be directed out of the region of interest. Severe collimation is usually required to meet this condition.

Broad-beam geometries are commonly encountered when radiation is used in medical and industrial situations. In such cases an appreciable contribution

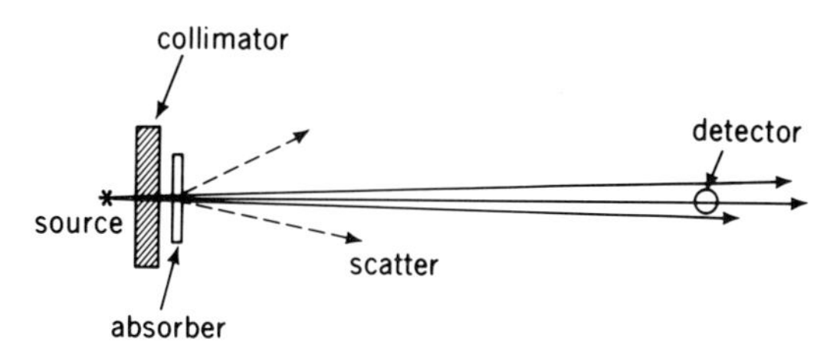

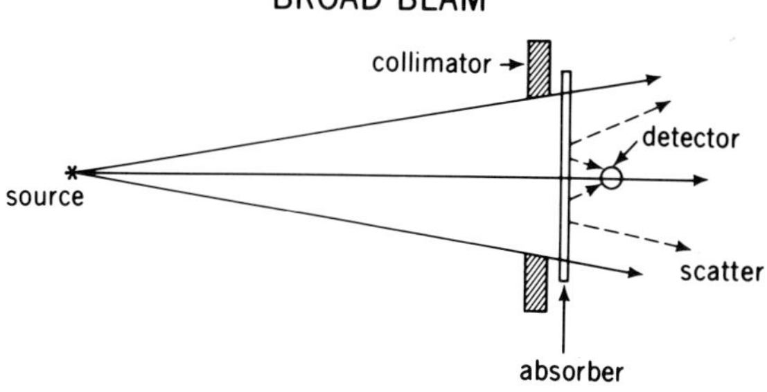

Figure 5.3 Under narrow beam conditions, the beam size at the absorber is much less than the distance from the absorber to the detector. Thus, scatter contributions are minimal. Under broad-beam conditions, scatter in the absorber contributes to the detector response. Adapted from W. R. Hendee, *Medical Radiation Physics* (Yearbook, Chicago, 1970).

from scattered and secondary radiation is present in the transmitted beam. The total fluence ϕ_t is given by

$$\phi_t = \phi_p + \phi_s + \phi_e > \phi_p \tag{5.13}$$

where ϕ_p is the fluence due to the primary beam; ϕ_s is the fluence due to the scattered radiation; and ϕ_e is the fluence emitted in the absorber as a secondary component (such as bremsstrahlung). The extra fluence, over and above transmitted primary radiation, is often referred to as scatter buildup radiation.

The effect of buildup radiation is illustrated in Figure 5.4, where the fractional transmission is graphed on semilogarithmic paper for a monoenergetic beam of radiation. The dashed line, which is expected for exponential attenuation, is located under the experimental result. The difference between the two curves on the figure is due to the scattered and emitted fluence. It is evident from Figure 5.4 that the ratio of buildup to primary radiation in the beam is an increasing function of the thickness of the absorber. It is also an increasing function of beam size.

A practical example of the effect of the scattered radiation occurs in radiography with a broad x-ray beam. The radiation scattered in the subject that reaches the film reduces the contrast in the image. When large fields and thick subjects are involved, the effect is very troublesome. Devices called grids are often installed between the subject and the film; they preferentially remove the scattered radiation (13) and produce an improved image.

In Figure 5.4, the absolute value of the slope of the broad-beam transmission curve increases from the initial value at small absorber thickness to a

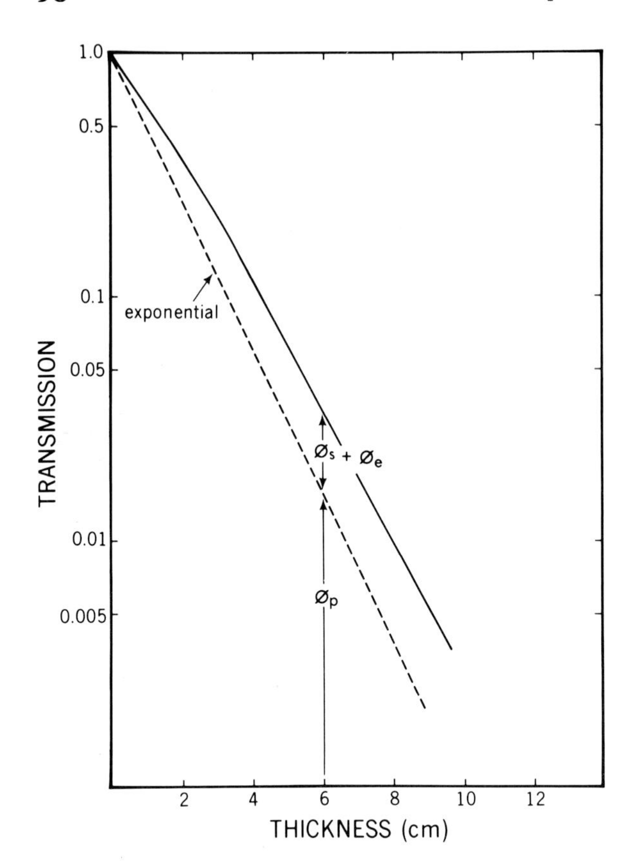

Figure 5.4 For broad-beam conditions, the scatter plus emitted fluence, $\phi_s + \phi_e$, is added to the primary fluence ϕ_p. This distorts the transmission so the exponential form does not occur.

larger value at somewhat greater thickness. After this initial increase, the slope is nearly independent of absorber thickness. It is generally true that broad-beam semilogarithmic transmission curves have approximately constant slope at very small transmission values (14, 15). Because of this fact, an equilibrium HVT for broad beams can be used in approximate shielding calculations for thick protective barriers.

If the incident radiation beam is not monoenergetic, the transmission curve generally will not have the form of a true exponential, even under narrow-beam conditions. The reason is because the attenuation coefficient is a function of energy for indirectly ionizing radiation. Figures 5.5 and 5.6 (16, 17) illustrate the dependence on energy for a photon and a neutron beam, respectively. The structures that depart from smooth dependence are the K and L edges for photons (Figure 5.5) and absorption resonances for neutrons (Figure 5.6). These phenomena will be discussed in upcoming chapters.

For photons, the attenuation coefficient is always large at low energies and generally decreases as the energy increases until a minimum value is attained. When an inhomogeneous beam with wide energy range (but maximum energy less than the energy of the minimum in the coefficient) is employed, the transmitted spectrum is changed after the addition of a thin absorber. The beam is

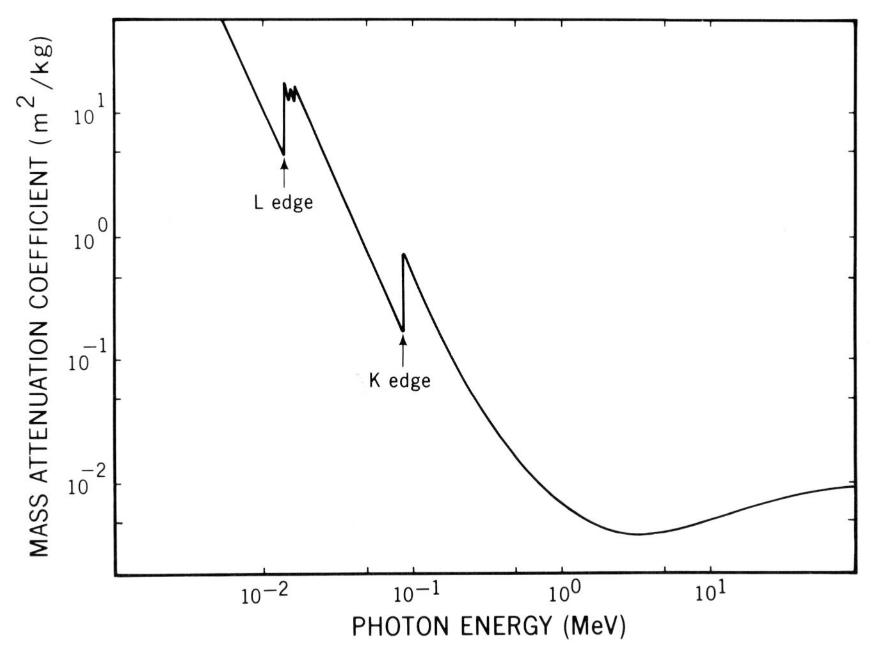

Figure 5.5 The mass attenuation coefficient for a lead absorber for photons has an absolute minimum at about 3 MeV (16).

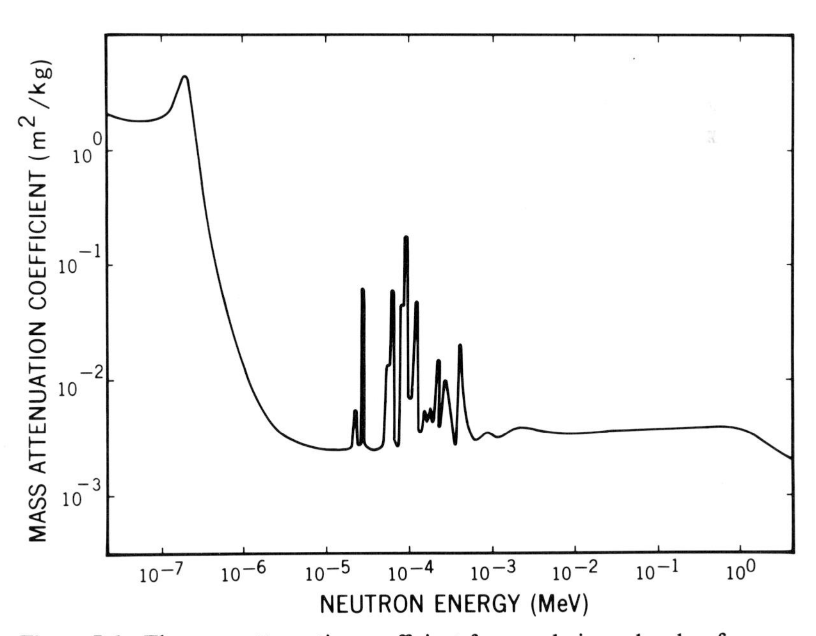

Figure 5.6 The mass attenuation coefficient for a cadmium absorber for neutrons has a large maximum at 0.3 eV (17).

said to be filtered, since the lowest energy photons are selectively absorbed out. It is evident that filtration leaves a more penetrating residual beam. Because of this, the average attenuation coefficient for the filtered beam is decreased and HVT is increased. The same general conclusion applies to a broad-spectrum neutron beam emerging from a thin cadmium absorber.

If one measures the narrow-beam transmission for an inhomogeneous beam using successively increasing absorber slab thicknesses, the deviation from the exponential relationship becomes apparent. Figure 5.7 shows such a transmission curve for photons produced by an x-ray generator operating at 250 kVp. In this circumstance one can measure the first and second half value thicknesses HVT_1 and HVT_2, respectively, and calculate the homogeneity coefficient η defined by

$$\eta = \frac{HVT_1}{HVT_2} \tag{5.14}$$

Narrow monoenergetic beams have $\eta = 1$; for inhomogeneous, narrow photon beams, $\eta < 1$.

Beams from diagnostic x-ray machines are routinely filtered with thin aluminum sheets. Figure 5.8 shows the transmitted beam spectrum from an

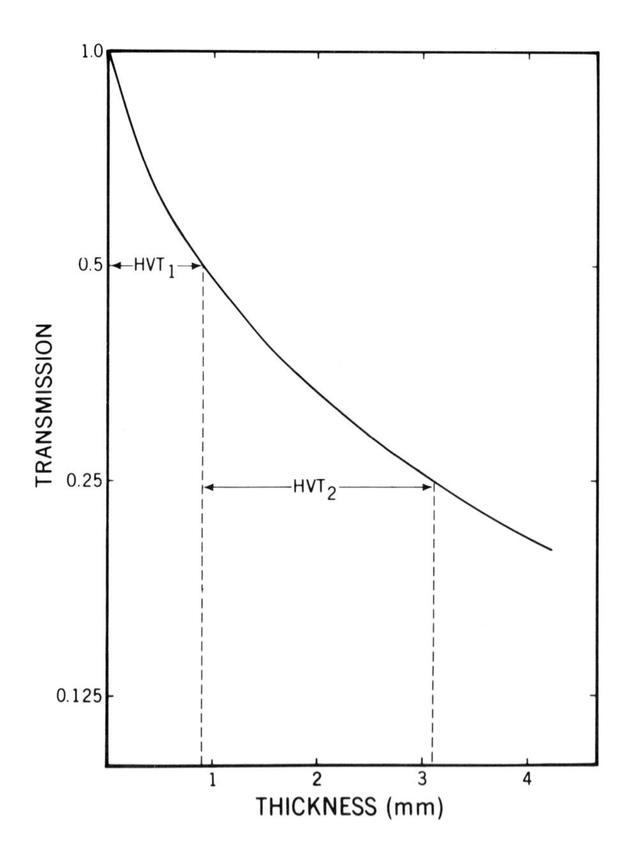

Figure 5.7 The transmission of a beam of x rays with a continuous spectrum has an HVT_2 larger than HVT_1.

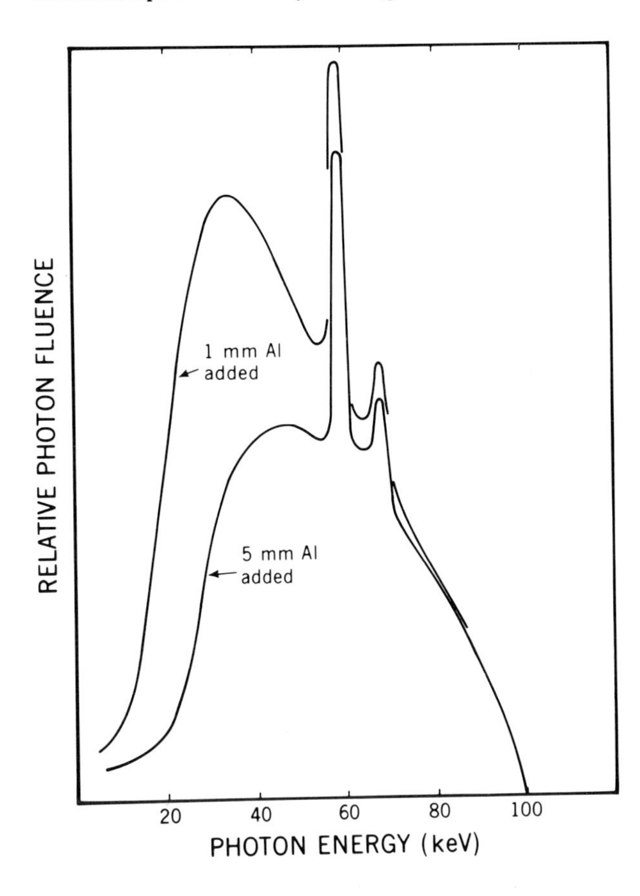

Figure 5.8 The addition of aluminum filters in the beam of an x-ray machine operating at 100 kVp preferentially removes low-energy photon components. Adapted from Ref. 19.

x-ray machine before and after the addition of several external filters (18, 19). Notice that the lower energy components are preferentially removed by the filter, although some x rays are lost at even the highest energies. The low-energy components, which are filtered out in great number, have very small probability of contributing to the image in the x-ray procedure. Yet they would add unwanted superficial patient dose.

For either photon or neutron beams with inhomogeneous spectra, it is sometimes useful to find the equivalent energy of the beam. The equivalent energy of an inhomogeneous beam is equal to the energy of a monoenergetic beam, which has the same transmission value for the particular absorber under consideration. In practice one can measure the effective attenuation coefficient μ_{eff} for the beam and estimate the corresponding photon energy from a table of attentuation coefficient values. μ_{eff} is defined by

$$\exp(-\mu_{\text{eff}}x) = \frac{\int \exp[-\mu(E)x]\phi(E)dE}{\int \phi(E)dE} = Tm, \quad \mu_{\text{eff}} = \frac{\ln(1/Tm)}{x} \tag{5.15}$$

where x is the absorber thickness, $\phi(E)dE$ is the photon fluence distribution, and Tm is the transmission of the inhomogeneous beam. The integration must include the complete spectrum of energies.

5.4 Kerma and Dose Relationships

The concept of charged-particle equilibrium will be discussed in this section, and a relationship between the absorbed dose and the incident fluence of photons or neutrons will be developed. In Section 5.5 these ideas will be extended so that the exposure for photon beams can be related to the dose for several practical situations.

As a first step, recall that the kerma is the charged particle kinetic energy transferred from indirectly ionizing beams per unit mass of absorber. If the Kerma K is expressed in grays,

$$K = \frac{dE_k}{dm} = \frac{dE_{tr}}{\rho A dx} = \phi \frac{dE_{tr}/N}{\rho dx} \tag{5.16}$$

Equation 5.16 relates to the case where N monoenergetic particles are in perpendicular incidence on the face of a small parallelepiped of mass ρAdx, where ρ is the density, A is the area of the face, and dx is the thickness. The incident fluence ϕ is therefore N/A. The average particle energy loss is such that dE_{tr}/N is transferred to charged-particle kinetic energy in the medium in distance dx. If the kerma is in rads, a factor of 100 must be included with the kinetic energy transferred dE_{tr}.

When the mass energy transfer coefficient ($\mu_{tr}\rho$) is applied (19),

$$K = \phi E \frac{dE_{tr}/EN}{\rho dx} = \phi E(\mu_{tr}/\rho) \tag{5.17}$$

For an inhomogeneous beam, with fluence spectrum $\phi(E)dE$,

$$K = \int E\phi(E)(\mu_{tr}/\rho)dE \tag{5.18}$$

Equations 5.17 and 5.18 clearly show that K is a function of incident particle energy E. Figure 5.9 shows the kerma per fluence for a neutron beam in water and muscle tissue (20). Figure 5.10 gives the kerma-to-fluence ratio for a photon beam in water (16). The kerma-to-fluence ratio for a photon beam in muscle tissue is quite similar to the water value. It is interesting to note that for 250-keV neutrons, a fluence of $10^{15}/m^2$ produces a tissue kerma of about 1 Gy. However, a fluence of $10^{15}/m^2$ photons gives 1 Gy at 2.5 MeV.

With the preceding equations in mind, the terms "radiation equilibrium" and "charged-particle equilibrium" can be discussed. These concepts have been presented previously in many places (21–23). Refer to Figure 5.11 where an element of mass m is shown in the large volume V. This mass is under irradiation. Radiation equilibrium is said to exist in m when every type of radiation with

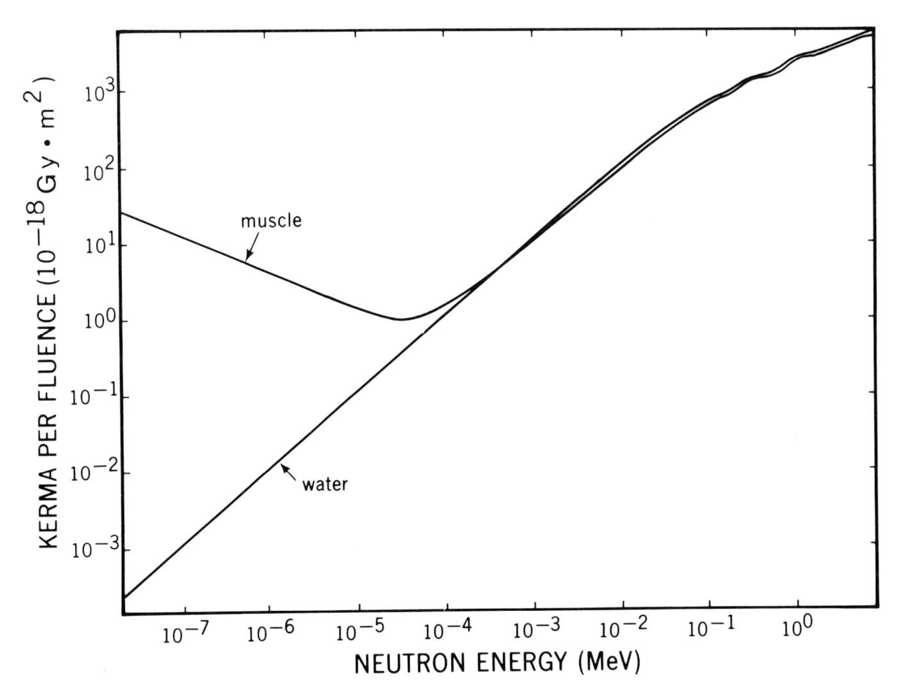

Figure 5.9 The kerma per fluence of neutrons increases with increasing energy for a water absorber. The large values at low energy for muscle tissue are due to neutron particle reactions in the nitrogen constituent (20).

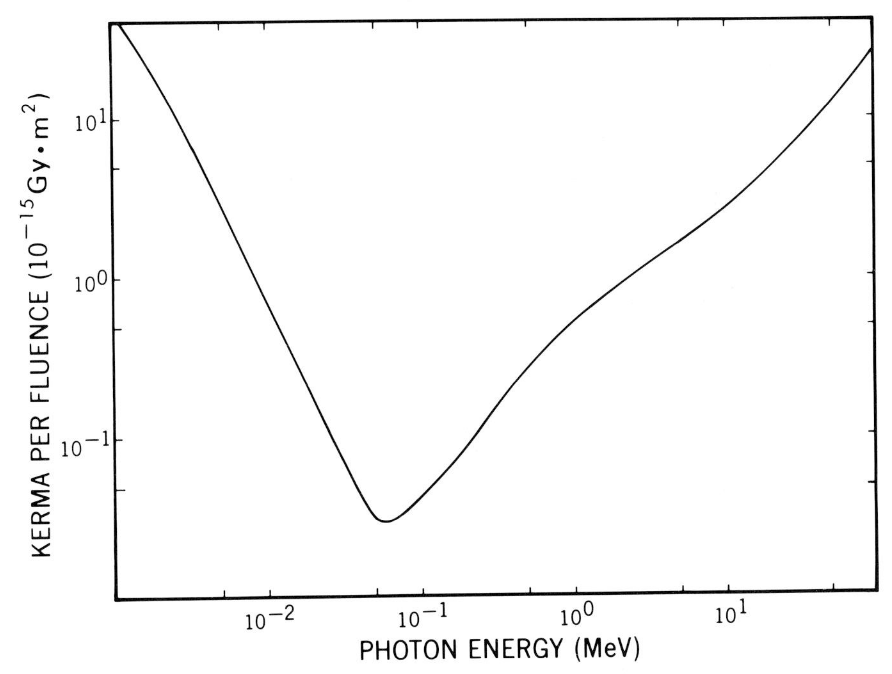

Figure 5.10 The kerma per fluence of photons has a minimum at about 60 keV in water (16). The values for muscle tissue are about 2% larger than those shown at lower energies and about 2% smaller than those shown at higher energies.

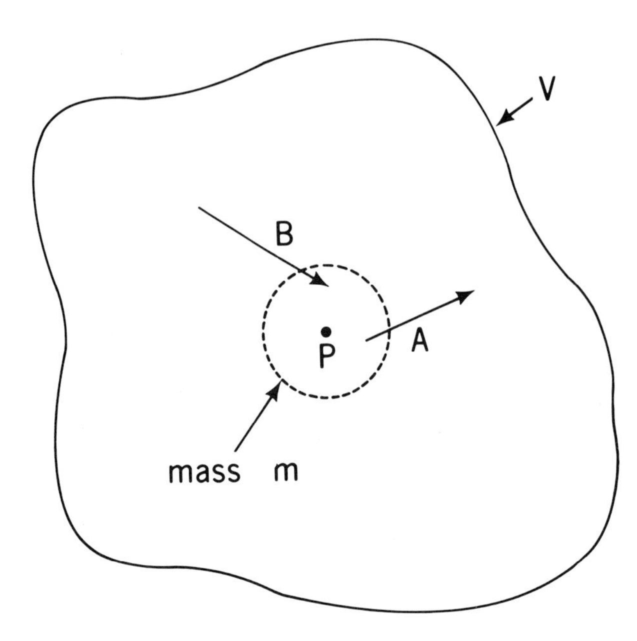

Figure 5.11 Radiation equilibrium is illustrated for mass m centered at point P in medium V. A and B are identical radiation particles paired so that the net energy transferred out of m is zero. Adapted from Ref. 23.

energy E that enters m can be matched with an identical radiation that leaves m with the same energy. If we suppose that A and B on the figure represent identical radiations and that if A enters the element, B leaves the element, then the two constitute a pair. If all radiations entering and leaving m can be paired in this way, radiation equilibrium exists in m.

Radiation equilibrium can occur in situations where radioactive materials are uniformly imbedded in a large volume of absorber; otherwise, it is difficult to achieve. When indirectly ionizing radiations are used, there are two radiation components in the absorber material: the uncharged primaries and the charged secondaries. Equilibrium of the secondaries may be considered exclusive of the primaries.

Charged-particle equilibrium is a special case of radiation equilibrium wherein the restrictions on the indirectly ionizing components are lifted. Charged-particle equilibrium is said to exist for element m when every charged particle with energy E that enters m can be matched with an identical particle that leaves m with the same energy.

Necessary conditions for charged-particle equilibrium during irradiation by indirectly ionizing beams include uniformity of the indirectly ionizing component and a homogeneous absorber (21). In practice, when external beams of neutrons or photons are employed, the conditions for charged-particle equilibrium can only be satisfied in the limit when the mean free path of the primary beam is much greater than the charged-particle range and the incident fluence is uniform over an area with dimensions much larger than the charged particle range. Table 5.2 gives some values for mean free path and secondary-particle range for several neutron and photon beams (23). For example, conditions for

Table 5.2 Mean Free Path (λ) and Maximum Range in Tissue

Photon or Neutron Energy (MeV)	*Photon λ (cm)*	*Secondary Electron Maximum Range (cm)*	*Neutron λ (cm)*	*Secondary Proton Maximum Range (cm)*
0.3	8	0.08	1.7	0.001
1.0	14	0.43	4.2	0.003
3.0	26	1.47	6.7	0.016
10.0	46	4.90	17.0	0.140

Data from Ref. 23.

charged-particle equilibrium are clearly violated in the region near the boundary between the materials of widely different composition.

Charged-particle equilibrium is a sufficient condition for measurement of the exposure, as defined in Chapter 1, with an air ionization chamber. All ions produced by electrons liberated in the chamber must be collected. Of course, some extra ions collected in the chamber are produced by electrons liberated outside the chamber. Similarly, some ions will not be collected because electrons liberated in the chamber volume are projected out of the collection region. If charged-particle equilibrium is ensured, a balance in charge lost and charge gained will occur, which allows an accurate measurement.

Consider again the absorbed dose in a mass m. Recall that

$$E_D = \sum E_{\text{in}} + \sum Q - \sum E_{\text{out}} \tag{5.19}$$

where E_D is the energy absorbed in the mass element m; ΣE_{in} is the sum of the energies carried into the element by radiation; ΣE_{out} is the sum of the energies carried out of the element by radiation; and ΣQ is the sum of the net energies released in the element due to mass conversion. Rest mass energies are excluded from incoming and outgoing sums and are included explicitly only when there is a net conversion of mass to energy. Using subscripts c and u to indicate charged and uncharged radiation contributions, respectively, means that

$$E_D = \sum (E_{\text{in}})_c + \sum Q_c - \sum (E_{\text{out}})_c + \sum (E_{\text{in}})_u + \sum Q_u - \sum (E_{\text{out}})_u \tag{5.20}$$

Suppose that charged-particle equilibrium holds; then

$$\sum (E_{\text{in}})_c = \sum (E_{\text{out}})_c,$$

$$E_D = \sum Q + \sum (E_{\text{in}})_u - \sum (E_{\text{out}})_u \tag{5.21}$$

If no bremsstrahlung escapes, the difference in the energy sums for the uncharged particles in the second of Equations 5.21 is E_K, the kinetic energy liberated. If bremsstrahlung is produced and particle induced mass change reactions do not occur, then

$$E_D = E_K - \sum (h\nu)_e = E_K(1 - g) \tag{5.22}$$

where $(h\nu)_e$ represents the energy of bremsstrahlung photons that escape. The fraction of the liberated kinetic energy that is radiated away as bremsstrahlung is given the symbol g, as was done in Equation 5.8.

Suppose these equations relating E_D and E_K in mass m hold in the infinitesimal limit; then dE_D and dE_K are energies for an absorber of mass dm. The dose D and kerma K can be substituted for the energy density:

$$D = (1 - g)K \tag{5.23}$$

Equation 5.23 is the basic result for charged-particle equilibrium with no added particle induced mass conversion. More generally, the kerma and the dose can be related by

$$D = bK \tag{5.24}$$

where b is the kerma-to-dose conversion factor, left unspecified.

The kerma-to-dose conversion factor for an external beam in an absorber is not a constant but depends on such parameters as beam quality, size of field, and the depth of penetration into an absorber (13). If we consider broad beams with flat fluence profiles and low-energy secondaries with very short range, the kerma and dose are nearly equal along the central axis at all depths. If the beam produces high-energy secondaries with appreciable range, the kerma-to-dose relationship is more complex. Figure 5.12 illustrates these two situations (24).

For high-energy indirectly ionizing external beams, the kerma-to-dose conversion factor is less than $(1 - g)$ near an air-absorber surface because of charged-particle buildup. Charged-particle buildup occurs because many of the charged secondaries are projected forward. Imagine that the absorbing material is divided into very thin layers 1, 2, 3, and so forth, with layer 1 on the air interface of the absorbing medium. Successive numbers indicate layers at successively greater depths in the absorber. Only a small number of secondaries pass through layer 1 because none are incident from previous layers. Layers 2, 3, and 4 receive an increasing number of charged particles, since those produced in more superficial layers are projected forward through these layers. Thus, the total number

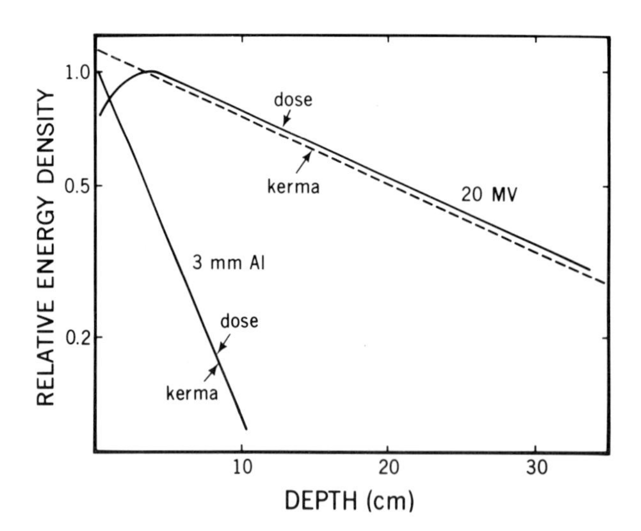

Figure 5.12 The central axis kerma and dose are shown as a function of depth for a superficial x-ray beam with HVT of 3 mm of aluminum and a 20-MV bremsstrahlung beam (24).

of projected charged particles increases with depth (in a limited region). In this situation,

$$\sum(E_{\text{in}})_c < \sum(E_{\text{out}})_c, \quad E_D < (1-g)E_K, \quad D < (1-g)K \qquad (5.25)$$

Charged-particle buildup ceases at depths approximately equal to the projected range of the particles. Beyond this point, secondaries produced in the most superficial layers are stopped and cannot contribute at greater depth. Furthermore, attenuation of the primary beam in more superficial layers means that fewer and fewer charged particles are liberated in successively deeper layers of comparable thickness. Beyond some depth the kerma-to-dose ratio increases to a value greater than $(1 - g)$:

$$\sum(E_{\text{in}})_c > \sum(E_{\text{out}})_c, \quad E_D > (1-g)E_K, \quad D > (1-g)K \qquad (5.26)$$

There are two depths of importance in this context: one is d_c, the depth for which the kerma-to-dose conversion factor equals $(1 - g)$; the second is d_m, where m indicates the dose maximum. The values d_c and d_m are usually related by $d_c \lesssim d_m$ as shown on the 20-MV curve on Figure 5.12 (13, 25). At depths considerably greater than d_m, the value of the kerma-to-dose conversion factor can be roughly constant if beam energy degradation is not severe. A transient equilibrium can exist (22), wherein the decreasing dose follows the decreasing kerma at increasing depths.

5.5 Exposure and Dose Relationships for Photons

For photon beams, conversion from exposure to dose is important because x-ray generators are often calibrated in terms of exposure rate. For further evaluations the dose rate can be calculated from the exposure rate.

Recall from Chapter 1 that exposure involves the charge generated per unit mass of air. The exposure X in coulombs per kilogram can be calculated from

$$X = dQ/dm_a = e(1-g)_a \frac{dE_{\text{tr}}/W_a}{\rho_a A dx} = \frac{\phi(1-g)_a}{(W_a/e)N} dE_{\text{tr}}/\rho dx \qquad (5.27)$$

For Equation 5.27, N monoenergetic photons are in perpendicular incidence on the face of a very small parallelepiped of mass $\rho_a Adx$, where ρ_a is the density of air, A is the area of the face, and dx is the thickness. The energy $h\nu$ is such that on the average dE_{tr}/N is transferred to electron kinetic energy per photon. A fraction of g of this kinetic energy is radiated out of the mass element as bremsstrahlung. The rest is dissipated in collisions that produce ions with charge e. W_a/e, the average energy per ion pair in air expressed in electron volts, is equal to 33.8 eV (26). The incident fluence ϕ is N/A.

Some algebraic manipulation of Equation 5.27 yields

$$X = \frac{\phi h\nu(1-g)_a}{W_a/e} \frac{dE_{\text{tr}}/h\nu N}{\rho_a dx} = \frac{\phi h\nu}{W_a/e} (\mu_{\text{en}}/\rho)_a \qquad (5.28)$$

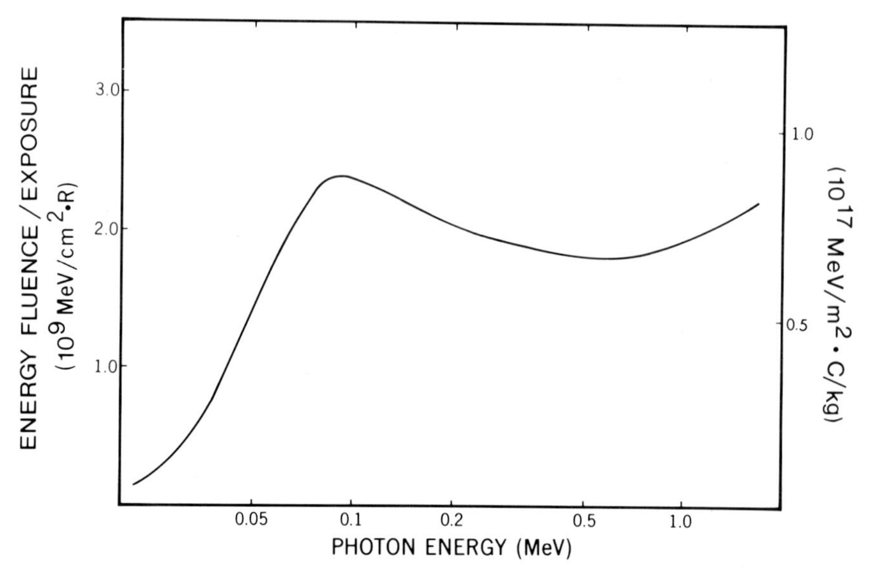

Figure 5.13 The energy fluence per unit exposure in air (16) increases to a maximum at about 90 keV photon energy and does not change rapidly at higher energies.

In Equation 5.28, the mass energy absorption coefficient μ_{en}/ρ has the subscript a for air. Solving for the product $\phi h\nu$:

$$\psi = \phi h\nu = \frac{X(33.8)}{(\mu_{en}/\rho)_a} = \frac{33.8X}{(\mu_{en}/\rho)_a} \tag{5.29}$$

Figure 5.13 is an illustration of the energy fluence ψ per coulomb per kilogram (16). Notice that above 100 keV this quantity is approximately equal to $(8)10^{16}$ $(\text{MeV/m}^2) \cdot (\text{C/kg})^{-1}$.

Equations 5.17 and 5.29 can be used to express the kerma K_m in some medium m in terms of the exposure X_m in the medium:

$$K_m = \phi h\nu(\mu_{tr}/\rho)_m = 33.8\frac{(\mu_{tr}/\rho)_m}{(\mu_{en}/\rho)_a}X_m = \frac{33.8}{(1-g)_m}\frac{(\mu_{en}/\rho)_m}{(\mu_{en}/\rho)_a}X_m \tag{5.30}$$

Using Equation 5.24,

$$D = b_m K_m = b_m \frac{33.8}{(1-g)_m}\frac{(\mu_{en}/\rho)_m}{(\mu_{en}/\rho)_a}X_m \tag{5.31}$$

The kerma-to-dose conversion factor for the medium is b_m.

Equation 5.31 is further simplified at depth d_c near a surface where $b_m = (1-g)_m$. At this depth the exposure-to-dose conversion factor f_m for medium m for monoenergetic photons is given by

$$D = 33.8\frac{(\mu_{en}/\rho)_m}{(\mu_{en}/\rho)_a}X_m = f_m X_m, \quad f_m = 33.8\frac{(\mu_{en}/\rho)_m}{(\mu_{en}/\rho)_a} \tag{5.32}$$

The units for f_m are grays per coulomb per kilogram. If a nonhomogeneous beam with photon fluence distribution $\phi(h\nu)dh\nu$ is used

$$\langle f_m \rangle = 33.8 \frac{\int (\mu_{en}/\rho)_m h\nu\phi(h\nu)dh\nu}{\int (\mu_{en}/\rho)_a h\nu\phi(h\nu)dh\nu} = 33.8 \frac{\int (\mu_{en}/\rho)_m \psi(h\nu)dh\nu}{\int (\mu_{en}/\rho)_a \psi(h\nu)dh\nu} \tag{5.33}$$

The exposure-to-dose conversion factor for air has the constant value of 33.8 since the integrals cancel. The factor depends on $h\nu$ for any other medium since the mass energy absorption coefficients are functions of $h\nu$.

Historically, the exposure-to-dose conversion factor has been given in terms of roentgens and rads rather than coulombs per kilogram and grays. Conversion to roentgens and rads in Equations 5.30 and 5.33 above is obtained by replacing 33.8 by 0.872. The exposure-to-dose conversion factor is displayed on Figure 5.14 for muscle and bone as a function of energy (13). Notice that the factor is approximately four times as large for bone as it is for muscle at low photon energies.

Large values of the exposure-to-dose factor for bone occur because μ_{en}/ρ is much larger for bone than for air at low energies. Bone contains substantial amounts of calcium and phosphorous (27) with atomic numbers of 20 and

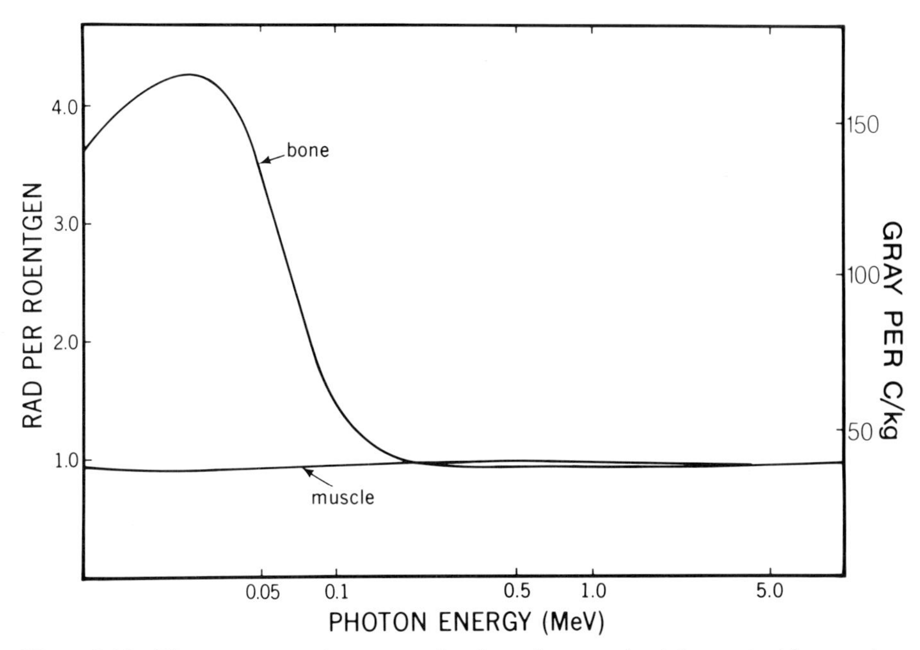

Figure 5.14 The exposure-to-dose conversion factor is approximately constant for muscle tissue. The factor is about four times larger for bone at photon energies below 100 keV (13).

15, respectively. Air and muscle tissue are composed almost entirely of lower atomic number materials. The mass energy absorption coefficient at low energies is mostly due to the photoelectric effect and increases with increasing atomic number. Photoelectric coefficients will be discussed in Chapter 7.

Equation 5.31 can be used to calculate the dose when the exposure is known in the medium. A situation often occurs wherein the exposure rate in air in a beam is measured at some standard distance F from a point source or target. Later, the dose rate in a medium m placed in the beam with surface at F can be calculated at the depth d_c where $b_m = (1 - g)_m$. Of course, $K_m(d_c)$, the photon kerma in the absorber at depth d_c, is different from $K_a(F)$, the kerma in air at the standard distance F. Differences are manifested by an attenuation factor for the indirectly ionizing radiation in the absorber, $A(d_c)$; a scatter buildup factor, $\mathrm{BF}(d_c)$, which is greater than one because scattered and emitted secondaries from the absorber add to the kerma; and a beam divergence factor, $(F/F + d_c)^2$. Then

$$K_m(d_c) = A(d_c)\mathrm{BF}(d_c)(F/F + d_c)^2 K_a(F) \tag{5.34}$$

This expression for the kerma can be substituted into Equation 5.31 and the dose can be written

$$D(d_c) = f_m A(d_c)\mathrm{BF}(d_c)(F/F + d_c)^2 X_a(F) \tag{5.35}$$

where $X_a(F)$ is for the measured value of the exposure in air at distance F.

5.6 Problems

1. Suppose that the mass energy absorption coefficient for 1-MeV photons in aluminum is 0.00269 m^2/kg. Calculate the mass energy transfer coefficient using the radiation yield expression from Chapter 3 under the assumption that 0.5 MeV is the average electron energy produced by photons in the aluminum.
2. The linear attenuation coefficient for 10 keV photons in lead is about 1460 cm^{-1}. Since this value is so much greater than one, how can it represent an interaction probability per unit length?
3. Calculate the cross-section for interaction of photons in lead at 140 keV (^{99m}Tc) and 1.25 MeV (^{60}Co). Use the mass attenuation coefficients given in Figure 5.5.
4. Calculate the cross-section for interaction of neutrons in cadmium at 0.02 eV and 1 MeV. Use the mass attenuation coefficients given in Figure 5.6.
5. The lead lining in the walls of a diagnostic x-ray room is 5 mm thick, equivalent to 10 HVT for the heavily filtered beam. It is necessary to increase the lead lining so that the amount of time the machine is on may be doubled, while preserving the same protection to the environs. What thickness of lead must be added?
6. The following data represent photon fluence values from an x-ray beam transmitted through an absorber:

Thickness	*Fluence*
0	100.0
1 mm Al	60.2
2 mm Al	41.4
3 mm Al	30.2
4 mm Al	22.4
5 mm Al	16.9

Plot the data on semilogarithmic paper and determine the HVT_1 and HVT_2 and the homogeneity coefficient.

7. Calculate the mean free path of a neutron in an absorber using the integrations suggested in Equation 5.10, with help from Equations 5.3 and 5.11.
8. Suppose a beam of photons containing a continuous distribution of energies up to 50 MeV, such as might occur in bremsstrahlung, is passed through a very thick absorber of lead. Which gamma-ray energy is likely to emerge with the largest fluence?
9. A narrow beam of gamma rays passes through 2.0 cm of lead. The incident beam consists of 30% 0.4-MeV photons and 70% 1.5-MeV photons. What fraction of the incident fluence is transmitted? Use Figure 5.5.
10. A narrow beam of neutrons passes through 2.0 cm of cadmium. The incident beam consists of 60% 0.02-MeV neutrons and 40% 0.5-MeV neutrons. What fraction of the incident fluence is transmitted? Use the information on Figure 5.6.
11. Show that the kerma-to-dose conversion factor is constant at depths well beyond the depth of maximum dose if the kerma and dose curves have an exponential form.
12. In Equation 5.23, the dose and kerma were related through the factor $(1 - g)$ for the case of charged-particle equilibrium and negligible mass conversion. Is charged-particle equilibrium the least restrictive condition on the energies that is necessary to obtain this relationship? If not, what other condition is useful?
13. Estimate the photon fluence per roentgen at 18 keV (Mo-K_α), 140 keV (^{99m}Tc), and 1.25 MeV (^{60}Co) from data on Figure 5.13. Why does the intermediate energy photon have the largest value of fluence per roentgen?
14. Calculate the dose for a 100-R exposure measured in muscle tissue and bone at 18 keV (Mo-K_α), 140 KeV (^{99m}Tc), and 1.25 MeV (^{60}Co) from the information on Figure 5.14. Assume that electronic equilibrium holds at the point of consideration.

5.7 References

1. N. Barash-Schmidt, A. Barbaro-Galtieri, C. Bricma, V. Chaloupka, R. L. Kelly, T. A. Lasinski, A. Rittenberg, M. Roos, A. H. Rosenfeld, P. Soding, and T. G. Trippe, Rev. Mod. Phys. Suppl. **45**, 515 (1973).
2. H. Frauenfelder and E. M. Henley, *Subatomic Physics* (Prentice-Hall, Englewood Cliffs, NJ, 1974), pp. 227–352.

3. M. O. Scully and M. Sargent III, Phys. Today **25**, 38 (1972).
4. W. T. Grandy, *Introduction to Electrodynamics and Radiation* (Academic, New York, 1970).
5. *Table of Isotopes,* edited by C. M. Lederer and V. S. Shirley (John Wiley & Sons, New York, 1978).
6. A. E. S. Green, Science **169**, 933 (1970).
7. A. P. Arya, *Fundamentals of Nuclear Physics* (Allyn and Bacon, Boston, 1966), p. 16.
8. R. M. Littauer, H. F. Schopper, and R. R. Wilson, Phys. Rev. Lett. **7**, 144 (1961).
9. R. D. Evans, *The Atomic Nucleus* (McGraw-Hill, New York, 1955), p. 422.
10. A. M. Weinberg and E. P. Wigner, *The Physical Theory of Neutron Chain Reactors* (University of Chicago, Chicago, 1958), p. 24.
11. F. Daniels and R. A. Alberty, *Physical Chemistry,* 3rd ed. (John Wiley and Sons, New York, 1966), p. 529.
12. *Radiation Quantities and Units, ICRU Report 33* (International Commission on Radiation Units and Measurements, Washington, 1980).
13. H. E. Johns and J. R. Cunningham, *The Physics of Radiology,* 3rd ed. (Charles Thomas, Springfield, MA, 1969).
14. *Structural Shielding Design and Evaluation for Medical Use of X Rays and Gamma Rays of Energies up to 10 MeV; NCRP Report 49* (National Council on Radiation Protection and Measurements, Washington, 1976), p. 103.
15. *Protection Against Neutron Radiation, NCRP Report 38* (National Council on Radiation Protection and Measurements, Washington, 1971), p. 106.
16. E. Storm and H. I. Israel, *Photon Cross Sections from 0.001 to 100 MeV for Elements 1 through 100, LA 3753* (University of California, Los Alamos, NM, 1967).
17. D. J. Hughes and R. B. Schwartz, *Neutron Cross Sections, BNL325,* 2nd ed. (Brookhaven National Laboratory, Upton, NY, 1958).
18. *Medical X-Ray and Gamma-Ray Protection for Energies up to 10 MeV, Equipment Design and Use, NCRP Report 33* (National Council on Radiation Protection and Measurement, Washington, 1971).
19. *Radiation Dosimetry: X Rays Generated at Potentials of 5 to 150 kV, ICRU Report 17* (International Commission on Radiation Units and Measurements, Washington, 1970).
20. *Neutron Fluence, Neutron Spectra and Kerma, ICRU Report 13* (International Commission on Radiation Units and Measurements, Washington, 1966).
21. *Physical Aspects of Irradiation, ICRU Report 10b* (National Bureau of Standards, Washington, 1964), p. 1.
22. J. Dutreix, A. Dutreix, and M. Tubiana, Phys. Med. Biol. **10**, 177 (1965).
23. W. C. Roesch and F. H. Attix, "Basic concepts of dosimetry," in *Radiation Dosimetry,* vol. 1, edited by F. H. Attix and W. C. Roesch (Academic, New York, 1968), p. 38.
24. M. Cohen, D. E. A. Jones, and D. Greene, Br. J. Radiol. Suppl. **11** (1972) pp. 12, 81.
25. D. W. Anderson, Phys. Med. Biol. **21**, 524 (1976).
26. *Average Energy Required to Produce an Ion Pair, ICRU Report 31* (International Commission on Radiation Units and Measurements, Washington, 1979).
27. P. Rubin and G. W. Casarett, "Concepts of clinical pathology," in *Medical Radiation Biology,* edited by G. V. Dalrymple, M. E. Gaulden, G. M. Kollmorgen, and H. H. Vogel, Jr. (W. B. Saunders, Philadelphia, 1973), p. 187.

6

Photon Scattering Reactions

6.1 Introduction

In the next few chapters, the most common reactions induced by photons and neutrons will be considered, one by one. We use the term "reaction" for extranuclear as well as nuclear processes. This chapter and the one following deal mostly with extranuclear photon reactions, although photonuclear reactions are considered in a later section of Chapter 7. Photon reactions can be either elastic or inelastic. Recall that the total kinetic energy remains unchanged in an elastic reaction. In an inelastic photon-induced reaction, an outgoing secondary particle with mass is always liberated. Inelastic reactions are especially important because the liberated charged secondaries deposit energy locally in an absorber.

In the present chapter, the scattering of photons will be considered. During a scattering process, the incoming photon undergoes an interaction that produces a change in the direction of travel. In addition, substantial energy degradation is involved for inelastic processes. The energy of the outgoing photon can take any one of a continuous set of values bounded by the incoming photon energy, since transitions between discrete states of the atomic system are usually not directly involved. The scattering process can be contrasted with a typical absorption process, which includes the complete disappearance of the incident photon.

Table 6.1 shows some helpful energy categories for photons; the energy intervals are useful only as guidelines for identification of regions of dominance for particular reactions. For example, elastic scattering, and in particular coherent scattering, is the dominant scattering reaction in the low-energy region. That process is still important at intermediate energies and until about 50 keV for high Z absorbers (1, 2). For an elastic reaction, the photon momentum is

Table 6.1 Photon Energy Ranges

Type	*Energy Range*
Nonionizing	$0 \lesssim h\nu \lesssim 10$ eV
Low energy	10 eV $\lesssim h\nu < 0.5$ keV
Intermediate energy	0.5 keV $\lesssim h\nu < 1$ MeV
High energy	1 MeV $\lesssim h\nu$

sufficiently small to allow a directional change with negligible energy transfer. Inelastic scattering dominates in the intermediate- and high-energy regions.

We will discuss elastic and inelastic scattering in sequence in the following pages, and contrasting aspects of the nature of electromagnetic radiation will be exhibited. In the discussion of elastic scattering given in Section 6.2, the description of electromagnetic radiation as a wave is emphasized, a useful procedure at sufficiently low energies. Conversely, in Section 6.3 inelastic Compton scattering will be discussed without reference to waves. Instead, the particulate nature of the photon will be exhibited, with momentum and energy conservation directly involved. This procedure is appropriate at high energies.

Elastic scattering from a single free electron is called Thomson scattering; inelastic scattering from a free electron is called Compton scattering or the Compton effect. A theoretical formulation that unifies photon scattering results was developed by Klein and Nishina using field theory techniques (3). In the low-energy limit, Klein-Nishina scattering formulas describe Thomson scattering. Klein-Nishina results are easily combined with Compton expressions at higher energy. Klein-Nishina scattering results are presented in Section 6.4 in a variety of differential cross section forms.

Elastic and inelastic scattering of photons by the atomic nucleus will not be discussed in this chapter, since these reactions occur relatively infrequently. They are nevertheless of considerable interest in specialty areas of physics. For example, nuclear resonance scattering is central in demonstrations of the Mössbauer effect (4, 5).

6.2 Elastic Scattering

Thomson scattering is the elastic scattering of an electromagnetic wave by a free charged particle, usually an electron. J. J. Thomson was the first to give the classical description of the process (6). Thomson scattering is a prime example of a reaction that can be understood from classical ideas of electromagnetic radiation, since consideration of the electric field wave is central. In the development to follow (7, 8), it will be assumed that the electric field from an incident electromagnetic wave sets the electron into oscillation and the oscillating electron subsequently radiates.

Electromagnetic waves are often represented as transverse plane waves oscillating in space and time. The properties of such waves have been discussed

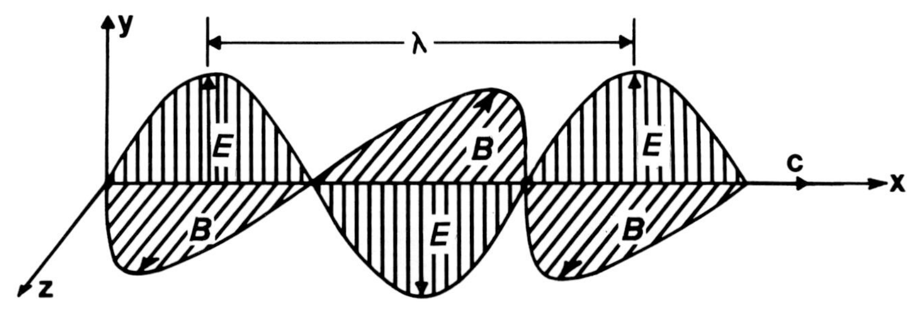

Figure 6.1 The spatial dependence of the fields of a plane electromagnetic wave traveling in the x direction is shown. ***E*** is the electric field vector, ***B*** the magnetic induction vector, and λ the wavelength.

in detail (9) and will not be reviewed extensively here. Figure 6.1 depicts the spatial oscillations of a polarized electromagnetic wave moving to the right (x direction). It is a plane wave if the spatial dependence of the amplitude in the yz plane can be ignored. For the elastic scattering problem, an electron in position on the x axis is assumed to be immersed in the electric field, which oscillates in time as the wave moves along in the x direction.

Figure 6.2 shows some geometrical parameters for the case where the electric field has components $\boldsymbol{E_y}$ and $\boldsymbol{E_z}$. If the field amplitude at the electron is $\boldsymbol{E_i}$, the time dependence can be written as

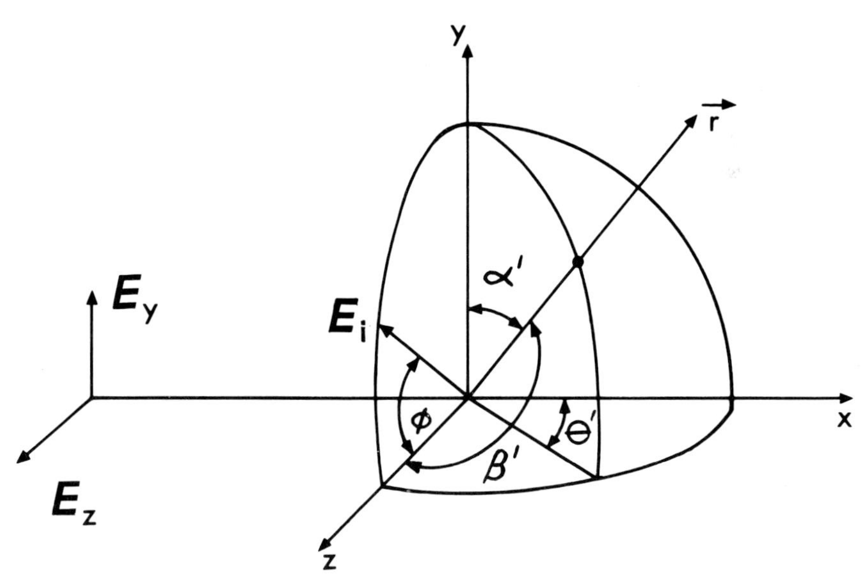

Figure 6.2 The electric field vector $\boldsymbol{E}_\mathrm{i}$ of a polarized electromagnetic wave is shown at the position of a scattering electron. The angles defining the exit direction along r for the scattered wave are shown. Primes are used on angles describing outgoing wave directions.

$$E_y = \hat{j} E_i \sin \phi \sin 2\pi\nu t, \quad E_z = \hat{k} E_i \cos \phi \sin 2\pi\nu t \tag{6.1}$$

where $\hat{j}$ and $\hat{k}$ are the unit vectors for the y and z directions, respectively, and ν is the frequency of the wave. The angle ϕ is used to give the appropriate projections of the polarization direction as shown on the figure. The average electromagnetic energy fluence rate or intensity, I_i, associated with the incoming wave is given by (8, 9)

$$I_i = c\epsilon_0 \langle (E_y{}^2 + E_z{}^2) \rangle = c\epsilon_0 E_i{}^2 \langle (\sin 2\pi\nu t) \rangle = \frac{c\epsilon_0 E_i{}^2}{2} \tag{6.2}$$

A factor of ½ was utilized in Equation 6.2 for the time average of the square of the sine function.

The driving force on the electron attributable to the electric field $\mathbf{E}_i$ produces a sinusoidal acceleration **a:**

$$\mathbf{a} = \frac{-eE_i}{m_e} \sin 2\pi\nu t \tag{6.3}$$

The oscillating electron subsequently radiates. It can be shown that the average electromagnetic intensity I_r radiated by the electron along the direction $\hat{r}$ of Figure 6.2 is given by (7, 8)

$$I_r = \frac{e^2}{16\pi^2 \epsilon_0 c^3 r^2} [\langle a_y{}^2 \rangle \sin^2 \alpha' + \langle a_z{}^2 \rangle \sin^2 \beta'] \tag{6.4}$$

The squares of the accelerations a in the foregoing equation are averaged. Each of the terms in the square brackets has the dipole shape (square of sine function) for the radiated intensity. Substituting from Equation 6.3 for a_y and a_z into Equation 6.4 yields

$$\begin{aligned} I_r &= \frac{e^2}{16\pi^2 \epsilon_0 c^3 r^2} \left[\frac{eE_i}{m_e}\right]^2 [\sin^2 \phi \sin^2 \alpha' + \cos^2 \phi \sin^2 \beta'] \langle (\sin 2\pi\nu t)^2 \rangle \\ &= I_i \left[\frac{r_0}{r}\right]^2 [\sin^2 \phi \sin^2 \alpha' + \cos^2 \phi \sin^2 \beta'] \end{aligned} \tag{6.5}$$

The expression for I_i from Equation 6.2 and the relationship $r_0 = e^2/4\pi\epsilon_0 m_e c^2$ for the classical electron radius were required for equation 6.5.

The polarized wave used in the foregoing development is unrealistic, since most beams include components with many polarization directions. Thus, an average over all incident polarization directions must be obtained. For nonpolarized waves, $\langle \sin^2 \phi \rangle = \langle \cos^2 \phi \rangle = ½$:

$$\langle I_r \rangle = \langle I_i \rangle \left[\frac{r_0}{r}\right]^2 \frac{(\sin^2 \alpha' + \sin^2 \beta')}{2} = \langle I_i \rangle \left[\frac{r_0}{r}\right]^2 \frac{(1 + \cos^2 \theta')}{2} \tag{6.6}$$

The expression on the right side of Equation 6.6 can be justified from Figure 6.2 because of axial symmetry about the direction of incidence for a nonpolarized beam. The $\sin \alpha'$ and $\sin \beta'$ terms can be evaluated in the xz plane; in that case, $\sin \alpha' = \sin \pi/2 = 1$, $\sin \beta' = \sin(\pi/2 - \theta') = \cos \theta'$.

The differential cross section per unit solid angle for Thomson scattering between angle θ' and $\theta' + d\theta'$ is symbolized by $d\sigma_T(\theta')/d\Omega'$. This cross section is defined as the probability for scattering into unit solid angle per electron per unit incident fluence. It is equivalent to the quotient of the average energy radiated into unit solid angle by the electron, divided by the incident energy fluence, since radiated and scattered energies are the same. Then

$$d\sigma_T(\theta')/d\Omega' = r^2\langle I_r\rangle/\langle I_i\rangle = (r_0^2/2)(1 + \cos^2\theta'),$$

$$\sigma_T = \int_0^{\pi} (d\sigma_T/d\Omega')d\Omega' = \pi r_0^2 \int_0^{\pi} (1 + \cos^2\theta')\sin\theta'\, d\theta' = 8\pi r_0^2/3 \tag{6.7}$$

Apart from the geometrical factor of 8/3, the Thomson cross section is equal to the classical electron cross-sectional area.

At low energies the photon wavelength is comparable to the diameter of the atoms in the absorber. In this case, all the electrons in a scattering atom are induced to oscillate and radiate in phase. When the entire atom acts as the scattering agent rather than individual electrons, the reaction is called coherent or Rayleigh scattering. In this situation a coherent scattering form factor $F(h\nu,\theta',Z)$ can be introduced (1, 2), and the angular distribution can be given by

$$d\sigma_{cs}(h\nu,\theta',Z)/d\Omega' = (r_0^2/2)(1 + \cos^2\theta')[F(h\nu,\theta',Z)]^2 \tag{6.8}$$

where cs stands for coherent scattering. The square of the form factor is the ratio of the differential elastic scattering cross section for an atom to the differential elastic scattering cross section for a free electron. The squared factor is proportional to Z^2 for the atom rather than Z because of the enhanced effectiveness of the coherent action (7, 10).

The form factor is often displayed as a function of $x = [\sin(\theta'/2)]/\lambda$, the momentum transfer variable. We utilize the variables $h\nu = hc/\lambda$ and θ' in our functional notation because they are more general. Figure 6.3 shows

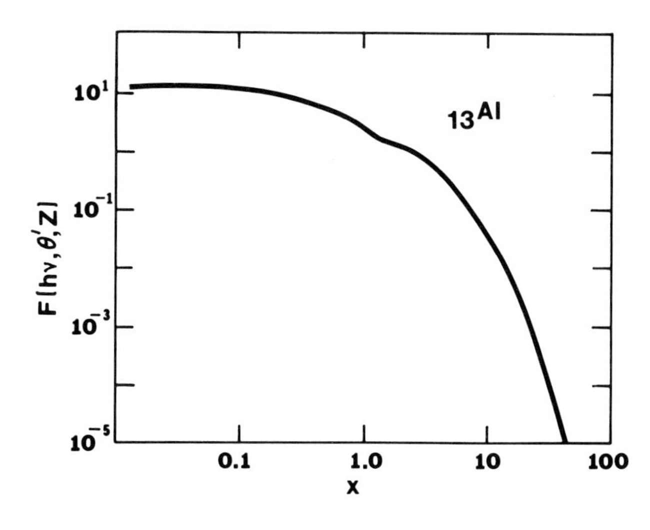

Figure 6.3 The coherent scattering form factor for aluminum is shown as a function of the $x = [\sin(\theta'/2)]/\lambda$, where the wavelength is in angstroms (10). Adapted from Ref. 11.

$F[h\nu,\theta',Z]$ for aluminum (11). Notice that the magnitude decreases drastically as x (proportional to $h\nu$) is increased. At constant $h\nu$ the form factor decreases with increasing scattering angle θ'. At 1 MeV most of the scattered radiation is confined to a cone with 10° half-angle (12).

The angular representation given by Equation 6.8 is valid for atoms acting independently. When scattering atoms are held in an array in a crystal, coherent scattering from crystal planes can occur if the monochromatic wavelength is of the order of the crystal lattice spacing. Coherent scattering of this sort can produce maxima in the angular distributions which are related to the crystal parameters. Scattering from crystal planes is called Bragg scattering or crystal diffraction (13). Crystal diffraction techniques are widely used to study the structure of complex molecules (14).

6.3 Compton Scattering Energy Values

Compton scattering or the Compton effect was named for A. H. Compton, who first described the process in 1923 (15, 16). Compton scattering is the inelastic scattering of a photon from a free electron and is the dominant photon scattering reaction at high energies. During Compton scattering the photon reacts much like a hard ball; a scattered photon with degraded energy and a fast electron, termed the Compton electron, result from the inelastic reaction. The Compton electron is responsible for local deposition of much of the energy from a high-energy photon beam.

Figure 6.4 is a schematic of a Compton scattering reaction. The parameters $h\nu$, $h\nu'$, T'_e, θ', and θ'_e are all used on the figure. Of course, $h\nu$ and $h\nu'$ are the energies of the incident and scattered photons, respectively. T'_e is the kinetic

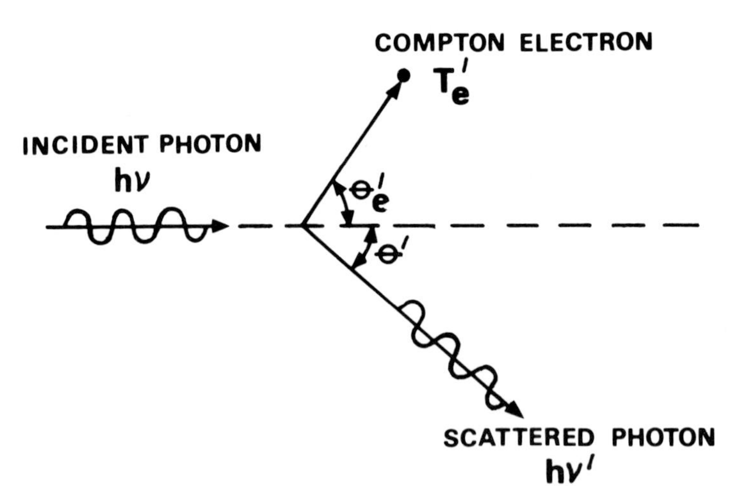

Figure 6.4 The variables for Compton scattering of an incident photon from a free electron are shown.

energy of the Compton electron; θ' and θ'_e are the scattering angles for the outgoing photon and the Compton electron, respectively. Notice that unprimed symbols are used to describe the incident beam, and primed symbols are used for emerging particles; this convention will be used throughout.

In our elementary considerations, the electron is assumed to be at rest prior to the collision. The parent nucleus for the electron will be completely ignored since nuclear recoil momentum can be neglected. The two outgoing particles are sufficient for simultaneous momentum and energy conservation. In addition, the electron binding energy BE can be ignored as long as $h\nu \gg \text{BE}$. Although these assumptions simplify the Compton scattering problem, inelastic scattering from inner-shell electrons does occur. The effects of electron binding are discussed in Section 6.5.

Consider the kinetics of Compton scattering. Application of momentum conservation to the setup shown in Figure 6.4 yields

$$h\nu/c = (h\nu'/c)\cos\theta' + p'_e \cos\theta'_e, \quad 0 = p'_e \sin\theta'_e - (h\nu'/c)\sin\theta'$$

$$p'_e c \cos\theta'_e = h\nu - h\nu' \cos\theta', \quad p'_e c \sin\theta'_e = h\nu' \sin\theta' \tag{6.9}$$

Here the momentum of the outgoing photon is $h\nu'/c$ and the outgoing electron momentum is p'_e. Squaring the final two equations in Equation 6.9 and adding them yields

$$(p'_e c)^2(\sin^2\theta'_e + \cos^2\theta'_e) = (h\nu)^2 - 2h\nu h\nu' \cos\theta' + (h\nu')^2(\sin^2\theta' + \cos^2\theta'),$$

$$(p'_e c)^2 = (h\nu)^2 - 2h\nu h\nu' \cos\theta' + (h\nu')^2 \tag{6.10}$$

In the algebraic manipulation preceding Equation 6.10, we have chosen to eliminate θ'_e rather than θ', although either operation is simple enough. The choice means that the scattered photon angle will be retained in the final algebraic expressions. This is often an appropriate choice because the scattered photon can initiate subsequent reactions at distant sites.

In the limit of negligible binding energy, conservation of energy for the Compton scattering process gives

$$h\nu = T'_e + h\nu', \quad T'_e = h\nu - h\nu' \tag{6.11}$$

Equations 6.11 can be used to obtain an independent expression for the electron momentum if it is substituted into the relativistic expression $p^2c^2 = T^2 + 2Tmc^2$:

$$(p'_e c)^2 = (T'_e)^2 + 2T'_e m_e c^2 = (h\nu - h\nu')^2 + 2(h\nu - h\nu')m_e c^2,$$

$$(p'_e c)^2 = (h\nu)^2 - 2h\nu h\nu' + (h\nu')^2 + 2(h\nu - h\nu')m_e c^2 \tag{6.12}$$

After subtracting the last part of Equations 6.10 from the last part of Equations 6.12, it is evident that

$$0 = -h\nu h\nu'(1 - \cos\theta') + (h\nu - h\nu')m_e c^2,$$

$$h\nu = h\nu'\left[1 + \frac{h\nu}{m_e c^2}(1 - \cos\theta')\right], \quad h\nu' = \frac{h\nu}{1 + \alpha(1 - \cos\theta')},$$

$$\alpha = h\nu/m_e c^2 \tag{6.13}$$

The ratio α will be called the collision energy parameter. An expression for the outgoing energy of the Compton electron can be obtained by substituting from Equations 6.13 into Equations 6.11:

$$T'_e = h\nu - \frac{h\nu}{1+\alpha(1-\cos\theta')} = \frac{h\nu\alpha(1-\cos\theta')}{1+\alpha(1-\cos\theta')} \tag{6.14}$$

Notice from Equations 6.13 that the Compton scattering reaction degrades the photon energy and that the fraction of energy scattered, $h\nu'/h\nu$, depends on the angle of scatter and the collision parameter. Because α is directly related to $h\nu$, it is evident that the Compton process becomes less efficient for degrading a photon beam as the photon energy decreases. In most cases photons degraded after a series of Compton collisions are absorbed in a photoelectric reaction before their energy is less than 10 keV (1). The photoelectric effect will be discussed in detail in Chapter 7.

For the Compton scattering problem, one must know two of the variables $h\nu$, $h\nu'$, T'_e, θ', or θ'_e to be able to calculate any of the others. Equations 6.13 and 6.14 include all of these variables except θ'_e. To complete the set of equations, we will explicitly consider the Compton electron angle next. From Equations 6.9,

$$\frac{p'_e c \cos\theta'_e}{p'_e c \sin\theta'_e} = \frac{h\nu - h\nu'\cos\theta'}{h\nu'\sin\theta'}, \qquad \cot\theta'_e = \frac{h\nu/h\nu' - \cos\theta'}{\sin\theta'} \tag{6.15}$$

Using Equations 6.13 and a trigonometric identity (17):

$$\cot\theta'_e = \frac{1+\alpha(1-\cos\theta') - \cos\theta'}{\sin\theta'}$$

$$= (1+\alpha)\frac{(1-\cos\theta')}{\sin\theta'} = (1+\alpha)\tan\theta'/2. \tag{6.16}$$

Thus, the angle of the outgoing electron and the angle of the outgoing photon are related by

$$\theta'_e = \cot^{-1}[(1+\alpha)\tan\theta'/2], \quad \theta' = 2\tan^{-1}[\cot\theta'_e/(1+\alpha)] \tag{6.17}$$

There are no constraints on θ' except those imposed by symmetry around the axis of the direction of incidence of the photon. All values of θ' are possible from $\theta' = 0$, which might be termed the no-scattering limit, to the 180° backscatter limit, $\theta' = \pi$. Values for the Compton electron angle decrease from $\theta'_e = \pi/2$, the limit for no scattering, to $\theta'_e = 0$ for 180° photon backscattering. The ranges for the angle variables are summarized by $0 < \theta' \leq \pi$, $\pi/2 > \theta'_e \geq 0$.

The range of allowed final energy values can be obtained from the limiting cases for scattering angle:

1. For the limit with no scattering:

   $\theta' \to 0, \quad \theta'_e \to \pi/2, \quad h\nu' \to h\nu, \quad T'_e \to 0$

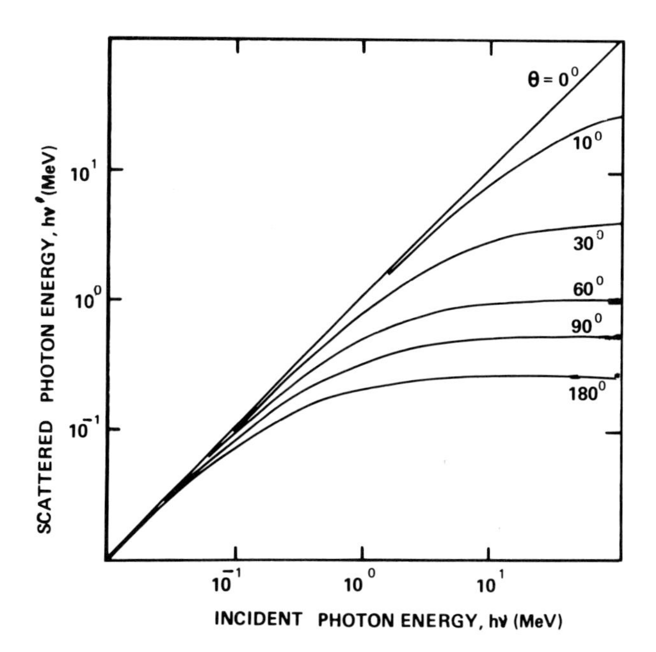

Figure 6.5 The scattered photon energy $h\nu'$ is shown as a function of the incident photon energy $h\nu$ for several scattering angles.

2. For the 180° backscatter limit:

$$\theta' \to \pi, \quad \theta'_e \to 0, \quad h\nu' \to \frac{h\nu}{1+2\alpha}, \quad T'_e \to h\nu \frac{2\alpha}{1+2\alpha}$$

Thus, the energy ranges can be written as $h\nu > h\nu' \geq h\nu/(1 + 2\alpha)$, $0 < T'_e \leq h\nu[2\alpha/(1 + 2\alpha)]$. Figure 6.5 shows that the energy of the scattered photon varies within the limits given above for a wide range of incident photon energy values and angles. Notice from this figure that the scattered photon energy changes very little when θ' is between 90° and 180°.

For very large $h\nu$,

$$\alpha \gg 1, \quad h\nu' \to \frac{h\nu}{\alpha(1 - \cos\theta')} = \frac{m_e c^2}{(1 - \cos\theta')},$$

$$\frac{\pi}{2} \leq \theta' \leq \pi, \quad m_e c^2 \geq h\nu' \geq m_e c^2/2. \tag{6.18}$$

No matter how large the primary photon energy may be, the energy of a backscattered photon ($\pi/2 \leq \theta' \leq \pi$) will not exceed 511 keV. In practical situations this means that shielding for backscattered photons requires minimal amounts of lead, even if the primary photon energy is very large (18).

Figure 6.6 shows that the number of scattered photons tends to peak around $h\nu/(1 + 2\alpha)$, the backscatter edge. Figure 6.7 shows that the number of Compton electrons tends to peak at $h\nu 2\alpha/(1 + 2\alpha)$, the Compton edge. As was mentioned earlier, the final photon energy is a slow function of angle

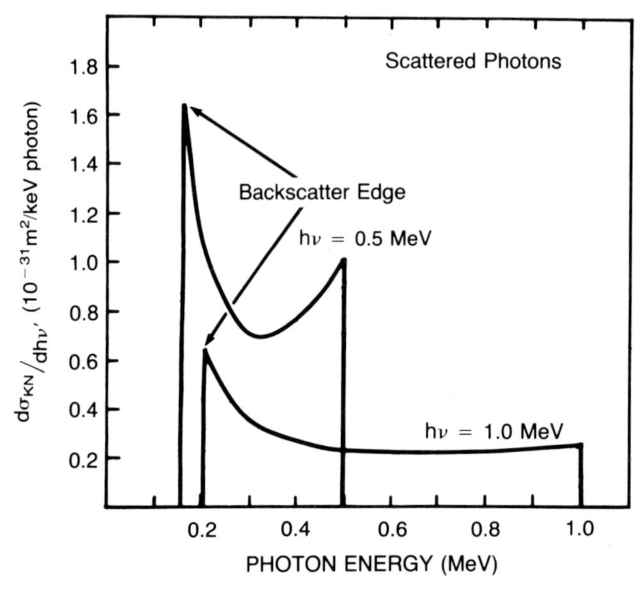

Figure 6.6 The spectra for photons emerging after Compton scattering at 0.5 MeV and 1 MeV are shown as a function of energy. The backscatter peaks are the prominent features. See Section 6.4 for explanation of Klein-Nishina calculations.

for backscatter conditions. Thus, backscattered photons (and Compton electrons) tend to pile up in a narrow interval at the edge.

The effect of the electron peak at the Compton edge is usually prominent on pulse-height spectra taken with sodium iodide scintillation detectors for high-energy gamma-ray sources. If shielding material is stacked around the scintillation detector, the backscatter peak produced by gamma rays scattered at a backward angle into the sodium iodide crystal may also be noticeable (19).

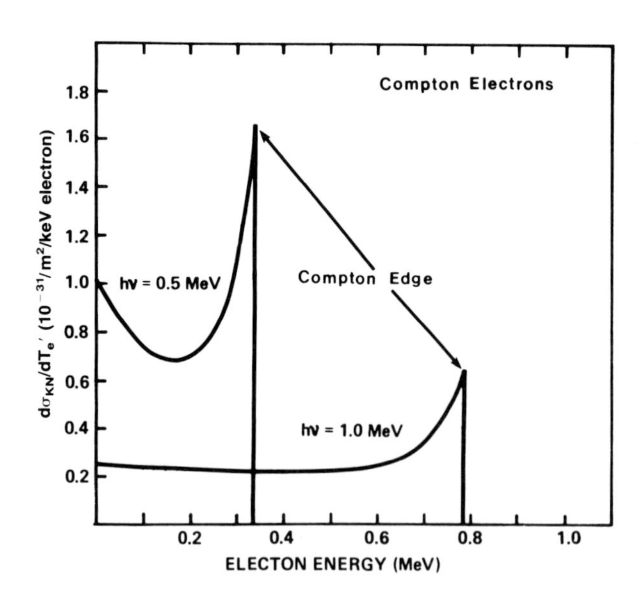

Figure 6.7 The spectra for Compton electrons emerging after scattering of 0.5-MeV and 1-MeV photons are shown as a function of energy. The Compton peaks are the prominent features. See Section 6.4 for explanation of Klein-Nishina calculations.

6.4 Klein-Nishina Differential Cross Sections

In Section 6.2 elastic scattering at low energies was described according to classical electromagnetic wave formulas; in Section 6.3 higher energy photons were described as particles that collide and rebound during Compton processes. The methods of quantum electrodynamics applied by Klein and Nishina in 1929 to photon scattering by free electrons (3) produced a common link between wave theory results and the photon particle concepts. The classical concept of a force between the particle and the field has little significance in the formalism of quantum electrodynamics. The exchange of energy is dominant, even the exchange of virtual photons (8), which do not actually appear in the laboratory system.

A basic result from the Klein-Nishina (KN) calculations was the free electron differential cross section per unit solid angle for scattering an unpolarized photon into the solid angle bounded by angles θ' and $\theta' + d\theta'$:

$$\frac{d\sigma_{KN}(h\nu,\theta')}{d\Omega'} = \frac{r_0^2}{2}\left\{\frac{1+\cos^2\theta'}{[1+\alpha(1-\cos\theta')]^2} + \frac{\alpha^2(1-\cos\theta')^2}{[1+\alpha(1-\cos\theta')]^3}\right\} \tag{6.19}$$

Of course, the scattering parameter $\alpha = h\nu/m_ec^2$ is the energy variable. The variable θ' is the scattering angle as defined in Section 6.3. At low energies, when $\alpha \rightarrow 0$, Equation 6.19 reduces to

$$\frac{d\sigma_{KN}(h\nu,\theta')}{d\Omega'} = r_0^2\frac{(1+\cos^2\theta')}{2} \tag{6.20}$$

This expression has the same form as Equation 6.7, the Thomson scattering angular distribution.

Figure 6.8 is a plot of the angular distribution given by Equation 6.19 as a function of α but normalized to 1.0 at $\theta' = 0$. Notice that at low energies (Thomson scattering) forward scattering and backscattering are equally probable. Lateral scattering is reduced. For successively higher energy incident photons,

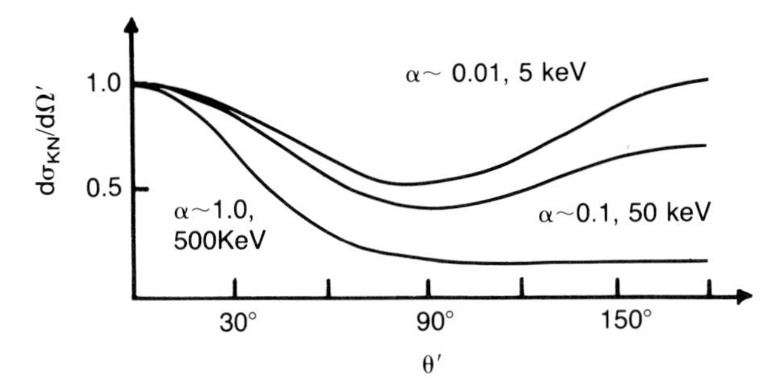

Figure 6.8 Klein-Nishina angular distributions for photon scattering into the solid angle between Ω' and $\Omega' + d\Omega'$ are shown as a function of the outgoing angle of the scattered photon. All results are normalized to 1 at $\theta' = 0$.

the forward directions become more favored. The average angle of scatter is reduced and backscatter becomes less significant.

The cross section per unit solid angle given by Equation 6.19 is easily related to measurements made with a small detector at some distance from the scattering target. The angle describing the detector position and the solid angle subtended by the detector are required to apply the relationship. However, other differential cross sections are useful in other situations. As mentioned in Section 6.3, $h\nu$, θ', θ'_e, T'_e, and $h\nu'$ are all basic variables of the scattering process. Cross-section derivatives with respect to several other variables will be obtained since they represent interesting angular and spectrum functions.

As a beginning, consider the relationship between the differential scattering angle and the differential solid angle illustrated in Figure 6.9:

$$d\Omega' = 2\pi \sin\theta'\, d\theta' \tag{6.21}$$

The Klein-Nishina free electron differential cross section per unit angle, for scattering an unpolarized photon through an angle between θ' and $\theta' + d\theta'$ is

$$\frac{d\sigma_{\text{KN}}(h\nu,\theta')}{d\theta} = 2\pi \sin\theta' \frac{d\sigma_{\text{KN}}(h\nu,\theta')}{d\Omega'}$$

$$= \pi r_0^2 \sin\theta' \left\{ \frac{1+\cos^2\theta'}{[1+\alpha(1-\cos\theta')]^2} + \frac{\alpha^2(1-\cos\theta')^2}{[1+\alpha(1-\cos\theta')]^3} \right\} \tag{6.22}$$

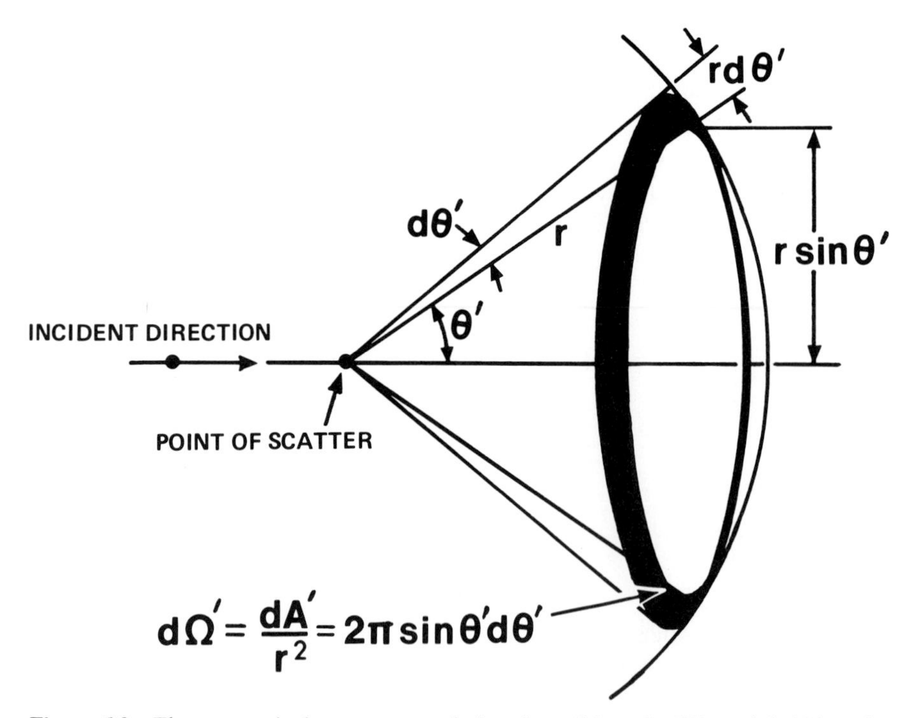

Figure 6.9 The geometrical parameters relating the solid angle differential $d\Omega'$ to the scattering angle θ' are shown.

Equation 6.22 was obtained by substituting Equation 6.21 into Equation 6.19.

Equation 6.22 is plotted on the *lower half* of Figure 6.10. Note that the distribution shown has cylindrical symmetry about the axis of the incident photon direction. Note also that for energies $h\nu > 1$ MeV, scatter occurs in predominantly forward directions (as discussed with reference to Figure 6.8), except that the number of photons scattered at $\theta' = 0$ is clearly shown to be zero. Since $\theta' = 0$ is the no-scattering limit, one should expect a vanishing scattering cross section at this angle. In this light it may be surprising that $d\sigma_{KN}/d\Omega'$ does not go to zero at $\theta' = 0$. In fact, the differential solid angle contains a $\sin\theta'$ factor, so the quotient represented by $d\sigma_{KN}/d\Omega'$ remains finite as θ' goes to zero.

The angular distribution for the Compton electrons is shown on the *top half* of Figure 6.10. The Klein-Nishina result in terms of electron angle θ_e' can be obtained from Equation 6.22 with

$$\frac{d\sigma_{KN}(h\nu,\theta_e')}{d\theta_e'} = \frac{d\sigma_{KN}(h\nu,\theta')}{d\theta'}, \quad \theta_e' = \cot^{-1}\left[\frac{(1+\alpha)\tan\theta'}{2}\right] \tag{6.23}$$

Notice that all electrons are expelled within a 90° interval, $\pi/2 > \theta_e' \geq 0$. Furthermore, the average electron emergence angle decreases (forward direc-

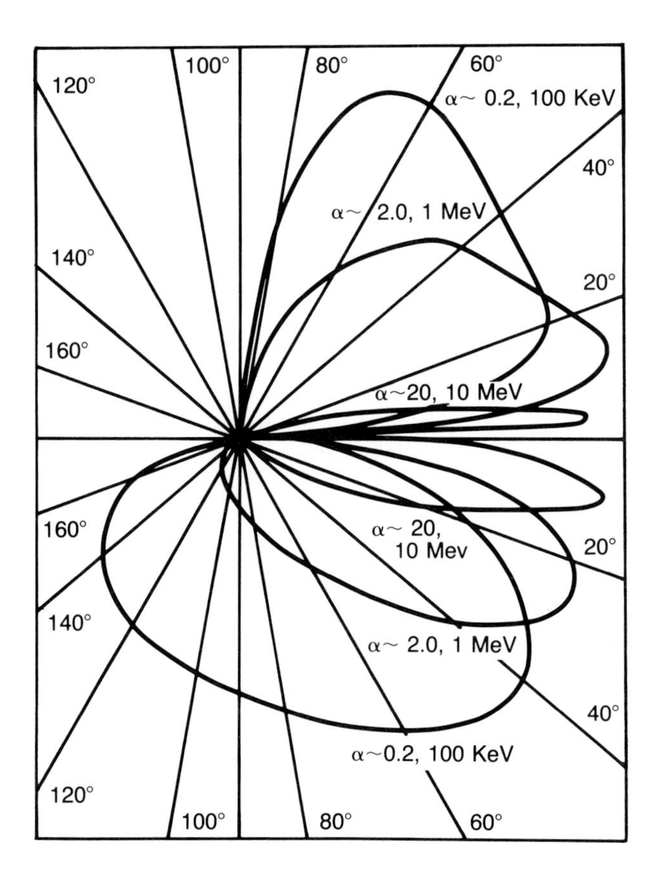

Figure 6.10 The differential cross sections per unit angle for KN scattering of photons are shown for the Compton electrons (*top*) and the scattered photons (*bottom*) for several energies. Adapted from W. R. Hendee, *Medical Radiation Physics* (Yearbook, Chicago, 1970).

tions) as the incoming photon energy increases. This tendency is related to the increasing amount of forward momentum carried by the incident photon.

The Klein-Nishina differential cross section per unit energy for expulsion of the Compton electron with energy between T'_e and $T'_e + dT'_e$ can be obtained after differentiating and rearranging Equation 6.14:

$$dT'_e = \frac{h\nu\alpha \sin\theta' \, d\theta'}{[1 + \alpha(1 - \cos\theta')]^2}, \quad \frac{d\theta'}{dT'_e} = \frac{m_e c^2}{\sin\theta'(h\nu - T'_e)^2},$$

$$1 + \alpha(1 - \cos\theta') = \frac{h\nu}{h\nu - T'_e}, \quad (1 - \cos\theta') = \frac{T'_e}{\alpha(h\nu - T'_e)} \tag{6.24}$$

Since $d\sigma/dT'_e = d\sigma/d\theta'(d\theta'/dT'_e)$,

$$\frac{d\sigma_{\mathrm{KN}}(h\nu, T'_e)}{dT'_e} = \frac{\pi r_0^2}{\alpha h\nu}\left\{2 + \frac{(T'_e)^2}{(h\nu - T'_e)^2}\left[\frac{1}{\alpha^2} + \frac{h\nu - T'_e}{h\nu} - \frac{2(h\nu - T'_e)}{\alpha T'_e}\right]\right\} \tag{6.25}$$

Equation 6.25 has been plotted for several photon energies on Figure 6.7. The Compton peak, which is clearly displayed, has already been discussed in Section 6.3. An expression for $d\sigma_{\mathrm{KN}}(h\nu,h\nu')/dh\nu'$ can be obtained from Equation 6.25 by substitution of $T'_e = h\nu - h\nu'$ and $dT'_e = -dh\nu'$. The distribution in scattered photon energy is shown on Figure 6.6. Evaluation of $d\sigma_{\mathrm{KN}}(h\nu,h\nu')/dh\nu'$ at several $h\nu$ values shows that the backscatter edge ($\theta' = 180°$) shown on Figure 6.6 occurs at energies less than $m_e c^2/2$.

The relationships used in the derivation of the Klein-Nishina cross section contain no provisions for electron bound states. The free electron procedure is usually valid for valence electrons but may be inadequate to describe collisions involving inner-shell electrons, especially when the binding energy is a considerable fraction of the energy transferred. In fact, even the most tightly bound electrons participate in scattering. When the effects of electron binding are included, the incoherent (KN) scattering cross section decreases and the coherent (Rayleigh) scattering cross section increases at lower incoming photon energies.

The Klein-Nishina free-particle results can be corrected for electron binding (1, 2) by multiplying by the incoherent scatter function $S(h\nu,\theta',Z)$. This function is usually given with $x = [\sin(\theta'/2)]/\lambda$ as a variable, but we prefer to use the related variables $h\nu$ and θ' because they are important in all scattering processes. The incoherent scatter function represents the probability that the atom will be excited or ionized after an impulse is applied to an atomic electron. Each of Equations 6.19, 6.20, 6.22, and 6.25 can be rewritten with

$$d\sigma_{\mathrm{is}} = d\sigma_{\mathrm{KN}} S(h\nu,\theta',Z) \tag{6.26}$$

Where $d\sigma_{\mathrm{is}}$ is the differential cross section for incoherent scattering. Incoherent scatter functions have been calculated for several elements (20, 21). They are found to increase with increasing $h\nu$ and to decrease with increasing Z. $S(h\nu,\theta',Z)$, the incoherent scatter function, and $F(h\nu,\theta',Z)$, the coherent scatter function, are often used together since incoherent and coherent scatter corrections must both be applied to the KN cross section at low energies. If increasing one of the variables $h\nu$, θ', or Z serves to increase $F(h\nu,\theta',Z)$, it will decrease $S(h\nu,\theta',Z)$ and vice versa. Figure 6.11 is included to illustrate the substantial

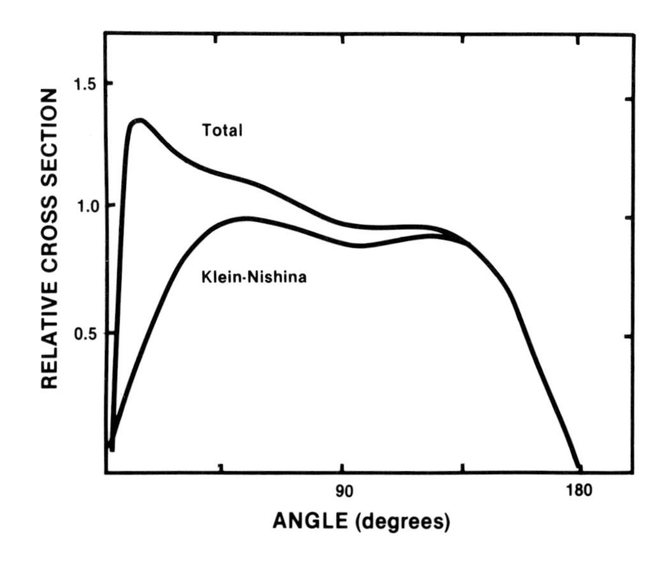

Figure 6.11 The Klein-Nishina differential cross section is compared to the total differential cross section, including coherent scattering. The results are for tissue at 20 keV. Adapted from Ref. 22.

corrections that must be made on $d\sigma_{KN}(h\nu,\theta')/d\theta'$ (22) when the photon energy is sufficiently low.

6.5 Attenuation Coefficients for Photon Scattering

The attenuation coefficients for photon scattering can be obtained from total cross-section expressions as outlined in Chapter 5. For incoherent scattering any of the Klein-Nishina differential cross-section expressions given in the previous section can be used if the effect of electron binding is ignored. We choose to integrate Equation 6.25 and include all possible values of the Compton electron energy from 0 to $(T'_e)_{max} = h\nu 2\alpha/(1 + 2\alpha)$.

It is convenient to rearrange Equation 6.25 before integrating:

$$\sigma_{KN}(h\nu) = \int \frac{d\sigma_{KN}}{dT'_e}\, dT'_e$$

$$= \frac{\pi r_0^2}{\alpha h\nu} \int \left[2 + \frac{(T'_e)^2}{\alpha^2 (h\nu - T'_e)^2} + \frac{(T'_e)^2}{h\nu(h\nu - T'_e)} - \frac{2T'_e}{\alpha(h\nu - T'_e)} \right] dT'_e \qquad (6.27)$$

Use of standard forms (17) for integrals with integrands $x^2 dx/(a + bx)^2$, $x^2 dx/(a + bx)$, and $x dx/(a + bx)$ will give

$$\sigma_{KN}(h\nu) = \frac{\pi r_0^2}{\alpha h\nu} \left\{ 2T'_e + \frac{1}{\alpha^2} \left[\frac{(h\nu)^2}{h\nu - T'_e} + 2h\nu \ln(h\nu - T'_e) - (h\nu - T'_e) \right] \right.$$
$$+ \frac{1}{h\nu} \left[2h\nu(h\nu - T'_e) - (h\nu)^2 \ln(h\nu - T'_e) - (h\nu - T'_e)^2 \right]$$
$$\left. - \frac{2}{\alpha} \left[(h\nu - T'_e) - h\nu \ln(h\nu - T'_e) \right] \right\}_0^{(T'_e)_{max}} \qquad (6.28)$$

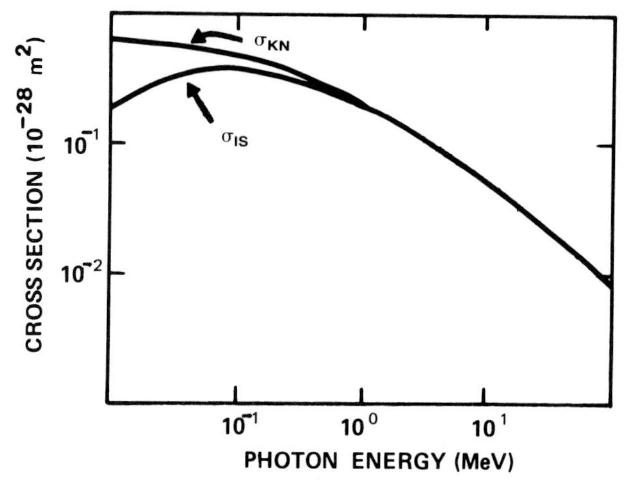

Figure 6.12 The total KN cross section for photon scattering, σ_{KN}, and the result for incoherent scattering, σ_{is}, including the correction for electron binding, are shown as a function of energy $h\nu$ for a lead absorber (1).

Application of the limit $(T'_e)_{\max} = h\nu 2\alpha/(1 + 2\alpha)$ gives

$$\sigma_{KN}(h\nu) = \pi r_0^2 \left\{ \frac{2(1+\alpha)}{\alpha^2} \left[\frac{2(1+\alpha)}{1+2\alpha} - \frac{\ln(1+2\alpha)}{\alpha} \right] + \frac{\ln(1+2\alpha)}{\alpha} - \frac{2(1+3\alpha)}{(1+2\alpha)^2} \right\} \tag{6.29}$$

This equation is plotted on Figure 6.12, where it is shown to be monotonically decreasing with increasing $h\nu$.

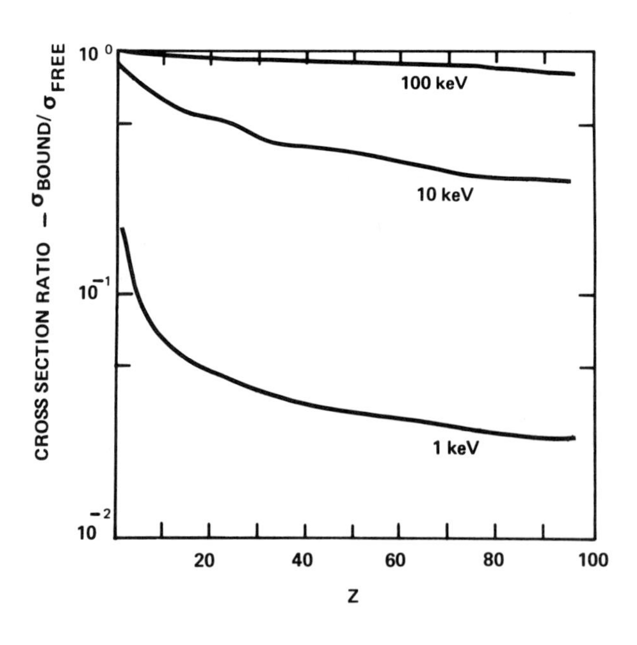

Figure 6.13 The ratio of the cross section for scattering by bound electrons divided by the KN free electron cross section is shown as a function of atomic number Z for several photon energies. The curve at 1 keV has been smoothed since fluctuations are substantial from element to element. Adapted from Ref. 11.

The effect of electron binding on the Klein-Nishina total scattering cross section is illustrated on Figures 6.12 and 6.13 (2). Results were obtained from the incoherent scattering cross section, but integration over angle θ' was used. There is considerable irregularity in the 1-keV result (Figure 6.13), which has been smoothed out in the figure. Electron binding reduces the cross section for incoherent scatter at low photon energies, especially for high Z materials. Notice from Figure 6.12 that the KN and the incoherent scatter cross sections converge at high energies. Beyond this energy any of the names incoherent scattering, KN scattering, and Compton scattering can be used. A cross-section expression for high energies can be obtained from Equation 6.29 in the limit where the collision energy parameter α is very large:

$$\sigma_{\text{is}}(h\nu) = \pi r_0^2 \left(\frac{1 + 2\ln 2\alpha}{2\alpha}\right), \quad \alpha \gg 1 \tag{6.30}$$

Photon scattering coefficients depend on the cross sections through

$$(\mu/\rho)_{\text{is}} = n_m \sigma_{\text{is}}, \quad (\mu/\rho)_{\text{cs}} = n_m \sigma_{\text{cs}} \tag{6.31}$$

where μ/ρ is the mass attenuation coefficient, n_m is the number of electrons per unit mass, and σ_{is} and σ_{cs} are the incoherent and coherent scattering cross sections, respectively. Some values of mass attenuation and mass energy absorption coefficients for photon scattering are given in Appendix 9 for several elements at a variety of energies.

If the effects of atomic binding can be ignored, the KN total scattering cross section can be used in Equation 6.31 for σ_{is}. Since σ_{KN} is independent of atomic number, the Z dependence in the mass attenuation coefficient comes through the factor n_m:

$$n_m = N_A(Z/M_m), \quad (\mu/\rho)_{\text{KN}} = N_A(Z/M_m)\sigma_{\text{KN}} \tag{6.32}$$

where N_A is Avogadro's number and M_m is the mass per mole for the absorber. In this case one can find the mass attenuation coefficient for Klein-Nishina scattering for element 2, from the coefficient for neighboring element 1 by

$$(\mu/\rho)_{\text{KN2}} = \frac{(Z/M_m)_2}{(Z/M_m)_1} (\mu/\rho)_{\text{KN1}}. \tag{6.33}$$

Of course, Equation 6.33 does not hold for coherent scattering. In that case a ratio involving (Z^2/M_m) instead of (Z/M_m) is often employed (7, 10).

The mass energy transfer coefficient for scattering can be obtained from the mass attenuation coefficients (23) using

$$(\mu_{\text{tr}}/\rho)_{\text{KN}} = (\mu/\rho)_{\text{KN}} \langle (T_e'/h\nu) \rangle \tag{6.34}$$

In this relationship, $\langle T_e'/h\nu \rangle$ is the average ratio of the Compton electron energy to the photon energy. Of course, $T_e' = 0$ for elastic processes, Rayleigh or Thomson scattering. Figure 6.14 shows the ratio $\langle T_e'/h\nu \rangle$ as a function of photon energy for KN scattering; the dependence can be calculated from the distribution given in Equation 6.25. This figure shows that the efficiency of scattering as an energy transfer mechanism increases with increasing energy.

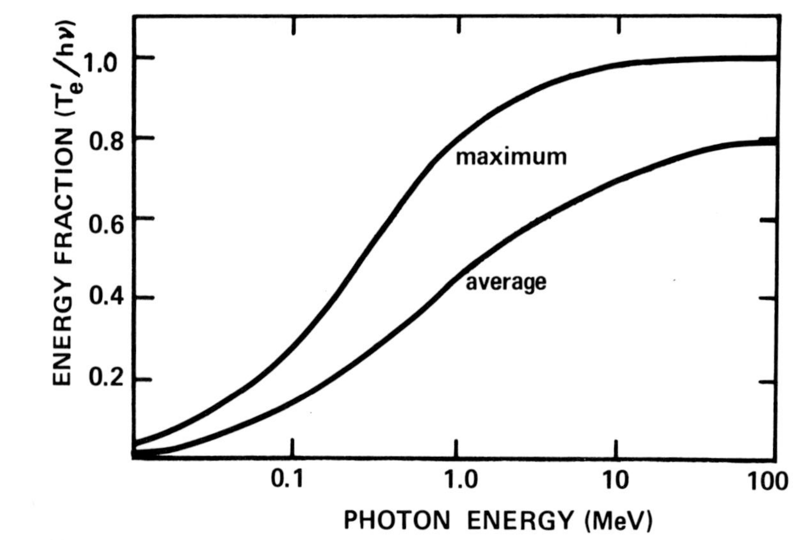

Figure 6.14 The average fraction of the total energy given to the Compton electron is shown for a wide range of incident photon energies. Adapted from H. E. Johns and J. R. Cunningham, *The Physics of Radiology,* 3rd Ed. (Charles C Thomas, Springfield, Ill., 1969).

The mass energy absorption coefficient for scattering is obtained by multiplying Equation 6.34 by the fraction of the Compton electron energy that is not radiated, $(1 - g)$. A first estimate of the fraction of energy radiated, g, can be obtained from the radiation yield expression in Chapter 3 if the mean Compton electron energy is utilized.

6.6 Problems

1. Evaluate the Thomson cross section for scattering from a free electron in square meters. Evaluate the Thomson cross section for scattering from a free proton in square meters.
2. For elastic scattering evaluate the ratio of the intensity radiated at 45° from the direction of incidence of a photon to the intensity radiated at 0°. Calculate the ratio of the intensity radiated at 90° to the intensity radiated at 0°.
3. Obtain the expression

$$T'_e = \frac{2h\nu \cos^2 \theta'_e}{\alpha\left[\left(\dfrac{1+\alpha}{\alpha}\right)^2 - \cos^2 \theta'_e\right]}$$

from Equations 6.9, 6.11, and 6.12. Hint: Eliminate θ' instead of θ'_e by squaring and adding several Equations in 6.9. Eliminate $(p_e c)^2$ from the result using the second equation of Equation 6.12. Square the result and use Equation 6.11.

4. What is the angle of scatter and the energy of a Compton electron when the incident photon energy is 140 keV and the angle of scatter of the photon is 60°?
5. Calculate the maximum energy that a Compton electron can have for each of the photons from ^{60}Co at 1.33 MeV and 1.17 MeV and for the ^{99m}Tc photon at 140 keV.
6. Calculate the minimum energy of a scattered photon for each of the photons from ^{60}Co at 1.33 MeV and 1.17 MeV and for the ^{99m}Tc photon at 140 keV.
7. Obtain the expression for the Klein-Nishina differential cross section $d\sigma_{KN}(h\nu,h\nu')/dh\nu'$ for scattering a photon by a free electron when the scattered photon energy is between $h\nu'$ and $h\nu' + dh\nu'$. Use Equations 6.11 and 6.25.
8. Integrate Equation 6.27 to obtain Equation 6.28 using tables for integrals with integrands $x^2dx/(a + bx)^2$, $x^2dx/(a + bx)$, and $xdx/(a + bx)$.
9. Calculate the mass attenuation coefficient for aluminum for scattering from the Klein-Nishina total cross section at 1.25 MeV.
10. Calculate the mass attenuation coefficient for scattering for lead from the Klein-Nishina total cross section at 1.25 MeV.
11. Calculate the mass attenuation coefficient for scattering from magnesium at 1.25 MeV in terms of the cross section for scattering from aluminum.
12. Calculate the mass energy transfer coefficient for scattering from the mass attenuation coefficient for scattering from aluminum at 140 keV and at 1.25 MeV.

6.7 References

1. E. Storm and H. I. Israel, *Photon Cross Sections from 0.001 to 100 MeV for Elements 1 through 100, LA 3753* (University of California, Los Alamos, NM, 1967).
2. J. H. Hubbell, *Photon Cross Sections, Attenuation Coefficients and Energy Absorption Coefficients from 10 KeV to 100 GeV, Report 29* (National Bureau of Standards, Washington, 1969).
3. K. N. Klein and Y. Nishina, Z. Physik **52**, 853 (1929).
4. R. L. Mössbauer, Naturwissenschaften **45**, 538 (1958).
5. R. L. Mössbauer, Z. Phys. **151**, 124 (1958).
6. J. J. Thomson, *The Corpuscular Theory of Matter* (Constable, London, 1907).
7. E. Segre, *Nuclei and Particles* (W. A. Benjamin, New York, 1965), p. 51.
8. R. D. Evans, *The Atomic Nucleus* (McGraw-Hill, New York, 1955), p. 819.
9. W. K. Panofsky and M. Phillips, *Classical Electricity and Magnetism* (Addison-Wesley, Reading, MA, 1955), p. 325.
10. D. R. White, Phys. Med. Biol. **22**, 219 (1977).
11. J. H. Hubbell, W. J. Veigele, E. A. Briggs, R. T. Brown, D. T. Cromer, and R. J. Howerton, J. Phys. Chem. Ref. Data **4**, 471 (1975).
12. A. T. Nelms and J. Oppenheim, J. Res. Natl. Bur. Stand. **55**, 53 (1955).

13. J. D. McGervey, *Introduction to Modern Physics* (Academic, New York, 1971), p. 453, 620.
14. E. Ackerman, *Biophysical Science* (Prentice-Hall, Englewood Cliffs, NJ, 1962), p. 267.
15. A. H. Compton, Phys. Rev. **21**, 483 (1923).
16. A. H. Compton and S. K. Allison, *X-Rays in Theory and Experiment,* 2nd ed. (van Nostrand, New York, 1935).
17. *Handbook of Chemistry and Physics,* edited by R. C. Weast and S. M. Selby (Chemical Rubber Co., Cleveland, 1966).
18. *Medical X-Ray and Gamma-Ray Protection for Energies up to 10 MeV, Structural Shielding Design and Evaluation, NCRP Report 34* (National Council on Radiation Protection and Measurements, Washington, 1971).
19. R. L. Heath, *Scintillation Spectrometry Gamma Ray Spectrum Catalogue, IDO-16880-1,* 2nd ed. (U.S. Atomic Energy Commission, Washington, 1964).
20. D. T. Cromer and J. B. Mann, J. Chem. Phys. **47**, 1892 (1967).
21. R. T. Brown, J. Chem. Phys. **55**, 353 (1971).
22. D. R. Dance, Phys. Med. Biol. **25**, 25 (1980).
23. *Radiation Quantities and Units, ICRU Report 33* (International Commission on Radiation Units and Measurements, Washington, 1980).

7

Photon Absorption Reactions

7.1 Energy and Momentum Relationships

In the present chapter, absorption reactions for photons are discussed. These reactions can take place in atomic, molecular, or nuclear systems. In an absorption reaction, the incident photon disappears. Other photons may subsequently emerge from the system, but they are the result of de-excitation involving rearrangement of the internal configuration of the absorber. Photon absorption reactions that include the emission of atomic or nuclear particles are inelastic. The liberated charged secondaries are important because they deposit most of the energy at local atomic sites in an absorbing medium.

In Chapter 6 four photon energy categories were given: nonionizing, low energy, intermediate energy, and high energy. In Section 7.2 we will be concerned with reactions initiated mainly by low-energy photons with $h\nu < 0.5$ keV. For this case the interactions considered will be those that usually involve valence or other weakly bound electrons. In Section 7.3 the atomic photoelectric effect will be discussed. The unique aspects of photoelectric reactions are most noticeable at intermediate energies, when electrons with substantial binding energies are involved. Section 7.4 deals with photonuclear reactions initiated by high-energy photons, $h\nu \geq 1$ MeV. In Section 7.5 pair production will be discussed, and in section 7.6 systematic relationships between attenuation coefficients are given so that one can compare the salient features of the different photon absorption processes.

Before the individual processes are discussed one by one, the basic energy and momentum equations, which are appropriate for photon absorption reactions, will be given. They will be adapted later to particular cases. One quantity of special interest is the threshold energy $h\nu_{\text{min}}$, the minimum possible input energy that will initiate a given reaction.

Let $X + \gamma \to b + Y$ symbolize a generalized photon absorption reaction on target system X with product system Y. Then the incoming photon γ dislodges the outgoing particle b. If X is initially at rest,

$$\mathbf{p}_\gamma = \mathbf{p}'_b + \mathbf{p}'_Y, \quad h\nu + M_X c^2 = E'_b + E'_Y = M_b c^2 + T'_b + M_Y c^2 + T'_Y,$$

$$h\nu = (M_b + M_Y - M_X)c^2 + T'_b + T'_Y = \text{BE}_b + T'_b + T'_Y \tag{7.1}$$

Here $\mathbf{p}$ symbolizes the momentum, M the mass, E' the total outgoing energy, and T' the outgoing kinetic energy. The subscripts identify the particle represented; BE_b is the binding energy for particle b.

The minimum possible photon momentum and energy are related by

$$p_\gamma = \frac{h\nu_{\text{min}}}{c} = (p'_b + p'_Y),$$

$$h\nu_{\text{min}} + M_X c^2 = M_b c^2 + M_Y c^2 + (T'_b + T'_Y)_{\text{min}} \tag{7.2}$$

The scalar momentum equation is appropriate at threshold since components of outgoing momentum that were not parallel to the photon direction would require extra kinetic energy. Apparently $(T'_b + T'_Y)_{\text{min}} > 0$ since $(p'_b + p'_y)_{\text{min}} > 0$ in the laboratory system.

Consider the same reaction in the center of mass (subscript cm) system where the total momentum is zero. The center of mass reference frame translates with speed v_{cm} with respect to the laboratory frame. At threshold in the center of mass system, the outgoing particles will be at rest, since $p'_b + p'_Y = 0$.

If b and Y are at rest in the center of mass system, each must be moving with speed $v = v_{\text{cm}}$ in the laboratory system. Thus, at threshold, in the laboratory system:

$$p'_b \propto M_b v_{\text{cm}}, \quad p'_Y \propto M_Y v_{\text{cm}}, \quad p'_b + p'_Y = p_\gamma,$$

$$p'_Y = \frac{M_Y}{M_b} p'_b = \frac{M_Y}{M_b}(p_\gamma - p'_Y), \quad p'_Y = \frac{M_Y}{M_b + M_Y} p_\gamma, \quad p'_b = \frac{M_b}{M_b + M_Y} p_\gamma \tag{7.3}$$

Equations 7.3 shows that at threshold in the laboratory system, the momentum of the photon is shared between outgoing particles in the direct ratio of their masses.

Suppose the results of Equations 7.3 are applied to the energy Equations 7.1 using the form $E' = \sqrt{(p')^2c^2 + M^2c^4}$:

$$h\nu_{\text{min}} + M_X c^2 = \sqrt{(p'_b)^2c^2 + M_b{}^2c^4} + \sqrt{(p'_Y)^2c^2 + M_Y{}^2c^4}$$

$$= \sqrt{\frac{M_b^2(h\nu_{\text{min}})^2}{(M_b + M_Y)^2} + M_b^2c^4} + \sqrt{\frac{M_Y^2(h\nu_{\text{min}})^2}{(M_b + M_Y)^2} + M_Y^2c^4}$$

$$= \sqrt{(h\nu_{\text{min}})^2 + (M_b + M_Y)^2c^4} \tag{7.4}$$

To solve for $h\nu_{\min}$, square Equation 7.4 and simplify:

$$2M_X c^2 h\nu_{\min} + M_X{}^2 c^4 = (M_b + M_Y)^2 c^4,$$

$$h\nu_{\min} = \frac{(M_b + M_Y)^2 c^4 - M_X{}^2 c^4}{2M_X c^2}$$

$$= (M_b + M_Y - M_X)c^2 \left[1 + \frac{(M_b + M_Y - M_X)c^2}{2M_X c^2} \right]$$

$$= \mathrm{BE}_b \left[1 + \frac{\mathrm{BE}_b}{2M_X c^2} \right] \tag{7.5}$$

Equation 7.5 shows that the threshold energy is larger than the binding energy. This is a direct result of the momentum of the incoming photon. However, in many practical situations, $\mathrm{BE}_b \ll 2M_X c^2$ and the correction term is negligible.

7.2 Photoabsorption at Low Energies

In this section we are primarily concerned with reactions initiated by photons with $h\nu < 0.5$ keV. In this case, absorption is governed by the interaction of the electric field of the incident photon and the electric dipole moment of the target system. Other electric and magnetic moments are much less effective (1).

Photoabsoprtion at low energies is discussed separately in this section because weakly bound or valence electrons are often involved. This means that the molecular configuration and physical state of the absorbing material is important, and a wide variety of phenomena can be included, each with distinctive properties. The overview given below is subdivided into photoabsorption in gaseous and liquid systems and photoabsorption in crystalline solids. The solid absorbers are discussed separately because of the distinctive electron energy bands associated with crystalline systems.

In gaseous and liquid absorbers, photoexcitation and photoionization are major subcategories of photoabsorption (1). For photoexcitation, the photon induces an electron transition between its initial bound energy state (indicated by subscript i) and the final energy state (indicated by subscript f), which is also bound. If target recoil is negligible, then

$$h\nu = E_f - E_i \tag{7.6}$$

For molecular absorbers, low-energy photoexcitation may involve dissociation of chemical bonds. Because of this, photons with appropriate energies may induce formation of radicals, i.e. particular molecular fragments with unpaired electrons.

When photon energies exceed the energy equivalent of the ionization potential, photoionization competes with photoexcitation. In a photoionization process, the atomic electron escapes from the parent atom as a free particle. Photoionization usually dominates when $h\nu$ is substantially greater than the binding

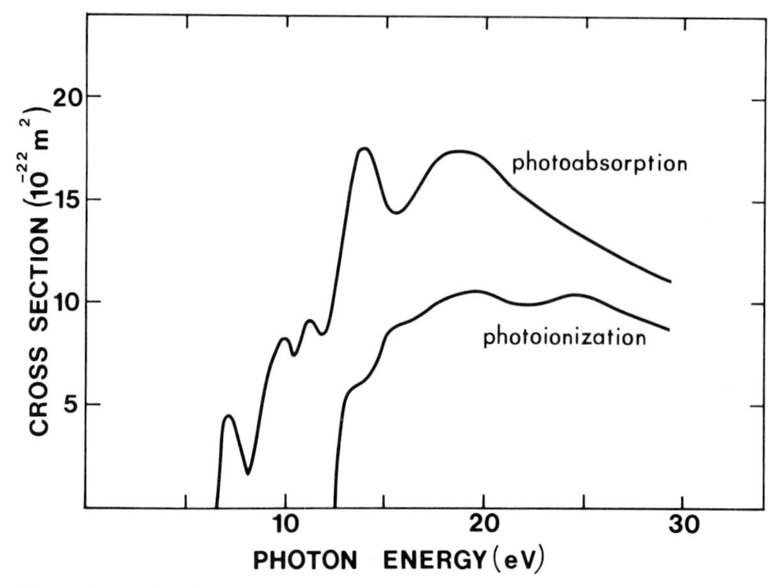

Figure 7.1 The photoabsorption cross section for water is shown as a function of photon energy. The photoionization cross section has a higher energy threshold but rises until it almost equals the photoabsorption cross section (2–4).

energy of electrons in the valence shell. However, multiple excitation processes are also possible at such energies.

Figure 7.1 illustrates the relationship between photoionization and photoabsorption cross sections for water vapor (2–4). The peaks or resonances between the excitation threshold at 6.6 eV and the ionization potential at 12.6 eV represent excitations associated with specific free radical products. Above 12.6 eV the molecule or the free radical fragments are often ionized (5).

Low-energy photoabsorption in liquid systems differs from the process in the gaseous state because of the perturbing effects of surrounding molecules. However, liquid-state solutions are of especial interest since photoabsorption in a solute can often be determined in the presence of the solvent. If the geometry is appropriate for exponential attenuation, the molar extinction coefficient ϵ can be used, as was discussed in Chapter 5. The solute concentration C_l, in moles per liter, and the vial thickness x are also important (6). For transmitted photon fluence ϕ,

$$\phi/\phi_0 = 10^{-\epsilon C_l x} = \exp(-2.30\epsilon C_l x) \tag{7.7}$$

where ϕ_0 is the incident fluence. When the transmission ϕ/ϕ_0 is measured for a monochromatic beam and when x and ϵ are known, the concentration C_l can be calculated.

Figure 7.2 shows the extinction coefficient for a ferric sulfate (Fe^{+++}) solution in water (6); the coefficient for a ferrous sulfate solution (Fe^{++}) is

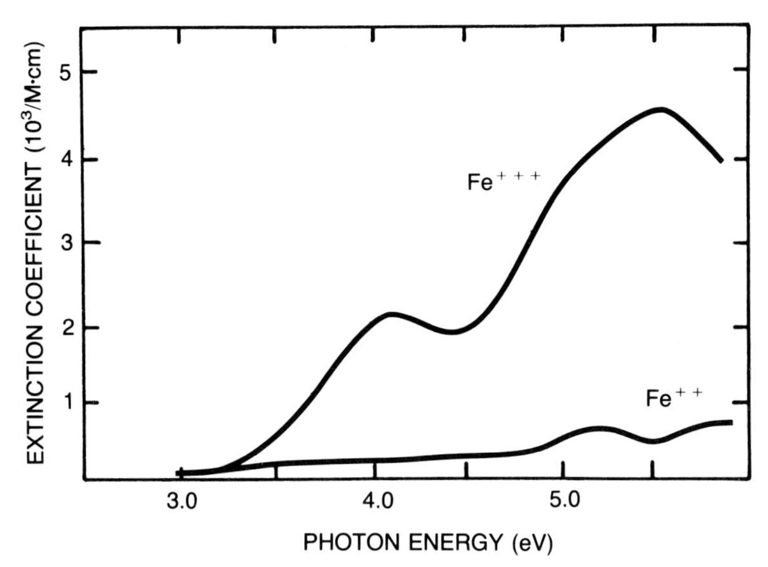

Figure 7.2 The extinction coefficient for Fe^{+++} ions in a water solution is compared to the coefficient for the Fe^{++} ion (6).

also shown, although it is quite small. Because of the difference in photoabsorption probability at values of 3040Å (4.1 eV) and 2240Å (5.5 eV), the concentration of Fe^{+++} can be measured in the presence of greater amounts of Fe^{++}. Ferrous sulfate solutions are used for dose-measuring systems (Fricke dosimeter) since radiation oxidizes the Fe^{++} ion to Fe^{+++}.

When atoms are held together in a regular array in a crystalline solid, electron energy levels for individual atoms are perturbed by the electric fields from neighboring atoms. External electrons are affected most noticeably. Their allowed energy levels broaden into bands of energy (7). This phenomenon is illustrated on Figure 7.3, and will be discussed in more detail in Chapter 12. For such a situation, the highest completely filled electron band is often called the valence band. The next band at higher energies is called the conduction band. Between the two bands a forbidden energy gap E_g occurs. Discussion of low-energy photoabsorption in crystalline solids usually involves these energy bands and the electron transitions between them that can be induced by the photon.

Interband excitations in crystalline solids are transitions involving a change in electron energy from one band to the next higher one. For interband photoabsorption the most common process is excitation of an electron with valence band energy into an empty conduction band state. The process leaves a vacancy or hole in the valence band. Thus, photoabsorption produces electron-hole pairs with

$$h\nu = E_e - E_h \geq E_g \tag{7.8}$$

where E_e is the electron energy and E_h is the hole energy. Incident photons with energy somewhat greater than E_g are often absorbed in a crystal; those with energy less than E_g will often be reflected from the crystal. Since the reflected photon energy determines the color perceived by an observer (7), the color of crystalline solids is often determined by E_g.

Figure 7.4 shows the mass attenuation coefficient for photoabsorption in CdS (Ref. 8). Below 2.4 eV photoabsorption is unlikely because the available energy is not sufficient to induce an interband transition. Above 2.5 eV the coefficient is large because a continuum of interband transitions is possible. The color of cadmium sulfide is yellow to orange, because green and blue photons are usually absorbed while yellow and red photons are reflected.

Intraband excitations occur when the electron is promoted to a previously empty state at higher energy in the same band. Since the gap is irrelevant in this situation, such transitions occur without a lower limit on energy. Intraband photoabsorption is especially important in metals. Figure 7.5 shows the mass attenuation coefficient for aluminum (9). Intraband processes are important at the lowest energies shown. Absorption at the higher energies shown on the abscissa is due to the atomic photoelectric effect, which will be discussed in the next section.

The surface photoelectric effect is a low-energy process closely related to the atomic photoelectric effect and occurs in solid absorbers at optical and ultraviolet photon energies. In the present context it can be viewed as a unique photoionization process because of the escape of the electron from the crystal surface.

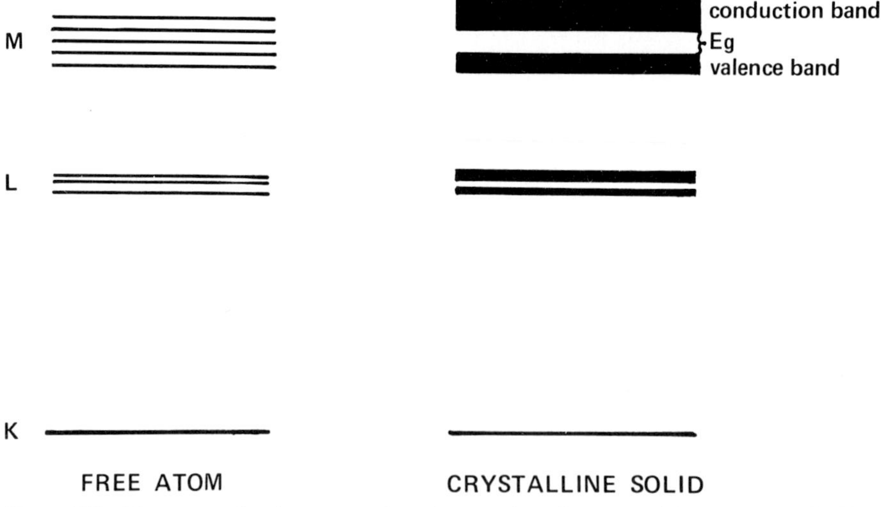

Figure 7.3 The energy level sequence for electrons in a free atom (*K*, *L*, and *M* shells) is compared to the level-band sequence for electrons in the crystalline atomic system.

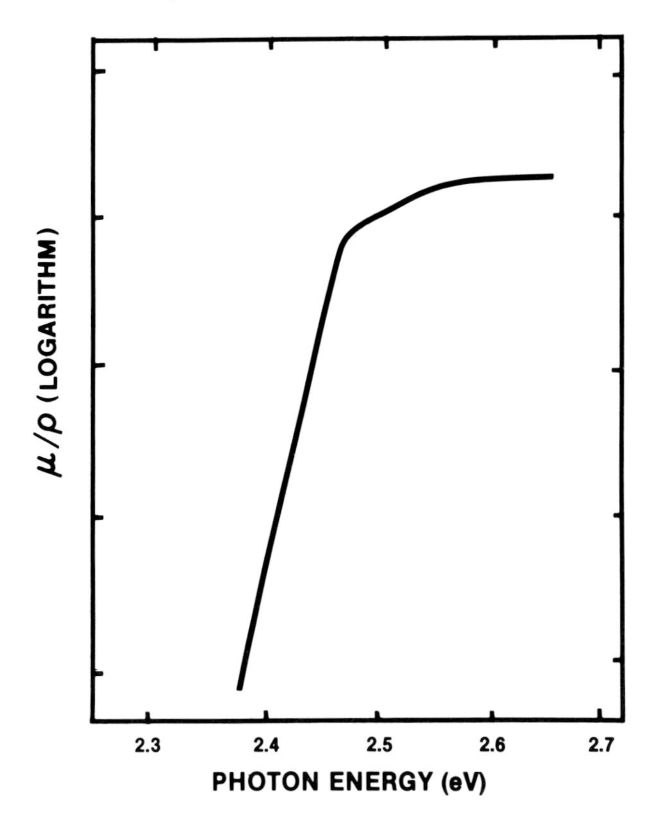

Figure 7.4 The mass attenuation coefficient for cadmium sulfide is shown as a function of energy at 300 K (8). The gap energy is about 2.4 eV.

The basic relationship for the surface photoelectric effect is

$$h\nu \geq T_s + \phi_W \tag{7.9}$$

where T_s is the kinetic energy of the electron after it has escaped from the surface, and ϕ_W is the work function for the photoelectric material. Note that the photon energy can be greater than $T_s + \phi_W$. Collisional energy loss by the electron is likely prior to its emission from the surface. The work function is due primarily to the electrostatic attraction between the escaping electron and the induced charge left on the surface. Some values for work functions are given in Table 7.1 (10).

Many important devices utilize the surface photoelectric effect, including photomultiplier tubes used in scintillation detectors for gamma rays and image intensifier tubes used for x-ray fluoroscopy. Cesium compounds such as cesium antimonide (Cs_3Sb) or cesium iodide (CsI) are often used as the photoelectric absorber (photocathode) in these devices.

7.3 The Atomic Photoelectric Effect

The atomic photoelectric effect is the ejection of an atomic electron from a bound shell after absorption of an incoming photon (11). The emitted electron

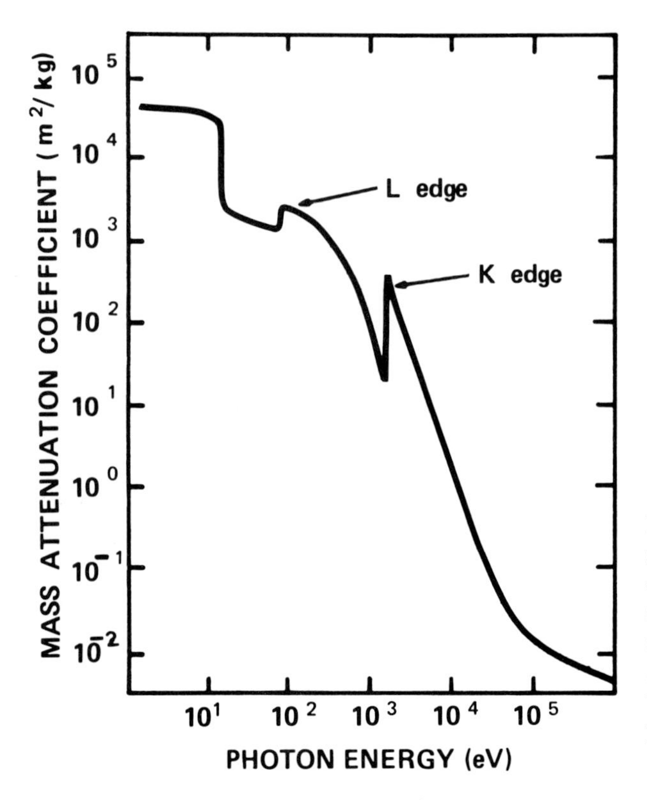

Figure 7.5 The mass attenuation coefficient for aluminum metal is shown as a function of photo energy. The K-edge and L-edge structures are due to the atomic photoelectric effect. Adapted from Ref. 9.

is called the photoelectron, as indicated on Figure 7.6. The atomic photoelectric effect is often the dominant photon absorption reaction in the intermediate-energy range, 0.5 keV $\leq h\nu \leq$ 1 MeV, although photoelectric processes are not limited to these energies. When the photon energy is sufficiently small and only valence electrons or loosely bound electrons are involved, the photoelectric effect cannot be distinguished from photoionization, discussed in the preceding section. In the intermediate-energy range, the atomic electrons involved are bound by substantial forces, so the physical state of the absorber and the molecular environment of the atoms have little effect. Because a vacancy is always

Table 7.1 Work Functions for the Surface Photoelectric Effect

Material	*Work function* ϕ_W *(eV)*
Aluminum (Al)	4.2
Silver (Ag)	4.5
Antimony (Sb)	4.6
Cesium (Cs)	1.9
Tungsten (W)	4.5
Cesium antimonide (Cs_3Sb)	2.1

Data from Ref. 10.

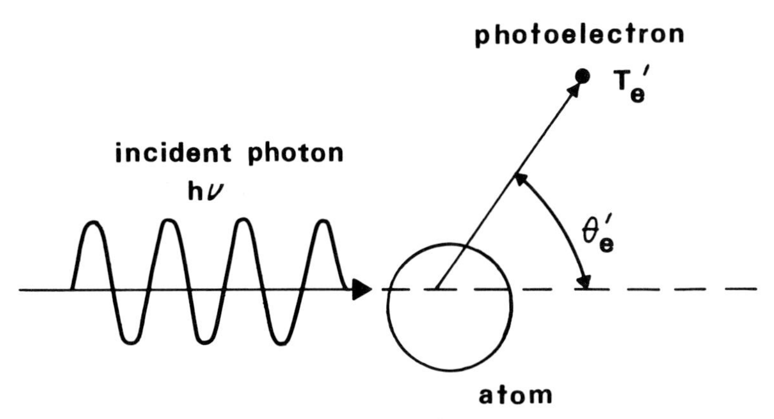

Figure 7.6 A schematic representation of the atomic photoelectric effect is shown. T'_e is the photoelectron kinetic energy.

produced in an atomic shell, de-excitation photons (fluorescence, phosphorescence) or electrons (Auger) are emitted after the photoelectric effect. These emissions are discussed in more detail in Chapters 11 and 12. In the following paragraphs, we will discuss the cross section and angular distribution for the atomic photoelectric effect and the origins of the absorption edges.

The atomic photoelectric effect is most easily understood as a reaction between a photon and an entire atom. This is appropriate because electron binding energy plays a significant role. Thus, the reaction shorthand is $X + \gamma \rightarrow e^- + X^+$, where X represents the target atom, X^+ the residual ion, and e^- the photoelectron. Equation 7.1 applied to this case gives

$$\mathbf{p}_\gamma = \mathbf{p}'_e + \mathbf{p}'_i, \quad h\nu = \mathrm{BE}_{nj} + T'_e + T'_i \tag{7.10}$$

where the subscripts γ, e, and i stand for the incoming photon, the photoelectron, and the recoil ion, respectively, and BE_{nj} stands for the binding energy of the atomic electron in the shell n and subshell j.

Equations 7.10 can be used to show that the atomic photoelectric effect must involve a bound electron. A hypothetical photoelectric absorption reaction with a free electron would occur with $p'_i = T'_i = \mathrm{BE}_{nj} = 0$. In these circumstances,

$$p_\gamma c = h\nu \stackrel{?}{=} p'_e c = \sqrt{(T'_e)^2 + 2m_e c^2 T'_e} > T'_e \tag{7.11}$$

The inequality in Equation 7.11 is incompatible with the energy conservation equation for a hypothetical free electron photoelectric reaction $h\nu = T'_e$. In reality, in the photoelectric process the electron must be bound and outgoing ionic momentum is necessary.

Equation 7.5 can be used to obtain the threshold for the atomic photoelectric effect. The subscripts b and Y must be replaced by e and i as indicated previously. The binding energy in this case refers to the atomic electron in the atom. Since the threshold for the atomic photoelectric effect is usually given with reference to a specific atomic shell n and subshell j,

$$h\nu_{\min,nj} = \mathrm{BE}_{nj}\left[1 + \frac{\mathrm{BE}_{nj}}{2M_X c^2}\right], \quad \mathrm{BE}_{nj} = E_{IP} + E^*_{nj} \tag{7.12}$$

The second equation in Equations 7.12 is a reminder that the residual ion is left in an excited state with energy E^*_{nj}. The value E_{IP} stands for the energy equivalent of the ionization potential, the minimum possible binding energy for the atomic or molecular target. Since $\mathrm{BE}_{nj} \ll 2M_X c^2$ for all cases, the threshold energy and the binding energy are used synonymously.

Whenever the photon energy is appreciably greater than the binding energy, the momentum and energy expressions in Equations 7.10 can be simplified. Consider first the ratio of momentum to kinetic energy for a particle of mass M and the ratio of momentum to energy for a photon:

$$p/T = \frac{\frac{1}{c}\sqrt{T^2 + 2TMc^2}}{T} = \frac{\sqrt{(2Mc^2/T) + 1}}{c}, \quad \frac{p_\gamma}{E_\gamma} = \frac{h\nu/c}{h\nu} = \frac{1}{c} \tag{7.13}$$

The massless photon has the smallest ratio of momentum to available energy. Furthermore, the more massive the particle the greater will be the ratio of momentum to kinetic energy.

Photon absorption by the photoelectric effect is most likely at energies where $2M_i c^2/T'_i \gg 2m_e c^2/T'_e \gg 1$. Under these conditions Equations 7.13 can be used to show that p_γ is negligible compared to p'_e and p'_i. Then from Equation 7.10,

$$\mathbf{p}'_e + \mathbf{p}'_i \simeq 0, \quad \mathbf{p}'_e \simeq -\mathbf{p}'_i, \quad T'_i = \frac{(\mathbf{p}'_i)^2}{2M_i} \simeq \frac{m_e}{M_i}\left[\frac{(p'_e)^2}{2m_e}\right] = \frac{m_e}{M_i}T'_e,$$

$$h\nu = \mathrm{BE}_{nj} + T'_e + T'_i \simeq \mathrm{BE}_{nj} + T'_e\left[1 + \frac{m_e}{M_i}\right] \simeq \mathrm{BE}_{nj} + T'_e \tag{7.14}$$

The final relationship in Equations 7.14 is sufficiently accurate in most situations.

In the preceding paragraphs, the energy and momentum relationships for the atomic photoelectric effect have been developed in terms of $h\nu$, T'_e, and T'_i. The geometric variables of the reaction are θ'_e and θ'_i, which define the outgoing directions of the photoelectron and the recoil ion with respect to the incoming photon direction. Since the kinetic energy of the ion is very small in most cases, T'_i and θ'_i are of little interest. In the next several paragraphs, cross-section expressions will be discussed in terms of variables $h\nu$, θ'_e, and T'_e.

The cross section for the atomic photoelectric effect can be evaluated using quantum mechanical techniques with approximate wave functions for the bound atomic electron and the emerging photoelectron. A theoretical result for the differential cross section per unit solid angle for the atomic photoelectric effect can be written as (12, 13)

$$\frac{d\sigma_K(h\nu, T'_e, \theta'_e)}{d\Omega'} = 4\sqrt{2}\,r_0^2\alpha^4 Z^5 (m_e c^2/h\nu)^{7/2} \frac{\sin^2\theta'_e}{[1 - (p'_e c/m_e c^2)\cos\theta'_e]^4} \tag{7.15}$$

where $p'_e c = \sqrt{(T'_e)^2 + 2m_e c^2 T'_e}$, r_0 is the classical electron radius, and α is the fine structure constant. The differential cross section given in Equation 7.15 is for the two electrons in the K shell of an atom with atomic number Z. It is an average value over all polarization directions and is therefore symmetrical about the direction of incidence of the photon. It is not relativistic. The angle θ'_e defines the photoelectron direction with respect to the incident photon direction as shown on Figure 7.6.

Figure 7.7 shows the angular distribution of photoelectrons for a low Z absorber BE $\ll h\nu$. Notice that the number emitted per unit solid angle tends to zero in the forward and backward directions. In fact, at low energies photoelectrons tend to come out at 90° with respect to the photon direction, with a distribution approaching the $\sin^2 \theta'_e$ shape. This distribution follows from the fact that the direction of the photon electric field vector is taken to be perpendicular to the incident direction. However, as the photon energy is increased, forward emission becomes more favored, as determined by the $\cos \theta'_e$ term in the denominator of the differential cross-section expression.

If one substitutes $d\Omega' = 2\pi \sin \theta'_e \, d\theta'_e$ in Equation 7.15, a nonrelativistic expression for the differential cross section per unit angle for ejection of K-shell electrons at angles between θ'_e and $\theta'_e + d\theta'_e$ can be written:

$$\frac{d\sigma_K(h\nu, T'_e, \theta'_e)}{d\theta'_e} = 8\sqrt{2}\,\pi r_0^2 \alpha^4 Z^5 (m_e c^2/h\nu)^{7/2} \frac{\sin^3 \theta'_e}{[1 - (p'_e c/m_e c^2)\cos \theta'_e]^4} \qquad (7.16)$$

This distribution is shown on Figure 7.8 for $h\nu = 0.02$ eV for a low Z (BE $\ll h\nu$) absorber. The preference for 90° electron emission is unmistakable at this energy. At higher values of $h\nu$, relativistic formulations (14) can be used to show that forward directions are favored.

When $p'_e \ll m_e c$, the $\cos \theta'_e$ term can be deleted in Equation 7.16 and the expression can be integrated over θ'_e. Then

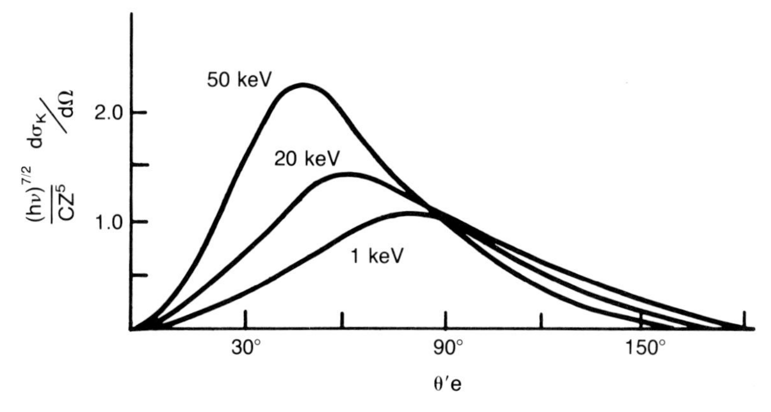

Figure 7.7 The angular distribution of emitted photoelectrons for low Z materials illustrates the shift to forward emission angles with increased energy. C is a constant which normalizes the curves to unity at 90°. Electron binding energy is neglected.

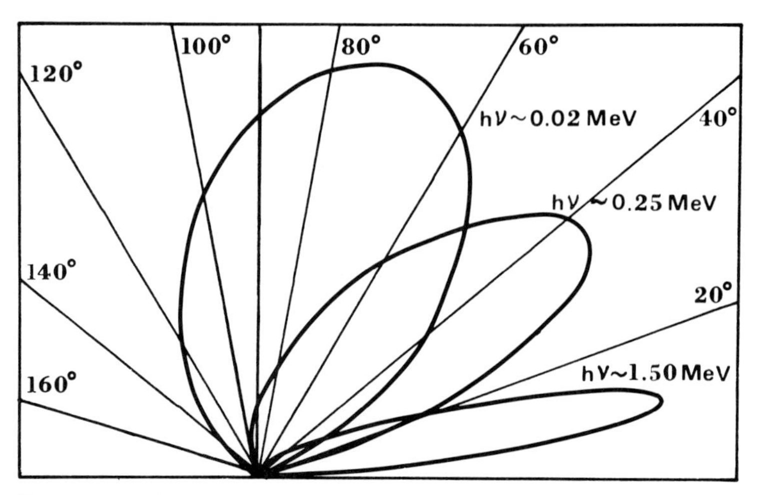

Figure 7.8 The differential cross section for emission of a photoelectron between θ'_e and $\theta'_e + d\theta'_e$ is shown for low Z materials.

$$\begin{aligned}\sigma_K(h\nu) &= 8\sqrt{2}\,\pi r_0^2\alpha^4 Z^5(m_ec^2/h\nu)^{7/2}\int_0^{\pi}\sin^3\theta'_e\,d\theta'_e \\ &= 4\sqrt{2}\,(8\pi r_0^2/3)\,\alpha^4 Z^5(m_ec^2/h\nu)^{7/2} \\ &= 4\sqrt{2}\,\alpha^4\sigma_T Z^5(m_ec^2/h\nu)^{7/2} \qquad (7.17)\end{aligned}$$

The symbol σ_T is for the Thomson cross section as defined in Chapter 6.

The total cross section per atom for the atomic photoelectric effect includes contributions from electrons in all occupied atomic shells. The ratio of the photoelectric cross section for the entire atom, σ, to the cross section for the two K-shell electrons, σ_K, has been evaluated just above the K-shell binding energy for many elements. This ratio can be calculated with sufficient accuracy ($\pm3\%$) for most purposes from (15):

$$\sigma/\sigma_K = 1 + 0.01481(\ln Z)^2 - 0.00079(\ln Z)^3 \qquad (7.18)$$

For most elements, 80%–90% of the photoelectric interactions occur with the two K-shell electrons when the photon energy is above the K-shell binding energy.

The calculations used to produce Equation 7.15 did not include relativistic formulations (16). Thus, the dependence of Equations 7.15 through 7.17 on Z and $h\nu$ is not exact. In the intermediate-energy range, because of screening of the nucleus by internal electrons, the atomic number dependence of the cross section varies with $h\nu$. The best average value for the Z exponent is generally between 4 and 5. Figure 7.9 shows that the exponent of the atomic number Z generally increases slightly with photon energy (17). As an example, the energy dependence of photoelectric processes for soft tissue has been calculated (18), with the Z exponent varying between 4.2 at 10 keV and 4.7 at 150 keV.

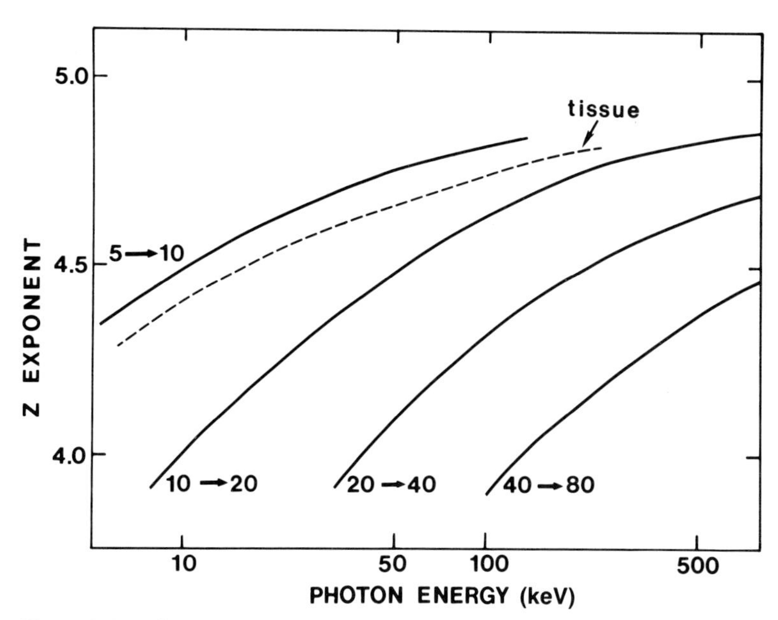

Figure 7.9 The atomic number exponent in the cross section per atom for the atomic photoelectric effect is shown as a function of Z and $h\nu$ (17). Only atoms with $BE_K < h\nu$ are utilized to construct the curves.

The dependence of the photoelectric cross section per atom on photon energy is illustrated in Figure 7.10 for two absorbers: iodine and lead (17). Apart from the discontinuities, the cross section decreases inversely with the energy to the third power at low energies rather than the $7/2$ power as given in Equation 7.17. At higher energies the dependence is approximately proportional to the inverse square of the energy. The rapid decrease in the cross section explains why photoelectric absorption is of little significance at energies greater than 1 MeV.

Figure 7.10 shows the smooth dependence of the photoelectric cross section on energy. It also shows several striking discontinuities occurring at 33 keV for iodine and 88 keV for lead. These structures are two examples of the K edge. A set of three secondary discontinuities called the L edges can also be seen for lead (L_I, L_{II}, and L_{III}). The K edge, L edges, M edges, and the like, occur for all absorbers with electrons populating the K, L, and M, shells and subshells, respectively. The edges occur at photon energies equal to the binding energies of electrons in the shells and subshells involved. Of course, when $h\nu < BE_{nj}$, photoelectric reactions can not occur in subshell of shell n. However, when the energy is sufficient, the cross section jumps to a large value. The jump ratio, is defined as the cross section just above the edge divided by the cross section just below it. The range of K-edge jump ratios varies from greater than 35 for light elements to less than 5 for heavy elements such as uranium. L- and M-edge jump ratios are smaller (19).

Electron binding energies are tabulated in Appendix 4. The general form

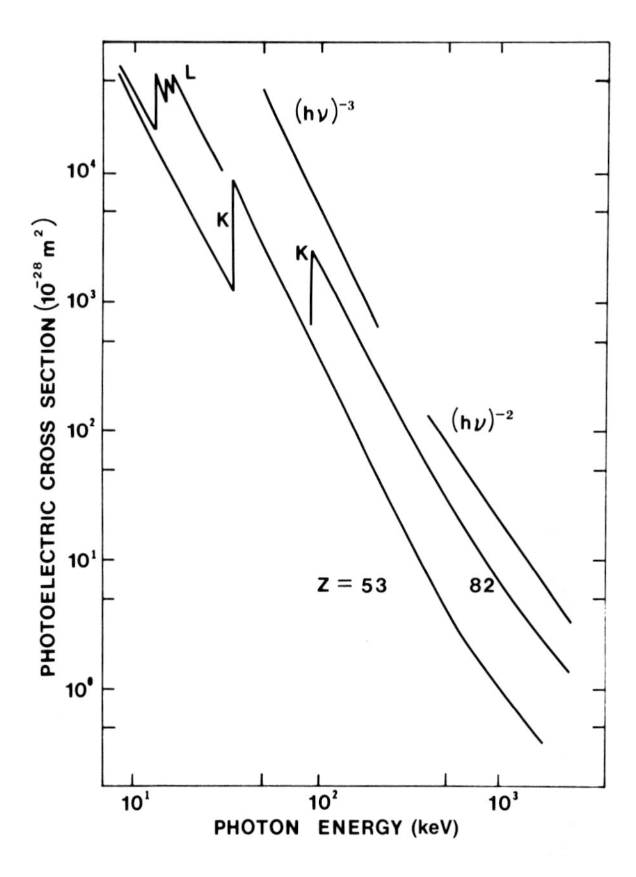

Figure 7.10 The photoelectric cross section for lead ($Z = 82$) and iodine ($Z = 53$) are compared (17).

of the expression describing these quantities is an extension of the Bohr energy expression for atomic electrons:

$$\mathrm{BE}_{nj} = \frac{\alpha^2 m_e c^2}{2} \frac{(Z - C_{nj})^2}{n^2} = h\nu_{\text{edge}} \tag{7.19}$$

where C_{nj} is the screening constant (20) for shell n and subshell j, included because of shielding of the nuclear charge by the innermost atomic electrons, and α is the fine structure constant. The coefficient $\alpha^2 m_e c^2/2$ equals 13.6 when evaluated in electron volts. Experimental determination of the location of an absorption edge is sometimes difficult because photon-induced transitions to bound states (photoexcitation) mask the precise value. Figure 7.11 shows the excitation peaks produced near the K edge for argon gas (21).

In closing, we note that the properties of the atomic photoelectric effect contribute greatly to the practical uses of photon beams. Radiographic images produced after transmission of x rays through inhomogeneous absorbers show superior contrast if photoelectric absorption is dominant. This is partly because the strong dependence of the photoelectric cross section on atomic number accentuates differences in x-ray transmission between structures with different elemental composition. In addition, competing scattering reactions may serve

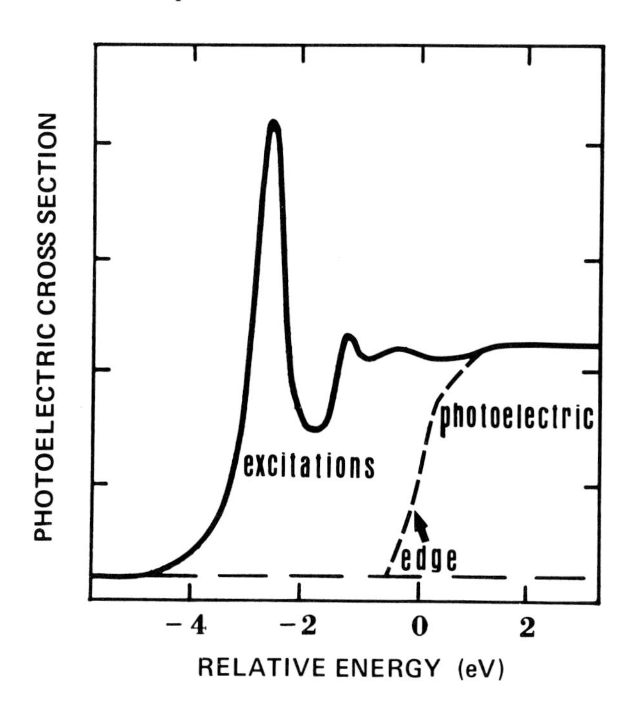

Figure 7.11 The K-shell photoelectric cross section for argon is shown near the K-edge energy. The peaks are due to photoexcitations (21).

to degrade the contrast by producing a relatively uniform scattered fluence over the entire radiograph.

Even the photoelectric absorption edges can be used to practical advantage. If the K edge for an absorber occurs at an energy just below the average photon energy in a beam, that absorber is especially opaque to the beam (see Figure 7.10). Thus, iodine or barium compounds introduced in specific body systems as contrast media are highly effective for medical fluoroscopic examinations with appropriate beam quality. For a similar reason, intensification screens used in x-ray cassettes are especially efficient in absorbing x rays if they are made with the rare earth elements gadolinium and lanthanum (22). The absorption edges for the rare earth elements occur at energies appropriate for maximum absorption of beams used in diagnostic radiology.

7.4 Photonuclear Reactions

A photonuclear absorption reaction is the ejection of at least one nuclear particle after absorption of a photon by a nucleus. The process is sometimes called nuclear photodisintegration or the nuclear photoeffect. A (γ,n) photonuclear reaction is illustrated on Figure 7.12, where the emitted neutron (photoneutron) moves away at angle θ'_n. Emission of a proton (photoproton) occurs in a (γ,p) process. Occasionally multiple nucleon emission follows absorption of a photon of sufficient energy. This process could be indicated by $(\gamma,2n)$, $(\gamma,2p)$, (γ,pn), and so forth, depending on the nucleons involved. Gamma-ray emission immedi-

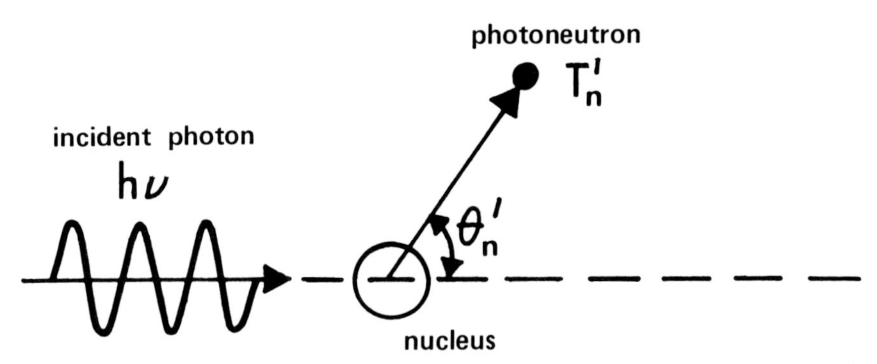

Figure 7.12 The schematic representation for a photonuclear reaction with the emission of a neutron is shown.

ately follows a photonuclear reaction if the residual nucleus is left in an excited configuration.

Although photonuclear reactions do not involve electrons directly, one often sees the notation $(e,e'n)$ or $(e,e'p)$ in the context of discussions of photonuclear reactions. These abbreviations stand for electronuclear reactions that are initiated with high-energy electron beams and do not involve photons directly. In an electronuclear reaction, part of the incident electron energy is transferred directly to the nucleus. The products of the process are similar to photonuclear products. Electronuclear reaction cross sections evaluated at energy T_e are smaller than analogous photonuclear reaction cross sections at energy $h\nu = T_e$ by a factor of $\alpha = 1/137$ (23, 24).

The momentum and energy conservation expressions for a photonuclear reaction can be taken directly from Equations 7.1. If the reaction is written $X + \gamma \rightarrow b + Y$, then

$$\mathbf{p}_\gamma = \mathbf{p}'_b + \mathbf{p}'_Y, \quad h\nu = \mathrm{BE}_b + T'_b + T'_Y \tag{7.20}$$

where BE_b is the binding energy of particle b in the target nucleus X. Recoil momentum of the product nucleus, $\mathbf{p}'_Y$, is necessary because the massless photon often does not carry sufficient momentum to balance the momentum of the emitted nuclear particle.

As a practical matter, the minimum binding energy of a nucleon in a nucleus can be obtained from atomic mass tables. If M_{gs} represents the ground-state nuclear mass in kilograms, the binding energy can be expressed in terms of $M_{\mathrm{gs}}c^2$.

$$\mathrm{BE}_b = (M_b + M_Y - M_X)_{\mathrm{gs}}c^2 + E_Y^* = \mathrm{SE}_b + E_Y^* \tag{7.21}$$

The residual excitation energy of the product nucleus is represented by E_Y^*, and the quantity SE_b, called the separation energy, is the minimum binding energy of b in nucleus X.

The ground-state nuclear mass excess (or mass deficit if negative) can be expressed in atomic mass units, Δm, or in megaelectron volts, ΔE. If the value must be obtained in megaelectron volts,

$$M_{gs}c^2/(1.6)10^{-13} = (A + \Delta m)931.5 = 931.5A + \Delta E \tag{7.22}$$

where A is the mass number. In this case 931.5 MeV has been used as the energy equivalent of a mass unit. Since the mass numbers for the nuclear particles in the photonuclear reaction are related by $A_X = A_b + A_Y$,

$$\begin{aligned} \text{SE}_b &= 931.5(A_b + A_Y - A_X) + \Delta E_b + \Delta E_Y - \Delta E_X \\ &= \Delta E_b + \Delta E_Y - \Delta E_X \end{aligned} \tag{7.23}$$

A table of mass excess (or mass deficit) values in kiloelectron volts is given in Appendix 5 for *atoms.* Ground-state nuclear masses can be obtained with sufficient accuracy from such tables by subtracting the appropriate number of electron rest mass energy values from the atomic mass values.

The threshold energy $h\nu_{\min}$ for a photonuclear reaction is the least possible photon energy that will initiate the reaction. Equations 7.5 give

$$h\nu_{\min,b} = \text{BE}_b\left[1 + \frac{\text{BE}_b}{2M_X c^2}\right], \quad \text{BE}_b = \text{SE}_b + E_Y^* \tag{7.24}$$

In the limit where $\text{BE}_b \ll 2M_X c^2$, the threshold equals the binding energy. Notice that in this case there is no specific notation for the shell and subshell of the bound particle as with the atomic photoelectric effect. However, reactions with nucleons bound in particular energy shells within the nucleus would be characterized by particular values of E_Y^*. When photonuclear thresholds are listed, they are usually values of SE_b ($E_Y^* = 0$). Table 7.2 shows some threshold energies for common photonuclear reactions. Only two reactions exist in which $h\nu_{\min} < 3$ MeV (25).

In the preceding section, the photon momentum-to-energy ratio was shown to be very small. Because of this, one can substitute for T_Y' in Equations 7.20. As long as the incoming energy is greater than the binding energy and $2M_b c^2/T_b' \gg 1$, $2M_Y c^2/T_Y' \gg 1$, then

$$\mathbf{p}_b' + \mathbf{p}_Y' \approx 0, \quad \mathbf{p}_b' \approx -\mathbf{p}_Y', \quad T_Y' \approx \frac{(p_Y')^2}{2M_Y} = \left[\frac{M_b}{M_Y}\right]\frac{(p_b')^2}{2M_b} = T_b'\left[\frac{M_b}{M_Y}\right]$$

$$h\nu = \text{BE}_b + T_b' + T_Y' \simeq \text{BE}_b + T_b'\left[1 + \frac{M_b}{M_Y}\right] \tag{7.25}$$

It is apparent that heterogeneous high-energy photon beams can release neutrons or protons from absorbing nuclei with a variety of energies. Figure 7.13 shows the spectrum of photoneutrons from tantalum produced by 30-MeV bremsstrahlung (26, 27).

The total photonuclear absorption cross section is the sum of all possible individual reaction cross sections, i.e. $\sigma_{\gamma,n} + \sigma_{\gamma,p} + \sigma_{\gamma,2n} + \sigma_{\gamma,2p} + \sigma_{\gamma,pn} + \ldots$. The giant resonance (GR) usually dominates the total absorption cross section. The photon energy that produces peak absorption $h\nu_p$ is about

Table 7.2 Photonuclear Threshold Energies

Nucleus	*Energy (MeV)*	
	(γ,n)	(γ,p)
^{2}H	2.22	2.22
^{9}Be	1.67	16.9
^{12}C	18.7	16.0
^{16}O	15.7	12.1
^{63}Cu	10.9	6.1
^{208}Pb	7.4	8.0

20 MeV for light targets and decreases with increasing mass number (25). For heavy nuclear targets, the GR cross-section function can be written as (28)

$$\sigma(h\nu)=\frac{h^2\alpha}{2M}\left(\frac{NZ}{A}\right)\frac{\Gamma/2\pi}{(h\nu-h\nu_p)^2+(\Gamma/2)^2} \tag{7.26}$$

where M is the nucleon mass, α is the fine structure constant, and N is the target nucleus neutron number. It is sometimes convenient to use $0.06NZ/A$ in megaelectron volt-barns for the initial constants. Γ is the level width and

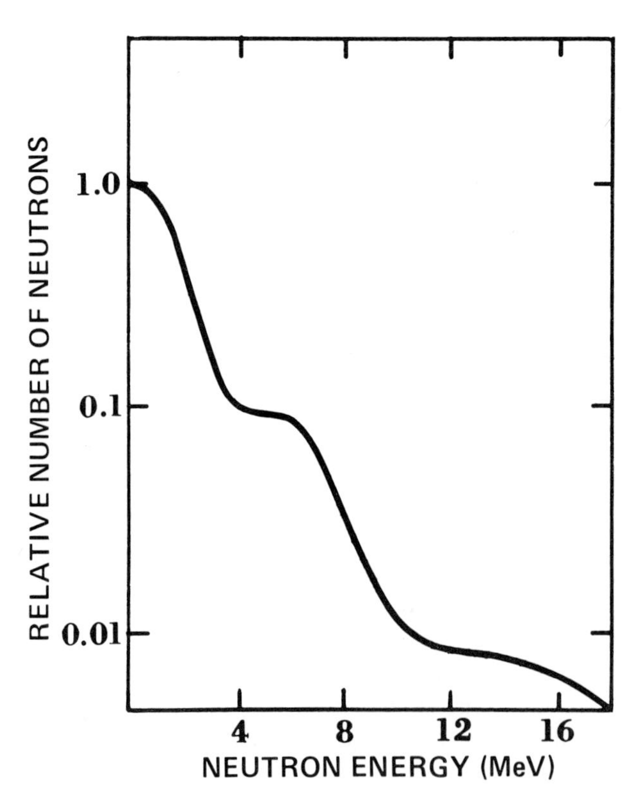

Figure 7.13 The energy spectrum of photoneutrons from tantalum is shown. The photoneutrons were produced by a bremsstrahlung beam with 30-MeV end-point energy (26, 27). Adapted from Ref. 26.

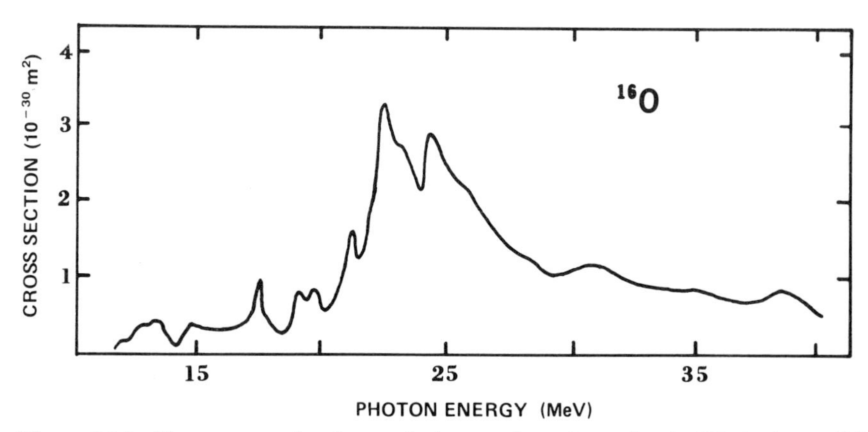

Figure 7.14 The cross section for total photonuclear absorption in ^{16}O is shown (31).

represents the full width at half maximum of the GR. In this equation the functional dependence on $h\nu$ was chosen to agree with the Breit-Wigner resonance line shape (29) to be discussed in Chapter 9, although other forms may also be appropriate (25, 30).

Peak values for the total cross sections fluctuate considerably around the smooth NZ/A dependence expected from Equation 7.26. In fact, the functional form of Equation 7.26 is not particularly appropriate for low- and medium-weight nuclei because the giant resonance has considerable resonance substructure. As an example, Figure 7.14 shows the photonuclear absorption cross section for an ^{16}O target (31).

Nuclear models are usually evoked to explain important aspects of photonuclear absorption. A nucleon fluid model is useful for explaining gross giant resonance properties, especially for medium and heavy nuclei (32). In this case the nucleus is described in terms of interpenetrating proton and neutron fluids. The center of mass of the proton fluid is capable of displacement with respect to the center of mass of the neutron fluid, thereby producing a dipole moment. The photon excites oscillations in the fluids when it is absorbed. GR substructure in light nuclei is usually explained with the aid of the nuclear shell model (33, 34).

Applications of photonuclear reactions are limited because of the relatively small cross sections involved. The magnitude of the peak photonuclear absorption cross section is never as much as 10% of the value for electronic absorption (15). Despite the small cross sections and the energies required, photonuclear activation of absorbers by high-energy bremsstrahlung is readily observed (35). A widespread manifestation of these reactions is the photoneutron component of the background radiation in accelerator rooms (36).

7.5 Pair Production

Pair production is the materialization of an electron-positron pair when a photon disappears in the vicinity of a nucleus (or atomic electron). The reaction was discovered by Anderson in 1932 (37). It is the dominant absorption process for high-energy photons. The pair production reaction can be indicated by $X + \gamma \rightarrow e^- + e^+ + X$, where e^- and e^+ stand for the electron and positron, and X stands for the nucleus (or atomic electron). Figure 7.15 is an illustration of pair production. One should remember that even when the reaction takes place near a nucleus, it is extranuclear; the particles produced are not nuclear emissions. Degradation of the kinetic energy of electrons or positrons and positron annihilation have already been discussed in Chapter 2.

Pair production occurs most often near a nucleus, but an atomic electron is another possible site. An electric field is always necessary. When the reaction takes place near an atomic electron, it is usually called triplet production. This name is understandable since three particles move away from the interaction site: the positron, the electron produced, and the atomic electron in recoil.

Momentum conservation and energy conservation equations for pair production are similar to Equations 7.1 except that two outgoing particles are involved:

$$\mathbf{p}_\gamma = \mathbf{p}'_- + \mathbf{p}'_+ + \mathbf{p}'_X, \quad h\nu + M_X c^2 = T'_- + m_e c^2 + T'_+ + m_e c^2 + T'_X + M_X c^2 \tag{7.27}$$

The subscripts $-$ and $+$ indicate the electron and positron, respectively. One need consider only the momentum components along the incident photon direction to show that the nucleus (or atomic electron) must be involved. Equations 7.27 can be modified with the use of the energy expression $E = T + m_e c^2 = \sqrt{(p'_e)^2 c^2 + m_e^2 c^4}$:

$$p_\gamma = \frac{h\nu}{c} = p'_- \cos\theta_- + p'_+ \cos\theta_+ + p'_X \cos\theta_X \Rightarrow h\nu \leq p'_- c + p'_+ c + p'_X c,$$

$$h\nu = \sqrt{(p'_-)^2 c^2 + m_e^2 c^4} + \sqrt{(p'_+)^2 c^2 + m_e^2 c^4} + T'_X \Rightarrow h\nu > p'_- c + p'_+ c + T'_X \tag{7.28}$$

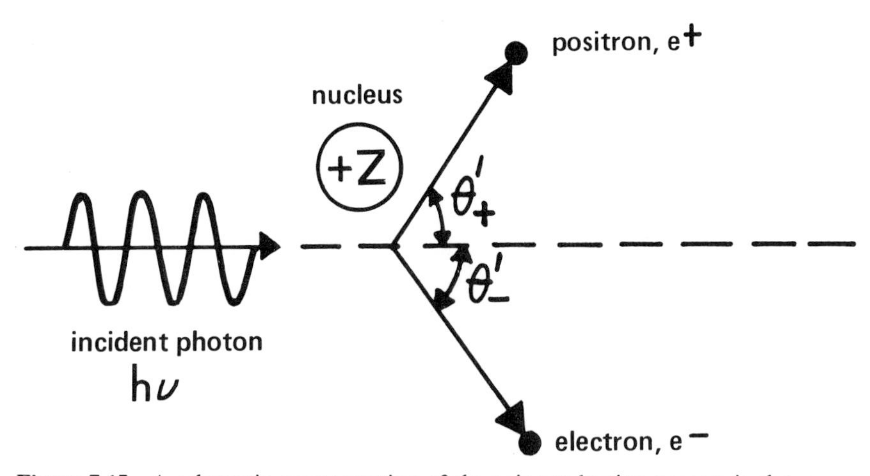

Figure 7.15 A schematic representation of the pair production process is shown.

The angles θ_-, and θ_+, and θ_X indicate directions of departure of the electron, positron, and nucleus (or atomic electron), respectively. The first inequality in Equations 7.28 occurs because $\cos\theta \leq 1$. The second is a result of the m_ec^2 terms. It is apparent that the two inequalities would be incompatible if the recoil particle X was not involved, i.e. if $p_X = T_X = 0$, both of the results in Equations 7.28 could not hold simultaneously.

The threshold energy for pair production near a nucleus can be obtained with the use of Equations 7.2 through 7.5. For threshold, Equations 7.27 can be written as

$$p_\gamma = p'_+ + p'_- + p'_X, \quad h\nu_{\min} = T'_- + T'_+ + T'_X + 2m_ec^2 \tag{7.29}$$

The minimum required energy occurs when the photon momentum is shared among the outgoing particles in the direct ratio of their masses. This leads to the analog of the result of Equations 7.5,

$$h\nu_{\min} = 2m_ec^2\left[1 + \frac{2m_ec^2}{2M_Xc^2}\right] \tag{7.30}$$

In this case $2m_ec^2$ is the mass difference between the outgoing particles and the initial nucleus (or atomic electron). When a nucleus is involved, $M_Xc^2 \gg m_ec^2$; therefore, the theshold energy reduces to $2m_ec^2 = 1.022$ MeV. When an atomic electron is involved, $2m_ec^2 = 2M_Xc^2$ and the threshold energy is $4m_ec^2 = 2.044$ MeV.

For pair production near a nucleus when $h\nu \gg 2m_ec^2$, the kinetic energy liberated is comparable to the photon energy. In this case the photon momentum can usually be neglected relative to the momentum of the outgoing particles and the nucleus will recoil as the electrons move away. Since the nucleus is very massive relative to the electron-positron pair, it will have relatively little recoil kinetic energy. In this case Equations 7.27 can be written

$$\mathbf{p}_\gamma = \mathbf{p}_- + \mathbf{p}_+ + \mathbf{p}_X \simeq 0, \quad h\nu = T_- + T_+ + T_X + 2m_ec^2$$
$$\mathbf{p}_X \simeq (\mathbf{p}_- + \mathbf{p}_+), \quad h\nu \simeq T_- + T_+ + 2m_ec^2 \tag{7.31}$$

The cross section for pair production near a nucleus was first evaluated in 1934 by Bethe and Heitler (38) using plane wave approximations in a quantum mechanical formulation. Since there is symmetry about the direction of the incoming photon, the variables in the problem are $h\nu$, θ'_-, θ'_+, θ'_X, T'_-, and T'_+. The nuclear kinetic energy can usually be ignored. Dependence on the angle of nuclear recoil θ'_X, on the angle of the outgoing electron θ'_-, and on the energy of the outgoing electron T'_-, can be eliminated using Equations 7.27. Then when $h\nu \gg 2m_ec^2$ and $\theta_+ \ll 1$, the differential cross section per unit solid angle per unit energy can be written as (38):

$$\frac{d^2\sigma(h\nu, T'_+, \theta'_+, Z)}{d\Omega_+ dT_+} = \frac{Z^2 r_0^2 \alpha}{\left[1 + \left(\dfrac{T'_+ + m_ec^2}{m_ec^2}\right)^2 \sin^2\theta_+\right]^2} f(h\nu, T'_+, \theta'_+, Z) \tag{7.32}$$

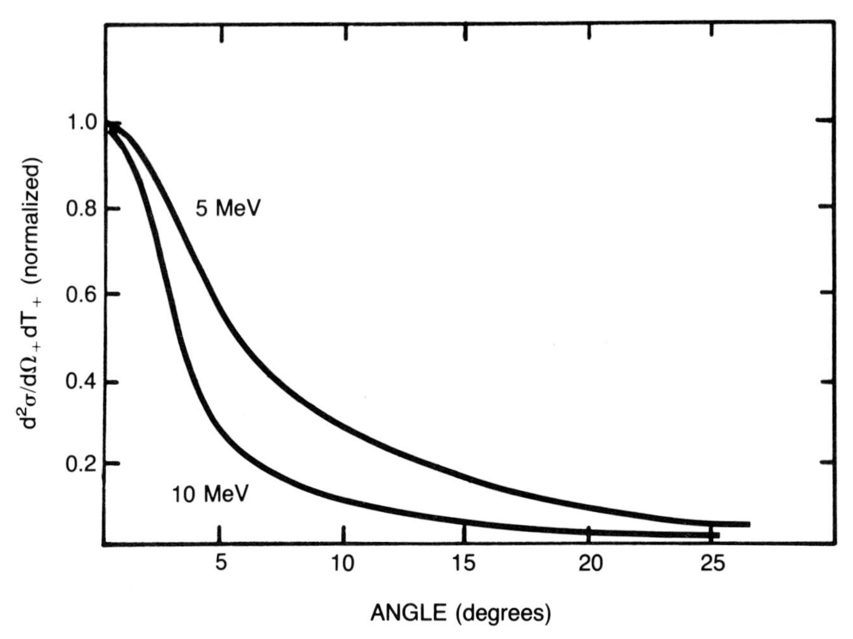

Figure 7.16 The angular distribution per unit solid angle and energy for production of a positron between angles θ'_+ and $\theta'_+ + d\theta'_+$ is shown (39).

In this case $f(h\nu, T'_+, \theta'_+, Z)$ is a rather complicated function. The dependence of the differential cross section on θ_+ is illustrated in Figure 7.16 for several values of T'_+. Notice that the distribution becomes more forward peaked as T'_+ increases. Generally, the root mean square angle between the incident photon and the outgoing electron is of the order of $m_ec^2/(T + m_ec^2)$ for relativistic electrons (39).

The result of integrating Equation 7.32 over solid angle $d\Omega_+$ can be written

$$\frac{d\sigma}{dT'_+} = Z^2 r_0^2 \alpha g(h\nu, T'_+, Z) \tag{7.33}$$

The function $g(h\nu, T'_+, Z)$ increases with increasing $h\nu$ and decreases only slightly with increasing Z. The differential cross section $d\sigma/dT'_+$ is shown on Figure 7.17 as a function of T'_+. This figure can be regarded as a display of the theoretical energy spectrum of the positrons (or the electrons) produced. Although the electron energy will differ from the positron energy in individual reactions, these results show that on the average the energy $h\nu - 2m_ec^2$ is evenly shared. However, experimental positron and electron spectra show that the actual energies are perturbed by the electric field of the nucleus. The positrons pick up extra energy from the nuclear coulomb force and the electrons lose energy by the same force; however, the change is small (11).

The differential cross section of Equation 7.33 can be integrated over T'_+ to yield

$$\sigma(h\nu,Z) = \int \frac{d\sigma}{dT'_+}\, dT'_+ = Z^2 r_0^2 \alpha \int g(h\nu, T'_+, Z)\, dT_+ = Z^2 r_0^2 \alpha h(h\nu, Z) \tag{7.34}$$

The function $h\ (h\nu,Z)$ is plotted on Figure 7.18 (17). Notice that the cross-section is monotonically increasing with increasing $h\nu$. At high energies the reaction can occur at a substantial distance from the nucleus, and atomic electrons tend to screen the nuclear charge (40). Then if $h\nu \gg m_e c^2$,

$$h(h\nu,Z) \simeq \left[\frac{28}{9} \ln \frac{183}{Z^{1/3}} - \frac{2}{27}\right] \tag{7.35}$$

The effects of pair production near atomic electrons (41) can be included by replacing Z^2 by $Z(Z + 0.8)$.

Dirac's interpretation (42) of the expression for total relativistic energy E can be used in the discussion of pair production. Recall that

$$E^2 = p^2c^2 + m_e{}^2c^4, \qquad E = \pm\sqrt{p^2c^2 + m_e{}^2c^4} \tag{7.36}$$

Dirac proposed that negative energy values given by the negative root should be associated with positrons much as the positive energy values are associated with electrons. Figure 7.19 is an illustration of this situation. The available free electron states are separated from the available free positron states by a gap of $2m_ec^2$. Thus, the photon energy required to produce a pair is $T'_+ + T'_- + 2m_ec^2$ and the threshold is $2m_ec^2$. According to Dirac the positron states are usually completely filled and are not normally observable. Small energy transfers between positrons are forbidden by the Pauli principle in this case.

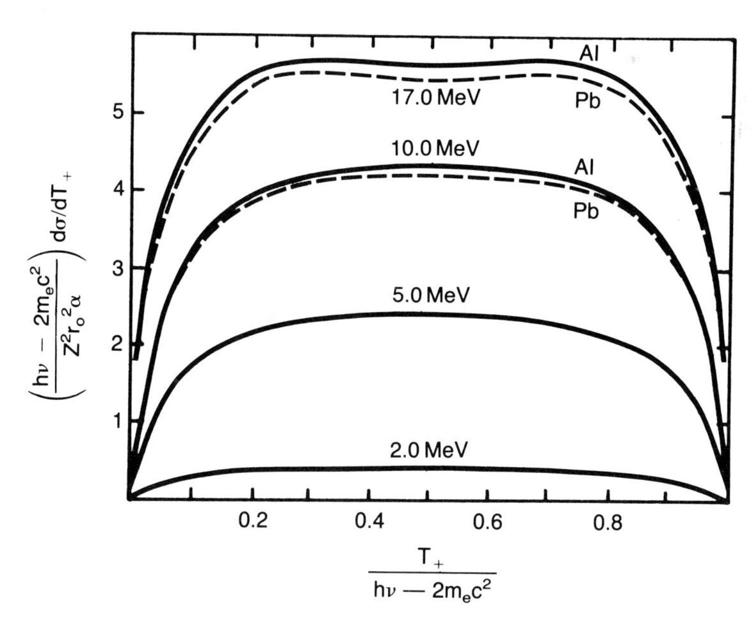

Figure 7.17 The energy spectrum of positrons from pair production is shown for several photon energies (38). Adapted from Ref. 11.

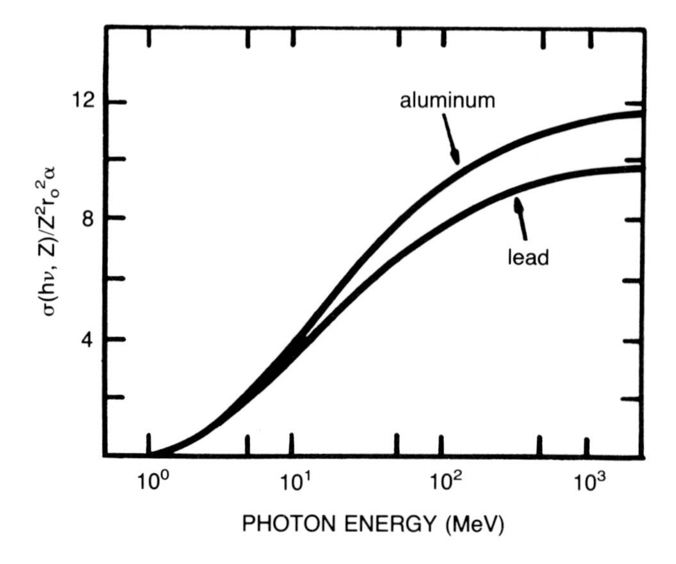

Figure 7.18 The pair production cross section is shown as a function of energy (17).

A positron and an electron are a specific example of a particle-antiparticle pair (43, 44). Charged particles (e.g. electrons) and their antiparticles (e.g. positrons) have the same physical properties except they have opposite charge. Thus, it is logical to treat them as different states of the same system. In such a scheme, the antiparticle can be identified with a particular manifestation of the particle, one which has opposite charge and moves backward in time. In addition, the available energy states belong to the negative square root in Equations 7.36.

Figure 7.20 shows several interactions between particles, antiparticles, and photons that occur in a space-time plane (45). In the *pair production* diagram,

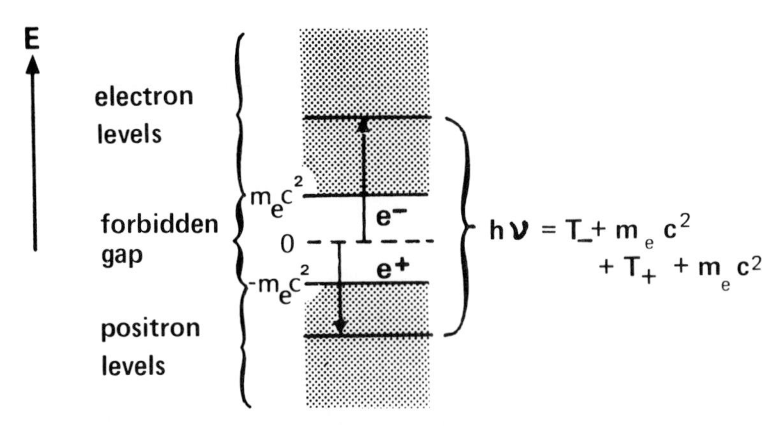

Figure 7.19 The energy level sequence for electrons and positrons in free space is shown. The electrons take positive total energy states in the continuum and the positrons take negative total energy states.

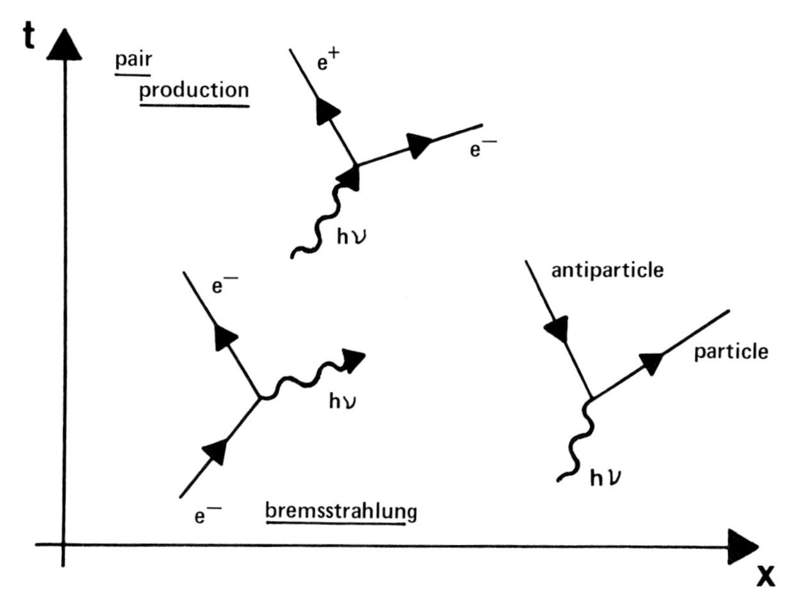

Figure 7.20 The schematic diagram for the pair production reaction is similar to the diagram for bremsstrahlung if both are considered to be special cases of an antiparticle-particle interaction with a photon.

the photon traveling forward in time interacts with an electron-positron pair, and the pair moves away in time. If a positron is equivalent to an electron moving backward in time, we can generalize from that representation to the particle-antiparticle representation in the figure. Positron time reversal is indicated by the reversed arrow direction. The *bremsstrahlung* representation is similar to the *pair production* diagram, when this time reversal is considered. Departure of the positron after pair production is analogous to the arrival of the electron prior to emitting bremsstrahlung. Since the photon is its own antiparticle, emission is analogous to absorption on such a figure.

This relationship between pair production and bremsstrahlung is evident in the cross-section dependence. Notice the factor $Z^2 r_0^2 \alpha$ occurs in the bremsstrahlung cross-section expression given in Chapter 3 and in the pair production cross sections shown in this section.

7.6 Attenuation Coefficients for Photon Absorption

The photon absorption reactions discussed in the preceding sections of this chapter include photoabsorption, the atomic photoelectric effect (pe), photonuclear reactions (pn), and pair production (pp). The letters in parentheses will serve as subscripts to identify the reactions in the next few paragraphs.

The mass attenuation coefficient for an absorption reaction, $(\mu/\rho)_a$, can be related to σ_a, the absorption reaction cross section per atom, by

$$(\mu/\rho)_a = n_m \sigma_a = (N_A/M_m)\sigma_a \tag{7.37}$$

where N_A is Avogadro's number, n_m is the number of targets per unit mass, and M_m is the mass per mole for the target atom. The equation $n_m = N_A/M_m$ is appropriate for photon absorption, since cross sections are usually given for an entire atom or for the nucleus. Recall that the relationship $n_m = N_A(Z/M_m)$ was used in the preceding chapter for photon scattering coefficients. It was correct in that context since scattering cross sections are usually given for a single electron rather than for an entire atom.

Figure 7.21 shows the mass attenuation coefficients for an iodine absorber as a function of photon energy for the atomic photoelectric effect and pair production. The mass attenuation coefficients for photon scattering are also shown for purposes of comparison. Notice that the photoelectric effect dominates at the lower energies shown; incoherent scattering dominates at intermediate energies; and pair production dominates at higher energies. This sequence holds for other absorbers, but the boundary energies between regions of dominance are a function of atomic number. Figure 7.22 shows the equal probability lines per unit mass of absorber for pairs of reactions as a function of Z and $h\nu$. As an example, at $h\nu = 0.1$ MeV photon scattering is most probable in carbon whereas photoelectric absorption dominates in lead. At $h\nu = 10$ MeV, photon scattering remains dominant in carbon, but pair production is most probable in lead.

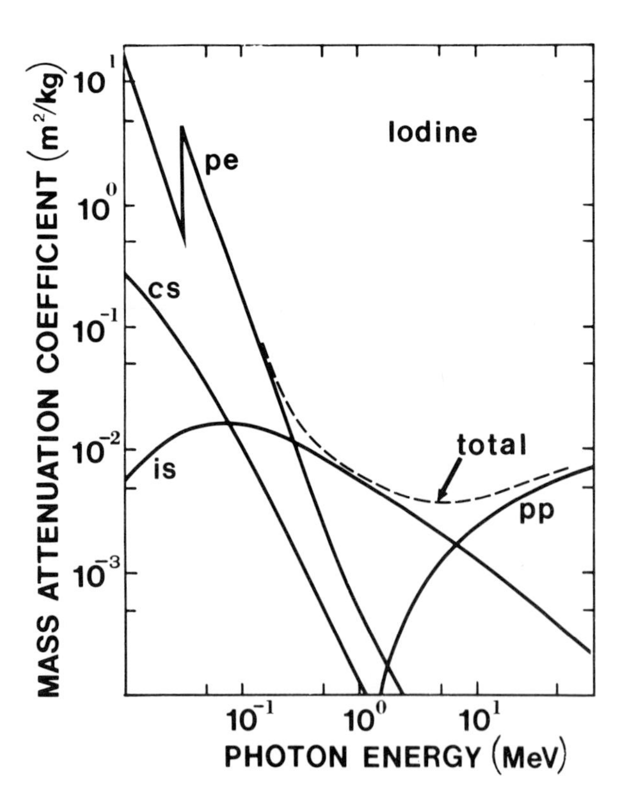

Figure 7.21 The mass attenuation coefficients for coherent scattering (cs), atomic photoelectric effect (pe), incoherent scattering (is), and pair production (pp) are compared for an iodine absorber (17).

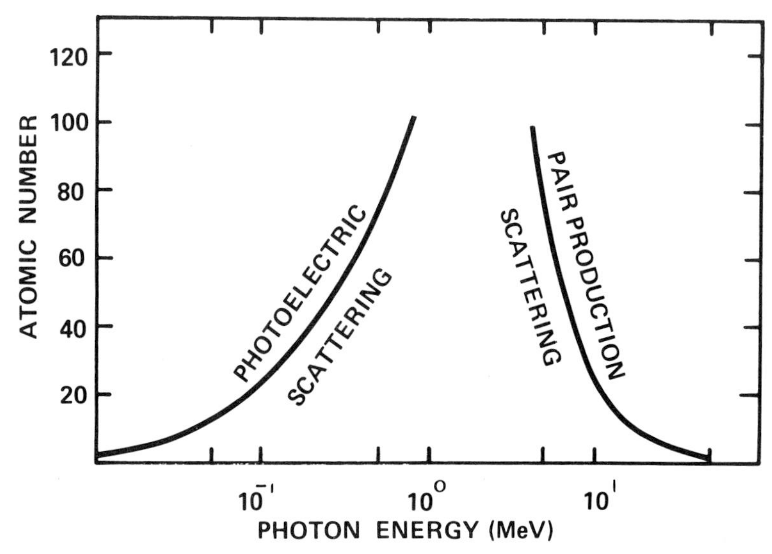

Figure 7.22 Equal reaction probability lines per unit absorber mass are shown in Z and $h\nu$ space for pairs of photon reactions (17).

In Section 7.3 the cross section per atom for the atomic photoelectric effect was shown to be proportional to Z^n where the exponent n is a function of Z and $h\nu$. Figure 7.9 can be used to estimate the exponent at energies above the K edge for particular elements. Thus, if the mass attenuation coefficient for the atomic photoelectric effect can be found in a table for element 1 at some energy well above the K edge, the photoelectric mass attenuation coefficient for neighboring element 2 at the same energy can be estimated from

$$(\mu/\rho)_{\text{pe2}} \simeq \frac{(Z^n/M_m)_2}{(Z^n/M_m)_1}(\mu/\rho)_{\text{pe1}} \tag{7.38}$$

This relationship is a special application of Equation 7.37 since the factor Z^n is attributable to the dependence of the photoelectric cross section on atomic number. If precise values are not necessary, one can substitute $n \Rightarrow 4$, $M_m \Rightarrow A \Rightarrow 2Z$. In this case the photoelectric absorption probability per unit mass of absorber varies approximately as Z^3.

Details of the internal arrangement of nucleons in the nucleus can affect photonuclear cross sections dramatically. Because of this variability, systematic relationships for photonuclear attenuation coefficients must usually include the giant resonance peak energies and widths (see Equation 7.26). For neighboring heavy elements, with similar GR widths, the average mass attenuation coefficient in the GR region is proportional to NZ/A. In this case a first-order estimate of the average photonuclear mass attenuation coefficient for element 2 can be obtained from the coefficient for neighboring element 1 by

$$\langle(\mu/\rho)_{\text{pn2}}\rangle \simeq \frac{(NZ/A)_2}{(NZ/A)_1}\langle\mu/\rho)\rangle_{\text{pn1}} \tag{7.39}$$

Since the neutron number $N \simeq Z \simeq A/2$, the photonuclear absorption probability per unit mass of absorber is approximately proportional to A.

In Section 7.5 the cross section per atom for nuclear pair production was shown to vary directly as Z^2 when the slow Z dependence in the function $h(h\nu,Z)$ was neglected. Under these circumstances the mass attenuation coefficient for pair production for element 2 can be estimated from the coefficient for pair production for neighboring element 1 using

$$(\mu/\rho)_{\text{pp}2} \simeq \frac{(Z^2/M_m)_2}{(Z^2/M_m)_1} (\mu/\rho)_{\text{pp}1} \tag{7.40}$$

If $M_m \simeq A \simeq 2Z$, the pair production absorption probability per unit mass can be shown to vary approximately as Z.

Mass energy transfer coefficients were defined in Section 5.2. They are proportional to the fraction of the energy lost by indirectly ionizing radiation which is converted into charged-particle kinetic energy. The mass energy transfer coefficient for the atomic photoelectric effect is given by (46)

$$(\mu_{\text{tr}}/\rho)_{\text{pe}} = (\mu/\rho)_{\text{pe}}\left[1 - \frac{\delta}{h\nu}\right] \tag{7.41}$$

where δ is the average energy emitted as fluorescent electromagnetic radiation. It is the average of the product of the atomic electron binding energy and the fraction of the binding energy radiated as characteristic photons (fluorescent yield). The mass energy transfer coefficient for pair production is given by

$$(\mu_{\text{tr}}/\rho)_{\text{pp}} = (\mu/\rho)_{\text{pp}}\left[1 - \frac{2m_ec^2}{h\nu}\right] \tag{7.42}$$

The fraction $2m_ec^2/h\nu$ is subtracted because the rest mass energy of the positron-electron pair is not directly converted to kinetic energy.

7.7 Problems

1. For the reaction $Y + \gamma \rightarrow b + Y$, show that the equations $p_b = p_\gamma[M_b/(M_b + M_Y)]$, $p_Y = p_\gamma[M_Y/(M_b + M_Y)]$ provide the minimum value of $\Delta = T_b + T_Y = (p_b{}^2/2M_b) + (p_Y{}^2/2M_Y)$ under the constraint that $p_\gamma = p_b + p_Y$. To accomplish this, differentiate Δ with respect to p_Y and set the result equal to zero.
2. Estimate the mass decrease because of the electron binding energy (13.6 eV) for the hydrogen atom. Is this mass difference likely to be directly measurable?
3. The transmission of a 1-cm cell containing a solution of Fe^{++} and Fe^{+++} ions (ferrous sulfate dosimeter) is only 50% of that value obtained before irradiation of the solution. The transmission was measured with a spectrophotometer at 3050Å, a wavelength in which Fe^{+++} absorption completely dominates. The molar extinction coefficient is 2197 per gram-mol$\cdot$l$^{-1}\cdot$cm^{-1}

for the Fe^{+++} ion at this wavelength. Calculate the concentration of the Fe^{+++} ion. If the solution density is 10^{+3} kg/m³ and the G value (reactions per 100 eV) for oxidation of Fe^{++} to Fe^{+++} is 15.5, calculate the dose to the solution.

4. Suppose a 140-keV photon undergoes photoelectric effect in a lead sheet with a K-shell electron.
   a. What is the kinetic energy liberated?
   b. If it is assumed that this is all photoelectron kinetic energy, calculate the electron momentum and the photon momentum and compare the two.
5. For a 30-keV photon, compute the ratio of the photoelectric interaction probabilities for K-shell electrons to that for L-shell electrons in iodine.
6. Calculate the threshold for these reactions: $^{9}Be(\gamma,n)^{8}Be$; $^{14}N(\gamma,n)^{13}N$; $^{12}C(\gamma,2n)^{10}C$. Use the mass tables in Appendix 5.
7. What is the mass attenuation coefficient for ^{16}O at about 22 MeV for photoneutron processes? Use the peak value from Figure 7.14.
8. Electric dipole absorption processes dominate when $kr = (2\pi\nu/c)r \ll 1$, where k is the wave number, ν is the incident photon frequency, and r is the average radius of the absorbing structure. Would you expect dipole absorption to dominate for 20-MeV photons in ^{16}O if the nuclear radius is given by $R = R_0 A^{1/3} = (1.4)10^{-15} A^{1/3}$ m?
9. Consider the pair production raction in the center of mass system.
   a. Calculate the speed of the center of mass, v_{cm}, in terms of $h\nu_{cm}$ and M_X, the target nuclear mass. At threshold in the center of mass system, the emitted pair and the nucleus are at rest; therefore, all outgoing particles must be moving with speed v_{cm} in the laboratory system.
   b. Calculate the laboratory system momentum values for p_+, $p_{\div}$, and p_Y using these facts.
   c. Utilize the relationship $p_Y + p_+ + p_- = p_\gamma = (h\nu/c)_{lab}$ for the laboratory system to calculate the relationship between the photon frequency in the center of mass system, ν_{cm}, and the photon frequency in the laboratory system, ν_{lab}.
10. Use the equation for the Lorentz force and the resulting acceleration $[m(d^2\mathbf{x}/dt^2) = \mathbf{v} \times \mathbf{B}]$ to show that a particle with mass m and charge $-q$ moving in magnetic induction $\mathbf{B}$ forward in time satisfies the same equation of motion as a particle with mass m and charge $+q$ moving backward in time.
11. Given that the mass attenuation coefficient for ^{63}Cu is 0.474 m²/kg at 40 keV (photoelectron dominates) and 0.0042 m²/kg at 2 MeV (incoherent scatter dominates), estimate the coefficient for ^{56}Fe at these energies.
12. What ratio of Z^n/M_m would be necessary so that some substance had the same photoelectric mass attenuation coefficient as water at 10 keV?
13. At what photon energy do the atomic photoelectric effect and the incoherent scattering have equal probability of occurrence in H_2O? At what photon energy is the kerma the same due to each process in H_2? Use the tables in Appendix 9.

7.8 References

1. L. G. Christophorou, *Atomic and Molecular Radiation Physics* (Wiley-Interscience, New York, 1971).
2. P. H. Metzer and C. R. Cook, J. Chem. Phys. **41**, 642 (1964).
3. K. Watanabe and A. S. Jursa, J. Chem. Phys. **41**, 1650 (1964).
4. C. E. Brion, Radiat. Res. **64**, 37 (1975).
5. I. G. Draganic and Z. D. Draganic, *The Radiation Chemistry of Water* (Academic, New York, 1971), pp. 29–32.
6. H. Fricke and E. J. Hart, "Chemical dosimetry," in *Radiation Dosimetry,* vol. 2, edited by J. H. Attix and W. C. Roesch (Academic, New York, 1966), pp. 167–239.
7. C. Kittel, *Introduction to Solid State Physics* (John Wiley & Sons, New York, 1966), pp. 255–297.
8. D. Dutton, Phys, Rev. **112**, 785 (1958).
9. H. R. Philipp and H. Ehrenreich, J. Appl. Phys. **35**, 1416 (1964).
10. A. H. Sommer, *Photoemissive Materials* (John Wiley and Sons, New York, 1968).
11. R. D. Evans, *The Atomic Nucleus* (McGraw-Hill, New York, 1955).
12. W. Heitler, *The Quantum Theory of Radiation* (Oxford University, London, 1954).
13. W. T. Grandy, *Introduction to Electrodynamics and Radiation* (Academic, New York, 1970), pp. 205–208.
14. F. Sauter, Ann. Phys. **11**, 454 (1931).
15. J. H. Hubbell, *Photon Cross Sections, Attenuation Coefficients and Energy Absorption Coefficients from 10 KeV to 100 GeV* (National Bureau of Standards, Washington, 1969), p. 27.
16. L. I. Schiff, *Quantum Mechanics* (McGraw-Hill, New York, 1955).
17. E. Storm and H. I. Israel, Nucl. Data Tables **1**, 565 (1970).
18. D. R. White, Phys. Med. Biol. **22**, 219 (1977).
19. W. J. Viegele, At. Data **5**, 55 (1973).
20. F. K. Richtmyer, E. H. Kennard, and J. N. Copper, *Introduction to Modern Physics,* 6th ed. (McGraw-Hill, New York, 1969), pp. 48–49.
21. L. G. Parratt, Phys. Rev. **56**, 295 (1939).
22. R. A. Buchanan, S. I. Finkelstein, and K. A. Wichershein, Radiology **105**, 185 (1972).
23. R. H. Dalitz and D. R. Yennie, Phys. Rev. **105**, 1598 (1950).
24. W. C. Barber, Phys. Rev. **111**, 1642 (1958).
25. E. G. Fuller, H. M. Gerstenberg, H. Vander Molen, and T. C. Dunn, *Photonuclear Reaction Data, NBS Special Publication 380* (National Bureau of Standards, Washington, 1973).
26. *Shielding for High Energy Electron Accelerator Installations, NBS Handbook 97* (National Bureau of Standards, Washington, 1964).
27. C. Cortini, C. Milone, A. Rubbino, and J. Ferrero, Nuovo Cimento **9**, 85 (1958).
28. J. S. Levinger, *Nuclear Photo-Disintegration* (Oxford University, London, 1960).
29. G. Breit and E. P. Wigner, Phys. Rev. **49**, 519 (1936).
30. R. L. Bramblett, S. C., Fultz, and B. L. Berman, "Systematic properties of the giant resonance current status," in *Proceedings of the International Conference on Photonuclear Reactions and Applications,* edited by B. L. Berman (E. O. Lawrence Livermore Laboratory, University of California, Livermore, CA, 1973), pp. 13–22, 40.
31. J. Ahrens, H. B. Eppler, H. Gim, H. Gundrum, M. Kronig, P. Riehm, G. Seta

Ram, A. Ziegler, and B. Ziegler, "Total photoneutron cross sections, particularly above 30 MeV," in *Proceedings of the International Conference on Photonuclear Reactions and Applications,* edited by B. L. Berman (E. O. Lawrence Livermore Laboratory, University of California, Livermore, CA, 1973), pp. 23–35.

32. H. Steinwedel and J. H. D. Jensen, Z. Naturforsch **59**, 413 (1950).
33. M. G. Mayer and J. H. D. Jensen, *Elementary Theory of Nuclear Shell Structure* (John Wiley & Sons, New York, 1955), p. 53.
34. G. E. Brown, L. Castillejo, and J. A. Evans, Nucl. Phys. **22**, 1 (1961).
35. H. V. Piltingsrud, Med. Phys. **10**, 147 (1983).
36. *Structural Shielding Design and Evaluation for Medical Use of X Rays and Gamma Rays of Energies up to 10 MeV, NCRP Report 49* (National Council on Radiation Protection and Measurements, Washington, 1976).
37. C. D. Anderson, Phys. Rev. **43**, 491 (1933).
38. H. A. Bethe and W. Heitler, Proc. R. Soc. London Ser A **146**, 83 (1934).
39. M. L. Ter-Mikaelian, *High-Energy Electromagnetic Processes in Condensed Media* (Wiley-Interscience, New York, 1972), pp. 29–31.
40. I. B. Bernstein, Phys. Rev. **80**, 995 (1950).
41. E. R. Gaerthner and M. L. Yeater, Phys. Rev. **78**, 621L (1950).
42. P. A. M. Dirac, Proc. R. Soc. **133**, 61 (1931).
43. E. C. G. Stueckelberg, Helv. Phys. Acta **15**, 23 (1942).
44. R. P. Feynman, Phys. Rev. **76**, 749 (1949).
45. H. Frauenfelder and E. M. Henley, *Subatomic Physics* (Prentice-Hall, Englewood Cliffs, NJ, 1974), pp. 166–170.
46. *Radiation Quantities and Units, ICRU Report 33* (International Commission on Radiation Units and Measurements, Washington, 1980).

8

Neutron Elastic Scattering Reactions

8.1 Introduction

Neutrons with substantial kinetic energy can initiate a variety of reactions in an absorber. In the discussions to follow, particular processes are grouped under two categories, scattering and absorption. In this chapter, neutron elastic scattering will be considered; the next chapter will introduce neutron absorption reactions. Division of neutron reactions into scattering and absorption categories is analogous to the division of photon reactions used in Chapters 6 and 7. When neutrons are considered, these categories must be clearly defined so that common terminology is not confusing.

For the purposes of this discussion, a neutron scattering reaction involves a change in direction of the incoming neutron due to an interaction with a target nucleus. Some degradation of the neutron kinetic energy is also implied. For elastic neutron scattering reactions, the incoming kinetic energy and the angle between the exit direction and the direction of incidence determine the exit kinetic energy.

The distinction between scattering and absorption reactions for neutrons can be made using the reaction notation introduced in Chapter 7. Suppose n indicates the incident neutron and X represents the target nucleus. Then if b represents the outgoing particle and Y is the symbol for the product nucleus, the reaction is $X + n \rightarrow b + Y$. If the target and product nuclides are not specified, the general reaction category can be indicated by (n,b). For example (n,n) represents neutron elastic scattering and (n,p) represents neutron absorption followed by proton emission.

Only reactions usually called elastic scattering and indicated by (n,n) are discussed in this chapter; inelastic scattering reactions indicated by (n,n') will not be included. The energy of the outgoing neutron from an (n,n') reaction depends on the residual nuclear configuration, since the product nucleus is left in an excited state (1). Inelastic neutron scattering reactions (n,n') are discussed in the next chapter since they fit more readily into that theoretical context.

Theoretical derivations for neutron reaction cross sections often utilize waves with reduced wavelength $\lambdabar$ to represent the neutrons (2, 3):

$$\lambdabar = \frac{\hbar}{p} = \frac{\hbar}{\sqrt{2M_n T}} \tag{8.1}$$

Of course, $\hbar$ is Planck's constant divided by 2π, p is the neutron momentum, M_n is the neutron mass, and T is the nonrelativistic kinetic energy of the neutron. In this representation, an elastic scattering process can be described as a reflection of the neutron wave by the nucleus (4, 5). One can consider attenuation of the neutron wave in the interior of the nucleus in a description of neutron absorption. Neutron wave representations will be used in explanations of the differential cross sections given in Section 8.3. The wave-particle dichotomy apparent in photon scattering is also evident for neutrons, since particle collision parameters will be used in Section 8.2 to develop energy expressions.

It is useful to subdivide the possible neutron kinetic energies into four ranges to facilitate discussion of elastic scattering and absorption of neutrons. Table 8.1 is a list of some commonly used energy ranges and the names applied (6, 7). The categories are useful since dominant reactions can often be identified in the regions given. The actual boundary energies do not imply a discontinuous change in properties and should not be emphasized in themselves.

The slow neutron category listed on the table includes several other well known subcategories. Among them are thermal neutrons and cold neutrons. Thermal neutrons are in thermal equilibrium with their surroundings. At 293 K the spectrum of thermal neutrons has a maximum at 0.025 eV. Cold neutrons have energies considerably less than 0.025 eV, often as low as 0.001 eV (7).

Usually, neutrons in an absorber are degraded in collisions with nuclear targets. After a number of collisions have occurred, the neutron kinetic energy approaches the average value of the vibrational kinetic energy of comparable nuclear targets. At this point it is possible for an individual neutron either to gain or lose energy in a given collision depending on the specific neutron-to-target kinetic energy ratio. When large numbers of collisions are involved, an equilibrium can be attained wherein the average neutron kinetic energy lost in collision is equal to the average neutron kinetic energy gained. When this equilib-

Table 8.1 Neutron Energy Ranges

Type	*Energy Range*
Slow neutrons	$0 \leq T < 1$ keV
Intermediate neutrons	1 keV $\leq T < 0.5$ MeV
Fast neutrons	0.5 MeV $\leq T < 10$ MeV
High-energy neutrons	10 MeV $\leq T$

rium is attained, the beam is said to be thermalized. On the average, cold neutrons experience a net gain in energy in collisions with room temperature targets until they are thermalized.

If the conditions for thermal equilibrium are fully satisfied, the group of neutrons will have a Maxwellian distribution in speed v that depends on the absolute temperature $T°$ of the medium (7). If N_0 is the total number of neutrons considered,

$$N(v)dv = 4\pi N_0\left[\frac{M_n}{2\pi kT^{\circ}}\right]^{3/2} v^2 \exp\left[-\frac{M_n v^2}{2kT^{\circ}}\right]dv. \tag{8.2}$$

where $N(v)dv$ is the number of neutrons with speed between v and $v + dv$ and k is the Boltzmann constant. Figure 8.1 is an illustration of the distribution of thermal neutrons at temperature T = 293K (8). The most probable speed for thermal neutrons is about 2200 m/s.

In the next section, energy relationships for elastic neutron-scattering reactions will be developed. The final equations will be given in laboratory system variables. In Section 8.3 the results of the partial wave treatment of neutron scattering will be discussed and some differential cross-section expressions will be given. Because elastic scattering is very important in energy degradation of a neutron beam, Section 8.4 includes a discussion of neutron degradation and

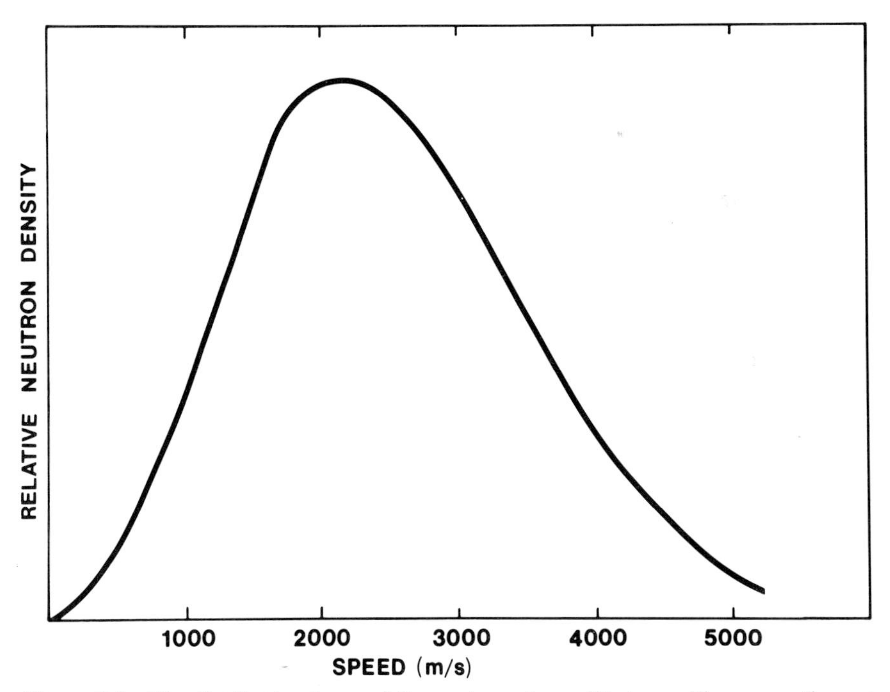

Figure 8.1 The distribution in speed for neutrons in equilibrium with surroundings at temperature of 293 K has a peak of 2200 m/s (8).

moderated spectra. Section 8.5 will relate the total cross section to the elastic scattering attenuation coefficient.

8.2 Energy and Momentum Relationships

In this section the equations governing the outgoing neutron kinetic energy and deflection angle for elastic scattering will be developed. The momentum and energy transferred to the nucleus and the energy loss fraction for a single collision will be obtained by application of conservation laws. The method will be sufficient for most neutron energies although relativistic formulations will not be used. The results will not be directly applicable to beams with an average kinetic energy approaching thermal values, since motion of target nuclei in the laboratory system will be neglected.

Collisional kinetics can often be described with simplified expressions in a reference frame moving so that the total momentum is zero. The center of mass (subscript cm) system, described briefly in Section 7.1, satisfies this description, and will be employed to describe a neutron-nucleus collision. Useful expressions can be obtained by comparing laboratory system velocities with center of mass system velocities. Final results will always be given in terms of laboratory system variables since they relate directly to measurable quantities.

To discuss an elastic scattering collision without confusion, variables that apply before and after the collision must be specified. Since two reference frames are used in the following, a double set of these variables is required. In the relationships that follow, the usual notation will be used for speed, momentum, and angle, except that these quantities will be represented by lowercase letters in the laboratory system and by capital letters in the center of mass system. Furthermore, the symbols for physical variables will be unprimed before and primed after the scattering collision. The latter convention was used in Chapters 6 and 7. When this notation proves unwieldy, symbols will be explicitly defined as used. In some circumstances, the subscript A will denote the nucleus with mass number A.

The scattering collision is easily understood with reference to Figure 8.2. Before the collision the neutron with mass M_n approaches the nucleus (at rest) with speed v in the laboratory system. After the collision the neutron leaves with speed v' at angle θ'. The position and speed of the center of mass is defined in the laboratory system by (9)

$$(M_A + M_n)\mathbf{r}_{cm} = M_A\mathbf{r}_A + M_n\mathbf{r},$$

$$v_{cm} = \frac{M_A v_A + M_n v}{M_A + M_n} = \frac{M_n v}{M_A + M_n} \simeq \frac{v}{A+1} \tag{8.3}$$

The second result in Equations 8.3 was obtained by differentiating the position vectors with respect to time. Note that $v_A = 0$, since the target is at rest before the collision in the laboratory system. The nucleus-to-neutron mass ratio is approximated by A:1 in the last relationship for simplicity.

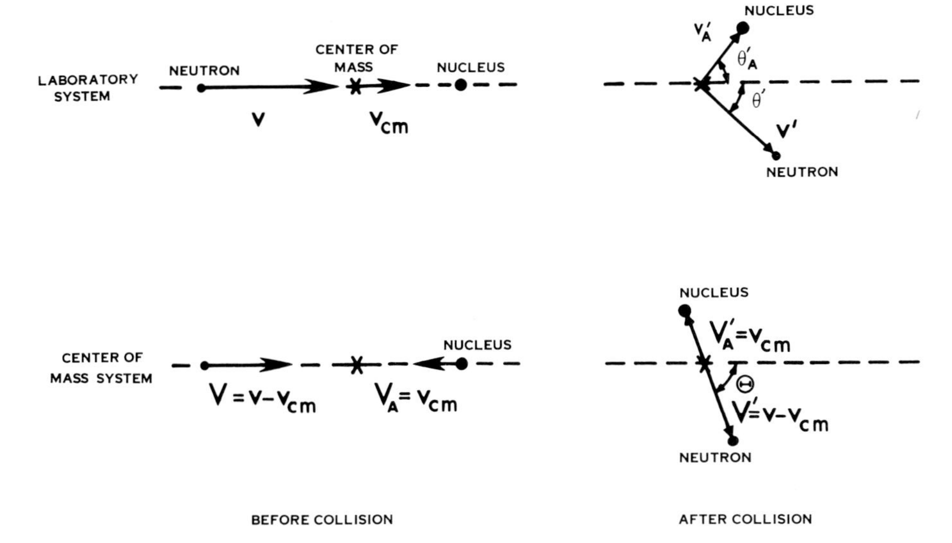

Figure 8.2 The collision between a neutron with speed v and a nucleus at rest is illustrated in two reference frames: the laboratory system and the center of mass system.

The collision as seen by an observer in the center of mass system is also illustrated on Figure 8.2. The initial speed of the neutron and the nucleus in the center of mass system can be obtained by subtraction of the center of mass speed from the particle speed in the laboratory system, since the incident motion is directed along the x axis. We must use the momentum conservation relationships to obtain the particle speed in the center of mass frame after the collision.

Since total momentum is zero in the center of mass system,

$$P + P_A = P' + P'_A = 0, \quad P_A = -P, \quad P'_A = -P' \tag{8.4}$$

Because the collisions described are elastic, the total kinetic energy is conserved:

$$\frac{P^2}{2M_n} + \frac{P_A{}^2}{2M_A} = \frac{(P')^2}{2M_n} + \frac{(P'_A)^2}{2M_A} \tag{8.5}$$

Substituting from Equations 8.4 into Equation 8.5 gives

$$P^2\left[\frac{M_A + M_n}{2M_nM_A}\right] = (P')^2\left[\frac{M_A + M_n}{2M_nM_A}\right], \quad P = P', \quad V = V' = v - v_{\text{cm}},$$

$$P_A{}^2\left[\frac{M_A + M_n}{2M_nM_A}\right] = (P'_A)^2\left[\frac{M_A + M_n}{2M_nM_A}\right], \quad P_A = P'_A, \quad V_A = V'_A = v_{\text{cm}} \tag{8.6}$$

The particle speeds in the center of mass system are unchanged by the elastic collision. This relationship is utilized on Figure 8.2.

In Figure 8.3 the velocity vectors after the collision in both reference frames are superposed on the same origin. The law of cosines (10) can be applied to the lower triangle giving

$$(v')^2 = (v - v_{cm})^2 + v_{cm}^2 - 2v_{cm}(v - v_{cm})\cos\chi' \tag{8.7}$$

Furthermore, trigonometric relationships for angles θ' and χ' on Figure 8.3 can be used to yield

$$v' \sin\theta' = (v - v_{cm})\sin\chi'$$

$$v' \cos\theta' = v_{cm} + (v - v_{cm})\cos(\pi - \chi') = v_{cm} - (v - v_{cm})\cos\chi' \tag{8.8}$$

for χ' between 0 and 180°.

Substitution for $-(v - v_{cm})\cos\chi'$ in Equation 8.7 from Equations 8.8 gives

$$(v')^2 = v_{cm}^2 + (v - v_{cm})^2 + 2v_{cm}(v' \cos\theta' - v_{cm}),$$

$$(v')^2 - 2v'v_{cm}\cos\theta' - v(v - 2v_{cm}) = 0 \tag{8.9}$$

Equation 8.3 can be used to eliminate v_{cm} in Equation 8.9:

$$(v')^2 - \frac{2v'v}{A+1}\cos\theta' - v\left[v - \frac{2v}{A+1}\right] = 0,$$

$$(v')^2 - \frac{2v'v}{A+1}\cos\theta' - v^2\frac{(A-1)}{(A+1)} = 0. \tag{8.10}$$

Application of the quadratic formula (10) to solve Equation 8.10 for v' yields

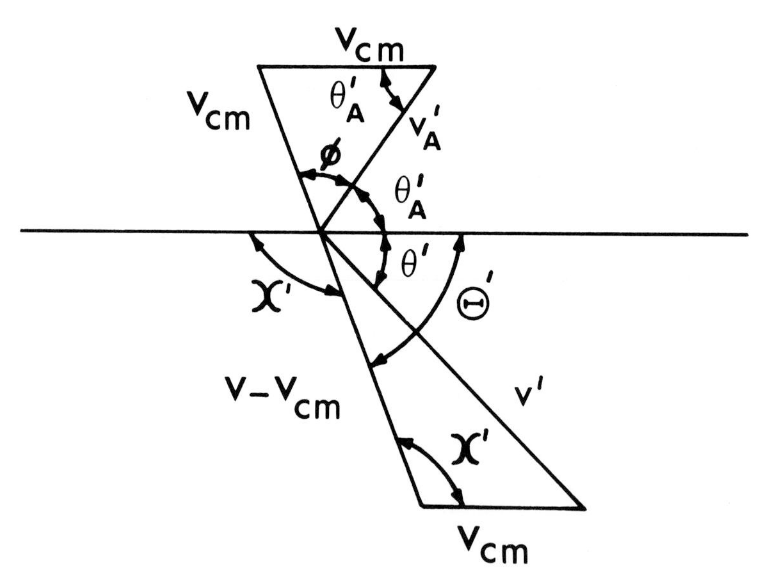

Figure 8.3 The vector triangles for the velocities in the neutron-nucleus collision are shown. Primes indicate values after the collision. The subscript A identifies the nuclear parameters; the subscript cm identifies center of mass values.

$$v' = \frac{v \cos \theta'}{A+1} \pm \frac{v}{A+1} [\cos^2 \theta' + (A-1)(A+1)]^{1/2},$$

$$v' = \frac{v}{A+1} [\cos \theta' + (A^2 - \sin^2 \theta')^{1/2}] \tag{8.11}$$

The negative sign before the square root in the first of Equations 8.11 is not used because the resulting value is not physically acceptable at $\theta' = 0$. Zero degrees is the no-scattering limit with $v' = v$.

The neutron kinetic energy in the laboratory system after the collision is $T' = M_n(v')^2/2$. That value can be related to the incident laboratory neutron kinetic energy, $T = M_n v^2/2$, using Equations 8.11 (1, 11):

$$T' = T \left[\frac{\cos \theta' + (A^2 - \sin^2 \theta')^{1/2}}{(A+1)} \right]^2 \tag{8.12}$$

The expression for the nuclear recoil kinetic energy in the laboratory system after the collision can be written as

$$T'_A = T - T' = T\left\{1 - \left[\frac{\cos \theta' + (A^2 - \sin^2 \theta')^{1/2}}{(A+1)} \right]^2 \right\} \tag{8.13}$$

Equations 8.12 and 8.13 are analogous to Equations 6.13 and 6.14 obtained for Compton scattering of photons.

Knowledge of any two of the elastic neutron collision variables, T, T', T'_A, θ', and θ'_A, is sufficient in principle to calculate any other variable. Energy and momentum conservation expressions assure it. A relationship giving θ'_A in terms of other variables is needed to complete the description.

It is evident that the uppermost triangle on Figure 8.3 has two equal length sides. Then

$$\chi' = \theta'_A + \phi = 2\theta'_A \tag{8.14}$$

By dividing the second of Equations 8.8 by the first, and substituting from Equation 8.14, one can obtain

$$\cot \theta' = \frac{[v_{cm}/(v - v_{cm})] - \cos \chi'}{\sin \chi'} = \frac{[v_{cm}/(v - v_{cm})] - \cos 2\theta'_A}{\sin 2\theta'_A}$$

$$\cot \theta' = \frac{1 - A \cos 2\theta'_A}{A \sin 2\theta'_A}$$

$$\theta' = \cot^{-1} \left[\frac{1 - A \cos 2\theta'_A}{A \sin 2\theta'_A} \right] \tag{8.15}$$

The mass number A enters because of substitution from Equations 8.3 for v_{cm}.

Apparently all values of θ' are possible from the limit for no scatter, $\theta' = 0$, to complete backscatter, with $\theta' = \pi$. However, as θ' approaches zero, θ'_A approaches the limit of $\pi/2$. Thus, the ranges on the angles are $0 < \theta' \leq \pi$,

$\pi/2 > \theta'_A \geq 0$. Consider two cases to obtain the range of values of the final energy:

1. For the limit with no scattering:

$$\theta' \to 0, \quad \theta'_A \to \pi/2, \quad T' \to T, T'_A \to 0$$

2. For the 180° backscatter limit:

$$\theta' \to \pi, \quad \theta'_A \to 0, \quad T' \to \left[\frac{A-1}{A+1}\right]^2 T = \alpha T, \quad T'_A \to (1-\alpha)T$$

Notice that the ranges for the kinetic energy variables are $T \geq T' \geq \alpha T$, $0 \leq T'_A \leq (1-\alpha)T$. We have used the collisional mass parameter defined by $\alpha = [(A-1)/(A+1)]^2$ above.

The important quantity in determining the energy limits for the scattered neutron and the nucleus is the mass number A of the nucleus. When A is small, a large fraction of the incident energy can be transferred in a single collision. This facet of the energy transfer characteristics of elastic scattering is discussed in more detail in Section 8.4 on neutron moderation.

8.3 Differential Neutron Cross Sections

Algebraic expressions for neutron scattering cross sections have been presented in many texts (1, 12–14). Derivation of the fundamental differential cross-section expression is a lengthy process. The calculations are usually made in the center of mass system with the incident neutrons represented by a plane wave and the scattered neutrons represented by a radial wave diverging from the nuclear scattering center. The incoming and outgoing waves are coherent, so interference effects occur. In the usual derivation, the incident neutron wave and the scattered wave have equal wavelength and momentum in the center of mass system. This guarantees that the reaction described is elastic scattering. It does not imply that neutron kinetic energy is unchanged in the laboratory frame after the collision. In fact, the target nucleus must absorb momentum in the laboratory system, so degradation of the incident neutron kinetic energy is assured.

Elastic scattering of neutron waves from a nucleus involves two different processes: potential (shape) scattering and resonance (compound) scattering (4, 5). Potential scattering is caused by the discontinuity in the neutron potential energy curve due to the nuclear force and can be considered a reflection at the nuclear surface interface. Resonance scattering occurs because part of the incident wave passing through the nuclear surface emerges as a radial wave. Resonance scattering occurs with large amplitude only for specific neutron wavelengths that produce transient nuclear states of well defined energy.

Total cross sections for elastic neutron scattering will be discussed in Section 8.5. The cross section (15) of Figure 8.4 for ^{27}Al is a good illustration since in the intermediate- and slow-neutron region for this nuclide, elastic scattering is dominant (7, 16). Potential elastic scattering produces the smooth compo-

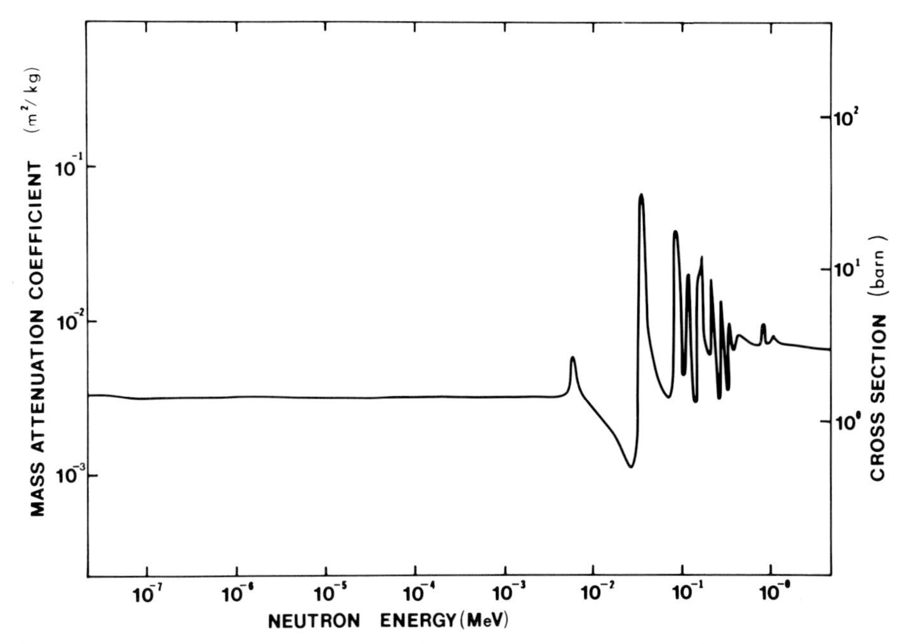

Figure 8.4 The total neutron cross section or mass attenuation coefficient for nucleus ^{27}Al has a flat low-energy region but shows resonances at higher energies (15). One barn $= 10^{-28}$ m^2.

nent of the curve at lower energies. The peaks occurring between 10 keV and 1 MeV are characteristic of resonance elastic scattering. At energies well above 1 MeV, absorption reactions occur with substantial probability for ^{27}Al. Neutron absorption reactions for targets other than ^{27}Al may dominate even at slow and intermediate energies (15).

It will be instructive to consider some differential cross-section results. First consider the effects of the potential of the nuclear force on the neutron wave. Suppose that the incident neutron wave, with center of mass wavelength Λ, can be represented by a sum of components of the general form $\sin(2\pi r/\Lambda - \phi_0)$. In this form r is the radial distance from the center of the spherical nucleus, and ϕ_0 is a phase factor. At distances well removed from the nuclear surface, the emerging neutron wave can be changed by the nuclear target in two ways: 1) the wave amplitude can be reduced by nuclear absorption; and 2) the phase of the wave can be shifted when compared to the phase of the incident wave.

Both effects of the nuclear center can be included in a complex wave perturbation factor η with amplitude factor N, which is real. Then

$$\eta = N \exp(2i\delta) = N(\cos 2\delta + i \sin 2\delta) \tag{8.16}$$

where $i = \sqrt{-1}$. The phase shift δ governs elastic scattering of the neutron waves. The amplitude factor N is especially important for neutron absorption in the nucleus. N is bounded such that $0 \leq N \leq 1$. When $N = 0$, absorption is at a maximum; when $N = 1$, only scattering occurs (6).

The effect of the phase shift on the scattering is more subtle than the amplitude effect. The shift occurs because the neutron wavelength inside the nucleus is altered from the free space wavelength (3). From Equation 8.1, if we use the total energy E as the constant of the motion,

$$r > R_A, \frac{\Lambda}{2\pi} = \frac{\hbar}{P} = \frac{\hbar}{\sqrt{2M_n E}}, \quad E = T_{\text{out}}, \quad V = 0;$$

$$0 \leq r \leq R_A, \frac{\Lambda}{2\pi} = \frac{\hbar}{P} = \frac{\hbar}{\sqrt{2M_n(E - V)}}, \quad E - V = T_{\text{in}}, \quad V < 0 \quad (8.17)$$

In these equations, R_A is the nuclear radius, T_{out} is the incident neutron kinetic energy in the center of mass system but outside of the nuclear force field, and T_{in} is the kinetic energy inside the region of the nuclear potential, V. Since the neutron wave must join smoothly through the nuclear surface, a phase shift δ in free space can occur as illustrated on Figure 8.5. In this case the total wave at large distances from the nucleus is given by sin $[(2\pi r/\Lambda) - \phi_0 + \delta)]$, instead of sin $[(2\pi r/\Lambda) - \phi_0]$ for the incident wave.

It is usual to write the total neutron wave as a sum of partial waves with index l, the orbital angular momentum quantum number. The allowed values of l are 0, 1, 2, 3, . . . corresponding to *s, p, d, f* . . . neutron waves, respectively. The differential cross section as a function of angle Θ' of the outgo-

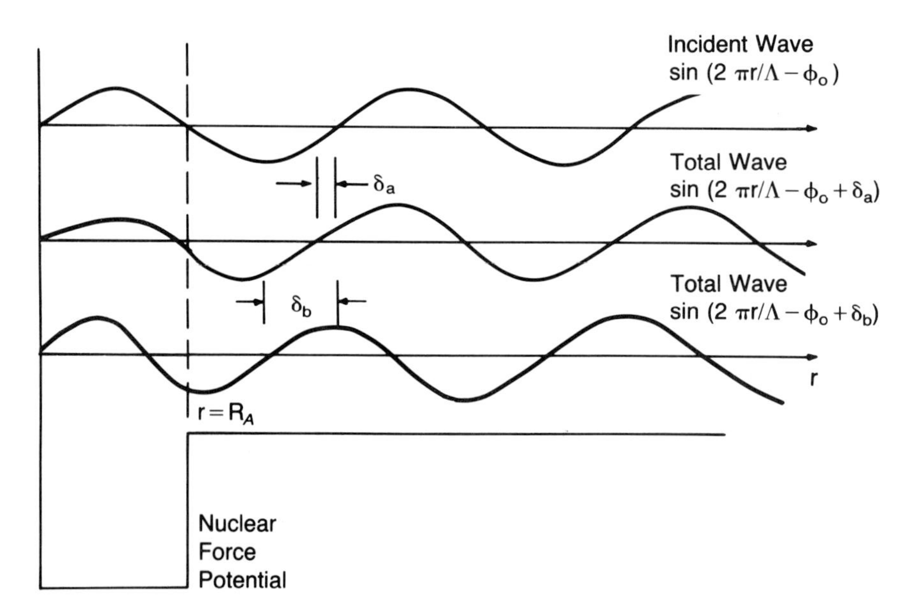

Figure 8.5 An incident neutron wave as well as several possible total waves are shown to illustrate the phase shift (13) for the case of small wave amplitude in the nucleus (δ_a) and the case of large wave amplitude in the nucleus (δ_b).

ing neutron is then written as a sum of cross-section values for each partial neutron wave indicated by l. These cross sections involve a set of wave perturbation factors η_l similar to the value given in Equation 8.16. Then (6)

$$\frac{d\sigma(\Lambda,\Theta')}{d\Omega'_{\text{cm}}} = \left[\frac{\Lambda}{2\pi}\right]^2 \left| \sum_{l=0}^{\infty} (1-\eta_l)\left[\frac{2l+1}{2}\right] P_l(\cos\Theta')\right|^2 \tag{8.18}$$

The symbol $d\sigma(\Lambda,\Theta')/d\Omega'_{\text{cm}}$ stands for the differential cross section per unit solid angle for neutrons leaving at angles between Θ' and $\Theta' + d\Theta'$. The functions $P_l(\cos\Theta')$ are the Legendre polynomials for center of mass system angles Θ' (17).

The sum in Equation 8.18 is formidable unless the number of terms can be limited with $\eta_l = 1$ (no absorption, no scatter) beyond some maximum l. Actual limitations depend on the neutron wavelength and the nuclear dimensions. Fortunately, important cases exist where only the $l = 0$ term (s wave) need be considered.

For our purposes the interaction probability is determined by the impact parameter b, which represents the perpendicular distance of closest approach between the particle center and the nuclear center. Since the angular momentum P of the incoming neutron about the nuclear center is quantized.

$$Pb \simeq \hbar\sqrt{l(l+1)}, \qquad l(l+1) \simeq \frac{P^2b^2}{\hbar^2} = \left[\frac{2\pi b}{\Lambda}\right]^2 \tag{8.19}$$

If the nucleus is spherical and the range of the nuclear force is small compared to the nuclear radius, the maximum impact parameter allowing efficient interaction is the sum of the neutron radius and the nuclear radius, $R = R_n + R_A$. Thus, reactions occur with high probability when $l(l+1) \leq [2\pi R/\Lambda]^2$. For larger l values the neutrons represented pass beyond the range of the nuclear force. The factor η must approach one for these cases since interactions are improbable. Although this picture is oversimplified, it is useful conceptually. The information on Table 8.2 can be used to demonstrate that for neutron beams with $T < 1$ MeV, $2\pi R/\Lambda < 1$. Thus, the s-wave ($l = 0$) term is likely to be dominant for intermediate and slow neutrons.

The center of mass system angular distributions resulting from Equation 8.18 can be represented as a power series in $\cos\Theta$, where the highest power

Table 8.2 Neutron Wavelength Versus Nuclear Radius

		$2\pi R_A/\Lambda$	
T_{cm} (eV)	$\Lambda/2\pi$ (m)	^{16}O	^{208}Pb
10^0	$(4.5)10^{-12}$	$(6.2)10^{-4}$	$(1.4)10^{-3}$
10^2	$(4.5)10^{-13}$	$(6.2)10^{-3}$	$(1.4)10^{-2}$
10^4	$(4.5)10^{-14}$	$(6.2)10^{-2}$	$(1.4)10^{-1}$
10^6	$(4.5)10^{-15}$	$(6.2)10^{-1}$	$(1.4)10^0$

$R_A = (1.1)10^{-15}A^{1/3}$ m (3).

that must be included is equal to twice the maximum value of l included in the partial wave formulation (18). This means that for s-wave scattering with $l = 0$, the angular distribution of emerging neutrons is symmetrical about the point of the collision in the center of mass system. If other values of l are included, the distribution is shifted toward the forward direction. Figure 8.6 shows two high-energy distributions in the center of mass system that show maxima and minima (19). For the energies shown, partial waves with $l > 0$ must be included. The qualitative pattern of these distributions is similar to that for light diffraction around the edge of an opaque object.

The distributions shown in Figure 8.6 occur when the nuclei act independently as scattering centers. When the target nuclei are held in a periodic array in a crystal, coherent scattering from the crystal planes can occur. The neutron wavelength must be of the order of magnitude of the crystal plane spacing (12). Thus, for common crystals coherent effects occur only for slow neutrons. Angular distributions for coherent neutron scattering from crystals have maxima and minima at angles governed by the lattice parameters and the wavelength, similar to Bragg x-ray scattering (6).

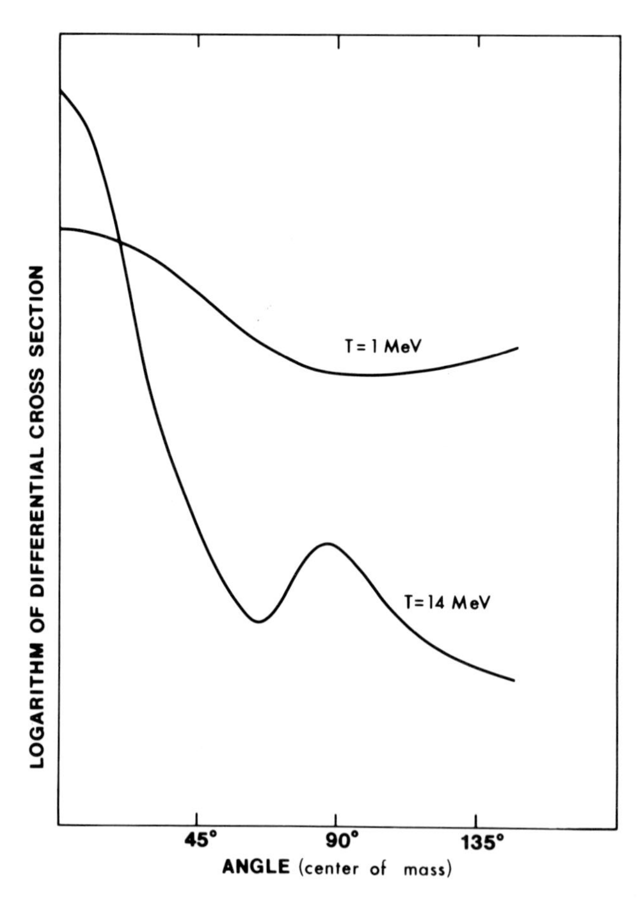

Figure 8.6 Angular distributions for elastic scattering of 1-MeV and 14-MeV neutrons are shown as a function of the center of mass scattering angle of the neutron for a ^{59}Co target (19).

In the following paragraphs, differential cross sections for low-energy elastic neutron scattering will be discussed, and laboratory system angular distributions will be obtained. In the low-energy limit, the neutron wavelength is considerably larger than the nuclear radius $R_A/\Lambda \ll 1$, and only s-wave scattering will be considered. Furthermore, at sufficiently low energies, the neutron wave does not penetrate into the nuclear surface but is reflected at the potential interface. Because of this, absorption is negligible and $N = 1$. When the neutron wave is excluded from the interior of the nucleus, the change in path of the wave must be R_A, the nuclear radius. The phase shift must be $\delta = 2\pi R_A/\Lambda \ll 1$. Under these conditions Equation 8.18 reduces to the s-wave result

$$\frac{d\sigma_0(\Lambda)}{d\Omega'_{\text{cm}}} = \left[\frac{\Lambda}{2\pi}\right]^2 \frac{|1-\eta_0|^2}{4}$$

$$= \left[\frac{\Lambda}{4\pi}\right]^2 |1 - \exp(4\pi i R_A/\Lambda)|^2 = R_A{}^2 = \frac{\sigma_0(A)}{4\pi} \tag{8.20}$$

The first part of Equation 8.20 follows directly from Equation 8.18 because $P_0(\cos\Theta) = 1$. One must expand the exponential in Equation 8.20 and keep the first two terms to obtain the final result. The symbol σ_0 for the s-wave ($l = 0$) total cross section can be used since the total solid angle is 4π sr. The dependence of σ_0 on mass number A is indicated because the radius R_A increases with A. Notice that σ_0 does not depend on angle or energy. Elastic s-wave scattering is isotropic in the center of mass system and the cross section versus energy is flat.

Equation 8.20 is a center of mass system relationship, but for our purposes expressions with the laboratory system variables are preferable. The second equation in Equations 8.8 can be differentiated to convert the s-wave cross section per unit solid angle to the laboratory frame.

$$\sin\chi'\,d\chi' = \frac{-v'\sin\theta'\,d\theta' + \cos\theta'\,dv'}{v - v_{\text{cm}}}, \tag{8.21}$$

$$\sin\chi'\,d\chi' = -\sin(\pi - \Theta')d\Theta' = -\sin\Theta'\,d\Theta' = \frac{-v'\sin\theta'\,d\theta' + \cos\theta'\,dv'}{v[A/(A+1)]}$$

The second of Equations 8.3, including the center of mass speed, was used above. In addition, it was necessary to utilize the fact that χ' and Θ' are supplementary angles. From Equations 8.11:

$$v' = \frac{v}{A+1}[\cos\theta' + (A^2 - \sin^2\theta')^{1/2}],$$

$$dv' = -\frac{v}{A+1}\sin\theta'\,d\theta'\left[1 + \frac{\cos\theta'}{(A^2-\sin^2\theta')^{1/2}}\right]$$

$$= -\frac{v}{A+1}\sin\theta'\,d\theta'\left[\frac{\cos\theta' + (A^2-\sin^2\theta')^{1/2}}{(A^2-\sin^2\theta')^{1/2}}\right] \tag{8.22}$$

Substituting the last of Equations 8.22 into the last part of Equations 8.21 yields

$$\sin\Theta'\,d\Theta' = \frac{\sin\theta'\,d\theta'}{A}\,[\cos\theta' + (A^2 - \sin^2\theta')^{1/2}]\left\{1 + \frac{\cos\theta'}{(A^2 - \sin^2\theta')^{1/2}}\right\}$$

$$= \frac{\sin\theta'\,d\theta'}{A}\,\frac{[\cos\theta' + (A^2 - \sin^2\theta')^{1/2}]^2}{(A^2 - \sin^2\theta')^{1/2}} \tag{8.23}$$

The solid angle Ω'_{cm} and wavelength Λ in the center of mass system determine the solid angle Ω' and energy T respectively in the laboratory frame (6, 11):

$$\frac{d\sigma_0(T,\theta')}{d\Omega'}\,d\Omega' = \frac{d\sigma(\Lambda,\Theta')}{d\Omega'_{\mathrm{cm}}}\,d\Omega'_{\mathrm{cm}},$$

$$\frac{d\sigma_0(\theta')}{d\Omega'} = \frac{\sigma_0(A)[\cos\theta' - (A^2 - \sin^2\theta')^{1/2}]^2}{4\pi A(A^2 - \sin^2\theta')^{1/2}} \tag{8.24}$$

This is one of the laboratory system results needed. Furthermore, the solid angle $d\Omega'$ involves angles between θ' and $\theta' + d\theta'$ such that $2\pi\sin\theta'\,d\theta' = d\Omega'$ in the laboratory system:

$$\frac{d\sigma_0(\theta)}{d\theta'}\,d\theta' = \frac{d\sigma(\theta')}{d\Omega'}\,d\Omega' = 2\pi\,\frac{d\sigma(\theta')}{d\Omega'}\,\sin\theta'\,d\theta',$$

$$\frac{d\sigma_0(\theta')}{d\theta'} = \frac{\sigma_0(A)[\cos\theta' + (A^2 - \sin^2\theta')^{1/2}]^2\sin\theta'}{2A(A^2 - \sin^2\theta')^{1/2}} \tag{8.25}$$

Note that Equations 8.24 and 8.25 are sufficient for the range $0 < \theta' \leq 180°$ for s-wave scattering. The angle θ' represents the half-angle of a cone with apex at the scattering center. The factor of $2\pi\sin\theta'$, which occurs in Equations 8.25 and not in Equations 8.24, means that the distribution per unit

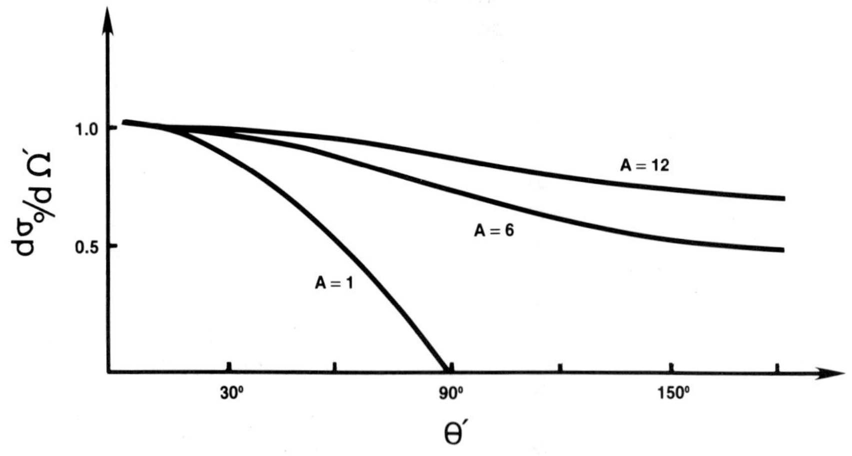

Figure 8.7 The angular distribution for s-wave scattering in the laboratory system is shown for scattering targets with several mass number values. The results are normalized to 1.0 at the maximum for comparison.

angle always goes to zero at 0° and 180°. The same effect occurs for Klein-Nishina scattering of photons (20), as described in Chapter 6. The s-wave laboratory system angular distribution given by Equations 8.24 is shown on Figure 8.7 for several values of A. The s-wave differential cross section per unit angle given by Equation 8.25 is shown on the *bottom half* of Figure 8.8 for several different target mass numbers. Forward angles are favored in the laboratory system, especially for light targets.

The differential cross section per unit angle for scattering with the nuclear recoil at an angle between θ'_A and $\theta'_A + d\theta'_A$ can be obtained with the relationships

$$\frac{d\sigma_0(\theta'_A)}{d\theta'_A}\,d\theta'_A = \frac{d\sigma_0(\theta')}{d\theta'}\,d\theta', \quad \theta' = \cot^{-1}\left[\frac{1 - A\cos 2\theta'_A}{A\sin 2\theta'_A}\right] \tag{8.26}$$

The second equation in Equations 8.26 was derived in Section 8.2. Radial plots obtained from Equations 8.26 for s-wave scattering are shown at the *top* of Figure 8.8 for several values of A.

The s-wave differential cross section given in terms of T' and T will be utilized in the discussion of neutron moderation in the next section. Since Equa-

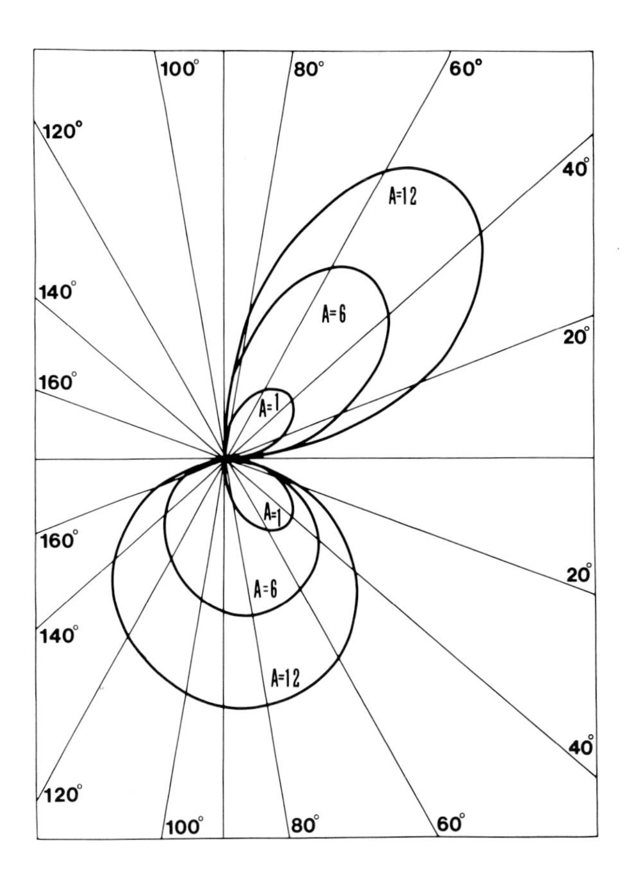

Figure 8.8 The relative differential cross section per unit angle for s-wave scattering in the laboratory system is shown for scattering targets with several mass number values. The lower half of the figure is shown for the outgoing neutrons, the upper half for outgoing nuclear targets.

tions 8.25 must be expressed in terms of energy T, it is useful to differentiate Equation 8.12:

$$dT' = -\frac{2T\sin\theta'\,d\theta'}{(A+1)^2}\left[\cos\theta' + (A^2-\sin^2\theta')^{1/2}\right]\left[1+\frac{\cos\theta'}{(A^2-\sin^2\theta')^{1/2}}\right]$$

$$\sin\theta'\,d\theta' = -\frac{(A+1)^2}{2T}\,\frac{(A^2-\sin^2\theta')^{1/2}dT'}{[\cos\theta'+(A^2-\sin^2\theta')^{1/2}]^2} \qquad (8.27)$$

The second of Equation 8.27 can be substituted into the last of Equation 8.25 to give

$$\frac{d\sigma(T',T)}{dT'}dT' = -\frac{d\sigma(\theta')}{d\theta'}d\theta' = \frac{\sigma_0(A)(A+1)^2}{4AT}dT' = \frac{\sigma_0(A)dT'}{(1-\alpha)T} \qquad (8.28)$$

The minus sign in Equation 8.28 occurs because T' decreases as θ' increases. The collision mass parameter α simplifies the expression. Equation 8.28 shows that the scattering probability is independent of T' and depends only on T and A. The illustration at the *top* of Figure 8.9 shows the differential cross section given in Equation 8.28 plotted against T'. Notice that it is flat in the allowed range where T' is between αT and T.

The cross section $d\sigma(T'_A,T)/dT'_A$ is similar to that given by Equation

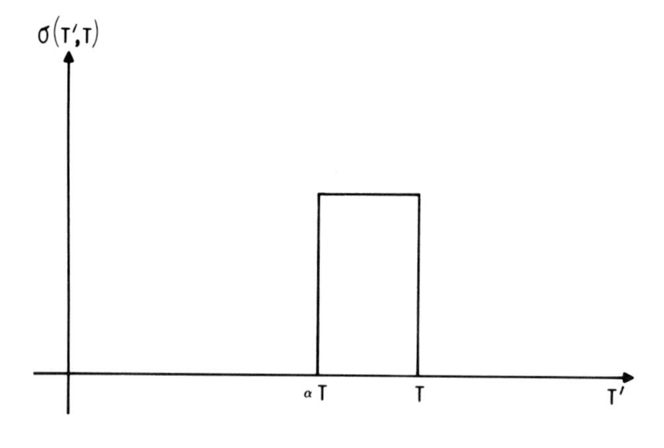

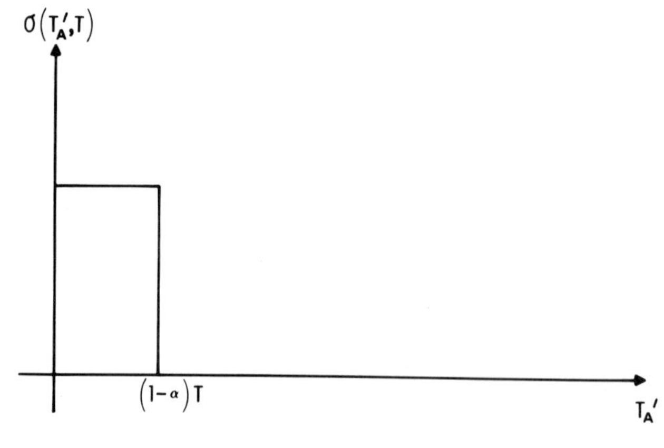

Figure 8.9 Differential cross sections per unit energy for s-wave scattering in the laboratory system are shown as a function of energy T' of the scattered neutron (*top*) and energy T'_A of the nucleus (*bottom*).

8.28 except that dT'_A should replace dT'. The allowed values in this case occur when T'_A is between 0 and $(1 - \alpha)T$. This quantity is shown at the bottom of Figure 8.9 when plotted against T'_A.

The energy distributions illustrated in Figure 8.9 are very simple. The *upper graph* is proportional to the scattered neutron spectrum; it should remind us that elastic scattering is very inefficient for energy degradation for heavy nuclei, with $A \gg 1$, $\alpha \to 1$. In this case, scattered neutrons have almost the same energy as the incident neutrons. The *lower graph* can be related to the pulse-height distribution obtained from elastic scattering in neutron scintillation detectors (21). The distribution is reminiscent of the Compton scatter plateau that occurs for photon beams except that there is no analog of the Compton peak.

8.4 Neutron Moderation

As a neutron beam is moderated, the kinetic energy of the individual neutrons is degraded, perhaps to thermal energies. In light nuclear materials, the process occurs predominantly through the agency of elastic scattering as long as the kinetic energy is less than 1 MeV. In medium to heavy nuclei, inelastic scattering often is significant in the degradation process. Nevertheless, elastic scattering usually dominates below 100 keV (1). A discussion of neutron moderation is included in this chapter because it is so closely associated with elastic scattering.

Neutron energy degradation by elastic scattering can be described with simple expressions because the average fractional energy loss in a single collision is independent of the incident energy. Equations 8.12 and 8.13, when divided by T, give the fractional energy retained and the fractional energy lost, respectively. These fractions are dependent on the angle of scatter and the mass number of the scattering nucleus, but are independent of incident kinetic energy. The fractional loss is not independent of incident energy when photons are degraded by scattering. The energy loss fraction for Compton scattering is a function of the incident photon energy through the factor $\alpha = h\nu/m_ec^2$.

The algebraic expressions obtained in this section will generally be valid in the region below 100 keV and well above thermal energies. Since the *s*-wave total cross sections will be used, the effects of higher angular momentum terms as well as inelastic scattering are not included. Furthermore, near thermal energies, chemical bond effects and nuclear motion may be important. The developments utilized previously have ignored these effects, and they will not be included in the discussions to follow.

Equation 8.28 can be utilized to calculate the average energy retained after an *s*-wave elastic collision between a neutron and a nuclear target. Since factors independent of T can be canceled,

$$\langle T' \rangle = \frac{\int_{\alpha T}^{T} T' \frac{d\sigma(T',T)}{dT'} dT'}{\int_{\alpha T}^{T} \frac{d\sigma(T,'T)}{dT'} dT'} = \frac{\int_{\alpha T}^{T} T'\, dT'}{\int_{\alpha T}^{T} dT'} = \frac{(1+\alpha)}{2} T \tag{8.29}$$

The average fraction $\langle T' \rangle / T$ depends only on $\alpha = [(A-1)/(A+1)]^2$. Consider neutron collisions with nuclei with $A = 1$, $A = 12$, $A = 208$, for elements hydrogen, carbon, and lead, respectively, to illustrate the dependence on A. The average fraction of energy retained after each collision is 0.50, 0.92, and 0.995, respectively. Light elemental targets are more efficient per collision for neutron energy degradation than those made from heavy elements.

Consider a series of moderating collisions $1 \rightarrow k$ for a single neutron. Suppose the initial neutron kinetic energy is T_0 and the final value is T_k. Intermediate kinetic energy values are T_1, T_2, T_3, and so forth, for k total collisions. One can write

$$T_0/T_k = (T_0/T_1)(T_1/T_2)(T_2/T_3) \ldots (T_{k-1}/T_k) \tag{8.30}$$

Average kinetic energy values are desired since one usually deals with a beam consisting of many neutrons. It is convenient to take the logarithm of each side of Equation 8.30 so that fractions can be separated before averaging. In this form one can make use of the rule that the average of a sum is equal to the sum of the averages. If $\langle \ \rangle$ indicates averages,

$$\langle \ln(T_0/T_k) \rangle = \langle \ln(T_0/T_1) \rangle + \langle \ln(T_1/T_2) \rangle + \ldots \langle \ln(T_{k-1}/T_k) \rangle,$$

$$\langle \ln T_0 \rangle - \langle \ln T_k \rangle = k \langle \ln(T/T') \rangle = k\xi \tag{8.31}$$

The last equation in Equations 8.31 represents the average result of k individual elastic scattering collisions. If the average value of the reciprocal degradation fraction ξ is known, the average value of the final kinetic energy can be estimated from the number of collisions and the initial energy.

The reciprocal degradation fraction for a beam of monoenergetic neutrons can be evaluated using Equation 8.28 when s-wave scattering is dominant:

$$\xi = \langle \ln(T/T') \rangle = \frac{\displaystyle\int_{\alpha T}^{T} \ln(T/T') \frac{d\sigma(T,T')}{dT'} dT'}{\displaystyle\int_{\alpha T}^{T} \frac{d\sigma(T,T')}{dT'} dT} = \frac{\displaystyle\int_{\alpha T}^{T} \ln(T/T') dT'}{(1-\alpha)T}$$

$$= 1 + \frac{\alpha \ln \alpha}{1-\alpha} \tag{8.32}$$

Evaluation of the integral in Equation 8.32 is straightforward (10). The result given by Equation 8.32 will hold whether or not the beam of neutrons is monoenergetic, since the reciprocal degradation fraction is the same for each component neutron energy.

Table 8.3 shows values of the average reciprocal degradation fraction for various targets. Notice that hydrogen represents a special case because Equation 8.32 is not defined when $\alpha = [(A-1)/(A+1)]^2 = 0$. The limit as α approaches zero is valid in this case. Table 8.3 also has a column showing the average final energies after 1-MeV neutrons have participated in 10 collisions. It is apparent that low A materials moderate neutrons with fewer collisions than high A materials.

Table 8.3 Properties of Moderating Nuclides

Nucleus	*Average Reciprocal Degradation Fraction*	*Average Final Energy*[a]
^{1}H	1.000	45 eV
^{2}D	0.726	703 eV
^{7}Li	0.260	74 keV
^{12}C	0.158	206 keV
^{16}O	0.120	301 keV
^{56}Fe	0.035	705 keV
^{208}Pb	0.010	905 keV

[a] $T = 1$ MeV, 10 collisions.

A quantity called the lethargy L (1, 7, 12) is defined by

$$L = \ln(T_0/T), \quad T_0 = 10 \text{ MeV} \tag{8.33}$$

It is often used as the variable for the abscissa of neutron spectrum plots. One advantage of this practice is that a wide range of energies can be covered in a single figure; another is that for equilibrium degradation spectra, equal lethargy intervals contain equal numbers of neutrons.

The form of the elastic scattering s-wave degradation spectrum (1, 22, 23) can be obtained for the case of a monoenergetic source emitting S neutrons per unit volume. The source is uniformly imbedded in moderating material with attenuation coefficient μ_{es} for elastic neutron scattering. Equation 8.32 shows that the average increase in lethargy for source neutrons per collision is a constant ξ for a given moderator. The number of elastic collisions in unit volume per unit lethargy change is also a constant, since it is related to the inverse of the lethargy increase. In the steady state, neutrons emitted from the source must pass through successive lethargy steps with lethargy gain of ξ per step. Therefore $n_L(L)$, the average number of collisions per unit volume per unit lethargy, is given by

$$n_L(L) = \frac{S}{\xi} \tag{8.34}$$

The average number of collisions per unit volume involving neutrons with initial kinetic energy between T and $T + dT$ can be called $n_T(T)dT$. This quantity must be equal to the product of the differential neutron fluence $\phi(T)dT$ and the attenuation coefficient μ_{es}:

$$n_T(T)dT = \phi(T)\,\mu_{es}dT, \quad \phi(T)\,\mathrm{d}T = \frac{n_T(T)dT}{\mu_{es}} \tag{8.35}$$

Distributions in lethargy and kinetic energy are related by Equations 8.33:

$$n_T(T)dT = -n_L(L)dL, \quad n_T(T) = -n_L(L)\frac{dL}{dT} = \frac{n_L(L)}{T} \tag{8.36}$$

Combining Equations 8.34 through 8.36 gives

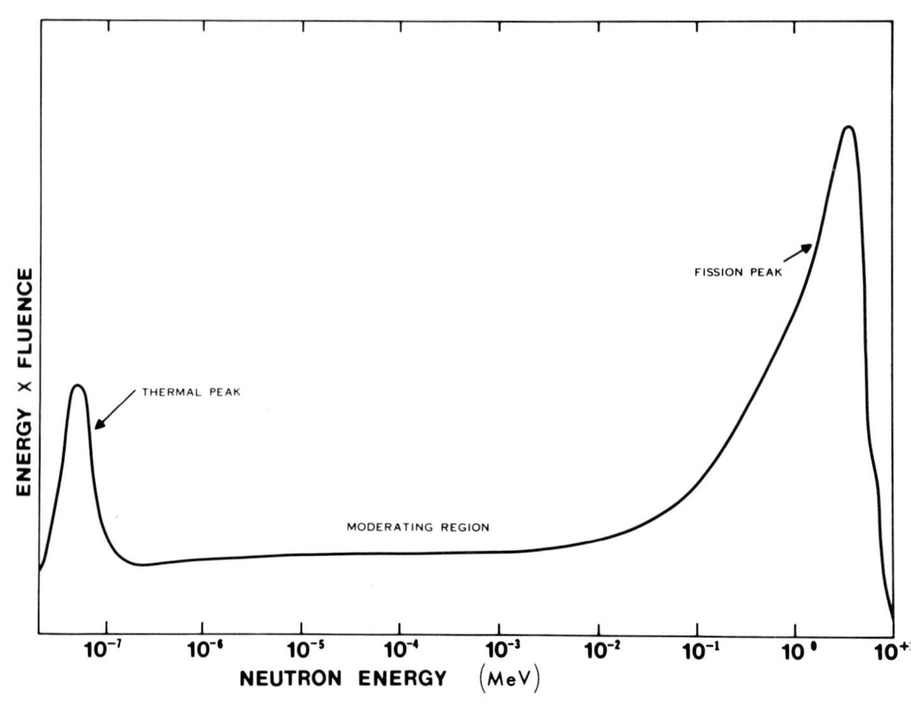

Figure 8.10 The product of the kinetic energy and the fluence for a fission reactor is shown. The fission peak and the thermal peak are apparent. The moderating region is quite flat. Adapted from Ref. 22.

$$\phi(T)\,\mathrm{d}T = \frac{n_L(L)dT}{\mu_{\mathrm{es}}T} = \frac{SdT}{\xi\mu_{\mathrm{es}}T} \tag{8.37}$$

Since μ_{es} is independent of T for s-wave elastic scattering, Equation 8.37 shows that the neutron fluence is inversely related to the kinetic energy. Thus

$$\Delta\Phi_{T_1,T_2} = \int_{T_1}^{T_2} \phi(T)\,dT = \frac{S}{\xi\mu_{\mathrm{es}}} \int_{T_1}^{T_2} \frac{dT}{T}$$
$$= \frac{S}{\xi\mu_{\mathrm{es}}}(\ln T_2 - \ln T_1) = \frac{S}{\xi\mu_{\mathrm{es}}}(L_1 - L_2) \tag{8.38}$$

Equation 8.38 shows that equal lethargy intervals contain equal numbers of neutrons.

Figure 8.10 is a plot of $T\phi(T)dT$ versus the logarithm of the energy for reactor neutrons (22). Notice that the curve is quite flat at intermediate energies well away from the fission peak and the thermal peak.

8.5 Attenuation Coefficients for Elastic Scattering

In the first part of this section, the differential cross-section expression given in Section 8.3 will be integrated and the total elastic scattering cross section will be obtained in laboratory system variables. Mass attenuation coefficients will then be related to the total cross sections. These values can be utilized in problems involving neutron elastic scattering in much the same way as the mass attentuation coefficients for incoherent scattering of photons are utilized in photon beam problems.

The total cross section for neutron interactions in the center of mass frame is most easily obtained by integrating the differential cross section of Equation 8.18 over the solid angle for a sphere. Because the functions $[(2l + 1)/2]^{1/2} P_l(x)$ are orthonormal on the interval $-1 \leq x \leq 1$ (17), the integration is simplified:

$$\sigma(\Lambda) = \int \frac{d\sigma(\Lambda,\Theta')}{d\Omega'_{cm}}\, d\Omega'_{cm}$$

$$= \left[\frac{\Lambda}{2\pi}\right]^2 \int_0^{\pi} \Bigg| \sum_{l=0}^{\infty} (1 - \eta_l)$$

$$\times \left(\frac{2l+1}{2}\right) P_l(\cos\Theta') \Bigg|^2 2\pi \sin\Theta'\, d\Theta',$$

$$\sigma(\Lambda) = \frac{\Lambda^2}{4\pi} \sum_{l=0}^{\infty} |1 - \eta_l|^2 (2l+1) \int_0^{\pi} \left(\frac{2l+1}{2}\right) P_l^2(\cos\Theta') \sin\Theta'\, d\Theta'$$

$$= \frac{\Lambda^2}{4\pi} \sum_{l=0}^{\infty} |1 - \eta_l|^2 (2l+1) \tag{8.39}$$

Equations 8.3 and 8.6 can be used to relate the momentum P of the neutron in the center of mass system to the momentum p in the laboratory system by the expression

$$P = M_n(v - v_{cm}) = p[1 - 1/(A+1)] = p\frac{A}{A+1} \tag{8.40}$$

Then from Equation 8.1,

$$\Lambda = \frac{h}{P} = \frac{h}{p}\left[\frac{A+1}{A}\right] = \frac{h}{(2M_nT)^{1/2}}\left[\frac{A+1}{A}\right], \quad \frac{\Lambda^2}{4\pi} = \frac{h^2}{8\pi M_n T}\left[\frac{A+1}{A}\right]^2 \tag{8.41}$$

Center of mass cross-section expressions can be evaluated in the laboratory system with the aid of Equations 8.41.

The maximum value for the elastic scattering cross section occurs when $N_l = 1$ and $\delta_l = \pi/2$ ($\eta_l = -1$). The waveform for this case is illustrated on the bottom of Figure 8.5, where the neutron wave amplitude is the same inside the nucleus as it is outside. This is a case where resonance scattering dominates:

$$\sigma_{max}(\Lambda) = \frac{\Lambda^2}{\pi} \sum_{l=0}^{\infty} (2l+1), \quad \sigma_{max}(T) = \frac{h^2}{2\pi M_n T}\left[\frac{A+1}{A}\right]^2 \sum_{l=0}^{\infty} (2l+1) \tag{8.42}$$

The first equation is for the center of mass system; the second is for the laboratory system.

The value for σ_{max} given above is the resonance peak value indicated on Figure 8.11. The conditions that produce cross-section resonances are discussed in some detail in the next chapter. For resonance scattering of neutrons, the usual shape of the peak is somewhat distorted because of interference effects (13). At energies just below the peak energy, potential and resonance scattering interfere destructively. At energies just above the peak energy, potential and resonance scattering interfere constructively. The resonance shape with interference effects is distinctive enough to be useful in identifying elastic scattering peaks on a cross-section plot (see Figure 8.4).

In the low-energy scattering limit, the neutron wave is excluded from the interior of the nucleus. Then, with $l = 0$, $N_0 = 1$, and $\delta_0 = 2\pi R_A/\Lambda \ll 1$, Equation 8.39 yields the s-wave elastic scattering result:

$$\sigma_0(A) = \frac{\Lambda^2}{4\pi} \sum_{l=0}^{\infty} |1 - \eta_l|^2(2l + 1) = \frac{\Lambda^2}{4\pi} \left| 1 - \exp\left(\frac{4\pi i R_A}{\Lambda}\right) \right|^2 = 4\pi R_A{}^2 \qquad (8.43)$$

The result was obtained by expanding the exponential. This value for σ_0 was anticipated in Equation 8.20. Although it was obtained from center of mass system expressions, it is valid for the laboratory frame also. Nonrelativistic total cross-section values are independent of the reference frame (6). The ^{27}Al cross section shown on Figure 8.4 is an example of an s-wave result and is nearly independent of the incident neutron energy in the slow-neutron range. Many other light elements have flat cross sections in this range although there are some notable exceptions (15). For cold neutrons the cross sections are not

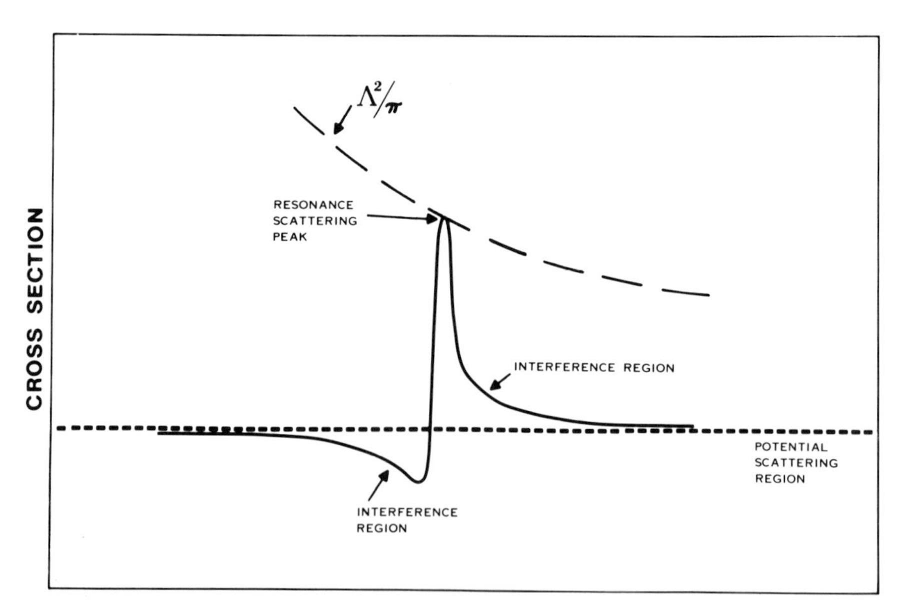

Figure 8.11 A typical elastic scattering resonance is shown. The characteristic interference effects are evident both before and after the peak energy.

constant when there are coherent interference effects in the sample crystal planes (12).

Figure 8.12 (24) shows that the values given by Equation 8.43 are approximate at best, since the measured cross sections are widely dispersed around the line given by $4\pi R_A{}^2$ for different Z values. Some of the differences between the data points and the line occur because Z is not a smooth function of A for the nuclear targets chosen. Furthermore, nuclei are generally not spherical, so the radius is not well defined (5, 6). A major reason for the differences is interference between potential and resonance scattering. The interference effects are very prominent if resonances exist in the energy region considered. The cumulative effects of clusters of distant resonances can also produce substantial deviations. (1).

The total cross section for neutrons incident on hydrogen from 0.1 eV to more than 1 MeV is shown on Figure 8.13. This figure is important because of the large amounts of hydrogen in tissue and the extensive use of hydrogenous material in neutron shielding material. The value of 20.7 barns (1 b = 10^{-28} m^2) at low energies is due almost entirely to elastic scattering, since the capture cross section is only 0.3 b at thermal energies. As shown on Figure 8.13, the cross section for a gaseous hydrogen target differs from the value for hydrogen in a water target at lowest energies. Furthermore, both cross sections increase

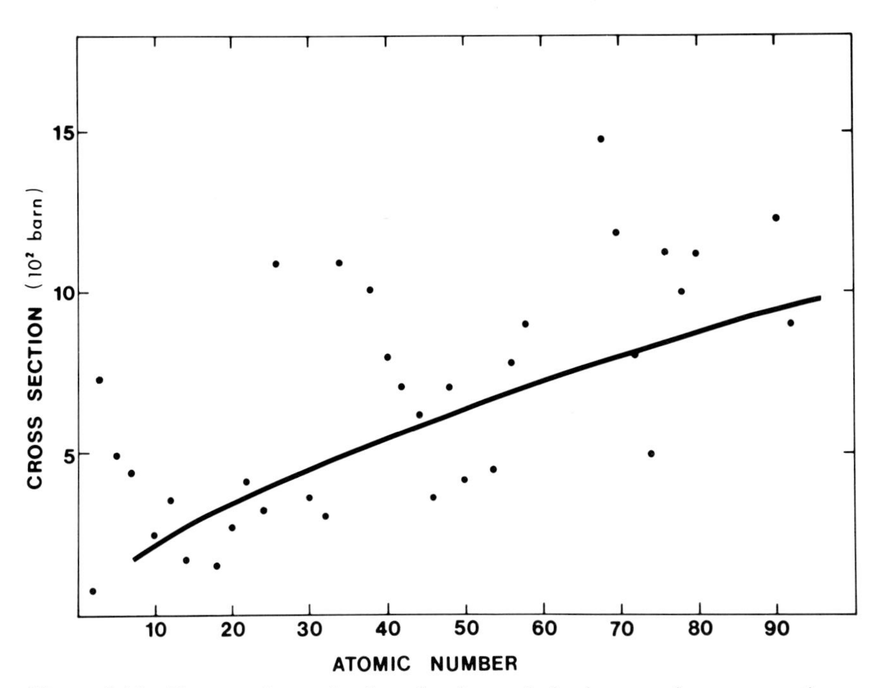

Figure 8.12 The experimental values for the total elastic scattering cross section are shown for slow neutrons incident on a variety of targets as a function of Z. The theoretical cross section values of $4\pi R_A{}^2$ are given by the solid line with $R_A = (1.4)10^{-15}A^{1/3}$ (1, 24). One barn = 10^{-28} m^2. Adapted from Ref. 1.

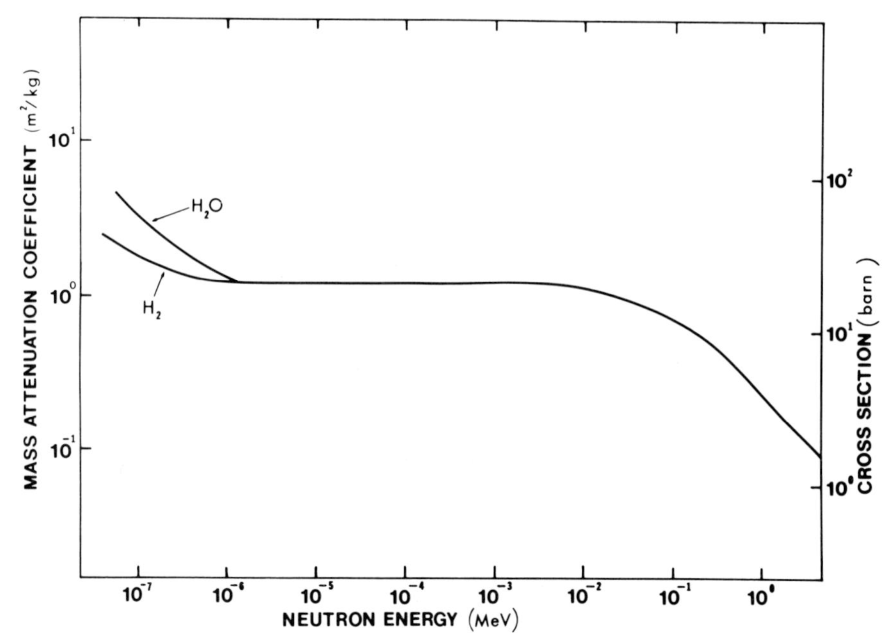

Figure 8.13 The cross section for neutrons incident on hydrogen is quite flat in the slow neutron region (8). It increases at lower energies because of the chemical bond effect (1, 12, 13).

as the neutron energy is decreased below 1 eV. These effects are attributable to chemical bonds: the neutrons can not transfer sufficient energy to sever bonds on the target atom or efficiently excite vibrational levels for the molecule. In this case the target proton does not act as a separate entity, and the entire molecule, with larger dimensions, becomes the scattering entity. these effects are also important for light nuclear targets other than hydrogen. Generally, when nuclear targets of mass number A are bound to a molecule (1, 12, 13), the cross section increases by a factor of $[(A + 1)/A]^2$ for neutrons with kinetic energy less than 10^{-2} eV.

The cross sections discussed above can be used to obtain the mass attenuation coefficient for elastic neutron scattering. As indicated in Chapter 5.

$$(\mu/\rho)_{es} = n_m \sigma_{es}. \tag{8.44}$$

where ρ is the density of the target material, and the number of target nuclei per unit mass is $n_m = N_A/M_m$, where N_A is Avogadro's number and M_m is the mass per mole. The subscript es denotes elastic scattering.

The mass energy transfer coefficient for elastic neutron scattering is given by

$$\left(\frac{\mu_{tr}}{\rho}\right)_{es} = \left(\frac{\mu}{\rho}\right)_{es}\left[\frac{\langle T'_A\rangle}{T}\right] = \left(\frac{\mu}{\rho}\right)_{es}\left[\frac{T-\langle T'\rangle}{T}\right] \tag{8.45}$$

It is assumed that the target nuclei are ionized because of the sudden momentum change (shake-off). This equation can be further evaluated in terms of mass

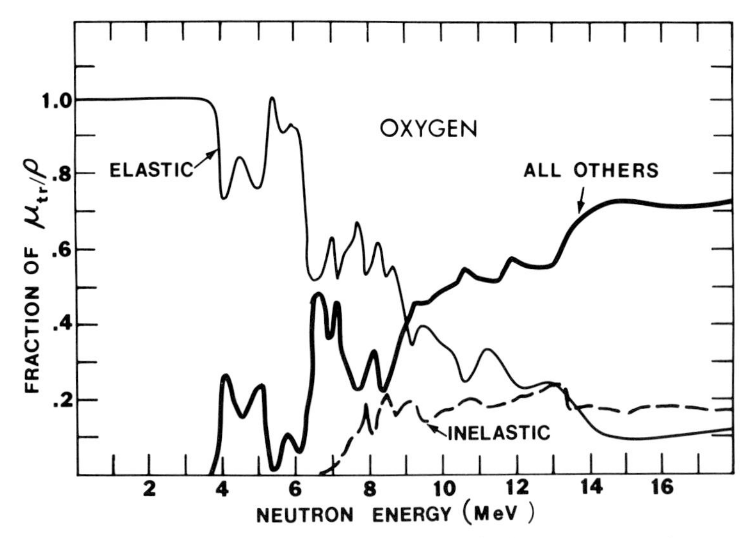

Figure 8.14 The fraction of the energy transfer from a neutron beam to an oxygen absorber is shown for various reactions as a function of neutron energy (25). Notice that the elastic scattering contribution dominates at lower energies. Adapted from Ref. 22.

number since the average value of T' can be obtained. For s-wave inelastic scattering, $\langle T' \rangle = [(1 + \alpha)/2]T$, so the quantity in brackets in Equation 8.45 is given by $[(1 - \alpha)/2]$ for all values of T. Appendix 10 contains values for the kerma per fluence for elastic scattering of neutrons for some materials.

Figure 8.14 shows that the mass energy transfer coefficient is determined primarily by the elastic scattering process for energies below 5 MeV for an oxygen target (25). However, at higher energies and /or for other targets, reactions such as (n,p) and (n,α) become important and cause considerable energy deposition. For nitrogen targets the (n,p) reaction is of great significance even at very low neutron energies. In all cases the mass energy absorption coefficient and the mass energy transfer coefficient are taken to be equal for neutron beams since the nuclei involved in the primary interactions radiate such little energy.

8.6 Problems

1. Calculate the most probable speed and kinetic energy and the average speed and kinetic energy for a Maxwellian distribution of neutrons with temperature T° = 293 K.
2. If neutrons in equilibrium with very cold surroundings have a most probable energy of 0.001 eV, what is the absolute temperature of the surroundings?
3. Calculate the wavelength of a 1-MeV neutron in free space. If a total energy of 1 MeV is a constant of the motion and the neutron is in a region of a potential well of depth 10 MeV, what is the wavelength?

4. What is the energy of 14-MeV neutrons after scattering from hydrogen atoms in paraffin when $\theta_A = 45°$?
5. Evaluate the fraction of the incident neutron energy remaining after a single maximum energy transfer collision. Consider hydrogen, carbon, and lead targets.
6. Substitute from Equations 8.3 and Equation 8.14 into Equation 8.7, and with the use of a trigonometric identity, show that $T'_A = T[4A \cos^2 \theta'_A / (A + 1)^2]$.
7. Calculate the mean free path for elastic scattering in water at 1 keV. Use the data on Figure 8.13 and assume that the elastic scattering cross section for oxygen is similar to that shown on Figure 8.4 for aluminum at this energy.
8. Carry out the integration in Equation 8.32 in detail to give the result indicated.
9. Use the result of the limit to evaluate Equation 8.32 for a hydrogenous scattering material.
10. Find the number of collisions (nearest integer) necessary to thermalize a beam of 1-MeV neutrons with deuterium (D), helium (He), beryllium (Be), carbon (C), and uranium (U) as elastic scattering materials.
11. Which elastic scattering material will give a smaller value of equilibrium neutron fluence at 1 keV, water or aluminum? Assume that a source with emission factor S per unit volume is present in each material. Use data from Figures 8.4 and 8.13.
12. Integrate the differential cross section per unit angle given in Equations 8.25 over angles between 0 and π rad and compare the result with Equation 8.43.
13. Integrate the differential cross section per unit energy given in Equation 8.28 over the allowed energy values and compare the result with Equation 8.43.
14. Express the energy fraction factor in Equation 8.45 as a function of mass number A. Calculate the appropriate conversion factor for a hydrogenous scatterer and an aluminum scatterer.

8.7 References

1. A. M. Weinberg and E. P. Wigner, *The Physical Theory of Neutron Chain Reactors* (University of Chicago, Chicago, 1958).
2. L. de Broglie, Philos. Mag. **47**, 446 (1924).
3. B. L. Cohen, *Concepts of Nuclear Physics* (McGraw-Hill, New York, 1971).
4. V. F. Weiskopf, Rev. Mod. Phys. **29**, 174 (1957).
5. W. E. Meyerhof, *Elements of Nuclear Physics* (McGraw-Hill, New York, 1967).
6. A. P. Arya, *Fundamentals of Nuclear Physics* (Allyn and Bacon, Boston, 1966).
7. I. Kaplan, *Nuclear Physics* (Addison-Wesley, Reading, MA, 1955).
8. D. J. Hughes, *Pile Neutron Research* (Addison-Wesley, Cambridge, MA, 1953).
9. H. Goldstein, *Classical Mechanics* (Addison-Wesley, Cambridge, MA, 1950), p. 85.

10. *Handbook of Chemistry and Physics,* edited by R. C. Weast and S. M. Selby (Chemical Rubber Co., Cleveland, 1966).
11. A. Foderaro, *The Elements of Neutron Interaction Theory* (The M.I.T. Press, Cambridge, MA, 1971), pp. 58–61, 349–393.
12. E. Segre, *Nuclei and Particles* (W. A. Benjamin, New York, 1965).
13. J. M. Blatt and V. F. Weisskopf, *Theoretical Nuclear Physics* (John Wiley and Sons, New York, 1952).
14. L. I. Schiff, *Quantum Mechanics,* 2nd ed. (McGraw-Hill, New York, 1955), p. 103.
15. D. J. Hughes and R. B. Schwartz, *Neutron Cross Sections, Report BNL 325,* 2nd ed. (Brookhaven National Laboratory, Upton, NY, 1958).
16. R. L. Henkel and H. H. Barschall, Phys. Rev. **80**, 145 (1950).
17. I. A. Stegun, "Legendre functions," in *Handbook of Mathematical Functions,* edited by M. Abramowitz and I. A. Stegun (National Bureau of Standards, Washington, 1964).
18. P. Stehle, *Quantum Mechanics* (Holden-Day, San Francisco, 1966), p. 196.
19. M. D. Goldberg, V. M. May, and J. R. Stehn, *Angular Distributions in Neutron Induced Reactions,* 2nd ed. (Brookhaven National Laboratory, Upton, NY, 1962).
20. R. D. Evans, *The Atomic Nucleus* (McGraw-Hill, New York, 1955), p. 690.
21. C. D. Swartz and G. Owen, "Recoil detection in scintillators," in *Fast Neutron Physics,* vol. 1, edited by J. B. Marion and J. L. Fowler (Interscience, New York, 1960).
22. *Neutron Fluence, Neutron Spectra and Kerma, ICRU Report 13* (International Commission on Radiation Units and Measurements, Washington, 1969).
23. E. Amaldi, "The production and slowing down of neutrons," in *Encyclopedia of Physics* (Springer-Verlag, Berlin, 1959).
24. K. K. Seth, *Conference on Neutron Physics by Time of Flight Methods, Report 2309* (Oak Ridge National Laboratory, Oak Ridge, TN, 1957), p. 5.
25. R. L. Bach and R. S. Caswell, Radiat. Res. **35**, 1 (1968).

9

Neutron Absorption Reactions

9.1 Energy Expressions and Q Relationships

In this chapter we will be concerned with absorption reactions for neutrons. These reactions involve the atomic nucleus since common neutron interactions occur through the agency of the nuclear force (1). In an absorption reaction, the incident neutron disappears inside the nucleus. Secondary radiations, with energies that are influenced by the intermediate excited nuclear configuration, are emitted after the absorption process. The secondaries may be photons from radiative capture reactions, (n, γ). They may be nuclear particles such as protons, neutrons, or other ions from particle emission reactions (n,b). If the secondary particle is another neutron, the reaction is often called inelastic scattering, (n,n'). Inelastic scattering can be included as a special case in the particle emission formalism since the incident neutron is actually absorbed into the nucleus as evidenced by the excited nuclear configuration which results (2). Neutron-induced fission, (n,f), is discussed as a special case because of the unusual reaction characteristics.

Energy categories for neutrons were listed in Chapter 8. They are useful because there is some correlation between energy of the neutron and the most probable type of reaction (3, 4). For example, radiative neutron capture reactions are the prevalent absorption reactions in most medium- and heavy-mass targets for incident neutron energies of less than 0.5 keV, the slow-neutron and intermediate-energy categories. These reactions will be discussed in Section 9.3. When $T \geq 0.5$ keV, particle emission reactions (n,b), with emitted protons, neutrons, or alphas, often dominate. These reactions are discussed in Section 9.4. Very light and very heavy nuclei provide exceptions to these generalizations. In some cases individual nuclear properties and individual threshold values are of overriding importance.

We will next develop equations to describe the energies for a neutron-induced particle emission reaction, $X + n \rightarrow b + Y$, and the results will be used in specific situations in later sections. In this notation, X indicates the absorbing target nucleus, n the incident neutron, b the emitted particle, and Y the product nucleus. Suppose X is initially at rest and that the incident neutron momentum and kinetic energy in the laboratory system are indicated by p and T, respectively. Momentum and energy conservation expressions are

$$\mathbf{p} = \mathbf{p}'_b + \mathbf{p}'_Y, \quad T + M_n c^2 + M_X c^2 = T'_b + M_b c^2 + T'_Y + M_Y c^2,$$

$$T = (M_b + M_Y - M_n - M_X)c^2 + T'_b + T'_Y = -Q + T'_b + T'_Y \tag{9.1}$$

The Q value used in Equation 9.1 is defined as the excess in kinetic energy liberated for the reation (5), where

$$Q = T'_b + T'_Y - T = M_X c^2 + M_n c^2 - M_b c^2 - M_Y c^2 \tag{9.2}$$

Since the Q can be written as a sum and difference of constituent nuclear mass energies, reaction equations are often written including Q. Since the Q value is the net released kinetic energy, two categories are possible: If Q for a reaction is positive, the reaction liberates energy (exoergic); such reactions can be initiated at any incident energy. If the Q value is negative, input kinetic energy is required (endoergic) to balance the increase in mass. In this case the neutron kinetic energy must exceed some value before the reaction will occur.

The minimum energy necessary to induce a reaction is usually called the threshold. A threshold value $T_{\min} > |Q|$ exists for neutron-induced reactions with negative Q values. Recall from Section 7.1 that at threshold, components of outgoing momentum perpendicular to the incident-particle direction are zero since they require extra energy. Furthermore, for minimum energy expended, the incident-particle momentum is shared between the outgoing particles in the ratio of their masses. Then at threshold for nonrelativistic neutron-induced reactions,

$$p'_b = \left[\frac{M_b}{M_b + M_Y}\right]p, \quad p'_Y = \left[\frac{M_Y}{M_b + M_Y}\right]p, \quad p = p'_b + p'_Y,$$

$$T'_b = \frac{(p'_b)^2}{2M_b} = \frac{M_b p^2}{2(M_b + M_Y)^2} = \frac{M_b M_n}{(M_b + M_Y)^2}T,$$

$$T'_Y = \frac{(p'_Y)^2}{2M_Y} = \frac{M_Y p^2}{2(M_b + M_Y)^2}$$

$$= \frac{M_Y M_n}{(M_b + M_Y)^2}T \tag{9.3}$$

When $Q < 0$ at threshold,

$$Q = T'_b + T'_Y - T_{\min} = \frac{M_n}{(M_b + M_Y)} T_{\min} - T_{\min},$$

$$T_{\min} = |Q| \frac{(M_b + M_Y)}{(M_b + M_Y - M_n)} = |Q| \left[\frac{1}{1 - M_n/(M_b + M_Y)}\right]$$

$$= |Q| \left[1 + \frac{M_n c^2}{M_X c^2 + |Q|}\right] \tag{9.4}$$

The relationship $(M_b + M_Y)_c{}^2 = (M_n + M_X)c^2 + |Q|$ from Equation 9.2 was substituted in Equations 9.4 to produce the final result.

The variables T, T'_b, T'_Y, θ'_b, and θ'_Y can be used to specify a neutron absorption reaction. The angles are defined on Figure 9.1. In principle, any one of these quantities can be calculated from two others if the mass values are known. The Q equation (not Q value) will be obtained in the following as an example of relationships involving angles. In particular we seek a relationship including only T, T'_b, θ'_b, and Q. Since nonrelativistic expressions are normally sufficient, momentum conservation components are

$$P_n = \sqrt{2M_n T} = \sqrt{2M_b T'_b} \cos\theta'_b + \sqrt{2m_Y T'_Y} \cos\theta'_Y,$$

$$0 = \sqrt{2M_b T'_b} \sin\theta'_b - \sqrt{2M_Y T'_Y} \sin\theta'_Y \tag{9.5}$$

Rearranging these equations and squaring gives

$$M_b T'_b \sin^2\theta'_b = M_Y T'_Y \sin^2\theta'_Y, \quad \cos^2\theta'_Y = 1 - \frac{M_b T'_b}{M_Y T'_Y} \sin^2\theta'_b,$$

$$M_Y T'_Y \cos^2\theta'_Y = (\sqrt{M_n T} - \sqrt{M_b T'_b} \cos\theta'_b)^2,$$

$$\cos^2\theta'_Y = \frac{(\sqrt{M_n T} - \sqrt{M_b T'_b} \cos\theta'_b)^2}{M_Y T'_Y} \tag{9.6}$$

Equating $\cos^2\theta'_Y$ in Equations 9.6 means that

$$1 - \frac{M_b T'_b}{M_Y T'_Y} \sin^2\theta'_b = \frac{(\sqrt{M_n T} - \sqrt{M_b T'_b} \cos\theta'_b)^2}{M_Y T'_Y} \tag{9.7}$$

Then

$$M_Y T'_Y - M_b T'_b \sin^2\theta'_b = M_n T - 2\sqrt{M_n M_b T T'_b} \cos\theta'_b + M_b T'_b \cos^2\theta'_b,$$

$$T'_Y = \frac{M_n}{M_Y} T + \frac{M_b}{M_Y} T'_b - \frac{2}{M_Y} \sqrt{M_n M_b T T'_b} \cos\theta'_b \tag{9.8}$$

Substitution of the second of Equations 9.8 into Equation 9.2 yields

$$Q = \mathrm{T}'_b + T'_Y - T$$

$$= T'_b \left[1 + \frac{M_b}{M_Y}\right] - T\left[1 - \frac{M_n}{M_Y}\right] - \frac{2\sqrt{M_n M_b T T'_b}}{M_Y} \cos\theta'_b \tag{9.9}$$

Equation 9.9 is the usual form for the Q equation. However, a similar equation could be written in terms of T, T'_Y, θ'_Y, and the Q value. The relationship

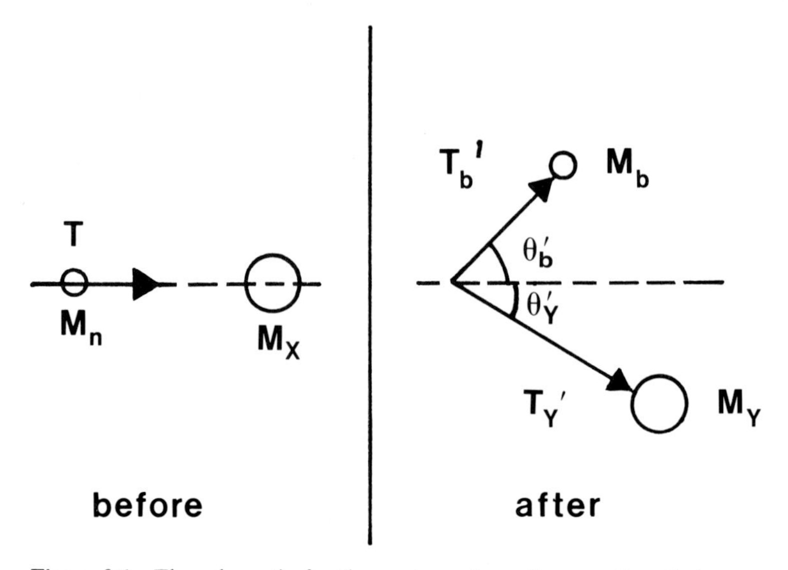

Figure 9.1 The schematic for the neutron absorption reaction $X + n \rightarrow b + Y$ is shown. The situation before absorption of the neutron and after emission of particle b is illustrated.

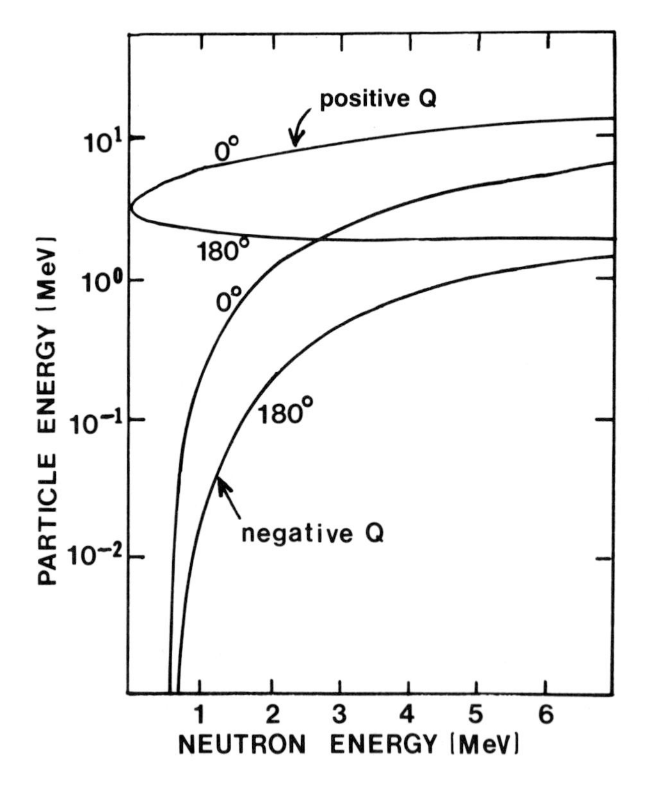

Figure 9.2 A plot of the emitted particle energy T_b' versus the incident neutron energy T is shown for the limits at 0° and 180° for the emitted particle angle θ_b'. Exoergic reactions take place in the region opening to the right labeled positive Q. Endoergic reactions take place in a horn-shaped region emanating from the threshold value and marked by negative Q. The Q values determine relative positions on the axes.

between the kinetic energy of the emitted particle and the neutron energy is indicated schematically in Figure 9.2 for angles $0 \leq \theta'_b \leq \pi$.

The utility of Equations 9.4 and 9.9 is greatly enhanced because the Q value for the ground-state reaction Q_{gs} can be obtained from tables of atomic masses. As was mentioned in Chapter 8, ground-state masses are given in terms of mass excess (or deficit) values in atomic mass units as Δm, or in energy units as ΔE. Then for energies in megaelectron volts,

$$M_{gs}c^2/(1.6)10^{-13} = (A + \Delta m)931.5 = 931.5A + \Delta E \tag{9.10}$$

where M_{gs} is the ground-state mass in kilograms. Substituting Equation 9.10 into Equation 9.2 gives an expression for Q value in megaelectron volts:

$$\begin{aligned} Q_{gs} &= (A_X + A_n - A_b - A_Y)931.5 + \Delta E_X + \Delta E_n - \Delta E_b - \Delta E_Y, \\ &= \Delta E_X + \Delta E_n - \Delta E_b - \Delta E_Y \end{aligned} \tag{9.11}$$

Equation 9.11 holds because $A_X + A_n = A_b + A_Y$. As has been mentioned, values for ΔE can be found in Appendix 5 for atomic masses (6). Since equations 9.1 through 9.11 involve nuclear masses, the appropriate number of electron mass values should be subtracted from the atomic values listed in the tables to give accurate Q values. In most circumstances this subtraction is not necessary since electron masses on the input side of the equation cancel those on the output side.

9.2 Resonances in Neutron Absorption Cross Sections

The cross sections for the neutron absorption reactions described in the following sections of this chapter have some common characteristics. The most striking is the appearance of resonances or cross-section peaks. Figure 9.3 shows the neutron cross section or neutron mass attenuation coefficient for cadmium as an example (7–9). Note that there are many resonances in the region around 100 eV and a dominating resonance is superposed on the smooth monotonic cross section at about 0.2 eV. We will show that these peaks are related to quasi-stable nuclear configurations formed after neutron absorption.

The compound nucleus is the intermediate system formed after absorption of a nuclear particle by a nucleus. The compound nucleus theory of nuclear reactions was first proposed by Bohr (10). In the following we consider neutron-induced reactions that occur via compound nucleus formation and exclude direct reactions that occur at high energies (11). During a compound nucleus reaction:

1. The absorbed particle and the target nucleus form the compound system, and the incident kinetic energy is shared among the system nucleons.
2. The compound nucleus decays into the final reaction products in a manner independent of the mode of formation but statistically dependent on the net excitation energy.

For neutron absorption this two-step reaction sequence can be included in the formalism by adding C, the compound nucleus symbol, in the reaction

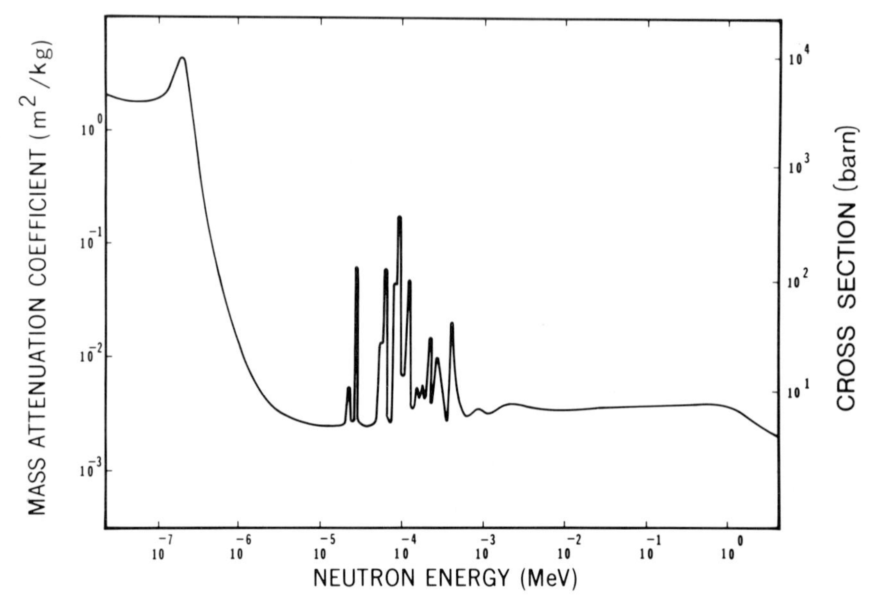

Figure 9.3 The total neutron mass attenuation coefficient and cross section for cadmium is shown (7–9) as a function of incident neutron energy.

equation $X + n \rightarrow C \rightarrow b + Y$. In this case the reaction cross section σ_{nb} is a product: $\sigma_{nb} = \sigma_c P_b$, where σ_c is the cross section for formation of the compound nucleus by neutron absorption and P_b is the probability it decays with emission of particle b.

A resonance peak in σ_{nb} indicates a particularly large value of the neutron absorption probability. It occurs when the input energy is appropriate for formation of an eligible excited state of the compound nucleus. Since the subscript C indicates compound nucleus values,

$$T + M_n c^2 + M_X c^2 = T'_C + M_C c^2 + E^*_C$$

$$E^*_C = T - T'_C + (M_n c^2 + M_X c^2 - M_C c^2) = T - T'_C + \mathrm{SE}_n \tag{9.12}$$

where E^*_C is the excited-state energy of the compound nucleus and SE_n is the neutron separation energy from that system. Note that T'_C is small compared to T for heavy nuclear targets, since it is primarily a recoil energy. Figure 9.4 illustrates the relationship between the total excess kinetic energy, $\Delta T = T - T'_C$, the separation energy, and the excited state energy.

Individual cross-section resonances for compound nucleus reactions are relatively narrow since the compound nucleus is relatively long lived. After the incident particle is captured into the compound nucleus, it is assumed to be reflected back and forth several times at the interior nuclear interface. Finally, the incident kinetic energy is dissipated in collisions with internal nucleons. Because of the reflections at the nuclear surface, the total elapsed time is much greater than the time necessary for a single traversal of the nucleus. For such

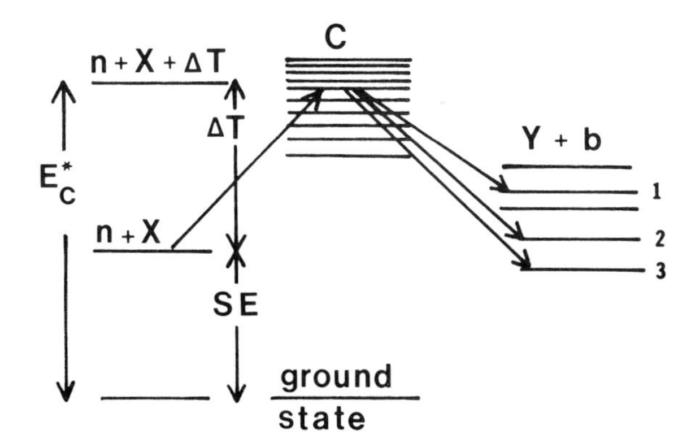

Figure 9.4 The excitation energy of the compound nucleus, E_c^*, is shown to be equal to a separation energy plus the excess input kinetic energy ΔT. Compound nucleus decay into several different configurations is also illustrated. Adapted from Ref. 17.

a process the uncertainty principle allows a relatively narrow minimum energy spread, $\Delta E \sim \hbar/\Delta t$.

The relationship between the energy level diagram and the cross section is shown on figure 9.5 (7–9, 12) for the nucleus ^{15}N. For each eligible level in the compound nucleus ^{15}N there corresponds a resonance peak in the cross

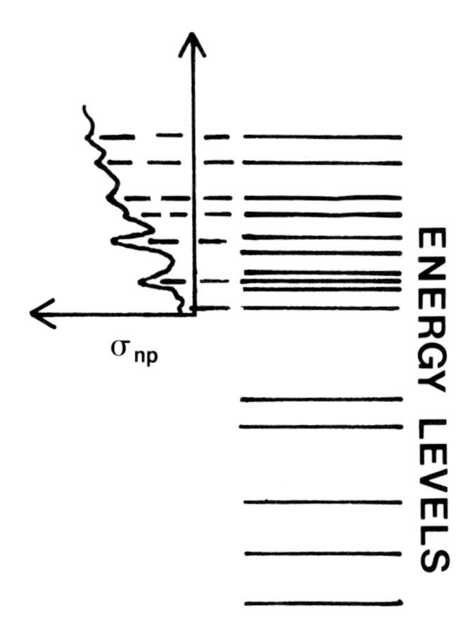

Figure 9.5 The energy level diagram for the excited compound nucleus ^{15}N is shown. The cross section for the $^{14}N + n \rightarrow {}^{15}N^* \rightarrow p + {}^{14}C$ reaction is also shown (7–9, 12). Resonance energies are compared with compound nucleus energy levels.

section for the reaction $^{14}N(n,p)^{15}C$. Notice that at lower energies the spacing of levels is greater than the width of the excited resonances. Consequently, the cross section shows more or less isolated resonances. At higher energies, the resonances overlap and the resulting cross-section curve appears nearly smooth, since it is the sum of the individual resonance curves.

The resonance line shape, often called the Breit-Wigner shape (13), can be obtained from the properties of the quasi-stable excited states of the compound system (2, 4, 14). The wave function used to represent such a state must have a time-dependent part that describes the decay as well as the usual oscillatory part. Suppose H is the total energy operator and $\psi(\mathbf{r},t)$ is the wave function with spatial factor $\phi(\mathbf{r})$ and time factor $f(t)$.

$$H\psi(\mathbf{r},t) = i\hbar \frac{\partial\psi(\mathbf{r},t)}{\partial t},$$

$$\psi(\mathbf{r},t) = \phi(\mathbf{r})f_j(t),$$

$$f_j(t) = \sqrt{\lambda}\exp(-iE_jt/\hbar)\exp(-\lambda t/2) \tag{9.13}$$

In this case E_j is the energy for a state designated by index j, and λ is the decay constant for that excited configuration.

The wave function in Equations 9.13 is appropriate for a nuclear state subject to decay. Recall that the probability of finding a system at a spatial position between $\mathbf{r}$ and $\mathbf{r} + d\mathbf{r}$ at time between t and $t + dt$ is proportional to the square of the absolute value of the normalized wave function (15, 16):

$$|\psi(\mathbf{r},t)|^2\, dVdt = |\phi(\mathbf{r})|^2\,\lambda\exp(-\lambda t)dVdt = |\phi(\mathbf{r})|^2\, dV[\lambda\exp(-\lambda t)dt] \tag{9.14}$$

where $dV = dxdydz$ is the volume element. If $\psi(\mathbf{r},t)$ is normalized, $|\phi(\mathbf{r})|^2$ must yield a value of one when integrated over all space, and $|f_j(t)|^2$ must yield a value of one when integrated over time:

$$\int_0^\infty |f_j(t)|^2 dt = \int_0^\infty \lambda\exp(-\lambda t)dt = 1 \tag{9.15}$$

We assume that the state was formed when $t = 0$.

Suppose that the nuclear system can be described as a harmonic oscillator with angular frequency ω and quantized energy $\hbar\omega$ (17). We ignore spin in this simple case. The Fourier transform operations outlined in Chapter 3 are useful. If we define $f(t)$ and $g(\omega)$ as a Fourier transform pair and let $E_j = \hbar\omega_j$ for all states j,

$$f(t) = 0,\ t < 0;$$

$$= \sqrt{\lambda}\exp[-(\lambda/2 + iE_j/\hbar)t] = \sqrt{\lambda}\exp[-(\lambda/2 + i\omega_j)t], \quad t \geq 0;$$

$$g(\omega) = 1/\sqrt{2\pi}\int_{-\infty}^{\infty} f(t)\exp(i\omega t)dt$$

$$= \sqrt{\lambda/2\pi}\int_0^\infty \exp\left\{-[\lambda/2 - i(\omega - \omega_j)]t\right\}dt \tag{9.16}$$

The integration in Equations 9.16 is straightforward:

$$g(\omega) = \sqrt{\frac{\lambda}{2\pi}} \frac{\hbar}{[\hbar\lambda/2 - i\hbar(\omega - \omega_j)]} \tag{9.17}$$

The squares of the absolute values of the transform functions yield equal results when integrated over the appropriate space (18):

$$\int_{-\infty}^{\infty} |f(t)|^2 dt = \int_{-\infty}^{\infty} |g(\omega)|^2 d\omega = \int_{-\infty}^{\infty} P(E^*, E_j^*) dE^* = 1 \tag{9.18}$$

$P(E^*, E_j^*)\ dE^*$ is called the resonance excitation function and represents the probability of finding the system with excitation energy between E^* and $E^* + dE^*$ when an energy level exists at E_j^*. A suitable form for this probability is

$$|g(\omega)|^2 d\omega = \frac{\hbar\lambda}{2\pi} \left[\frac{\hbar d\omega}{(\hbar\lambda/2)^2 + (E^* - E_j^*)^2} \right]$$

$$= \frac{1}{2\pi} \left[\frac{\Gamma dE^*}{(\Gamma/2)^2 + (E^* - E_j^*)^2} \right] = P(E^*, E_j^*) dE^* \tag{9.19}$$

The level width, $\Gamma = \hbar\lambda$, and excitation energy, $E^* = \hbar\omega$, were utilized in Equation 9.19 to produce the customary form for the Breit-Wigner resonance line shape. Figure 9.6 is a plot of Equation 9.19 as a functin of E^*. As seen on this figure, Γ is the full width at half maximum of the resonance line.

The Breit-Wigner resonance line shape can be incorporated into an expression for σ_C, the cross section for formation of the compound nucleus. We start with the developments in Section 8.5. From Equation 8.39 the neutron cross section is given by

$$\sigma(\Lambda) = \frac{\Lambda^2}{4\pi} \sum_l |1 - \eta_l|^2 (2l + 1), \quad \eta_l = N_l \exp(2i\delta_l) \tag{9.20}$$

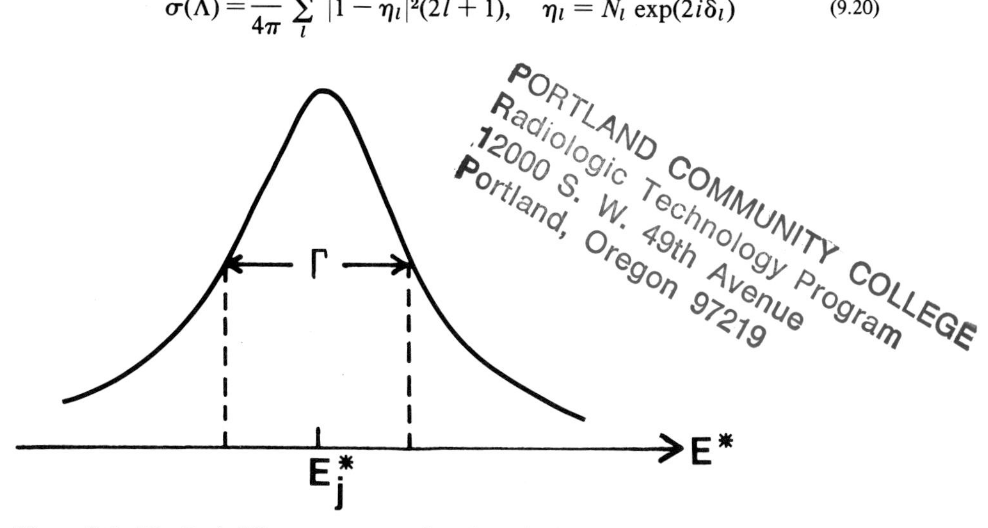

Figure 9.6 The Breit-Wigner resonance line shape is shown (13). The level width Γ is the full width at half maximum.

The variable Λ in the equation is the incident neutron wavelength in the center of mass system. The index l is for the partial waves, and η_l is the neutron wave perturbation factor given in terms of the wave amplitude factor N_l and the phase shift δ_l.

Suppose the target nucleus were to absorb neutrons with maximum efficiency. For this special case $N_l = 0$, so $\eta_l = 0$:

$$\sigma_{\max}(\Lambda) = \frac{\Lambda^2}{4\pi} \sum_l (2l+1),$$

$$\sigma_{\max}(T) = \frac{h^2}{8\pi M_n T}\left[\frac{A+1}{A}\right]^2 \sum_l (2l+1) \tag{9.21}$$

The first equation is given in terms of center of mass system variables, the second is given in terms of laboratory system variables as indicated in Equation 8.41.

In many cases Equations 9.21 are inadequate. Usually only part of the incident neutron wave is transmitted through the nuclear surface while part is reflected at that interface. Furthermore, the effect of energy levels in the compound nucleus and the resulting resonances must be considered. With appropriate modification, the compound nucleus formation cross section for the laboratory system can be written as

$$\sigma_C(T) = \frac{h^2}{8\pi M_n T}\left[\frac{A+1}{A}\right]^2 \sum_l (2l+1) Tm[D\,P(E^*, E_j^*)] \tag{9.22}$$

In this case Tm is the transmission of the neutrons at the nuclear interface, and D, the average spacing between levels, is the weighting factor that ensures that the average value obtained from many resonances is correct. It is necessary when many resonances are interspersed over a substantial energy range (4).

Equation 9.22 will be used in subsequent sections. Generally, the functional form is written utilizing E^*, the excitation energy of the compound nucleus system. However, if T_j is the incident kinetic energy that produces excitation E_j^*, then for specific reaction conditions, $E^* - E_j^* \simeq T - T_j$. Thus, any cross-section resonance can be displayed as a function of the kinetic energy of the incident neutron.

9.3 Radiative Capture Reactions

Radiative neutron capture is the absorption of a neutron followed by the emission of one or more gamma rays. Figure 9.7 is an illustration of a radiative capture reaction. The reaction is often indicated by (n, γ), which is shorthand for the sequence $X + n \rightarrow Y^* \rightarrow Y + \gamma$. The secondary gamma rays have energies characteristic of the levels of the product nucleus Y since they are emitted from Y^*, the excited compound nucleus. Notice that no nuclear particles are emitted in a radiative capture reaction, only photons. Neutron-induced reactions

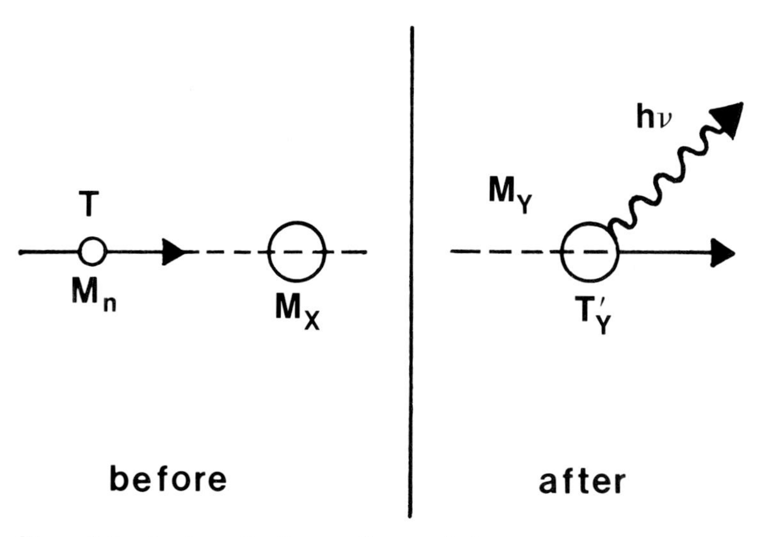

Figure 9.7 A schematic diagram for a radiative neutron capture reaction is illustrated.

can occur in which nuclear particles are emitted as well as gamma rays. They are classified as particle emission reactions and are discussed in the next section.

Radiative neutron capture reactions occur for most stable nuclear targets at low incident energies (19). Radiative capture is usually the dominant absorption reaction for slow neutrons and intermediate-energy neutrons incident on intermediate and heavy nuclear targets (4). In some special cases, neutron-induced fission dominates near the thermal energy. The most probable absorption reaction is not so easy to predict for light nuclear targets, since particle emission reactions with positive Q values may be competitive.

In general, radiative capture is an inelastic process and is exoergic. Momentum and energy expressions for the two-step reaction can be written

$$\mathbf{p} = \mathbf{p}'_{Y*}, \quad \mathbf{p}'_{Y*} = \mathbf{p}'_Y + \mathbf{p}_\gamma,$$

$$T + M_n c^2 + M_X c^2 = T'_{Y*} + M_Y c^2 + E^*_Y = T'_Y + M_Y c^2 + \Sigma h\nu,$$

$$T'_{Y*} + E^*_Y = (M_n + M_X - M_Y)c^2 + T = \mathrm{SE}_n + T,$$

$$\Sigma h\nu = \mathrm{SE}_n + T - T'_Y \tag{9.23}$$

where subscript Y^* indicates the excited compound nucleus, and E^*_Y is the excitation energy of the compound system prior to emission of gamma rays. The separation energy SE_n for the neutron in the compound nucleus is between 7 and 9 MeV for most nuclei (17). Since the nuclear kinetic energy is usually small, even thermal neutrons can induce reactions which proceed via highly excited nuclear states. When highly excited states are involved, de-excitation proceeds via a cascade of capture gamma rays. If T is small, the total gamma-ray energy $\Sigma h\nu$ approaches the value of the separation energy. The angular distribution of the emitted gamma rays is isotropic in the center of mass system

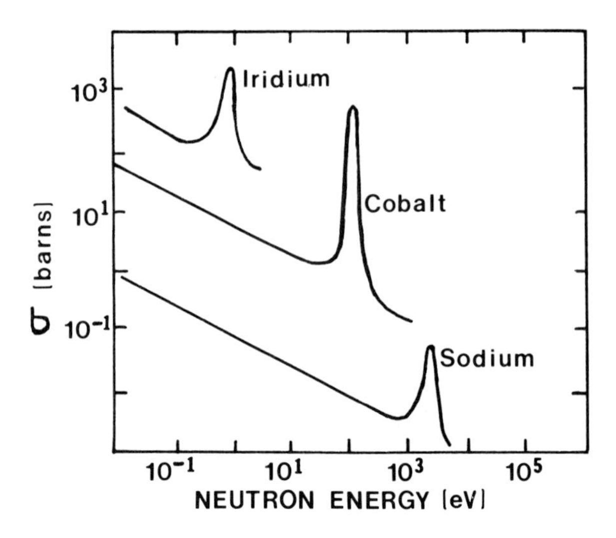

Figure 9.8 Neutron cross sections are compared for iridium, cobalt, and sodium in the slow- and intermediate-energy regions (7–9). Note the $1/v$ dependence near thermal energies.

as long as the mode of decay of the compound nucleus is independent of the mode of formation.

The cross section for radiative capture of neutrons can be discussed conveniently using Equation 9.22. Figure 9.8 shows that a cross-section resonance often occurs somewhere in the low-energy region. The resonance can be superposed on a region of smooth energy dependence.

From Equation 9.22 it is apparent that the smooth dependence at low energies must be obtained from the factors multiplying the resonance excitation function. Thus,

$$\sigma_C(T) = \frac{h^2}{8\pi M_n T}\left[\frac{A+1}{A}\right]^2 Tm\, D\, P(E^*,E_j^*) = \sigma_{1/v}\, D\, P(E^*,E_j^*)$$

$$\sigma_{1/v} = \frac{h^2}{8\pi M_n T}\left[\frac{A+1}{A}\right]^2 Tm \tag{9.24}$$

for s-wave absorption. The transmission, of neutrons, Tm, at the nuclear potential interface must be evaluated before $\sigma_{1/v}$ can be obtained.

Refer to Figure 9.9 (4,17,19) to obtain an expression for Tm. Suppose that the nuclear potential for s-wave incident neutrons has the form shown, with depth V_0 and interface at $x = 0$. Notice that no barrier is included for this simple case since s-wave neutrons have no angular momentum or charge. Nevertheless, some of the incoming wave will be reflected at the nuclear surface. We consider the one-dimensional problem for simplicity, although radial waves would be more realistic (4).

In region I, outside of the influence of the nuclear potential, an incident traveling wave and a reflected traveling wave can be represented by

$$\psi_I(x,t) = A \exp[i(kx - \omega t)] + B \exp[-i(kx + \omega t)],$$

$$k = \frac{2\pi}{\lambda_I} = 2\pi \frac{p_I}{h} = \frac{\sqrt{2M_n T}}{\hbar} \tag{9.25}$$

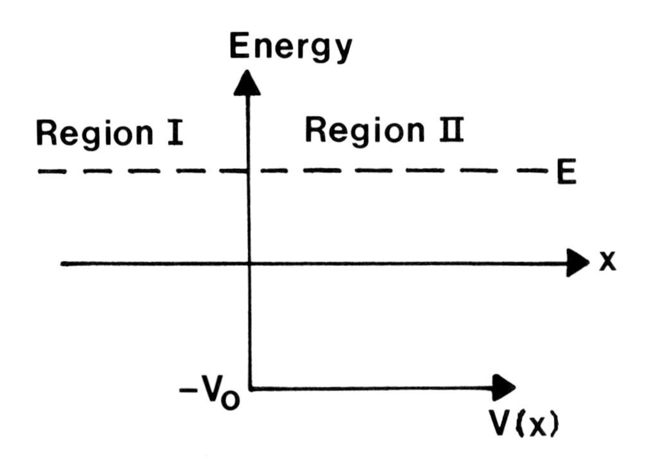

Figure 9.9 The change in potential at the interface between region *I* and region *II* causes partial reflection of an incident wave.

where factors A and B are the amplitudes of the incident wave, traveling in the $+x$ direction, and the reflected wave traveling in the $-x$ direction, respectively. The wave number k is related to the free-space neutron kinetic energy T.

In region *II* only the wave representing transmitted neutrons exists:

$$\psi_{II}(x) = C \exp[i(Kx - \omega t)],$$

$$K = \frac{2\pi}{\lambda_{II}} = 2\pi \frac{p_{II}}{h} = \frac{\sqrt{2M_n(E - V)}}{\hbar} = \frac{\sqrt{2M_n(T + V_0)}}{\hbar} \tag{9.26}$$

E is the sum of the kinetic and potential energy and is unchanged in regions *I* and *II*. C is the amplitude of the wave traveling in the $+x$ direction. The wave number K can be much increased over the free-space value, since the magnitude of V_0 is usually several tens of megaelectron volts for the attractive nuclear force.

If this wave is to represent a physical situation, it must be continuous at x = 0. Then

$$\psi_I(0,t) = \psi_{II}(0,t), \quad A + B = C,$$

$$\left.\frac{d\psi_I}{dx}\right|_{0,t} = \left.\frac{d\psi_{II}}{dx}\right|_{0,t}, \quad ikA - ikB = iKC \tag{9.27}$$

The transmission at the interface is the ratio of the incident fluence rate divided by the transmitted fluence rate. It is proportional to the ratio of the absolute value of the square of the amplitudes, $|C|^2/|A|^2$. We obtain C/A by substituting from Equations 9.27:

$$B = C - A, \quad kA - k(C - A) = KC \quad 2kA = (K + k)C, \quad C/A = \frac{2k}{K + k},$$

$$Tm = \frac{K|C|^2}{k|A|^2} = \frac{4kK}{(K + k)^2} \tag{9.28}$$

K/k was included in the equation for the transmission since the fluence rate ratio must include a factor involving the ratio of particle speeds.

Substituting the results of Equations 9.28 for s-wave neutron transmission into the last of Equations 9.24 gives

$$\sigma_{1/v} = \frac{h^2}{8\pi M_n T}\left[\frac{A+1}{A}\right]^2 Tm = \frac{h^2}{2\pi M_n}\left[\frac{A+1}{A}\right]^2 \frac{kK}{T(K+k)^2},$$

$$\sigma_{1/v} \simeq \frac{h^2}{2\pi M_n}\left[\frac{A+1}{A}\right]^2 \frac{k}{KT} = \frac{h^2}{2\pi M_n}\left[\frac{A+1}{A}\right]^2 \frac{\sqrt{2M_n T}}{T\sqrt{2M_n(T+V_0)}}, \quad k \ll K,$$

$$\sigma_{1/v} \simeq \frac{h^2}{2\pi M_n (V_0)^{1/2}}\left[\frac{A+1}{A}\right]^2 \frac{1}{T^{1/2}}, \quad T \ll V_0 \tag{9.29}$$

Equation 9.29 is an expression of the $1/v$ law for s-wave neutron absorption, since $T^{-1/2}$ is proportional to v^{-1}. The dependence is valid in the low-energy limit $k \ll K$ and $T \ll V_0$ for our development.

The $1/v$ law for neutron absorption was discussed in 1935 by Fermi and co-workers (20). It is generally applicable for the smooth dependence for radiative capture reactions for slow neutrons. The $1/v$ dependence is also apparent for particle emission reactions with positive Q values. Of course, the resonance excitation function tends to dominate the energy dependence of the cross section near a resonance peak.

Table 9.1 shows some radiative capture thermal neutron cross-section values and energies of principal gamma rays for interesting materials (21, 22). The data in this table illustrate the large variations in thermal capture cross sections for neighboring nuclear absorbers. Figure 9.8 indicates that part of this variation can be traced to resonances occurring in the cross section for formation of the compound nucleus. Notice that large thermal neutron cross sections often occur when the resonance energy is close to $T = 0.025$ eV (19). For iridium the resonance occurs at 0.9 eV; for cobalt the closest resonance peak is at about 100 eV.

Table 9.1 Radiative Capture Reactions

Reaction X(n,γ)	*Thermal Cross Section*	*Highest Energy Gamma Ray (MeV)*	*Average Number of Gamma Rays*
$_{1}$H	0.33	2.23	1.0
$_{11}$Na	0.53	6.41	2.0
$_{26}$Fe	2.53	10.16	1.7
$_{27}$Co	37.0	7.49	
$_{45}$Rh	156	6.79	
$_{48}$Cd	2450	9.05	4.1
$_{49}$In	196	6.38	3.3
$_{62}$Sm	5600	7.89	5.6
$_{64}$Gd	46000	7.78	3.9
$_{79}$Au	98.8	6.49	3.5
$_{82}$Pb	0.17	7.38	

Data from Refs. 21 and 22.

The wide distribution in thermal neutron cross sections is related to the fluctuating positions of levels at about 7–9 MeV excitation energy in the compound nucleus for different nuclides. At these energies the level spacing is about 30–50 eV on the average (4, 19). The thermal cross section σ_{th} will be very large if by chance a compound nucleus level occurs at an energy that produces a resonance with incoming neutron energy just above 0.025 eV. The resonance excitation function $P(E^*, E_j^*)$ is a fast function of the energy difference $E^* - E_j^*$. Because of this, σ_{th} will drop off rapidly as $|E^* - E_j^*|$ increases.

The use of absorbers with large thermal capture cross sections will be discussed further in the next chapter. Since thermal neutrons are an important component of background spectra, radiative capture reactions are very important when materials are selected for shielding against a neutron background. These reactions often produce the terminal event in a neutron degradation sequence. The nuclide ^{113}Cd is especially useful for this purpose because of its large cross section ($\sigma_{th} = 20{,}000$ b). The high-energy gamma-ray secondaries must always be considered in shielding situations.

9.4 Particle Emission Reactions

A particle emission reaction is the emission of a nucleon or combination of nucleons after absorption of an incident neutron. Figure 9.1 can be used to illustrate this reaction. For compound nucleus reactions, the sequence $X + n \rightarrow C \rightarrow b + Y$ with abbreviated notation (n,b) is appropriate. The outgoing particle indicated by b might be a proton, another neutron, a deuteron, or an alpha particle shown as (n,p), (n,n'), (n,d), and (n,α), respectively. Other reactions, including the emission of several nucleons, are $(n,2n)$, $(n,2p)$, and (n,np).

After emission of the secondary particle or particles, the residual nucleus may be left in an excited state. In this case, emission of gamma rays can follow, either singly or in cascade. One group of reactions that always includes gamma-ray emission is called inelastic scattering, (n,n'). In spite of the name, inelastic scattering reactions are included in the group of particle emission reactions because the emitted neutron can be considered to be a secondary particle (2).

The (n,b) reactions are important in the fast-neutron region and usually dominate in the high-energy region with $T \geq 10$ MeV. Some particle emission reactions may dominate in the slow- and intermediate-energy region if they are exoergic, i.e. $Q > 0$. This situation occurs mainly for light targets. Exoergic particle emission reactions with cross sections that are inversely proportional to the incident neutron speed ($1/v$ dependence) in the slow-neutron region compete favorably with the (n,γ) reactions in some cases. Several important exoergic neutron-particle reactions are listed in Table 9.2.

The $^{14}N(n,p)^{14}C$ reaction cited in the table is a means of production of naturally occurring ^{14}C, which can be ingested into living systems. It occurs in the atmosphere and is induced by neutrons of cosmic origin. Radiocarbon dating techniques utilize ^{14}C originally produced in this manner. The $^{10}B(n,\alpha)^7Li$ and the $^6Li(n,\alpha)^3H$ reactions are the basis of neutron detecting and dosimetry

Table 9.2 Exoergic Neutron-Particle Reactions

Reaction	Q (*MeV*)	*Thermal Cross Section* (*b*)
$^{3}He(n,p)^{3}H$	0.77	5330
$^{6}Li(n,\alpha)^{3}H$	4.64	940
$^{10}B(n,\alpha)^{7}Li$	2.78	3840
$^{14}N(n,p)^{14}C$	0.63	1.8
$^{33}S(n,p)^{33}P$	0.75	0.002
$^{35}Cl(n,p)^{35}S$	0.62	0.5

Data from Refs. 6 and 21.

systems. The boron is used in boron trifluoride (BF_3) gaseous detector tubes, the lithium in neutron scintillation detectors made with lithium iodide-europium [LiI(Eu)] crystals or in lithium fluoride (LiF) thermoluminescent dosimeters. Since the (n,α) reaction cross section of ^{10}B is so large at thermal energies, boron-loaded plastics are used for neutron shielding (21).

Generally, particle emission reactions are characterized by negative Q values. Equations 9.4 show that the threshold for an endoergic (n,b) reaction is given by

$$T_{\min} = |Q|\left[1+\frac{M_n c^2}{M_X c^2 + |Q|}\right] \simeq |Q|\left[1+\frac{M_n c^2}{M_X c^2}\right] \simeq |Q|\left[\frac{A+1}{A}\right] \quad (9.30)$$

The approximations utilized in Equation 9.30 are based on the assumptions that $|Q| \ll M_X c^2$ and $M_n/M_X \approx 1/A$. Notice that the value of $|Q|$, which is the major factor in the threshold energy, is determined by nuclear binding forces. The correction, $M_n c^2/M_x{}^2$ is included because of momentum conservation considerations. Table 9.3 shows threshold values for several important neutron-particle reactions with negative Q values.

The threshold energies for inelastic scattering reactions (n,n') vary widely from target to target. They depend on the energy of the first excited state of the product nucleus. From Equations 9.1 and 9.30 with $n \equiv b$ and $X \equiv Y$,

Table 9.3 Endoergic Neutron-Particle Reactions

Reaction	Q (*MeV*)	*Threshold* (*MeV*)	*Cross Section 14 MeV* (*b*)
$^{12}C(n,\alpha)^{9}Be$	−5.7	6.2	
$^{16}O(n,p)^{16}N$	−9.6	10.2	
$^{16}O(n,d)^{15}N$	−9.9	10.5	0.04
$^{16}O(n,\alpha)^{13}C$	−2.2	2.3	
$^{32}S(n,p)^{32}P$			0.23
$^{54}Fe(n,p)^{54}Mn$			0.35
$^{56}Fe(n,p)^{56}Mn$			0.11

Data from Refs. 6–9 and 21.

$$Q = T'_b + T'_Y - T = M_X c^2 + M_n c^2 - M_b c^2 - (M_{Y*})c^2$$
$$= M_Y c^2 - (M_{Y*})c^2 = -E^*_Y,$$
$$T_{\min} = |Q| \left[1 + \frac{M_n c^2}{M_X c^2 + |Q|}\right] \simeq E^*_Y \left[1 + \frac{M_n c^2}{M_X c^2}\right] \simeq E^*_Y \left[\frac{A+1}{A}\right] \quad (9.31)$$

where M_{Y*} stands for the mass of the excited compound nucleus. Table 9.4 shows the energy of the first excited state and the threshold energy for some common target nuclides for inelastic scattering. They extend over an energy range in excess of 5 MeV.

The cross section for a particle emission reaction is often written with the use of the resonance excitation function (Equations 9.19 and 9.22):

$$\sigma_{nb}(T) = \sigma_C(T) P_b =$$
$$\frac{h^2}{8\pi M_n T}\left[\frac{A+1}{A}\right]^2 \sum_l (2l+1)\left\{\frac{Tm\, D\, \Gamma\, P_b}{2\pi[(\Gamma/2)^2 + (E^* - E^*_j)^2]}\right\} \quad (9.32)$$

where $\sigma_C(T)$ is the cross section for formation of the compound nucleus, and P_b is the probability for emission of particle b from the excited system. We will develop expressions to replace P_b, Tm, and D in the following paragraphs.

There are usually several possible modes of decay (emission of different particles) from each highly excited compound nucleus state. If particle b is emitted,

$$\lambda_b = \hbar\Gamma_b, \quad P_b = \frac{\lambda_b}{\sum_i \lambda_i} = \frac{\hbar\Gamma_b}{\sum_i \hbar\Gamma_i} = \frac{\Gamma_b}{\Gamma} \quad (9.33)$$

The relationship between level width Γ and the decay probability λ was also used in Equation 9.19. In Equations 9.33, Γ_i is the partial width for emission of particle i and λ_i is the partial decay probability for emission of particle i.

Table 9.4 Energy of First Excited States

Nucleus	*First Excited-State Energy (MeV)*	*Inelastic Scattering Threshold (MeV)*
^{6}Li	2.19	2.56
^{10}B	0.72	0.79
^{12}C	4.43	4.80
^{14}N	2.31	2.48
^{16}O	6.06	6.44
^{27}Al	0.84	0.87
^{54}Fe	1.41	1.44
^{56}Fe	0.84	0.86
^{65}Cu	1.12	1.14
^{208}Pb	2.61	2.62
^{235}U	0.01	0.01

Data from Refs. 2 and 12.

The transmission Tm can be eliminated from Equation 9.32 using the decay probability for neutron emission. The assumption is that the barrier transmission is independent of the direction of incidence (19). A neutron in the nucleus usually has substantial kinetic energy, so it will cross the nuclear diameter many times during the lifetime of a compound nucleus. Then

$$\lambda_n = \frac{\Gamma_n}{\hbar} = Tmf_n, \quad Tm = \frac{\Gamma_n}{\hbar f_n} \tag{9.34}$$

In this case f_n is the striking frequency of the neutron on the nuclear surface.

Suppose that the nuclear energy level sequence can be approximated by a harmonic oscillator energy level spacing (4). Then

$$E_j^* = E_0 + jD \tag{9.35}$$

where E_0 is the ground-state energy, j is the level index, and D is the average level spacing. The wave function for the compound nucleus can be expanded in a linear combination of wave functions for configurations indicated by j. In this case, decay of the levels can be ignored since time intervals that are small compared to the lifetime of the compound nucleus are considered:

$$\begin{aligned}\psi(\mathbf{r},t) &= \sum_j a_j\psi_j = \sum_j a_j\phi(\mathbf{r})\exp(-iE_j^*t/\hbar) \\ &= \exp(-iE_0t/\hbar)\sum_j a_j\phi(\mathbf{r})\exp[-i(jDt/\hbar)]\end{aligned} \tag{9.36}$$

Under these circumstances, $|\psi(\mathbf{r},t)|^2 = |\psi(\mathbf{r},t + 2\pi\hbar/D)|^2$ since the wave functions in the summation differ by only a phase factor, $\exp[-i(j2\pi)]$, when t is advanced by $2\pi\hbar/D$. This means that if a particle appears at the nuclear surface at time t and is reflected, on the average it will reappear again in a time interval $2\pi\hbar/D$. Thus, the system has a period of $2\pi\hbar/D$ and a frequency of $D/2\pi\hbar$. In this case the transmission can be written

$$Tm = \frac{\Gamma_n}{\hbar f_n} = \frac{2\pi\Gamma_n}{D} \tag{9.37}$$

When a resonance occurs at excitation energy E_j^*, Equations 9.32, 9.33, and 9.37 give

$$\begin{aligned}\sigma_{nb}(T) &= \frac{h^2}{8\pi M_n T}\left[\frac{A+1}{A}\right]^2 \sum_l (2l+1)\left\{\frac{Tm\,D\,\Gamma\,P_b}{2\pi[(\Gamma/2)^2 + (E^* - E_j^*)^2]}\right\} \\ &= \frac{h^2}{8\pi M_n T}\left[\frac{A+1}{A}\right]^2 \sum_l (2l+1)\left[\frac{2\pi\Gamma_n}{D}\right]\left\{\frac{D\Gamma(\Gamma_b/\Gamma)}{2\pi[(\Gamma/2)^2 + (E^* - E_j^*)^2]}\right\} \\ &= \frac{h^2}{8\pi M_n T}\left[\frac{A+1}{A}\right]^2 \sum_l (2l+1)\left\{\frac{\Gamma_n\Gamma_b}{(\Gamma/2)^2 + (E^* - E_j^*)^2}\right\}\end{aligned} \tag{9.38}$$

The last of Equations 9.38 is the cross section expression for emission of particle b after absorption of a neutron of energy T in the laboratory system. The effect of nuclear spin has been ignored in this relationship, but multiplicative

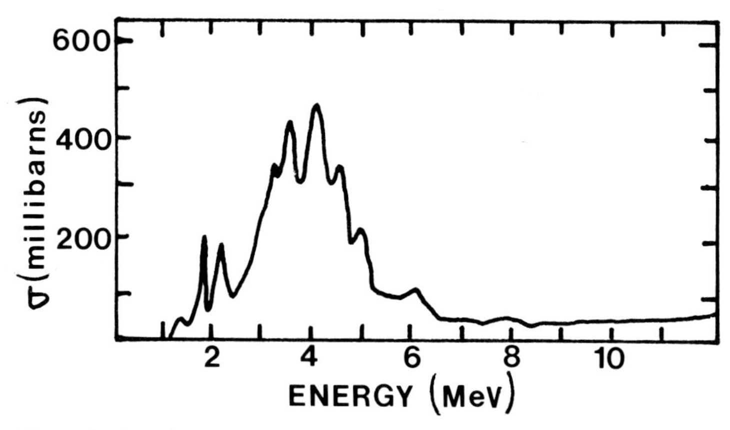

Figure 9.10 The cross section for the reaction $^{14}N(n, \alpha)^{11}B$ is shown as a function of incident neutron kinetic energy. Adapted from Ref. 23.

spin factors can be included easily (4, 19). The particular decay mode is indicated by the partial width Γ_b. For example, the cross section for neutron radiative capture can be described using Γ_γ. The resonance scattering cross section includes Γ_n, and the cross section for proton emission includes Γ_p. Thus, Equation 9.38 is quite general and can be applied to resonance processes discussed previously. It includes the effects of all angular momentum components, but *s*-wave or *p*-wave expressions can be obtained from the sum (4, 19), using the term with $l = 0$ or $l = 1$, respectively.

Figure 9.10 is given (23) as an example of resonances in a cross section for neutron-particle reactions. Note that the $^{14}N(n, \alpha)^{11}B$ reaction shown has resonance peaks that are well separated and resolved at lower energies. At higher energies the individual peaks start to overlap and merge together. Listings of many particle emission reaction cross sections have been published (21–25).

The partial level width values in Equations 9.38 must include a barrier penetration factor to describe particle emission completely. Figure 9.11 indicates that a charged nuclear particle encounters a coulomb barrier before escaping from a compound nucleus. In addition, any nuclear particle with angular momentum quantum number $l > 0$ encounters a centrifugal barrier, also illustrated on the figure. Thus, only *s*-wave neutron emission can be described without reference to a barrier. Figure 9.12 shows the barrier transmission values for protons and neutrons with $l = 0$ and $l = 1$. Note that neutron re-emission is greatly favored at low energies for high Z targets.

9.5 Neutron-Induced Fission

Several names are given to reactions that result in the breaking up of a heavy nucleus, and the names relate to the size of the nuclear pieces (19). A fragmentation reaction results in a group of lightweight nuclear fragments. A spallation

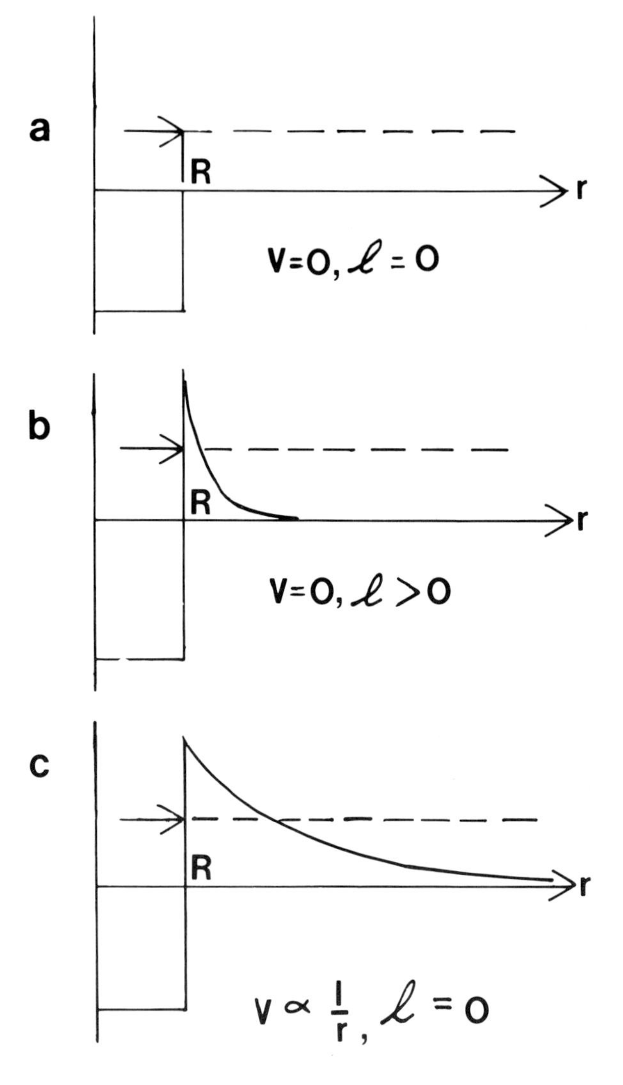

Figure 9.11 The situations that can occur for particle emission from a compound nucleus include: a) neutron emission for *s* wave, no barrier; b) neutron emission for angular momentum quantum number $l > 0$, centrifugal barrier; c) charged particle emission for *s* wave, coulomb barrier. A combination of barriers shown in b and c would occur for $l > 0$ emission of charged particles. V is the nuclear potential energy shown as a function of the radial coordinate r; R is the nuclear radius.

reaction results in one relatively heavy fragment and several much lighter weight products. Fission is the splitting of the nucleus into two fragments of roughly equal mass. Neutron-induced fission occurs frequently in only a few heavy nuclei, but it is significant because of the large amount of energy released in the process.

Spontaneous fission, as a nuclear decay mode, occurs frequently in the artificially produced nuclides with $Z > 92$. High nuclear charge is necessary since the coulomb repulsive force between the nuclear protons provides impetus for the process. Neutron-induced fission, (n,f), is possible in many heavy nuclides. Fission is the dominant thermal neutron absorption process in ^{235}U and ^{239}Pu. The (n,f) cross sections are several hundreds of barns in these nuclides at thermal energies. Figure 9.13 shows the cross section for ^{235}U neutron-induced fission (7–9).

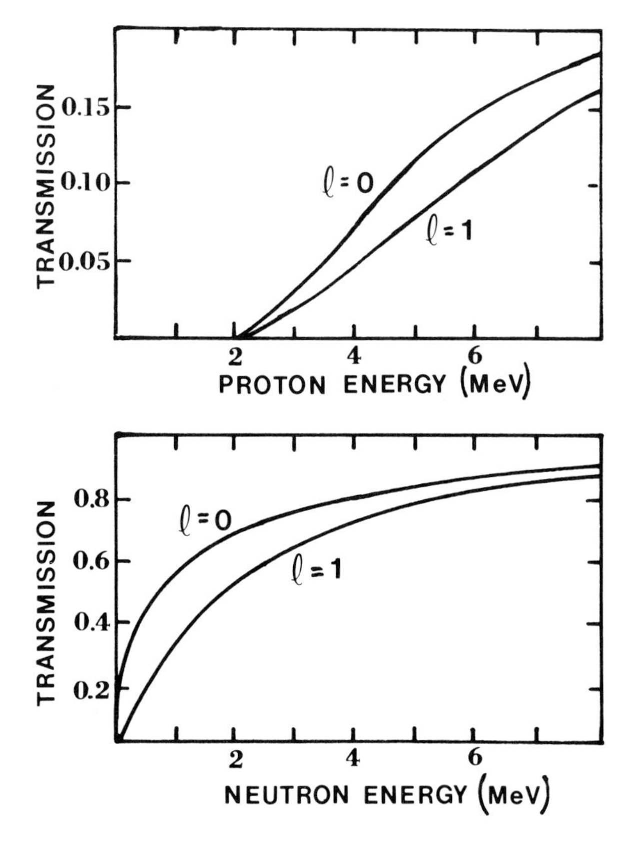

Figure 9.12 The transmission of the barrier for proton (*top*) and neutron (*bottom*) emission from a nucleus of radius (5) 10^{-15} m and $Z = 20$ is shown. Adapted from Ref. 4.

Q values are exceptionally large for fission of the very heavy nuclides. Figure 9.14 shows the binding energy per nucleon in the nucleus for stable nuclides (19). Notice that nucleons in heavy nuclei ($A > 50$) are less tightly bound than those in nuclei with $A \simeq 50$. Suppose a compound nucleus with mass number $235 + 1 = 236$ were to split into two equal mass fragments. From this figure the binding energy difference in this situation can be estimated as 1 MeV per nucleon; this corresponds to about 236 MeV released. Table 9.5 shows some data for the products of thermal neutron fission of ^{235}U. In fact, the total energy released is 195 MeV per reaction.

The Bohr and Wheeler liquid drop model for fission was published in 1939 (26). This model can be understood with reference to Figure 9.15. Prior to absorption of the neutron, the nucleus is assumed to be nearly spherical. Because of the excitation caused by absorption of the neutron, the compound nucleus becomes ellipsoidal and begins to oscillate along the axis of elongation. When the oscillation amplitude is large, a neck forms between the two end masses. When the coulomb repulsive force is sufficient, the masses are pushed completely apart. Prompt neutrons, which often come from the neck, are emitted immediately as each fragment reforms. Prompt gamma rays are emitted from the separated fragments, about $(3)10^{-14}$ s after neutron absorption. The fragments

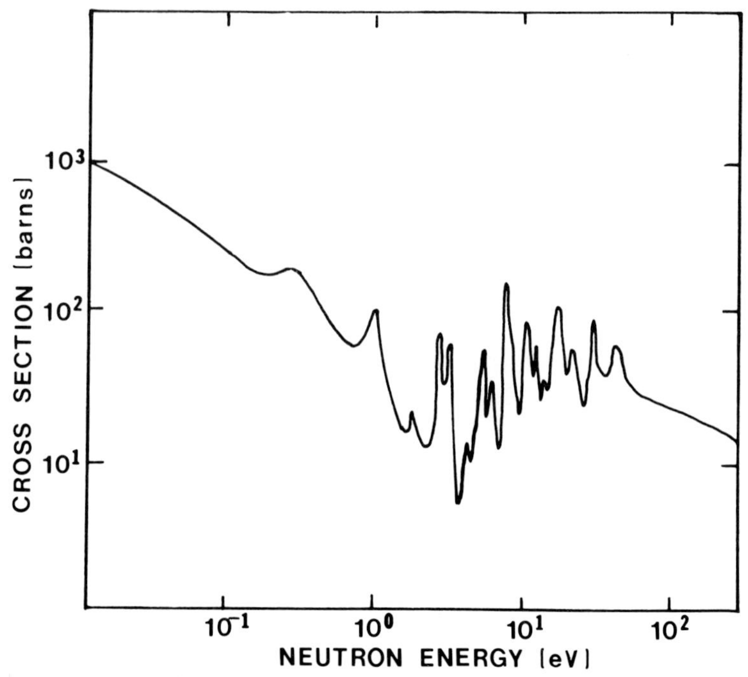

Figure 9.13 The fission cross section σ_{nf} is shown as a function of energy for ^{235}U (7–9).

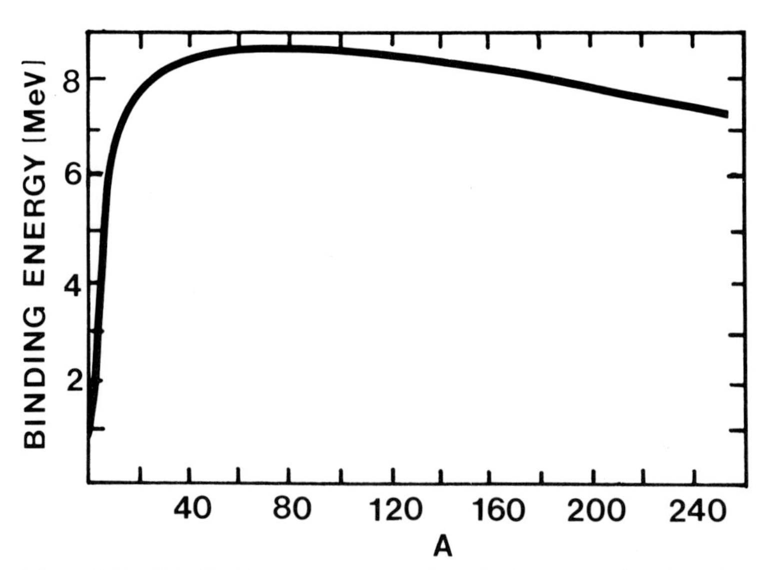

Figure 9.14 The binding energy per nucleon is shown as a function of mass number A (6, 19). Fluctuating individual values have been smoothed to obtain the curve shown.

Table 9.5 Energy Release in ^{235}U Thermal Neutron Fission

Energy release	*MeV*
1. Fragment kinetic	162
2. Neutron kinetic	5
3. Prompt gamma ray	7
4. Fragment decay	
beta particle	5
gamma ray	5
5. Neutrino	11
Total	195

Data from Ref. 17.

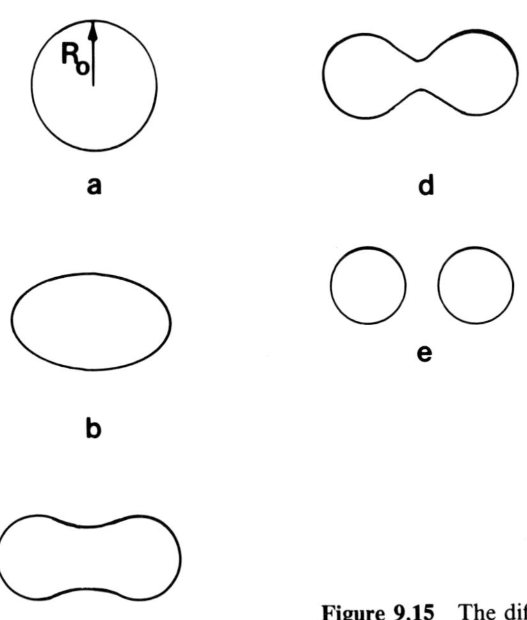

Figure 9.15 The different shapes of the compound nucleus during the fission process are shown.

are themselves usually neutron rich and so will decay by either delayed neutron emission or β^- decay (19).

The fission barrier can be discussed with reference to Figure 9.16. Consider three situations that might exist if two nuclear fragments were brought together from afar. At large distances the coulomb repulsive force would surely dominate, and it would increase if the separation was decreased. If the surfaces of the fragments came into contact, the attractive nuclear force would tend to draw them together and they would coalesce.

For case (a) on the figure, the attractive nuclear forces are so much stronger than the repulsive coulomb force that a maximum in the potential energy curve is produced near the point of contact. The potential energy decreases as the fragments coalesce. This produces a potential barrier of substantial height. Intermediate case (b) on the figure applies when the nuclear charge is very large and the repulsive energy of the merging fragments almost balances the nuclear binding force; there is still a fission barrier but reseparation is possible with slight excitation. This situation is appropriate for some fissionable heavy nuclei such as ^{235}U. For case (c), the coulomb repulsion is so large that there is no net binding force, and certainly no barrier. A nucleus in this category is not stable and will fission immediately.

As an example, consider the coulomb repulsive energy for equal mass fragments (symmetric fission), each with atomic number $Z/2$ and mass number $A/2$. The potential energy at contact, E_c, is given by

$$E_c = \frac{q^2}{4\pi\epsilon_0 d} = \frac{(Z/2)^2 e^2}{4\pi\epsilon_0 (2R)} = \frac{Z^2 e^2}{32\pi\epsilon_0 R} \tag{9.39}$$

where $q = +Ze/2$ is the fragment nuclear charge and $d = 2R$ is the separation distance when the fragments with radius R touch. For spherical fragments $R = R_0(A/2)^{1/3}$. Thus,

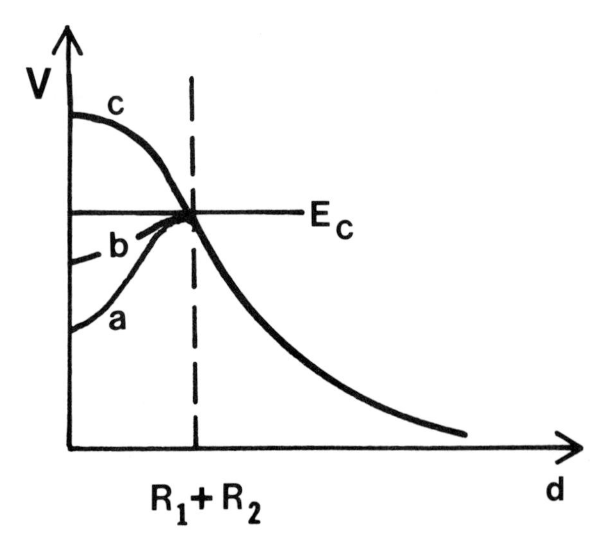

Figure 9.16 The fission barrier is illustrated for (a) low or medium Z nuclide; (b) high Z nuclide fissionable after absorption of a thermal neutron; and (c) high Z nuclide which fissions instantaneously. V is the potential energy, R the radial separation between fragment centers, and E_c the coulomb repulsive energy at contact. Adapted from Ref. 17.

$$E_c = \frac{Z^2e^2}{32\pi\epsilon_0 R_0 (A/2)^{1/3}} = \frac{2^{1/3}e^2}{32\pi\epsilon_0 R_0}\left[\frac{Z^2}{A^{1/3}}\right] \tag{9.40}$$

Although Equation 9.40 is strictly appropriate only for the special case of symmetric fission, the factor $Z^2/A^{1/3}$ is an important parameter in any case. Large values of $Z^2/A^{1/3}$ indicate a high probability of fission.

About 2.5 neutrons are emitted (19) per fission process. The spectrum of these liberated neutrons is shown on Figure 9.17; fast neutrons clearly dominate (27).

The yield for the nuclear fragments from neutron-induced fission of ^{235}U is shown on Figure 9.18. Notice the double hump with centroids at $A \simeq 95$ and $A \simeq 135$. The distribution in atomic number for the fission fragments has a low-mass hump centered around $Z = 38.5$ and a high-mass hump centered around $Z = 54.5$ (7–9). The asymmetry thereby represented is related to the fact that fission fragments with a magic number core of 50 protons are particularly stable. Since symmetric fission would require two fragments, each with $Z < 50$, it is not favored.

9.6 Attenuation Coefficients for Neutron Absorption Reactions

The neutron absorption reactions discussed in previous sections include radiative capture, $n\gamma$; particle emission reactions, *nb;* and neutron-induced fission, *nf.*

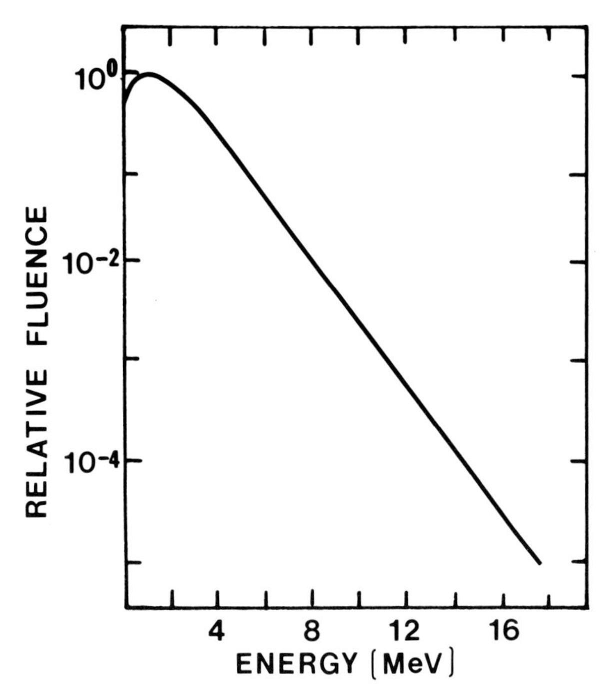

Figure 9.17 A typical energy distribution of fission neutrons is shown (27).

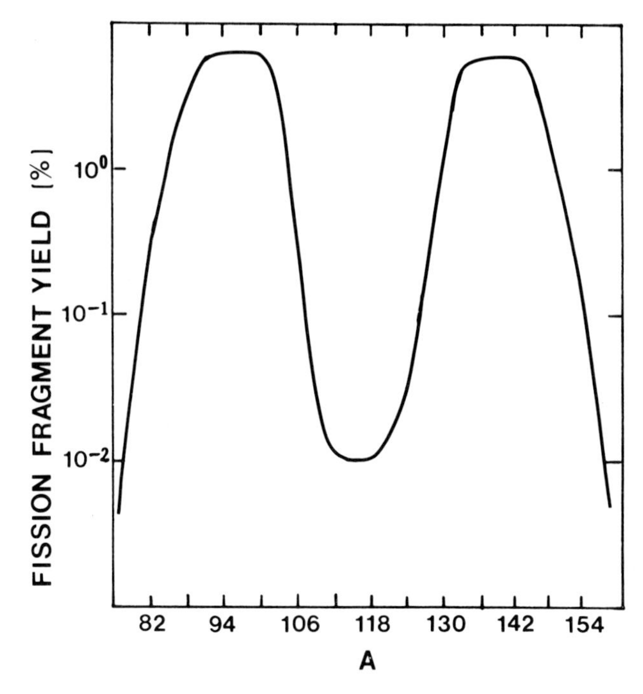

Figure 9.18 The mass yield curve for fragments from neutron-induced fission in ^{235}U is shown (2). Fluctuating individual values have been smoothed to obtain the curve shown.

The symbols given will be used as subscripts to specify particular coefficients as they are presented in the succeeding paragraphs.

The mass attenuation coefficient for a neutron absorption process, $(\mu/\rho)_a$, can be related to the reaction cross section by

$$(\mu/\rho)_a = n_m \sigma_a = (N_A/M_m)\sigma_a \tag{9.41}$$

where n_m is the number of targets per unit mass, N_A is Avogadro's number, and M_m is the mass per mole.

Figure 9.19 is a schematic illustration of the neutron mass attenuation coefficients for a medium weight nuclear target. Notice that usually radiative capture (subscript $n\gamma$) and elastic scattering (subscript es) are important at lowest energies. Elastic scattering often dominates at somewhat higher energies until the cumulative contribution of the particle emission reactions takes over. More precise regions of dominance for neutron reactions are difficult to define. The occurrence of resonances, which is highly variable from nuclide to nuclide, is a major factor. Thus, Figure 9.19 is intended to be representative but may not apply in all particular cases (light targets especially).

Values of the kerma per fluence for neutrons have been reported in several places (24, 28–30). The appropriate mass energy transfer coefficients can be obtained from the kerma per fluence by dividing by the incident kinetic energy. In the following discussion, we assume that the product nucleus is ionized by shake-off of valence electrons (31, 32) because of the nuclear recoil.

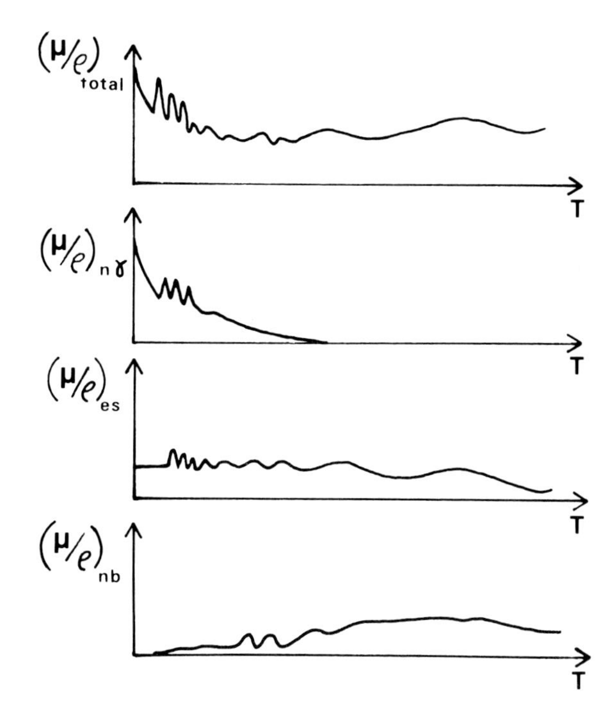

Figure 9.19 An illustration of the relationship between neutron attenuation coefficients for a typical medium mass nucleus is shown. Notice that the curve for the total coefficient is the sum of the other curves. Adapted from Ref. 11.

For radiative capture reactions, the mass energy transfer coefficient can be written as

$$(\mu_{tr}/\rho)_{n\gamma} = (\mu/\rho)_{n\gamma}[\langle T'_Y \rangle / T] \tag{9.42}$$

In this case, the average charged-particle kinetic energy liberated, $\langle T'_Y \rangle$, can be related to T through Equations 9.23. Momentum conservation insures that (T'_Y / T) will be small when heavy nuclear targets are involved. Since the liberated gamma-ray energies are not included in the mass energy transfer coefficients, they will not directly contribute to the neutron kerma per fluence. They will contribute to the total kerma and dose in an extended medium, however.

For charged-particle emission reactions with liberated kinetic energy $T'_b + T'_Y$,

$$(\mu_{tr}/\rho)_{nb} = (\mu/\rho)_{nb}[\langle T'_b + T'_Y \rangle / T] = (\mu/\rho)_{nb}[(T + Q)/T] \tag{9.43}$$

When Q is positive, this equation indicates that the transfer coefficient is greater than μ/ρ, an anomalous situation. In most cases Q is negative, so the factor multiplying the mass attenuation coefficient is less than one. Figure 8.14 shows the relative contributions of particle emission reactions to μ_{tr}/ρ for an oxygen target (27). Note that particle emission reactions do not contribute substantially to the charged-particle energy released until $T \simeq 4$ MeV. In nitrogen-containing substances, particle emission reactions contribute to the kerma even at thermal energies because the (n,p) reaction is exoergic.

9.7 Problems

1. Calculate the Q values and threshold energies for the $^{16}O(n,\alpha)^{13}C$ and $^{16}O(n,p)^{16}N$ reactions. Use the mass tables in the appendixes.
2. Use the Q equation to calculate the maximum energy of the protons emitted after 100-keV neutrons are absorbed in ^{14}N.
3. Consider the case of a 1-MeV neutron incident on a nucleus of mass number $A = 100$. Using $R = R_0A^{1/3} = (1.1)10^{-15}A^{1/3}$ m, calculate the time required for a single traversal of the nuclear diameter by the neutron. Assume a square well nuclear potential with depth of 10 MeV. Compare this time with an estimate of the compound nucleus lifetime from the uncertainty principle for a resonance with level width $\Delta E = 0.1$ eV.
4. Use the wave function of Equations 9.13 and substitute it into the time-dependent Schrödinger equation, $H\psi = i\hbar\frac{\partial\psi}{\partial t}$. Show that the time differentiation indicated results in an energy expression with real part E_n, which is measurable, and an imaginary part related to the decay constant.
5. Integrate the resonance excitation function of Equation 9.19 over the variable E^* from zero to infinity. Show that it is normalized. In addition, show that the level width Γ is in fact the full width at half maximum of the resonance line.
6. What thickness of cadmium is necessary to remove 99.9% of a beam of thermal neutrons? Assume negligible scatter.
7. Use the resonance excitation function of Equation 9.19 to calculate the absorption probability at energy $E^* = 0.025$ eV due to a resonance peak at $E_j^* = 20$ eV. In addition, calculate the same probability for a resonance peaking at 0.025 eV. Assume the resonance width is about 0.1 eV in both cases. Compare the values.
8. Show that the resonance level width in the laboratory system, Γ_{lab}, is larger than the level width in the center of mass system, Γ_{cm}, by

$$\Gamma_{\text{lab}} = \Gamma_{\text{cm}}\left[\frac{M_n + M_X}{M_X}\right]$$

9. What thickness of ^{10}B will remove 95% of a beam of 100-eV neutrons?
10. If the cross section for capture of thermal neutrons is 760 b for naturally occurring boron, what is the cross section for ^{11}B?
11. Calculate the fission barrier height $E_c - Q$ for symmetric neutron-induced fission of ^{235}U.
12. What is the relative probability of production of ^{131}I with respect to production of ^{137}Cs in thermal neutron fission of ^{235}U? Use the double-hump curve on Figure 9.18. What is the relative probability of production of ^{99}Mo with respect to production of ^{137}Cs?

9.8 References

1. H. Frauenfelder and E. M. Henley, *Subatomic Physics* (Prentice-Hall, Englewood Cliffs, NJ, 1974).

2. A. M. Weinberg and E. P. Wigner, *The Physical Theory of Neutron Chain Reactors* (University of Chicago, Chicago, 1958).
3. W. J. Hornyak, T. Lauritsen, P. Morrison, and W. A. Fowler, Rev. Mod. Phys. **22**, 291 (1950).
4. J. M. Blatt and V. J. Weisskopf, *Theoretical Nuclear Physics* (John Wiley, New York, 1952).
5. R. D. Evans, *The Atomic Nucleus* (McGraw-Hill, New York, 1955), pp. 441–469.
6. L. Mattauch, J. Thiele and A. Wapstra, Nucl. Phys. **67**, 1 (1965).
7. D. J. Hughes and R. B. Schwartz, *Neutron Cross Sections, Report BNL 325,* 2nd ed. (Brookhaven National Laboratory, Upton, NY, 1938).
8. D. J. Hughes, B. A. Magurno, and M. K. Brussel, *Neutron Cross Sections, Report BNL 325,* 2nd ed., Suppl. 1. (Brookhaven National Laboratory, Upton, NY, 1960).
9. J. R. Stehn, M. D. Goldberg, B. A. Magurno, and R. Wiener Chasman, *Neutron Cross Sections, Report BNL 325,* 2nd ed., Suppl. 2. (Brookhaven National Laboratory, Upton, NY, 1964).
10. N. Bohr, Nature **137**, 344 (1936).
11. W. E. Meyerhof, *Elements of Nuclear Physics* (McGraw-Hill, New York, 1967), pp. 172–73.
12. F. Ajzerberg and T. Lauritsen, Rev. Mod. Phys. **24**, 321 (1952).
13. G. Breit and E. P. Wigner, Phys. Rev. **49**, 519 (1936).
14. E. Segre, *Nuclei and Particles* (W. A. Benjamin, New York, 1965).
15. M. Born, Z. Phys. **37**, 863 (1926).
16. L. I. Schiff, *Quantum Mechanics,* 2nd ed. (McGraw-Hill, New York, 1955).
17. A. P. Arya, *Fundamentals of Nuclear Physics* (Allyn and Bacon, Boston, 1966).
18. R. V. Churchill, *Fourier Series and Boundary Value Problems* (McGraw-Hill, New York, 1941), p. 92.
19. B. L. Cohen, *Concepts of Nuclear Physics* (McGraw-Hill, New York, 1971).
20. E. Amaldi, O. D'Agostino, E. Fermi, B. Pontecorro, F. Rasetti, and E. Segre, Proc. R. Soc. London Ser. A **149**, 522 (1935).
21. *Handbook on Nuclear Activation Cross Sections, Technical Report Series No. 156* (International Atomic Energy Agency, Vienna, 1974).
22. *Protection Against Neutron Radiation, NCRP Report No. 38* (National Council on Radiation Protection and Measurements, Washington, 1971).
23. J. A. Auxier, W. S. Snyder, and T. D. Jones, "Neutron interactions and penetration in tissue," in *Radiation Dosimetry,* vol. 1, edited by F. H. Attix and W. C. Roesch (Academic, New York, 1968), p. 275.
24. *Neutron Dosimetry for Biology and Medicine, ICRU Report No. 26* (International Commission on Radiation Units and Measurements, Washington, 1977).
25. W. E. Alley and R. M. Lessler, Nucl. Data Tables **11** (1973).
26. N. Bohr and J. A. Wheeler, Phys. Rev. **56**, 426 (1939).
27. B. E. Watt, Phys. Rev. **87**, 1037 (1952).
28. R. L. Bach and R. S. Caswell, Radiat. Res. **35**, 1 (1968).
29. R. H. Bassel and G. H. Herling, Radiat. Res. **69**, 210 (1977).
30. P. J. Dimbylow, Phys. Med. Biol. **25**, 637 (1980).
31. J. S. Levinger, Phys. Rev. **90**, 11 (1953).
32. T. A. Carlson and M. O. Krause, Phys. Rev. Sect. A **140**, 1057 (1965).

10

Photon and Neutron Energy Deposition in Depth

10.1 Introduction

Energy deposition in an absorber due to photon or neutron fluences will be discussed in this chapter. The mathematical analysis used will be sufficiently general to include both of these radiations. From the point of view of energy transfer, photons and neutrons both participate in scattering and absorption processes, and both generate charged and uncharged secondaries. Figures 10.1 and 10.2 show the kerma per fluence separated into components due to scattering and absorption (1–4) for a small mass of tissue. In both figures, it is evident that at intermediate energies (about 1 MeV), scattering reactions transfer most of the energy. After degradation, at lowest energies, absorption reactions dominate the energy transfer. The situation for neutrons is unique in that a substantial portion of the available energy is transferred to gamma rays. Furthermore, the shape of the neutron absorption curve in tissue is not representative for all low Z absorbers. Soft tissue (76% O, 11% C, 10% H, 3% N) contains (5) a substantial amount of nitrogen by weight, and low-energy neutron absorption in tissue is directly dependent on the nitrogen content.

Energy degradation of photons and neutrons can involve many processes. Figures 10.3 and 10.4 show that a great variety of secondary radiations can be involved. More than 30 separate interaction steps may be needed to convert the electromagnetic energy of a 1-MeV photon into kinetic energy (6) in tissue. Fewer interactions would be required in heavy absorbers because of the increased probability of the photoelectric effect. Degradation of the energy of a 1-MeV neutron requires fewer than 30 elastic scattering events for light absorbers. Considerably more elastic scattering events are required for heavy absorbers, but

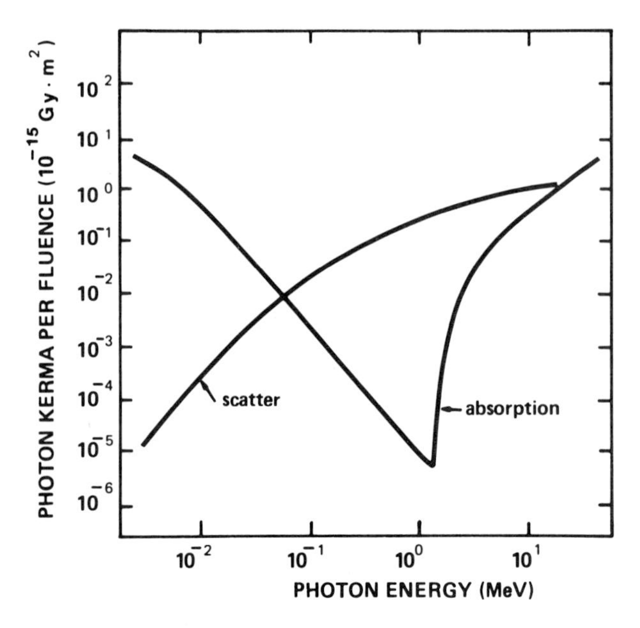

Figure 10.1 The curves show the kerma per fluence as a function of photon energy for a small tissue mass. The contribution from scatter reactions is shown separately from the contribution from absorption reactions (1).

inelastic scattering and particle emission reactions aid in the neutron degradation process in that case.

Since neither photons nor neutrons have a net charge, the pattern of energy deposition in depth for these radiations is different than was described in Chapter 4. The concept of path length, valid for charged particles, is not very useful for photons or neutrons. A charged particle usually participates in a multitude

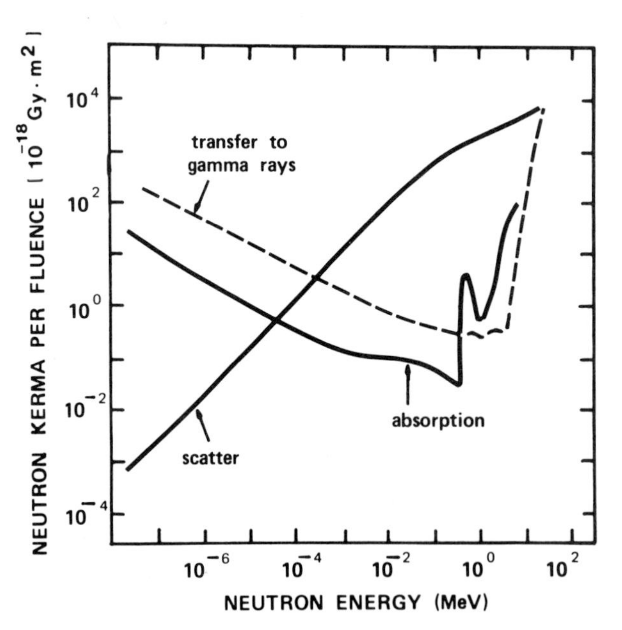

Figure 10.2 The curves show the kerma per fluence as a function of neutron energy for a small tissue mass. The dashed line is not actually a kerma curve but represents the energy per unit mass per unit fluence transferred to gamma rays (2–4). Adapted from Ref. 3.

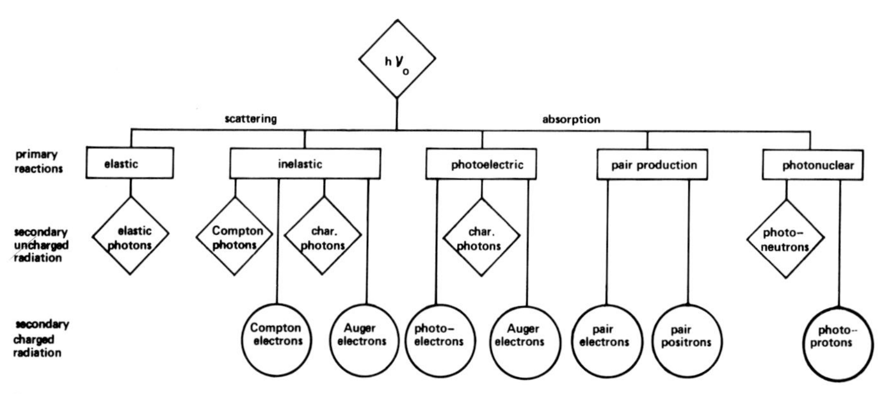

Figure 10.3 The block diagram shows primary reactions and secondary radiations for a photon indicated by $h\nu_0$ in an absorber.

of discrete events spaced at relatively regular intervals before it is stopped. Thus, a plot of the dose deposited versus penetration depth falls to zero abruptly when the kinetic energy is dissipated. Monoenergetic segments of a primary beam of photons or neutrons participate in events governed by the exponential attenuation law. Thus, many of these radiations travel relatively short distances before absorption, while other identical radiations by chance may penetrate to much more substantial depths. Because of this property, curves of dose versus depth maintain a gradually diminishing magnitude for indirectly ionizing species.

In the following section, the origins and composition of charged secondary radiations will be discussed in some detail for both photon and neutron primaries. The charged secondaries will be regarded as a connecting link between the kerma and the dose. In Section 10.3 the effect of buildup of uncharged secondary and higher order products on the depth dose will be discussed, and the barrier transmission will be defined.

In Section 10.4 we outline a method requiring the use of the absorbed

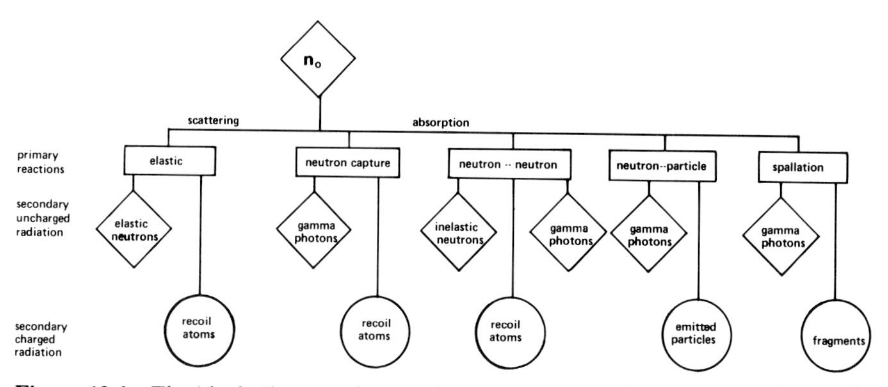

Figure 10.4 The block diagram shows primary reactions and secondary radiations for a neutron indicated by n_0 in an absorber.

fractions for calculating internal doses from deposited radionuclides. In Section 10.5 an analytical description of the central axis depth-dose curve for an incident beam of indirectly ionizing radiation will be developed in terms of an effective attenuation coefficient and a scatter buildup factor. The algebraic description given is not necessarily optimal for an efficient description of experimental data, but it is conceptually useful because major effects are clearly delineated.

Section 10.6 is included to acknowledge that indirectly ionizing radiation beams are often used to produce radioactive samples. Some fundamental expressions involving cross sections and irradiation times are given.

10.2 Charged Secondary Radiations

The chain of energy transfer for photon or neutron beams leads through the primary reactions to charged-particle secondaries. It can then be traced through ionization and excitation processes to local atomic sites. The kerma is a convenient intermediate quantity for use in a discussion of energy deposition by indirectly ionizing radiation because it is related to the kinetic energy given to charged secondaries. It is applicable to both photon and neutron beams and is directly related to the source fluence. In addition, it represents the released energy per unit mass and therefore can be expressed in grays, the unit for absorbed dose.

The total kerma K_t has three major components in an absorber under irradiation:

$$K_t = K_p + K_s + K_e \tag{10.1}$$

The primary kerma K_p is attributable to photons or neutrons coming directly from the source with no intervening events. The kerma from scattered radiation, K_s, is released by indirectly ionizing components degraded by one or more scattering events. For example, Compton scattered photons would contribute in an incident photon beam. K_e is the kerma from radiation emitted in the absorber but initiated by the primary radiation. It can include energy released by secondary components, such as characteristic photons, bremsstrahlung, annihilation photons, and even photoneutrons from high-energy photon sources. For neutron sources the emitted secondaries might include capture gamma rays and de-excitation gamma rays emitted after inelastic reactions. An array of scattered and emitted secondaries is indicated in Figures 10.3 and 10.4.

The kinetic energy released in the absorber by a photon beam is given to photoelectrons, Compton electrons, and electrons and positrons from pair production for the most part. The properties of these secondaries have been discussed in Chapter 6 and 7. If they have sufficient energy, they will carry energy away from the site of liberation to well removed atomic sites. Because of this transport, the secondary energies and emission angles are of significance.

Recall that the angular distributions for photoelectrons, Compton electrons, and electron-positron pairs generally favor forward directions when the incoming photon energy is substantial. Figure 10.5 shows some evidence of

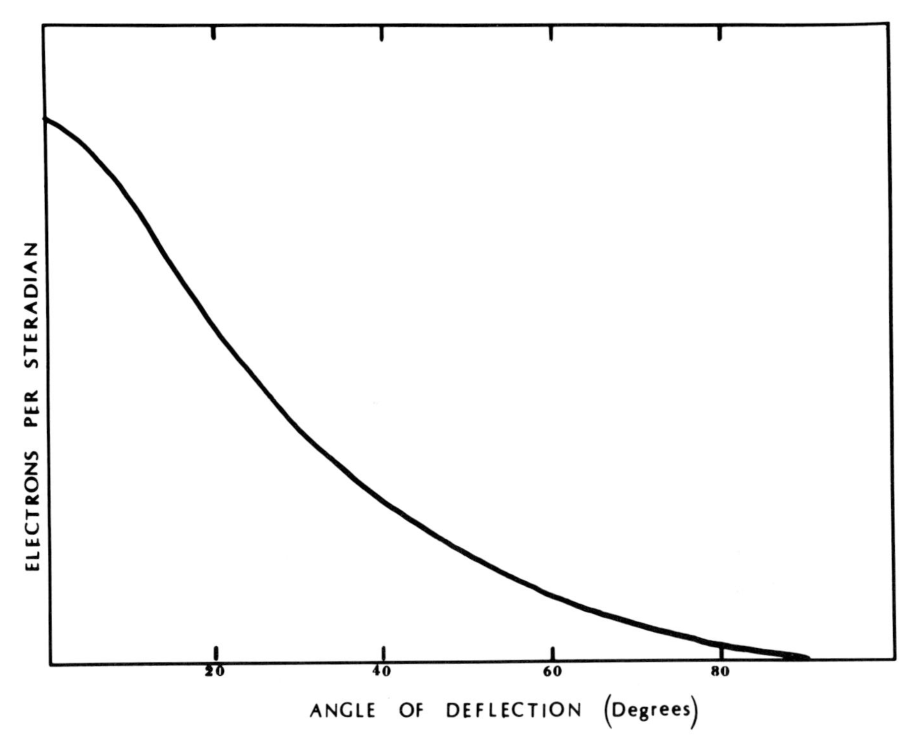

Figure 10.5 The angular distribution of electrons emitted from a 6-kg/m² aluminum absorber by ^{60}Co gamma rays is shown (7).

this fact for the case of ^{60}Co photons incident on a 6-kg/m² aluminum absorber (7). Thus, at sufficiently high incident beam energies, the dominant projection of the charged-particle momentum is along the incoming direction, and the dose deposited in depth is displaced from the kerma in the forward direction. On the average the distance the energy will be carried from the point of release can be no greater than the range of the most energetic secondaries. Table 5.2 gives some pertinent electron ranges for comparison purposes.

Auger (pronounced "OJ") electrons comprise a component in the array of charged-particle secondaries that has not been discussed yet. The Auger effect is the emission of an orbital electron after the production of an inner-shell vacancy. Thus, it is a possible de-excitation process after photoelectric absorption. Generally, there are three atomic subshells involved in an Auger process, as illustrated in Figure 10.6. They need not all be different subshells, however. As an example, a *KLM* Auger process means that a *K*-shell vacancy was filled by a transition from an *L*-shell electron, which resulted in emission of an *M*-shell electron (Auger electron) from the atom.

The Auger effect is governed by coulomb interactions of the electrons (8), and electrons from all subshells satisfying the energy considerations can participate. For a general example, consider the filling of a vacancy in subshell

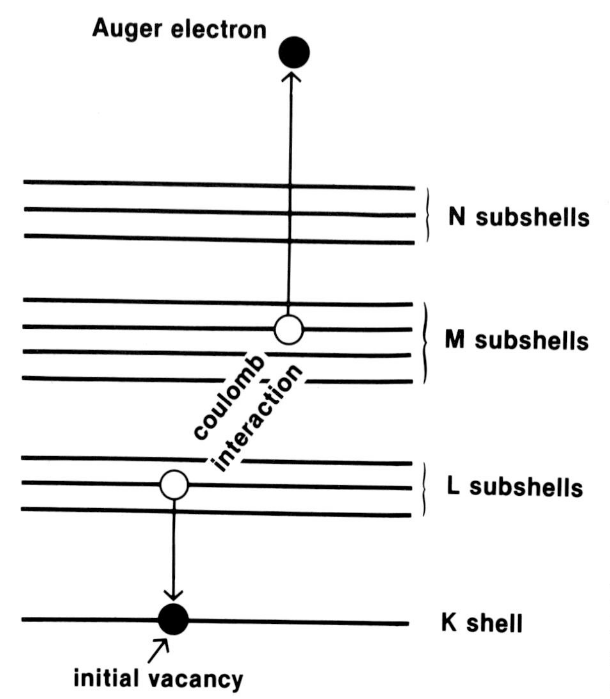

Figure 10.6 An energy level scheme is shown to illustrate the Auger effect.

Q by an electron transition from subshell R followed by emission of an electron from subshell S, where $BE_Q - BE_R \geq BE_S$. Each of the letters Q, R, and S could stand for subshells K, L_I, L_{II}, L_{III}, M_I, M_{II}, M_{III}, and so forth, in principle. With this notation the kinetic energy T of the Auger electron is related by

$$BE_Q - BE_R = T + BE_S, \quad T = BE_Q - BE_R - BE_s \tag{10.2}$$

where BE represents the electron binding energy for the subshell indicated. In atomic systems, valence electrons are partially shielded from the nuclear charge by inner-shell electrons. In the case of ions with single inner-shell vacancies, the effective nuclear charge for binding is about $+\frac{1}{2}e$ larger than for the neutral atom because of reduced shielding. Since an Auger electron is released from a charged ion, BE_S should be increased from the neutral atom value. An average of the appropriate binding energy in the neutral atom and the value for an atom with the atomic number increased by one can be used as an estimate in circumstances with singly charged ions.

In tissue constituents the energies of Auger electrons are relatively small, and they are of minor significance in the electron milieu in an absorber. Auger emission may have a pronounced effect on the particular molecule directly involved, however. This is because a single inner-shell vacancy can lead to two other vacancies, and so forth. For intermediate and heavy atoms, the potential for a shower process exists. Chemical bonds in an involved molecule are likely to be severely disrupted in such a circumstance (9).

For neutron beams the kinetic energy released can be given to a variety of ions, depending on the absorbing material. In hydrogenous absorbers, protons from elastic scattering are important. Other species such as helium ions from (n, α) reactions may also be present as well as residual ions from such reactions. Figure 10.7 shows the kerma for a 5-MeV neutron beam as a function of depth in a tissue absorber (4); notice that components from protons dominate at all depths shown. However, the component attributable to electrons becomes important at large depths. For neutron kinetic energies below about 10 keV, the electron kinetic energy is the most important component in tissue absorbers.

The radiative capture reaction $H(n, \gamma)D$ is a significant intermediary for production of fast electrons in hydrogenous absorbers (3, 4). The 2.2-MeV capture gamma ray comprises a substantial part of the radiation fluence in these absorbers when the incident neutron kinetic energy is low. Generally, gamma rays emitted from nuclei excited during various neutron reactions are present when neutron beams are absorbed. The resulting tertiary electron fluence consists of photoelectrons, Compton electrons, electron-positron pairs, and the like, set free by interactions of the secondary gamma rays.

The directions of the fast-proton secondaries emerging after elastic neutron scattering reactions have substantial forward components, as illustrated in Figure 8.8. Because of this forward projection, the dose distribution for high-energy neutron beams is displaced from the kerma distribution in the direction of the incident radiation. This is similar to the situation for photon beams except that the ranges of the secondary protons are much smaller than the ranges of secondary electrons of comparable energy from photon beams. Thus, the displacement between release and energy deposition for the charged secondaries is comparatively small for neutron beams.

The most striking examples of the effect of forward projection of secondary charged particles occur in depth-dose distributions near the entry surface of a

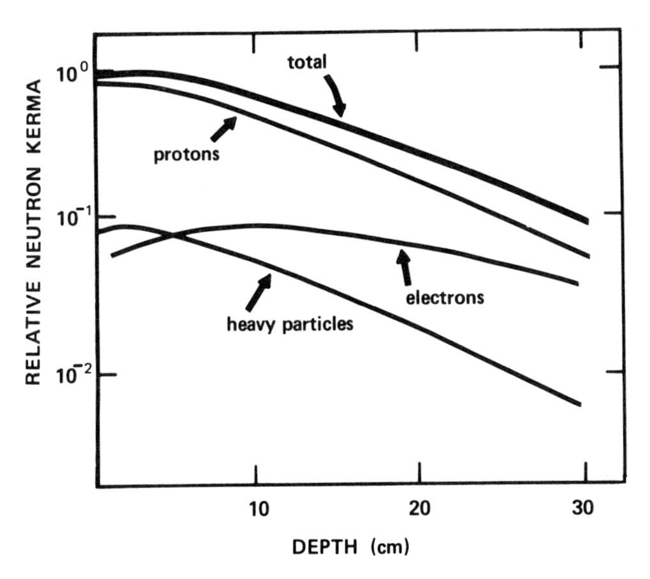

Figure 10.7 The total kerma in tissue can be divided into contributions given to protons, heavy ions and electrons for a 5MeV neutron beam. Adapted from Ref. 4.

beam into an absorber. The dose builds up to a maximum at some depth d_m under the surface as the secondary charged-particle fluence builds up. This phenomenon has already been mentioned in Chapter 5. At greater depth, beam attenuation causes a more gradual decrease in dose with increasing depth. Of course, the thickness of the charged-particle buildup layer depends on the beam quality. Figure 10.8 shows some values of the buildup thickness for a water absorber (10). Generally, the higher the primary energy, the larger the buildup depth.

Table 10.1 gives some results for dose measured in the charged-particle buildup region of inhomogeneous photon or neutron beams (11–13). The headings for the columns indicate the beam quality. For example, 25 MV is used to denote the inhomogeneous bremsstrahlung beam produced by 25-MeV electrons. For neutrons the heading 50 MV denotes the inhomogeneous neutron beam produced by incident 50-MeV charged particles. For neutron beams, several incident ions and targets can be used. Thus, the reaction $^{9}Be(d,n)^{10}B$ where d (deuteron) and n (neutron) indicate the incident and outgoing particles, respec-

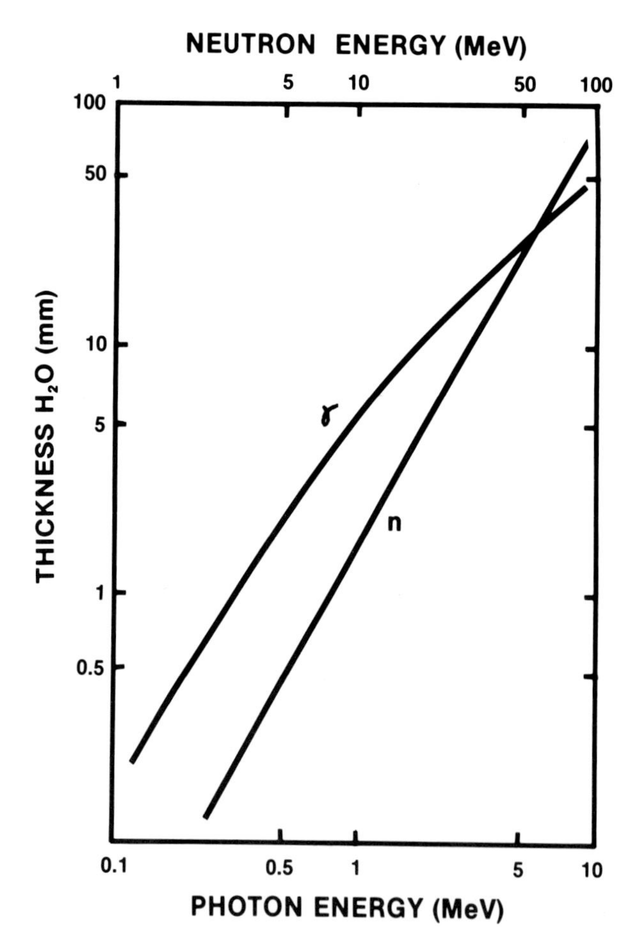

Figure 10.8 The water thickness necessary to achieve complete charged-particle buildup is shown (9).

Table 10.1 Relative Dose in Charged-Particle Buildup Region

Depth (mm)	Photon Beams		Neutron Beams	
	^{60}Co	25 MV	16-MV $^{9}Be(d,n)^{10}B$	50-MV $^{9}Be(d,n)^{10}B$
1	70	28	87	72
2	90	40	99	86
3	98	47	100	91
4	100	55		96
5		61		98
6		66		99
8		73		100
10		79		
15		89		
20		95		
25		99		
30		100		
35				

Data from Refs. 11–13. Values are expressed in percent of maximum value.

tively is also specified for this column. For inhomogeneous beams the maximum dose occurs at a depth related to the average range of the secondaries. Thus, low-energy x-ray machines produce no measurable buildup region, but high-energy beams from accelerators produce a substantial region of charged-particle buildup.

10.3 Scatter Buildup Effects

In many practical situations, the dose expected from a radiation source must be estimated prior to actual use of the source. When shielding barriers surround the source, barrier transmission values are helpful in making the estimate. We will discuss barrier transmission in the present section, with particular emphasis on the effects of scattered radiation. The principles introduced in this section will be further demonstrated in development of the schemata for calculating doses from distributed sources and calculating central axis depth-dose values for radiation beams.

As a first step, consider a monoenergetic, isotropic point source of indirectly ionizing radiation. The primary source fluence ϕ_p at a distance r is related directly to N, the total number of source events producing at least one emission, and to the branching ratio w, which is the fraction of emission events producing radiation of the type and energy under consideration:

$$\phi_p = \frac{wN}{4\pi r^2} \exp(-\mu r) \tag{10.3}$$

The factor $1/4\pi r^2$ occurs because the emitted particles intercept a spherical shell of area $4\pi r^2$ at distance r. The exponential factor is due to attenuation of the primary fluence in a surrounding absorber with attenuation coefficient μ.

The energy deposited in the absorbing material because of the indirectly ionizing radiation depends on the fluence ϕ and the energy E of the radiation emitted by the source. We will use the kerma in this section because it is directly related to these two variables and is expressed in units of energy density. In Chapter 5 the kerma K was given as

$$K = E(\mu_{tr}/\rho)\phi \tag{10.4}$$

where μ_{tr}/ρ is the mass energy transfer coefficient. The kerma per fluence was shown on Figures 10.1 and 10.2 for a photon and a neutron beam, respectively, in a tissue absorber. Table 10.2 shows specific values of the kerma per fluence for several common photon and neutron sources (1, 4, 15–16). An average energy and average kerma per fluence is listed for the heterogeneous neutron sources.

Combining Equations 10.1, 10.3, and 10.4 gives

$$K_t = K_p + K_s + K_e = K_p\text{BF}(r), \quad \text{BF}(r) = \left[1 + \frac{K_s + K_e}{K_p}\right]$$

$$K_t = \frac{wNE}{4\pi r^2}(\mu_{tr}/\rho)\text{BF}(r)\exp(-\mu r) \tag{10.5}$$

The scatter buildup factor BF(r) corrects the kerma expression for energy released from scattered and emitted radiation in the absorber. The scatter buildup referred to here for *indirectly ionizing components* should not be confused with *charged-particle* buildup discussed in Section 10.2.

Kerma and dose values from isotropic point sources are relatively easy to calculate. Line sources, cylindrical sources, and spherical sources can be treated as point sources if the distance from the center of the source to the point of interest is much greater than the source dimensions and if self-absorption is negligible (17). When cylindrical and spherical sources with appreciable self-absorption are utilized, primary attenuation and scatter buildup cause complications. Table 10.3 gives some useful relationships for several source geometries

Table 10.2 Fluence-to-Kerma Conversion Factors for Radioactive Sources

	Average Values		
		$\langle K/\phi \rangle (Gy \cdot m^2)$	
Source Type	$\langle E \rangle$ *(MeV)*	*Air*	*Tissue*
^{99m}Tc photons	0.14	0.68	0.73 $\times 10^{-16}$
^{137}Cs—^{137}Ba photons	0.66	2.92	3.22 $\times 10^{-16}$
^{60}Co—^{60}Ni photons	1.25	5.36	5.90 $\times 10^{-16}$
Am—Be neutrons	4.3		3.8 $\times 10^{-15}$
Pu—Be neutrons	4.1		3.7 $\times 10^{-15}$
Ra—Be neutrons	4.5		3.8 $\times 10^{-15}$

Data from Refs. 1, 4, 14–16.

Table 10.3 Approximate Transmission Tm for Several Source Geometries[a]

Expressions	*Coefficients*
A. Cylindrical source (length l, radius R) 1. Along axis: $Tm = \exp(-\mu' l/2)$ 2. Normal to axis a. No extra cylindrical shielding: $Tm = \exp(-8\mu' R/3\pi)$ b. Extra cylindrical shielding; shell of thickness Δr: $Tm = \exp(-8\mu' R/3\pi)\exp(-1.2\mu'\Delta r)$ B. Spherical source (radius R): $Tm = \exp(-3\mu' R/4)$	Effect of degraded radiation excluded: $\mu' = \mu$ Effect of radiation proportional to energy released: $\mu' = \mu_{tr}$ Effect of radiation possible only after decreased energy: $\mu' < 0$

Data from Ref. 18.
[a] Conditions: 1) source to detector distance must be much greater than any source dimension; 2) transmission must be greater than 0.9.

(18); notice that several special cases can be considered. If the effect of the degraded radiation on the system of interest can be neglected, the attenuation is best described using the total coefficient for the primary beam. If the effect of the radiation on the system of interest is proportional to the charged-particle energy released, one can approximate the effective coefficient with the energy transfer coefficient $\mu \simeq \mu_{tr}$.

When the released kinetic energy is of utmost importance, the product $\mathrm{BF}(r)\exp(-\mu r)$ for the point source in Equations 10.5 can be approximately replaced by $\exp(-\mu_{tr} r)$. The rationale for this substitution is that the energy transfer coefficient does not include scattered energy and energy emitted in secondary processes. Treating the scattered and emitted fractions as though they were transmitted primary tends to compensate for the buildup factor. As Figure 10.9 shows, this approximation is valid only for small absorber thickness (19); it fails when the absorber thickness approaches a mean free path length. In reality, the degraded secondary radiation is more readily absorbed than the primary component. Because of this, the measured kerma at some substantial distance is less than the value calculated using μ_{tr}.

When an energy-sensitive detector is located in an infinite absorbing medium at distance r from a central source emitting photon energy $h\nu$, the buildup factor can be approximated by

$$\mathrm{BF}(h\nu,r) = 1 + \beta(h\nu,Z)\mu r = 1 + \beta(h\nu,Z)r/\lambda \qquad (10.6)$$

The linear buildup constant $\beta(h\nu,z)$, appropriate for this geometry, has been tabulated for various photon energies and absorbers (20). As a first approximation, $\beta(h\nu,Z)$ can be set equal to one for photons with $h\nu \simeq 2$ MeV and for low Z absorbers. In this case the radial distance expressed in mean free path units, $\mu r = r/\lambda$, is the important parameter. Table 10.4 shows some values which approximate the buildup factor $\mathrm{BF}(h\nu,r)$ for photon point sources in water (21). Notice that the buildup factor becomes very large when the absorber thickness is equivalent to many mean free path lengths.

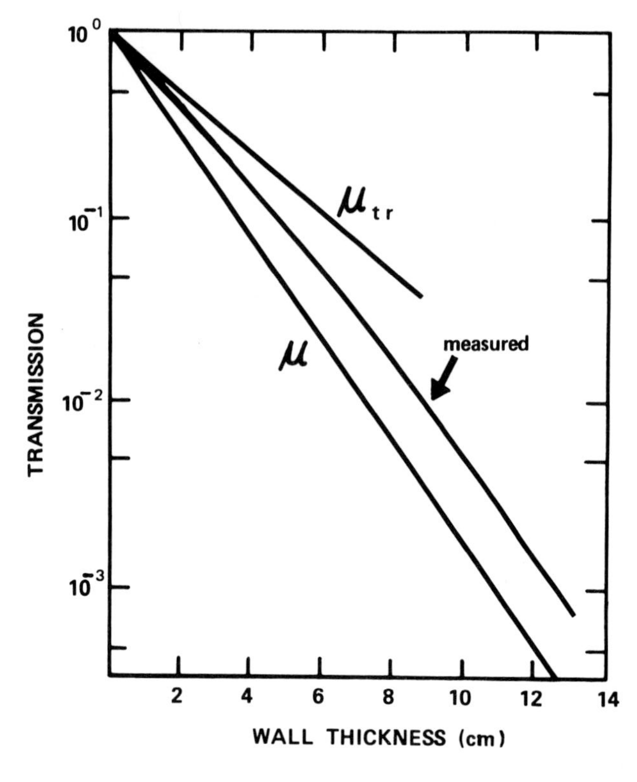

Figure 10.9 The measured curve is the transmission of ^{60}Co gamma rays through a surrounding lead shield (19). The exponential curve using μ_{tr} approximates the measured curve for small thicknesses.

Consider the case when a shielding barrier is interposed between an indirectly ionizing radiation source and the area of interest. Suppose that the barrier surface is at a distance F from the source that is much greater than any source dimension. Of interest is the kerma, $K_t(d,F,A/P)$, at a depth d within the barrier

Table 10.4 Buildup Factors for Photon Beams

Material	*Energy (MeV)*	*Number of Mean Free Paths*			
		1	*2*	*4*	*10*
Water	0.255	3.09	7.14	23.0	166
	0.5	2.52	5.14	14.3	77.6
	1.0	2.13	3.71	7.68	27.1
	2.0	1.83	2.77	4.88	12.4
	4.0	1.58	2.17	3.34	6.94
	10.0	1.33	1.63	2.19	3.72
Lead	0.5	1.24	1.42	1.69	2.27
	1.0	1.37	1.69	2.26	3.74
	2.0	1.39	1.76	2.51	4.44
	4.0	1.27	1.56	2.25	5.44
	10.0	1.11	1.23	1.58	4.34

Data from Ref. 21, calculated for dose build-up.

and on a line from the source that is perpendicular to the barrier surface (central axis). If $K_p(O,F)$ is the primary kerma incident on the barrier surface, then

$$K_t(d,F,A/P) = K_p(O,F)\mathrm{BF}(d,A/P)\left[\frac{F}{F+d}\right]^2 \exp(-\mu_{\mathrm{eff}}d) \qquad (10.7)$$

This result is easily understood with reference to Equation 10.5. The inverse square factor $[F/(F + d)]^2$ occurs because of continued beam divergence as it travels from the surface of the barrier to depth d. An effective attenuation coefficient μ_{eff}, which is a function of the average energy of the primary fluence, is used in the exponential. The coefficient itself will be an implicit function of depth if spectrum filtration effects are important. The buildup factor is given as a function of depth d and A/P, the area-to-perimeter ratio for the radiation field at the absorber surface. The area-to-perimeter ratio is utilized when collimated sources are involved. It has been shown to be an efficient parameter for use in describing scatter buildup effects in such situations (22).

The National Commission on Radiation Protection and Measurements (NCRP) has defined the transmission as the ratio of detector response behind a shield to the detector response at the same location without a shield (23). Presumably, either particle-sensitive or energy-sensitive detectors could be employed. At this point we restrict ourselves to energy-sensitive detectors and choose a kerma response ratio rather than a dose ratio, so that complications from projected secondary charged particles are avoided. In this case the transmission $Tm(d,A/p)$ for a barrier of thickness d is the ratio of the total kerma at depth d in a thick barrier to the total kerma at the same point in space if the initial absorber thickness d were removed. Using Equation 10.7,

$$Tm(d,A/P) = \frac{K_t[d,F,(A/P)_F]}{K_t[O,(F+d),(A/P)_{F+d}]} = \frac{\mathrm{BF}[d,(A/P)_F]}{\mathrm{BF}[O,(A/P)_{F+d}]}\exp(-\mu_{\mathrm{eff}}d) \qquad (10.8)$$

The subscripts on the area-to-perimeter ratios give the distances for evaluation of these ratios. With the definition given in Equation 10.8, the kerma due to radiation scattered backward from the barrier cancels out of the transmission expression and only forward scatter components are included. Thus, the transmission from Equation 10.8 is a special case of the NCRP definition.

Generally, the scatter buildup factor for collimated indirectly ionizing beams approaches a limiting value as the field dimensions on the absorber surface become large compared to the mean free path of the radiation. This occurs because radiation scattered and emitted from points well removed in a very large field are likely to be absorbed before they reach the volume of interest. The broad-beam limit, defined by

$$\mathrm{BF}(d,A/p) \rightarrow \mathrm{BF}_{\max}(d), \quad A/p \gg \lambda \qquad (10.9)$$

is useful in design of shielding barriers for radiation-generating equipment before actual installation. In this limit one obtains the maximum transmission that could occur.

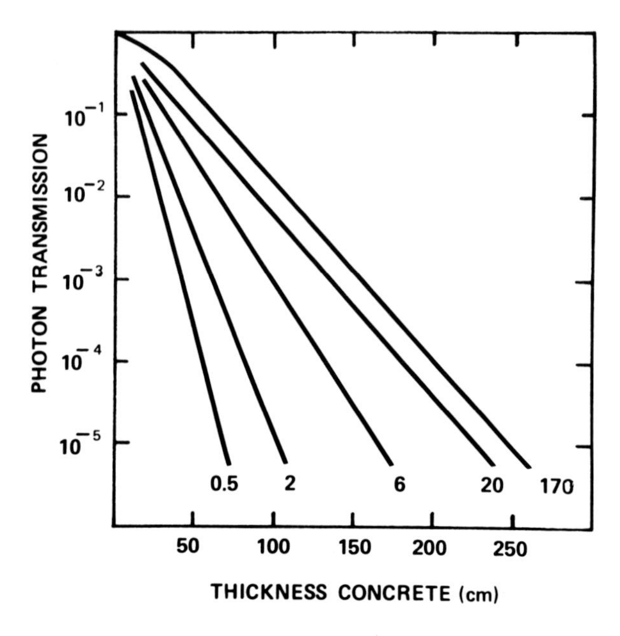

Figure 10.10 The broad-beam transmission of several bremsstrahlung beams through concrete is shown (23). The numbers are end-point energies in megaelectron volts. Adapted from Ref. 23.

Broad-beam transmission curves for concrete are shown on Figures 10.10 and 10.11 for several photon and neutron beams, respectively (5, 23). The dashed curve on Figure 10.11 includes the kerma contribution due to secondary gamma rays as well as neutrons. Note that both the photon and the neutron curves initially deviate from the exponential shape because of changes in the buildup

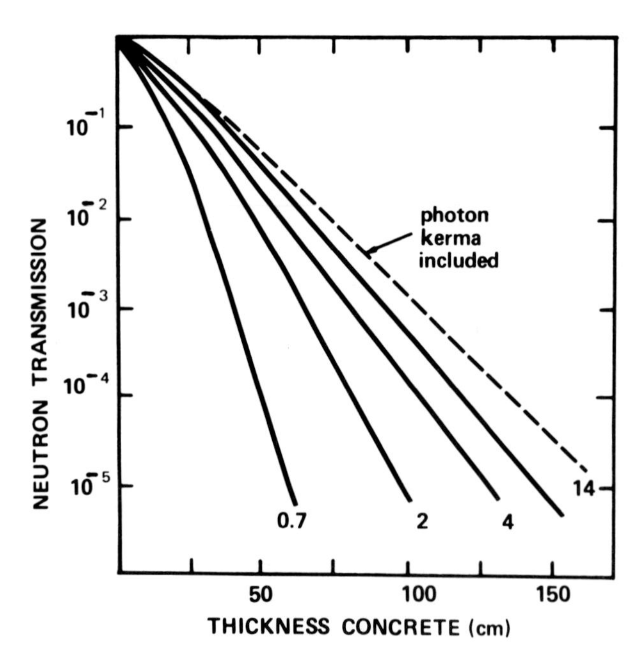

Figure 10.11 Several broad-beam transmission curves for monoenergetic neutrons in concrete are shown (5). The numbers are neutron energies in megaelectron volts. The dashed curve is the sum of the transmitted kerma for 14-MeV neutrons and the kerma due to secondary photons.

factor and effective attenuation coefficient with depth. Nonetheless, after depths greater than several mean free path lengths are attained, the dependence on depth can be approximated by an exponential. This exponential dependence is suggested by Equation 10.8 but has not been deduced a priori.

When the transmission curve is approximately exponential at large depths, one can obtain an equilibrium tenth-value thickness (TVT) and a kerma removal coefficient μ_{rm}.

$$\exp[-\mu_{rm}(\text{TVT})] = 1/10, \quad \mu_{rm} = \frac{2.303}{\text{TVT}} \tag{10.10}$$

The removal coefficient can be used to extrapolate measured transmission data to greater barrier thickness if necessary.

The effects of buildup in a collimated beam are largely due to radiation scattered at relatively small angles. Indirectly ionizing components scattered at large angles can produce a radiation background in regions removed from the primary beam. For example, the radiation energy scattered from the surface of a wall due to a horizontal radiation beam may be significant at locations well removed from the primary beam path. In this context, the name "albedo" has been given to the fraction of the incident radiation reflected or backscattered from a surface (24). The reflection coefficient $\alpha(\theta_i,\theta_s)$, which is generally a function of angle of incidence θ_i and angle of scatter θ_s for the radiation is often more useful than the total albedo. For example, the kerma at a point removed 1 m from a scattering area A in square meters can be obtained from

$$K = K_0\alpha(\theta_i,\theta_s)A \tag{10.11}$$

If K_0 is the kerma incident on the scattering surface, α is the fraction reflected from unit area struck by the beam. Figures 10.12 and 10.13 show calculated values which approximate $\alpha(\theta_i,\theta_s)$ for concrete (23).

10.4 Internal Doses from Distributed Sources

In this section we will develop some expressions that can be used to calculate the dose to an absorbing volume containing a uniformly distributed radioactive source. This problem can be viewed as an example of the use of Equations 10.5. It is of some practical importance because in some circumstances radioactive materials can be deposited in body organs. A first estimate of the radiation effect in such a situation requires knowledge of the absorbed dose for the organ.

The formulation we will use can be found in several references (25, 26). It is usually applied to photon sources, but in principle, doses from neutron sources could be calculated also. For simplicity, assume that the radioactive material is uniformly distributed throughout the homogeneous volume of interest, V, which has mass density ρ. The total activity, $A = dN/dt$, is contained within this volume. The branching ratio or fraction of emission events falling into the category under discussion is w, and the average energy of the radiation emitted per decay is E. Then ϕ_{af}, the absorbed fraction, is the fraction of emitted

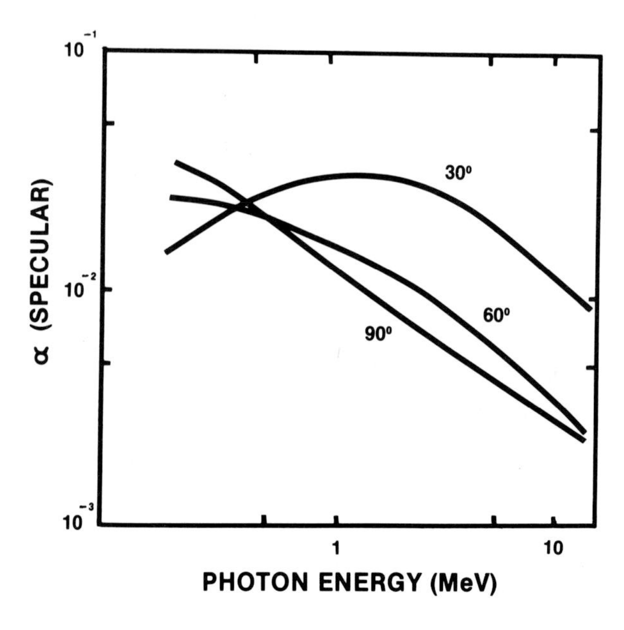

Figure 10.12 The specular reflection coefficient for several pairs of angles is shown for incident photons. Adapted from Ref. 23.

energy that is self-absorbed in the volume. The subscript af was attached to the absorbed fraction so that it would not be confused with the particle fluence. The dose rate dD_E/dt in the volume is a direct result of the activity and refers only to energy component E. Then for dose D_E in grays and activity A in becquerels,

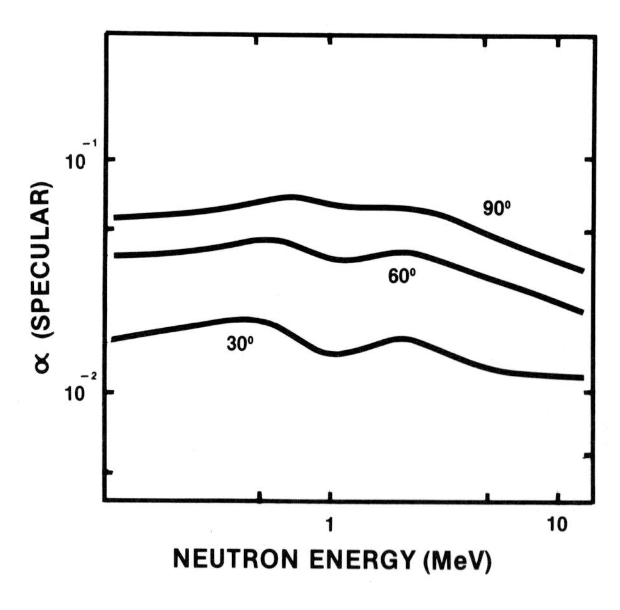

Figure 10.13 The specular reflection coefficient for several pairs of angles is shown for incident neutrons. Adapted from Ref. 23.

$$\frac{dD_E}{dt} = \left[\frac{dN}{dt}\right]\frac{wE}{\rho V}\phi_{af} = \frac{A}{\rho V}(wE)\phi_{af} = C\Delta\phi_{af} \tag{10.12}$$

where $C = A/\rho V$ is the uniform activity concentration in the mass in becquerels per kilogram and $\Delta = wE$ is the absorbed dose constant in joules.

The total dose is obtained by integrating the dose rate over the duration of the irradiation. In this case, after initial deposition of the radionuclide, the dose rate decreases in time because of physical decay, indicated by decay constant λ_p. If the volume of interest is an organ in a living system, biological processes may act to clear the radionuclide for elimination. Thus, one can define a biological clearing probability per second per molecule, λ_b, in analogy with the physical decay constant. The use of a single value for λ_b means that exponential biological clearance of the radionuclide is assumed. This is sufficient for a first approximation.

Assuming instantaneous filling, at time $t = 0$, the rate of change of activity in the volume is given by

$$\frac{dA}{dt} = -(\lambda_p + \lambda_b)A, \quad A = A_{max}\exp[-(\lambda_p + \lambda_b)t] \tag{10.13}$$

where A_{max} is the total activity contained in the volume at $t = 0$. Combining Equations 10.12 and 10.13 and integrating yields

$$D_E = \int_0^t (dD_E/dt')dt' = C_{max}\Delta\phi_{af}\int_0^t \exp[-(\lambda_p + \lambda_b)t']dt'$$

$$= \frac{C_{max}\Delta}{(\lambda_p + \lambda_b)}\phi_{af}\{1 - \exp[-(\lambda_p + \lambda_b)t]\} \tag{10.14}$$

Equation 10.14 was developed for a single component of radiation with energy *E*. The total volume dose is the sum of contributions from all radiation components. The summation over individual components, with index *i*, can be put in a standard form:

$$D = 1.44T_{1/2e}C_{max}[1 - \exp(-0.693t/T_{1/2e})]\sum_i \Delta_i(\phi_{af})_i \tag{10.15}$$

where $T_{1/2e}$ is the effective half-life, and $1.44T_{1/2e} = 1/(\lambda_p + \lambda_b)$.

Use of SI units in Equation 10.15 will ensure a result in grays. Other units commonly used together include the effective half-life $T_{1/2e}$ in hours, the maximum radionuclide concentration C_{max} in microcuries per gram (μCi/g), and the absorbed dose constant Δ_i in gram-rads (absorbed dose) per microcurie hour [g·rads/μCi·hr)]. Absorbed dose constants in these units have been tabulated for many photon-emitting radionuclides (14, 27).

Usually radiations are put into one of two categories for purposes of internal dose calculations: 1) nonpenetrating radiations, which are assigned a value $\phi_{af} = 1.0$ for the source volume; and 2) penetrating radiations, which have $0 < \phi_{af} < 1.0$. The nonpenetrating radiations are charged particles or low-energy indirectly ionizing radiations that generally penetrate less than 1 cm (25). In fact, absorbed fractions are never exactly one, even for an entirely uniform radionuclide deposition confined within a well defined volume, since energy must always escape from the boundary surfaces. This means that the dose is never exactly uniform in the volume. Nevertheless, the two categories are sufficient for large volumes with a large ratio of volume to surface area. If nonuniform depositions are made, charged-particle penetration must be considered and dose reduction coefficients must be applied (28).

The absorbed fraction discussed in the foregoing development will depend on the containing volume and the attenuation coefficient of the medium involved. In general, a complete calculation of ϕ_{af} is nontrivial. However, it is relatively simple to obtain an integral expression for the absorbed fraction for the case of uniform deposition in a sphere of homogeneous composition.

If we ignore the escape of secondary charged particles from the spherical surface, the absorbed fraction for indirectly ionizing radiation is equal to the fraction of the total emitted energy transferred to charged particles within the volume V. Then

$$\phi_{af} = \frac{\iint \rho \, dV_{tr} \, dK_e}{wNE} \tag{10.16}$$

where dK_e stands for the differential kerma caused by photons emitted in a differential element of the volume, and dV_{tr} is the elemental volume for energy transfer to charged particles. Thus, the subscripts e and tr stand for radioactive emission and energy transfer to charged particles, respectively. The symbols w, N, and E are for the branching ratio, the total number of emission events, and the emitted energy, respectively, as defined in Section 10.3.

From Equation 10.5 the differential kerma for dN emissions must be given by

$$\begin{aligned} dK_e &= (dN) wE(\mu_{tr}/\rho) \frac{\mathrm{BF}(r)\exp(-\mu r)}{4\pi r^2} \\ &= \left[\frac{N}{\rho V}\right] \rho \, dV_e wE(\mu_{tr}/\rho) \frac{\mathrm{BF}(r)\exp(-\mu r)}{4\pi r^2} \end{aligned} \tag{10.17}$$

In this case ρdV_e is the mass element for emission of the radiation, and $N/\rho V$ is the average number of emissions per unit mass; thus, $dN = (N/\rho V)\rho dV_e$. The gamma ray is assumed to be emitted in dV_e, and the energy is transferred to charged particles liberated in dV_{tr} at a distance r away. The situation is illustrated in Figure 10.14. The value $(\mu_{tr}/\rho)\mathrm{BF}(r)\exp(-\mu r)/4\pi r^2$ is the fraction of the photon energy transferred. Using $V = (4/3)\pi R^3$ and combining Equations 10.16 and 10.17 gives

$$\phi_{af} = \frac{\iint \rho \, dV_{tr} \, dK_e}{wNE} = \frac{3\mu_{tr}}{16\pi^2 R^3} \int dV_{tr} \int dV_e \frac{\mathrm{BF}(r_e)\exp(-\mu r_e)}{r_e^{\,2}} \tag{10.18}$$

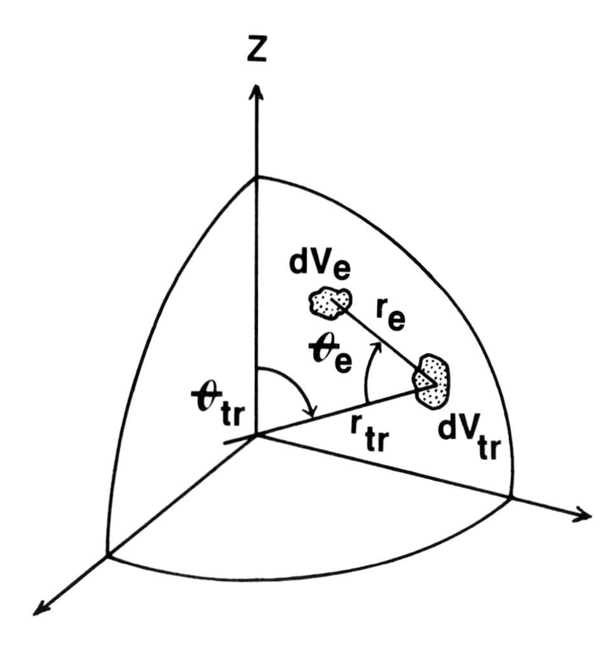

Figure 10.14 Elements for integration are shown in a spherical volume. Radiations are presumed to be emitted in dV_e at a position given by r_e and θ_e, and energy is transferred to charged particles in dV_{tr} at r_{tr} and θ_{tr}.

If the absorber is small ($\lambda \gg R$), Equation 10.18 can be simplified by substituting $\exp(-\mu_{tr}r)$ for $\mathrm{BF}(r)\exp(-\mu r)$:

$$\phi_{af} = \frac{3\mu_{tr}}{16\pi^2 R^3} \int dV_{tr} \int dV_e \frac{\exp(-\mu_{tr} r_e)}{r_e^2} \tag{10.19}$$

Since the exponential factor carries the fractional attenuation of the emitted photons, it is clear that after integration to radius R the absorbed fraction will be a function of $\mu_{tr}R = R/\lambda_{tr}$, where λ_{tr} is the mean free path for energy transfer for the indirectly ionizing component. This dependence ensures that the absorbed fraction will increase as the radius is increased. It will approach one when R is much greater than the mean free path for energy transfer. Figure 10.15 illustrates the specific dependence of ϕ_{af} on $\mu_{tr}R$ for a spherical absorber.

In practical situations the geometry is often quite complicated. In all cases the absorbed fraction increases as the volume dimensions increase relative to the mean free path for the radiation. However, use of the energy transfer coefficient in the exponential is not always sufficient, and a Monte Carlo procedure may be necessary to obtain the absorbed fractions. Table 10.5 shows some results for small spheres with radioactive photon emitters (29). Calculations for absorbed fractions of body organs are available (30).

10.5 Expressions for Depth-Dose Curves

In this section we will confine our attention to the dose deposited in depth in an absorber by indirectly ionizing beams. The word "beam" will indicate radia-

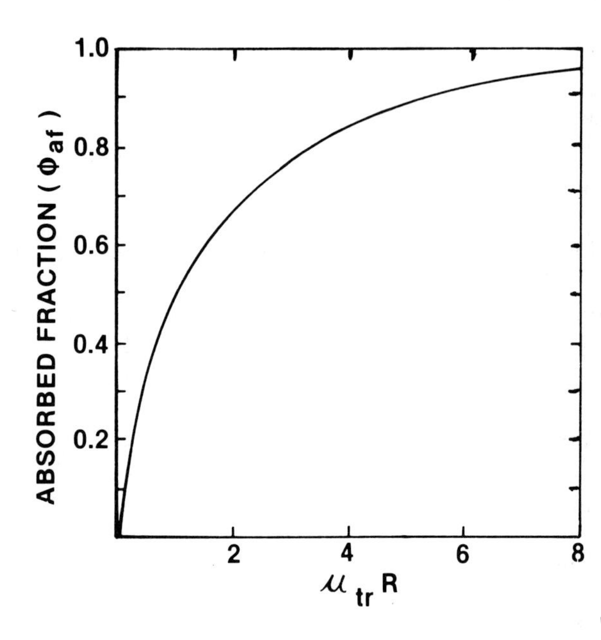

Figure 10.15 The absorbed fraction for a uniform sphere is shown as a function of $\mu_{tr}R$, where R is the radius. Adapted from Ref. 27.

tion diverging from a collimated source incident in air on a distant absorber. We will give analytical expressions with several parameters for the percent depth dose in the absorber along the central axis of the beam and incident perpendicular to the absorber surface. The expressions utilized relate directly to basic principles but are not necessarily efficient mathematical forms to use in curve fitting. A discussion of isodose curves follows development of the central axis expressions.

It is convenient to begin with an expression for the kerma at various central axis depths. Equation 10.7 relates the total kerma $K_t(d,F,A/P)$ to the most important parameters. We use the kerma-to-dose conversion ration $b(d)$, expressed as a function of depth, to obtain the expression for the central axis dose $D(d,F,A/P)$. From Equation 10.7,

Table 10.5 Absorbed Fractions for Uniformly Distributed Photon Sources in Unit Density Spheres

	Photon Energy (MeV)						
Mass (g)	*0.030*	*0.050*	*0.100*	*0.140*	*0.364*	*0.662*	*1.460*
1	0.050	0.011	0.009	0.010	0.011	0.011	0.010
2	0.064	0.014	0.012	0.012	0.014	0.014	0.012
4	0.081	0.019	0.015	0.016	0.018	0.018	0.016
10	0.111	0.027	0.021	0.022	0.025	0.024	0.021
20	0.139	0.035	0.027	0.028	0.031	0.031	0.027
40	0.174	0.046	0.036	0.036	0.039	0.038	0.033
100	0.306	0.087	0.067	0.066	0.072	0.070	0.061

Data from Ref. 29.

$$D(d,F,A/P) = b(d)\ K_t(d,F,A/P)$$

$$= b(d)\ K_p(O,F)\ \mathrm{BF}(d,A/P) \left[\frac{F}{F+d}\right]^2 \exp(-\mu_{\mathrm{eff}} d) \quad (10.20)$$

At the depth d_m, where the dose is a maximum, Equation 10.20 becomes

$$D(d_m,F,A/P) = b(d_m)\ K_p(O,F)\ \mathrm{BF}(d_m,A/P) \left[\frac{F}{F+d_m}\right]^2 \exp(-\mu_{\mathrm{eff}} d_m) \quad (10.21)$$

The kerma-to-dose conversion ratio is often approximately constant beyond the depth of the maximum for radiation beams producing exponential depth-versus-dose curves (31). If $b(d_m)$ and $b(d)$ are equal, and $d \geq d_m$,

$$D(d,F,A/P) = D(d_m,F,A/P) \frac{\mathrm{BF}(d,A/P)}{\mathrm{BF}(d_m,A/P)} \left[\frac{F+d_m}{F+d}\right]^2 \times \exp[-\mu_{\mathrm{eff}}(d-d_m)] \quad (10.22)$$

The percent depth dose $P(d,F,A/P)$ was defined on Chapter 4 in the context of charged-particle beams. For indirectly ionizing radiation, it depends on the quality of the radiation, the absorber characteristics, and the parameters listed, i.e. depth d, source distance F, and field-area-to-perimeter ratio A/P at the surface. The percent depth dose is the ratio of the dose at depth d to the dose at the maximum, expressed as a percent. When $d \geq d_m$:

$$P(d,F,A/P) = 100 \frac{D(d,F,A/P)}{D(d_m,F,A/P)}$$

$$= 100 \frac{\mathrm{BF}(d,A/P)}{\mathrm{BF}(d_m,A/P)} \left[\frac{F+d_m}{F+d}\right]^2 \exp[-\mu_{\mathrm{eff}}(d-d_m)], \quad (10.23)$$

Figure 10.16 shows central axis percent depth dose curves for an assortment (12, 32, 33) of photon and neutron beams for a water absorber. The charged—particle buildup region at the surface mentioned in Section 10.2 is evident for the high-energy beams. The curve shape well beyond the maximum is nearly exponential in many cases. Appendix 11 gives measured values of percent depth dose for several photon beams.

Equation 10.23 is useful because the consecutive factors remind one that scatter buildup, beam divergence, and absorber attenuation are three major effects to be considered in estimating the dose in depth. However, the buildup factor and the effective attenuation coefficient generally must be determined as a function of depth before percent depth-dose values can be calculated. The effective attentuation coefficient for the primary beam has a constant value for a monoenergetic gamma-ray beam. It cannot be expected to remain constant at different depths when the beam has a continuous energy distribution. Figure 10.17 shows a photon spectrum from a linear accelerator and a spectrum of neutrons from a cyclotron (34, 35). Because of the effect of filtration in the absorber, the spectrum of the primary beam will change with depth, causing a change in effective attenuation coefficient.

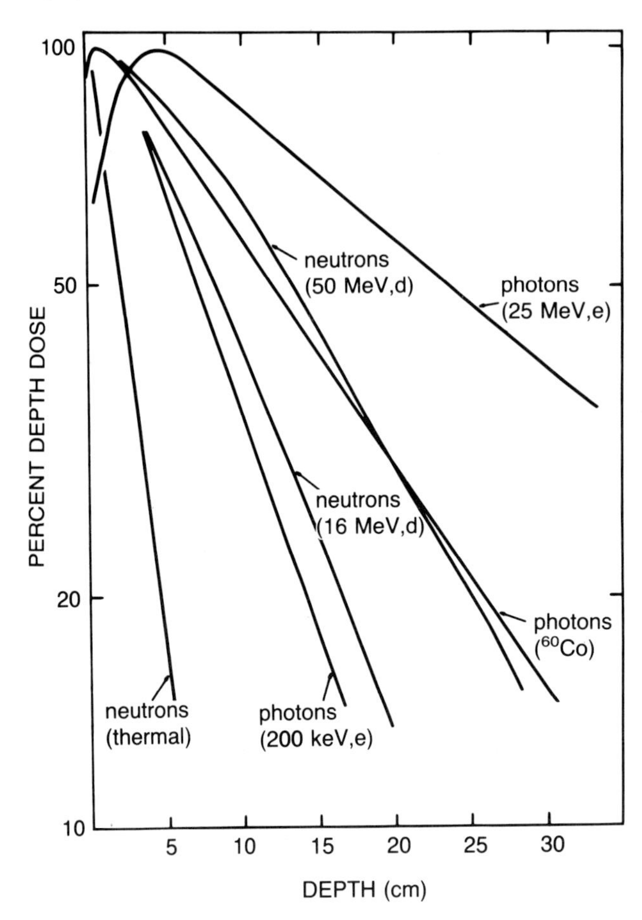

Figure 10.16 Central axis percent depth-dose curves for field sizes of about 100 cm² are shown for a variety of beams in water (12, 32, 33). Where energies are indicated, they refer to the incident kinetic energy of the deuterons [neutron curves (d)] or electrons [bremsstrahlung curves (e)]. The neutron curves are for source-to-surface distances of 125 cm (16 MeV) and 150 cm (50 MeV). The photon curves are for 50-cm (200 keV), 80-cm (^{60}Co), and 100-cm (25 MeV) distances.

A second quantity of practical value for collimated beams is the tissue-air ratio $\mathrm{TAR}[d,(A/P)_d]$. This ratio depends on the depth d and on the area-to-perimeter ratio evaluated at depth d for the field. It is commonly used when radiation beams incident from several directions are made to intersect in a common volume. The tissue-air ratio is defined as the dose to the center of an equilibrium tissue mass in depth in an absorber, divided by the dose delivered to the same mass at the same position but in air (no surrounding absorber). If the kerma-to-dose conversion factor is constant for $d \geq d_m$, then

$$\mathrm{TAR}[d,(A/P)_d] = \frac{D[d,F,(A/P)_d]}{D_a[F+d,(A/P)_d]} = \frac{b(d)K_t[d,F,(A/P)_d]}{b(d_m)K_p(O,F)\left[\dfrac{F}{F+d}\right]^2 \exp(-\mu d_m)}$$

$$= \mathrm{BF}[d,(A/P)_d]\exp[-\mu_{\mathrm{eff}}(d-d_m)] \qquad (10.24)$$

Equation 10.20 was used to obtain Equation 10.24 above. The phrase "equilibrium tissue mass" means that complete charged-particle buildup is assumed. Since an absorber of thickness d_m is required to produce this, the exponential is included in the denominator of Equation 10.24. The dose in air D_a, is propor-

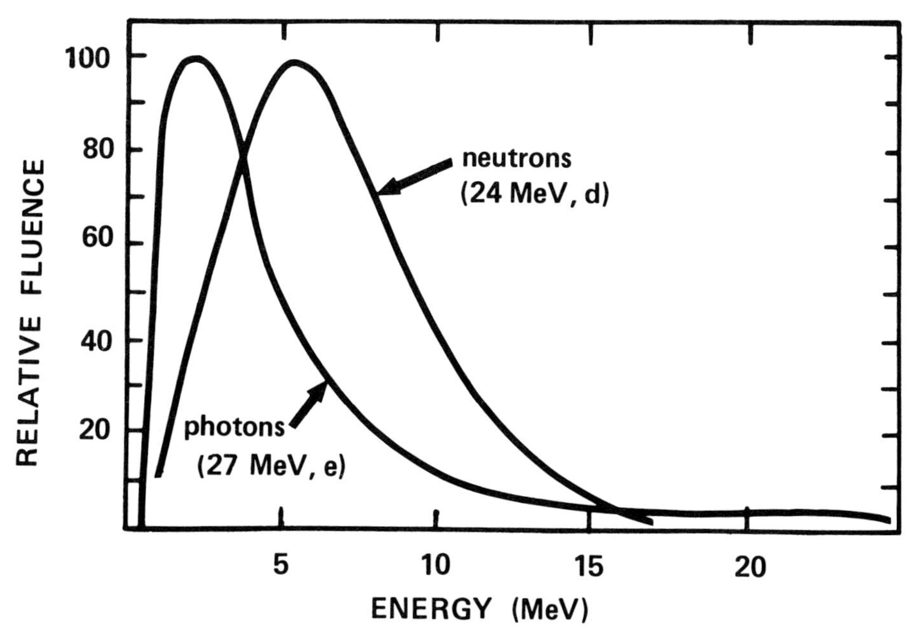

Figure 10.17 The spectrum of photons produced by 27-MeV electrons (e) incident on a tungsten target (34) and the spectrum of neutrons produced by 24-MeV deuterons (d) incident on a beryllium target (35) are shown.

tional to the primary kerma, since scatter buildup is assumed to be negligible in the equilibrium mass. The tissue-air ratio can be related to the percent depth dose using Equations 10.23 and 10.24.

The tissue-air ratio is of special interest because it is directly proportional to the scatter buildup factor for the beam and independent of distance from the source. Figure 10.18 shows the tissue-air ratio for ^{60}Co beams of several sizes (32). Notice that at a given depth, the ratio increases when the field size (A/P) is increased. This increase is directly related to an increase in the scatter buildup factor, and occurs primarily because the absorber volume available to contribute to scatter processes increases with increasing field size. Notice also that the curves on the figure tend to level out for very large fields. This feature can be related to self-absorption of the scattered radiation, which limits the contribution from distant parts of the volume. Local self-absorption is expected to be limiting when the field dimensions are larger than the mean free path for the secondary radiation.

A third quantity relating doses at different depths is TPR, the tissue-phantom ratio (36). The tissue-phantom ratio is the quotient of the dose at some depth in an absorber divided by the dose delivered to the same position but at a standard depth in the absorber. The tissue-phantom ratio is called the tissue-maximum ratio (TMR) if the standard depth is d_m. The tissue-maximum ratio has the advantage of being easily measured for any radiation quality. The difficulty in measuring the dose in air in an equilibrium tissue mass is

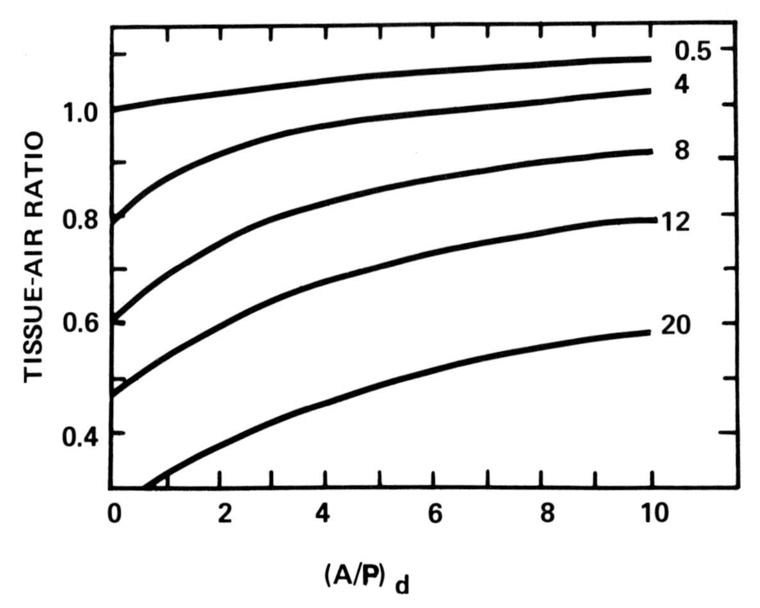

Figure 10.18 The TAR is shown for ^{60}Co radiation as a function of area-to-perimeter ratio of the field (32). Adapted from Ref. 6.

bypassed. If kerma-to-dose conversion factors are assumed constant, and $d \geq d_m$,

$$\begin{aligned}\mathrm{TMR}[d,(A/P)_d] &= \frac{D[d,F,(A/P)_d]}{D[d_m,F+d-d_m,(A/P)_d]} \\ &= \frac{b(d)K_t[d,F,(A/P)_d]}{b(d_m)K_t[d_m,F+d-d_m,(A/P)_d]} \\ &= \frac{\mathrm{BF}[d,(A/P)_d]}{\mathrm{BF}[d_m,(A/P)_d]}\exp[-\mu_{\mathrm{eff}}(d-d_m)] \end{aligned} \qquad (10.25)$$

The tissue-maximum ratio is independent of the distance from the radiation source and depends on scatter buildup and primary attenuation. In these properties it is very similar to the tissue-air ratio, the difference being that the scatter buildup factor at d_m is divided out of the TMR. The tissue-maximum ratio is widely used for high-energy photon beams.

The foregoing discussion was developed around central axis relationships. Scatter doses are somewhat reduced in magnitude in regions near the edge of a collimated field, but they remain significant. The overall effect is best displayed on an isodose curve. Recall that an isodose contour is a curve of constant percent depth dose. An isodose curve is a composite of such contours in a common plane containing the central axis. A variety of isodose curves is shown on Figure 10.19 (12, 37, 38) for both photon and neutron beams.

The shape of an isodose contour away from the central axis of the beam is related to a variety of phenomena. Several of them are listed below:

1. beam divergence,
2. penumbra,
3. scattered radiation,
4. source emission distribution.

The divergence of the edges of the isodose contours with increasing depth occurs because of the divergence of the incident fluence as produced by the source and collimators. Notice that beam divergence is more pronounced for the low-energy x-ray source in Figure 10.19 (*top left*) than for other sources; this is because the source-to-surface distance is substantially less (50 cm) for this beam than for the other beams (100–150 cm).

The sharp field edges for the low-energy x-ray beam can be contrasted with the rounded edges of the isodose lines for the ^{60}Co radiation (*top right*). Penumbra produced by the source collimator combination is a major cause of the gradual dose change near the field edge for the relatively large ^{60}Co source. A ^{60}Co teletherapy source has a face diameter of several centimeters. Within a substantial region near the lateral field edges, the collimator edge intercepts only part of the radiation emitted. Thus, the penumbra region is rather large for ^{60}Co teletherapy beams. Most x-ray sources have a focal spot with dimensions much less than a centimeter, and thus the beams produced have relatively sharp edge contours.

The concave shape of the isodose contours is related to several factors.

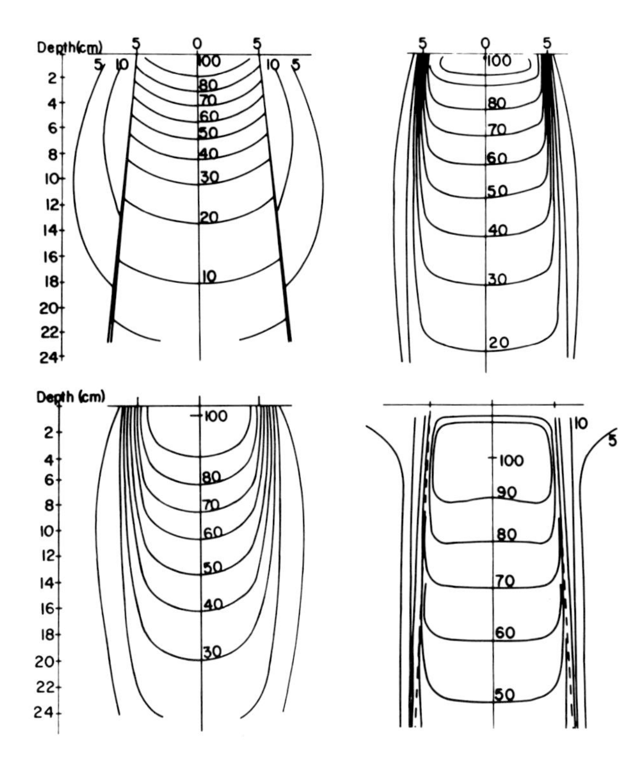

Figure 10.19 Isodose curves (12,37,38) for a 200-kVp x-ray beam (*top left*), a ^{60}Co γ-ray beam (*top right*), a neutron beam produced by 50-MeV deuterons (*bottom left*), and a 25-MV x-ray beam are shown. Adapted from Ref. 25.

Scattered and emitted radiation have an effect because field center is an especially favorable position and receives secondary radiation from all sides. A point at the field edge receives scatter from only a fraction of the surrounding solid angle; thus, the buildup factor is diminished at the edges. For smaller source-to-surface distances, radial divergence also contributes substantially to the concave shape of the isodose contours.

The angular distribution of the emitted radiation is another factor to be considered, especially at high energies. Generally, very high energy radiations tend to emerge from the target with preferred directions nearly parallel to the incident direction. This forward peaking can cause an unacceptable amount of curvature in the isodose curves for clinical accelerators. For high-energy accelerators used in radiation therapy, a cone-shaped field-flattening filter is made to intercept the beam and thereby absorb radiation near the beam axis preferentially. When a field flattener is used, nearly flat isodose contours can be obtained. The *bottom right* curve on Figure 10.19 was obtained from an accelerator with such a filter.

10.6 Activation by Photon and Neutron Beams

In this section we will discuss activation of an absorber during irradiation. When photon beams with sufficiently high energy are employed, radioactivity is induced in photonuclear reaction products. Neutron-beam activation can proceed from a variety of neutron absorption reactions. Generally, neutron-induced reactions have larger cross sections than the photonuclear reactions and produce sources with greater specific activity. Many widely used radioactive nuclides can be produced by neutron activation. For example, ^{60}Co and ^{137}Cs sources are often produced by using radiative capture reactions in the stable ^{59}Co and ^{136}Cs samples, respectively. In addition, ^{99}Mo, the precursor of ^{99m}Tc, and ^{131}I can be separated from other nuclei produced during neutron-induced fission. All four of these radioactive species have extensive medical applications.

Suppose that activation by fluence ϕ is governed by cross section σ. If N_a is the number of activated targets distributed over area A in the irradiating beam, the expression for the cross section can be arranged so that

$$\sigma = \frac{-d\phi/\phi}{n_v dx} = \frac{N_a/A}{\phi n_v dx} = \frac{N_a}{\phi n_v A dx} = \frac{N_a}{\phi N}$$

$$N_a = N\sigma\phi, \qquad \frac{dN_a}{dt} = N\sigma\frac{d\phi}{dt} = P_a \tag{10.26}$$

where N is the total number of nuclei under irradiation in the target, and P_a is the production rate of radioactive nuclei.

Since the reaction product is itself radioactive, Equations 10.26 must be modified for product decay. In this case the number of radioactive nuclei present at time t is governed by

$$\frac{dN_a}{dt} = P_a - \lambda_a N_a; \qquad N_a = 0, \quad t = 0 \tag{10.27}$$

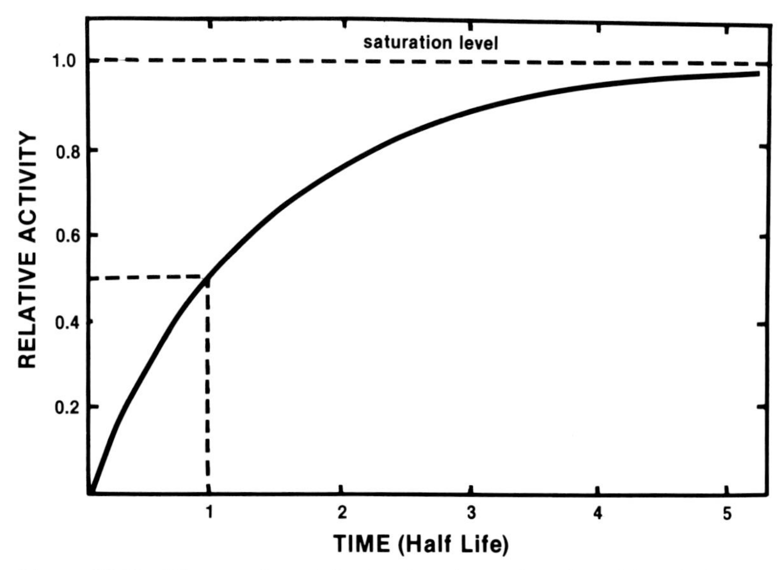

Figure 10.20 The sample activity is plotted as a function of irradiation time for constant production rate.

Equations 10.27 show that the net rate of increase of radioactive nuclei is the difference between the production rate P_a and the decay rate $\lambda_a N_a$. Integrating the first of Equations 10.27 over the irradiation time t_i yields a result valid for constant P_a:

$$\int \frac{dN_a}{N_a - P_a/\lambda} = -\lambda \int dt + C, \quad \ln\left(N_a - \frac{P_a}{\lambda}\right) = (-\lambda t_i + C),$$

$$N_a - \frac{P_a}{\lambda} = \exp(-\lambda t_i + C), \quad \exp(C) = -\frac{P_a}{\lambda} = (N_a)_{\max}$$

$$N_a = (N_a)_{\max}[1 - \exp(-\lambda t_i)] \tag{10.28}$$

The constant of integration C was evaluated using the boundary condition $N_a = 0$, $t = 0$.

It is evident from Equations 10.28 that the number of radioactive nuclei increases to a saturation maximum $(N_a)_{\max} = P_a/\lambda = N\sigma(d\phi/dt)/\lambda$ under steady irradiation. Furthermore, the rate of increase in activated nuclei diminishes continuously; thus, in practice one never reaches the saturation activity. After irradiating for a time equal to several half-life values of the product radionuclide, further irradiation is relatively inefficient. Figure 10.20 is an illustration of the buildup of activity in the sample with irradiation time.

As a special case, suppose the fluence of neutrons in a reactor with speed between v and $v + dv$ is given by $\phi(v)dv$. If the $1/v$ reaction cross-section dependence holds, $\sigma_{1/v} = \sigma_{\text{th}}(v_{\text{th}}/v)$, where thermal values are indicated by the subscript th:

$$dP_a = N\sigma_{1/v}\frac{d\phi(v)}{dt}dv = N\frac{(\sigma_{\rm th}v_{\rm th})}{v}\frac{d\phi(v)}{dt}dv$$

$$= N(\sigma_{\rm th}v_{th})\frac{d\phi(v)}{vdt}dv = N\sigma_{\rm th}v_{\rm th}\rho_v(v)dv \tag{10.29}$$

In this case $\rho_v(v)dv$ is the number of neutrons per unit volume with speeds between v and $v + dv$. Thus

$$P_a = N(\sigma_{\rm th}v_{\rm th})\int \rho_v(v)dv = Nn_v\sigma_{\rm th}v_{\rm th} \tag{10.30}$$

where the integral represents the total number of neutrons per unit volume. If the number of target nuclei N and the number of neutrons per unit volume n_v can be stipulated for a given experimental situation, the sample activity can be estimated using Equations 10.28, 10.29, and 10.30. Of course, resonance absorption has been neglected in these equations.

10.7 Problems

1. Identify the neutron-initiated absorption reaction that provides the largest part of the kerma per fluence at low neutron energies ($\ll$1 MeV) in tissue.
2. Is the use of water as a tissue-equivalent material justified for neutron beam dosimetry at all energies?
3. Compute the energies of M-shell Auger electrons emitted after formation of K-shell vacancies in calcium and oxygen atoms. Use the energy values given in Appendix 4 and ignore shells above M.
4. A beam of neutrons from a beryllium target produced by 24-MeV deuterons is incident on issue. Estimate the depth for the charged-particle (proton) buildup maximum. Use data from Appendix 6 and the spectrum from Figure 10.17. Compare with values from Table 10.1 for similar beams.
5. Why do ion chambers, used for measurements on ^{60} Co beams, have 0.4- to 0.5-cm walls?
6. a. A narrow beam of 1-MeV photons with fluence $10^{10}/m^2$ is incident on a tissue absorber. What is the kerma in grays at the surface?
   b. A narrow beam of 1-MeV neutrons of fluence $10^{10}/m^2$ is incident on a tissue absorber. What is the kerma in grays at the surface?
7. Express Δ_i and $C_{\rm max}$ in Equation 10.14 in terms of w, E, $A_{\rm max}$, and ρV.
8. Using the information given below for ^{99m}Tc, calculate the gamma- and x-ray dose rate to a 40-g thyroid if 1 mCi of ^{99m}Tc is uniformly distributed therein.

	E_i (*MeV*)	$\Delta_i[g \cdot rads/(\mu Ci \cdot h)]$	ϕ_{af}
Gamma 1	0.0021	0.000	0.974
Gamma 2	0.1405	0.263	0.029
Gamma 3	0.1426	0.000	0.029
$K_{\alpha 1}$	0.0183	0.002	0.479
$K_{\alpha 1}$	0.0182	0.001	0.480

9. Using Figures 10.10 and 10.11, estimate the transmission of bremsstrahlung beams of end-point energy 2 MeV and 20 MeV through a 36 in. concrete wall. Repeat the process for neutron beams of energy 2 MeV and 14 MeV (with and without secondary photon contribution).
10. What concrete wall thickness ($\rho = 2.35$ g/cm^3) is necessary to reduce the kerma from a 6-MV bremsstrahlung beam to 10^{-8} of the incident value?
11. The linear attenuation coefficient for ^{60}Co radiation in water is 6.5 m^{-1}.
    a. Calculate the dose at points at depths 0.01 m, 0.05 m, 0.1 m, and 0.2 m along the central axis for F of 0.8 m. Assume the maximum dose is 100 rad. Ignore scatter.
    b. Compare your calculations with the measured values in Appendix 11 for a 10 × 10-cm field. Calculate the dose attributable to scatter and the buildup factor at each depth.
12. Express the tissue-air ratio in terms of the percent depth dose using Equations 10.23 and 10.24. Calculate the tissue-air ratio at a depth of 10 cm for a ^{60}Co beam with dimension 10 × 10 cm at the surface. Use the percent depth-dose values from Appendix 11.
13. Determine the treatment time necessary to deliver a dose of 2 Gy to a tumor 8 cm below the surface for a ^{60}Co beam with surface dimensions 10 × 10 cm. Use percent depth-dose data from Appendix 11 and a dose rate at the maximum of 100 rads/min to a small mass of tissue.
14. Using an exponential probability of interaction for radiation in an absorber, calculate the probability for an interaction occurring between depth d_1 and d_2. Are small distance intervals more likely than large distance intervals? Why?
15. When ^{60}Co is produced by irradiation of ^{59}Co in a reactor, the irradiation time for the sample may be only 1 yr. What fraction of the maximum possible activity is obtained in this time?
16. For a steady irradiation, how many half-life times must elapse before the induced sample activity is 87.5% of the maximum activity?
17. Use Equations 10.26 and 10.28 to calculate the fraction of sample nuclei converted from ^{113}Cd to ^{114}Cd after a 1-h thermal neutron irradiation with thermal fluence rate 10^{+12} neutrons/(cm^2 · s).

10.8 References

1. E. Storm and H. I. Israel, Nucl. Data Tables **7**, 565 (1970).
2. R. L. Bach and R. S. Caswell, Radiat. Res. **35**, 1 (1968).
3. J. A. Auxier, W. S. Snyder, and T. D. Jones, "Neutron interactions and penetration in tissue," in *Radiation Dosimetry,* vol. 1, edited by F. H. Attix and W. C. Roesch (Academic, New York, 1968).
4. *Neutron Fluence, Neutron Spectra and Kerma, ICRU Report 13* (International Commission on Radiation Units and Measurements, Washington, 1969).
5. *Protection Against Neutron Radiation, NCRP Report 38* (National Council on Radiation Protection and Measurements, Washington, 1971), p. 11.
6. H. E. Johns and J. R. Cummingham, *The Physics of Radiology* (Charles Thomas, Springfield, IL, 1969), p. 186.
7. P. J. Ebert and A. F. Lauzon, IEEE Trans. Nucl. Sci. **13**, 735 (1966).

8. D. Chattorji, *The Theory of Auger Transitions* (Academic, New York, 1976).
9. T. A. Carlson, Radiat. Res. **64**, 53 (1975).
10. *Radiobiological Dosimetry, National Bureau of Standards Handbook 88* (U.S. Department of Commerce, Washington, 1962), p. 9.
11. D. E. Velkey, D. J. Manson, J. A. Purdy, and G. D. Oliver, Jr., Med. Phys. **2**, 14 (1975).
12. A. R. Smith, P. R. Almond, J. B. Smathers, and V. A. Otte, Radiology **113**, 187 (1974).
13. D. T. Goodhead, R. J. Berry, D. A. Bance, P. Gray, and B. Stedeford, Phys. Med. Biol. **23**, 144 (1978).
14. L. T. Dillman, J. Nucl. Med. Suppl. **4**, 7 (1970.
15. J. DePangher and E. Tochilin, "Neutrons from accelerators and radioactive sources," in *Radiation Dosimetry,* edited by F. H. Attix vol. 3, and E. Tochilin (Academic, New York, 1969), p. 329.
16. D. Nachtigall, Health Phys. **13**, 213 (1967).
17. R. D. Evans, *The Atomic Nucleus* (McGraw-Hill, New York, 1955), pp. 728–742.
18. R. D. Evans and R. O. Evans, Rev. Mod. Phys. **20**, 305 (1948).
19. A. Morrison, Nucleonics **5**, 19 (1949).
20. K. Z. Morgan and L. C. Emerson, "Dose from external sources of radiation," in *Principles of Radiation Protection,* edited by K. Z. Morgan and J. E. Turner (John Wiley and Sons, New York, 1967), p. 268.
21. H. Goldstein and J. H. Wilkins, Jr., *Calculations of the Penetration of Gamma Rays, Report NYO-3075,* (Nuclear Development Associates, 1954).
22. D. E. Wrede, Phys. Med. Biol. **17**, 548 (1972).
23. *Radiation Protection Guidelines for 0.1–100 MeV Particle Accelerator Facilities, NCRP Report 51* (National Council on Radiation Protection and Measurements, Washington, 1977).
24. J. Thewlis, *Concise Dictionary of Physics and Related Subjects* (Pergamon, New York, 1973).
25. W. R. Hendee, *Medical Radiation Physics* (Year Book, Chicago, 1970).
26. E. R. Powsner and D. E. Raeside, *Diagnostic Nuclear Medicine* (Grune and Stratton, New York, 1971).
27. *Methods of Assessment of Absorbed Dose in Clinical Use of Radionuclides, ICRU Report 32* (International Commission on Radiation Units and Measurements, Washington, 1979).
28. V. A. Brookeman, L. T. Fitzgerald, and R. L. Morin, Phys. Med. Biol. **23**, 852 (1978).
29. W. H. Ellett and R. M. Humes, J. Nucl. Med. Suppl. **5**, 27 (1971).
30. W. S. Synder, M. R. Ford, G. G. Warner, and H. L. Fisher, Jr., J. Nucl. Med. Suppl. **3**, 5 (1969).
31. J. Dutreix, A. Dutreix, and M. Tubiana, Phys. Med. Biol. **21**, 524 (1976).
32. M. Cohen, D. E. A. Jones, and D. Greene, Br. J. Radiol. Suppl. **11**, 1 (1972).
33. N. A. Frigerio, Phys. Med. Biol. **6**, 541 (1962).
34. L. B. Levy, R. G. Waggener, W. D. McDavid, and W. H. Payne, Med. Phys. **1**, 62 (1974).
35. E. Tochilin and G. D. Kohler, Health Phys. **1**, 332 (1958).
36. *Determination of Absorbed Dose in a Patient Irradiated by Beams of X or Gamma Rays in Radiotherapy Procedures, ICRU Report 24* (International Commission on Radiation Units and Measurements, Washington, 1976).
37. S. O. Fedoruk, H. E. Johns, and T. A. Watson, Radiology **60**, 348 (1953).
38. K. C. Tsien, Br. J. Radiol. **31**, 32 (1958).

11

Ionization and Excitation in Gaseous Systems

11.1 Introduction

In earlier chapters, processes initiated by fast charged particles and photons and neutrons have been discussed in some detail. In many cases in these chapters, emphasis was given to changes produced in the properties of the particles in a beam. Less discussion was aimed at the effect of the transferred energy in altering the properties of the atomic and molecular constituents of the absorbing medium. In this chapter and those to follow, the emphasis will shift to specific effects on the irradiated medium. For example, consideration will be given to the spectrum of liberated charge carriers and to perturbations of the atomic and molecular absorbing systems. De-excitation processes in the absorbing atoms and the division of the energy between constituents in the medium will also be considered. Insofar as possible, the discussions are designed to give background information about phenomena that can be applied to dosimetric or imaging devices, or those that have biological significance.

The sequence of topics to follow begins with phenomena in irradiated gaseous systems (present chapter), proceeds to solid systems (Chapter 12), and follows with effects in liquids and molecular solutions (Chapter 13). The final chapter, on radiation effects on living systems, follows (Chapter 14) because cells can be viewed as highly organized molecular solutions.

The measurement of radiation fluences or doses in an absorbing medium is a basic problem, encountered whenever radiation is utilized. The measurement process may require insertion of a device, often called a dosimeter, into the medium in the radiation field. The quantity measured by the device is assumed to be proportional to one of the basic radiation quantities. An ionization chamber is a common example of such a measuring device. Since the active volume of the chamber is a gas-filled cavity, a discussion of the basic radiation-induced phenomena in a gas-filled chamber is appropriate in this chapter.

When a directly ionizing beam is used, the ions produced in a gaseous cavity are due to both primary charged particles and secondary electrons passing through the cavity. When an indirectly ionizing beam is used, the ionization generated is due mainly to the secondary charged particles liberated either in a surrounding medium or in the cavity chamber walls.

Suppose $\sigma_i(T)$ is the cross section for ionization induced by charged particles with kinetic energy T in a gaseous material. The ionization probability P_i per gaseous molecule in the cavity is given by

$$P_i = \int_{E_{IP}}^{T_{max}} \sigma_i(T)\phi(T)dT \tag{11.1}$$

where $\phi(T)dT$ represents the fluence of charged particles with energy between T and $T + dT$, and E_{IP} is the energy equivalent of the ionization potential. Because Equation 11.1 is quite general, it serves to emphasize that the ionization produced depends on the energy spectrum of charged particles involved as well as the properties of the absorbing atoms.

Because the charged-particle energy distribution is directly involved, a discussion of electron spectra in irradiated media will be given in Section 11.2. In Section 11.3 the magnitude of the charges and currents liberated in a gas-filled volume will be considered. A general theory for cavity response will be given in Section 11.4, and in Section 11.5 the discussion will center around the response of some ion chamber systems.

11.2 Electron Spectra in Irradiated Media

A variety of processes that liberate atomic electrons have been mentioned in preceding chapters. In Chapters 5 and 10 some consideration was given to the properties of the charged particles liberated by indirectly ionizing beams and to the charged-particle buildup under an air-medium interface. In this section we will discuss the energy spectrum of the electrons present in an absorber under irradiation. A procedure for calculating the equilibrium spectrum of degraded primary and secondary electrons will be outlined as a first step. Some spectra will be provided in the figures, including illustrations of the effects of energy degradation of primary components and of secondaries.

Fano (1) and Spencer and Fano (2) and Spencer and Attix (3) have developed analytical procedures that can be used to calculate the equilibrium electron spectrum that occurs when source electrons are degraded in a medium. Dependence on spatial variables was not involved since the radiation field was assumed to be uniform in a homogeneous medium of substantial extent. A simplified method of computing the electron spectrum in uniformly irradiated media was developed by Birkhoff et al. (4, 5), under the condition that the energy lost by electrons due to bremsstrahlung emission is small compared to collisional energy losses. In this method the spectrum of electrons being degraded is explicitly divided into two components, one for source electrons and the other for atomic electrons ejected during ionizing collisions in the medium. An outline of this analytical technique will be given in the following paragraphs.

Consider first the equilibrium spectrum of degraded source electrons in an irradiated medium. If the source is a negative beta emitter, the initial spectrum function, ${}_s n_v(T')dT'$, represents the number of beta particles released per unit volume with energies between T' and $T' + dT'$ and the subscript s refers to the source. If the electrons are produced by photon beams, the source electron spectrum is the sum of the photoelectron, Compton electron, and pair production electron spectrum components liberated during these interactions. In any case, it is assumed that the source electrons lose energy in a continuous manner in the process of slowing down in the medium.

Suppose that $\phi(T)dT$ represents the equilibrium degradation fluence with electron energy between T and $T + dT$. The average energy per unit volume lost by equilibrium electrons with kinetic energy between T and $T + dT$ is given by

$$\langle dE_v \rangle = \phi(T)dT\langle(-dT/dx)_c\rangle = \phi(T)S_c(T)dT \tag{11.2}$$

where the function $S_c(T)$ is the stopping power for collisional loss. Since the density of electrons is assumed uniform throughout the homogeneous region surrounding the volume of interest, there is no spatial dependence in Equation 11.2. During degradation the number of electrons passing through the interval between T and $T + dT$ is equal to the integral of the source electron spectrum from T to $T_{\max}$. Thus, the average energy per unit volume deposited by electrons with energy in an infinitesimal interval is the product of the source electron density integral and the width of the energy interval:

$$\langle dE_v \rangle = dT \int_T^{T_{\max}} {}_s n_v(T')dT' \tag{11.3}$$

where T' is an integration variable for kinetic energy.

Elimination of $\langle dE_v \rangle$ in Equations 11.2 and 11.3 gives

$$\phi(T) = \frac{\int_T^{T_{\max}} {}_s n_v(T')dT'}{S_c(T)} \tag{11.4}$$

Equation 11.4 shows that the equilibrium fluence of uniformly degraded source electrons with kinetic energy T is inversely proportional to the stopping power at T. This dependence is reasonable since electrons with large stopping powers are less likely to be in a fixed energy acceptance interval than those with small stopping powers.

Equation 11.4 does not provide for contributions from atomic electrons ejected from atoms in ionizing coulomb collisions. Although the average energy loss per ion pair is small, in a substantial number of collisions, relatively large energy losses occur with large amounts of energy transferred to the atomic electrons involved. Because two electrons cannot be distinguished a priori, the electron emerging from the collision site with the larger kinetic energy is assumed to be the primary, and the electron with the smaller energy is assumed to be the ejected atomic electron. According to this convention, the maximum energy transferred by an electron of energy T' is $T'/2$.

The probability per unit pathlength that an electron of energy T' transfers energy between E and $E + dE$ to an atomic electron is given by $n_v[d\sigma(T',E)/dE]dE$ where $n_v = N_A(Z/M_m)\rho$ is the number of atomic electrons per unit volume. A complete expression for $d\sigma(T',E)/dE$ can be obtained from the relativistic Möller cross section, given in Equation 3.1. In this case the immediate aim is to calculate the number of electrons ejected with energy greater than T because of collisions with source electrons of energy T'. If $K(T',T)$ is the probability per unit path length that a secondary electron with energy in the interval from $E_{\min}$ to $E_{\max}$ is set free by a source electron of energy T', then

$$K(T',T) = N_A(Z/M_m)\rho \int_{E_{\min}}^{E_{\max}} \frac{d\sigma(T',E)}{dE}\, dE,$$

$$E_{\min} = T, \quad E_{\max} = T'/2 \tag{11.5}$$

The total number of ejected electrons per unit volume with energy greater than T is obtained by multiplying $K(T',T)$ by the fluence of electrons with energy between T' and $T' + dT'$ and integrating over all possible values of T' from $2T$ to the maximum value in the spectrum, $T_{\max}$. Thus, the spectrum of ejected electrons (per unit volume) before their degradation is given by

$$_e n_v(T) = \int_{2T}^{T_{\max}} \phi(T')K(T',T)dT' \tag{11.6}$$

The lower limit is $2T$ in the integral because electrons with smaller energies are considered to be the ejected electrons. Equation 11.6 must be set equal to zero for $T > T_{\max}/2$. If this is not done, source electrons would be added to the numbers of ejected electrons.

The final expression for the total electron fluence between T and $T + dT$ is obtained by adding the ejected electron contribution to the expression for source electrons in Equation 11.3. The analog of Equation 11.4 for the total spectrum is

$$\phi(T)dT = [S_c(T)]^{-1} \left[\int_{T}^{T_{\max}} {}_v n_s(T')dT' + \int_{2T}^{T_{\max}} \phi(T')K(T',T)dT' \right] dT \tag{11.7}$$

This equation can be used in calculations if the integrals are approximated by finite sums and calculations proceed from intervals at large T to those at small T. When T is nearly as large as $T_{\max}$, the fluence contains contributions from source electrons only, as obtained from the first term in the brackets, since the second term is set equal to zero. Ejected electron fluences at energies $T < T_{\max}/2$ can be obtained by substituting high-energy $\phi(T')dT'$ values obtained from prior source term calculations into the second integral in the brackets.

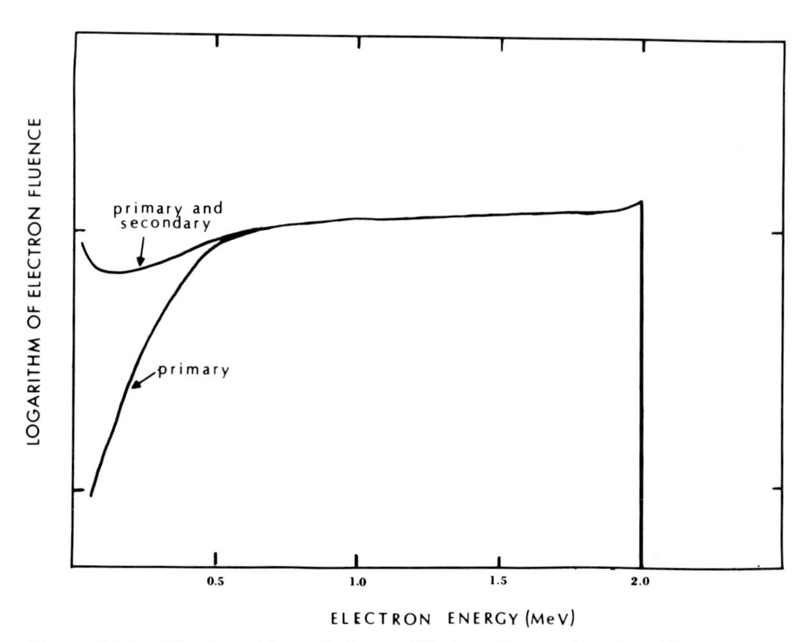

Figure 11.1 The logarithm of the equilibrium fluence for 2-MeV source electrons uniformly degraded in aluminum is shown as a function of energy (2).

In Equation 11.7 the overriding energy dependence at high kinetic energy is directly related to $[S_c(T)]^{-1}$, which in turn is approximately proportional to $(v/c)^2 = \beta^2$. Thus, the equilibrium fluence from a monoenergetic source is expected to be nearly constant at relativistic energies with $v \simeq c$. It will decrease with decreasing energy when $T < m_e c^2$ and finally increase with decreasing energy when the number of ejected electrons becomes important. Figure 11.1 shows the degradation spectrum in aluminum for a 2-MeV monoenergetic electron source (2) under uniform irradiation conditions. Notice that the curve labeled *primary* on this figure follows the reciprocal of the stopping power as described above. The contribution from the ejected electrons becomes appreciable at energies below about 500 keV, and the two curves diverge at this point. The bremsstrahlung process has little effect on the spectrum for low atomic number materials at these energies and can be ignored. If the absorber atomic number or electron energy is large, the effect of bremsstrahlung must be included (2).

Figure 11.2 shows a comparison between the beta spectrum from ^{32}P and the spectrum that results (4) from uniform degradation of beta-source electrons, including contributions from ejected atomic electrons. Notice that the average electron energy is shifted to smaller values in the degradation process, as would be expected. The ejected electrons dominate in the spectrum below about 40 keV.

Equation 11.7 has been used to calculate the electron fluence per unit

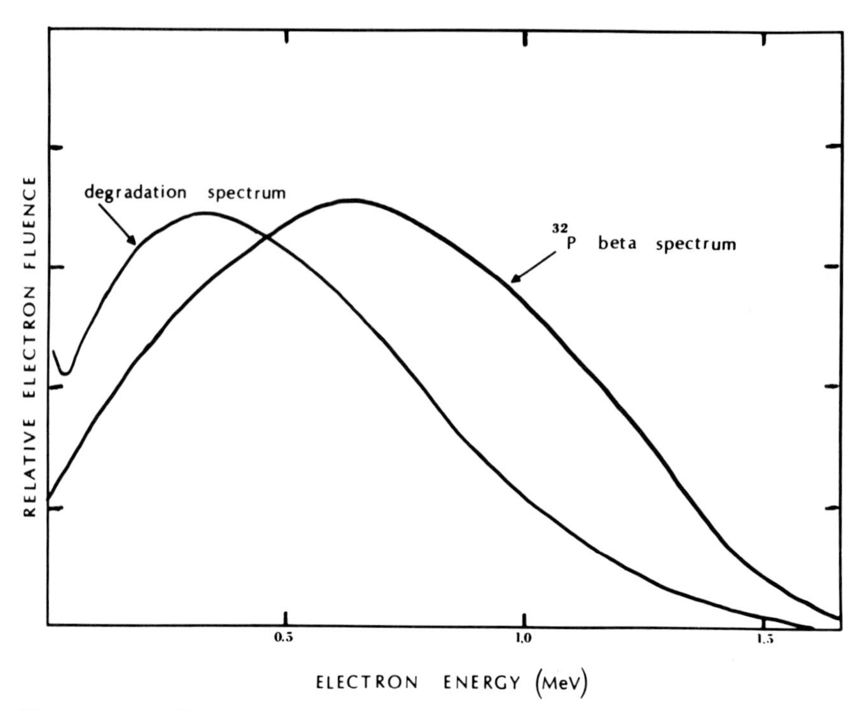

Figure 11.2 A ^{32}P beta-emission spectrum and the spectrum of ^{32}P electrons uniformly degraded in Bakelite are shown. The curves have been normalized so that the areas are equal. Adapted from Ref. 4.

energy (6,7). The results agree well with experimental results above 10 keV for uniformly embedded beta sources. Below 10 keV the stopping power expressions available are usually not sufficiently accurate. Furthermore, contributions to the electron spectrum from collisions with inner-shell electrons and the resulting Auger electron cascades must be considered (8) at lower energies. An experimental electron spectrum obtained with ^{198}Au beta rays embedded in aluminum is compared with the calculated curve in Figure 11.3.

Direct applications of the calculation scheme outlined in the preceding part of this section are somewhat restricted because of the assumption of radiation uniformity. In practical situations, external unidirectional photon beams are often employed. The radiation is not uniform in these situations, because source electron velocity components along the direction of incidence dominate. Furthermore, the beam is attenuated as the depth increases.

Figure 11.4 shows an experimental electron degradation spectrum produced by a narrow beam of ^{60}Co photons incident on a flat piece of aluminum (9). The thickness of aluminum used to obtain this spectrum was sufficient for complete charged-particle buildup. Notice that the maximum kinetic energy of the electrons was 1.12 MeV, the energy of the Compton edge electron from the 1.33-MeV photon. The step at 0.95 MeV is due to the Compton edge of the 1.17-MeV ^{60}Co photon. The onset of dominance of the ejected atomic elec-

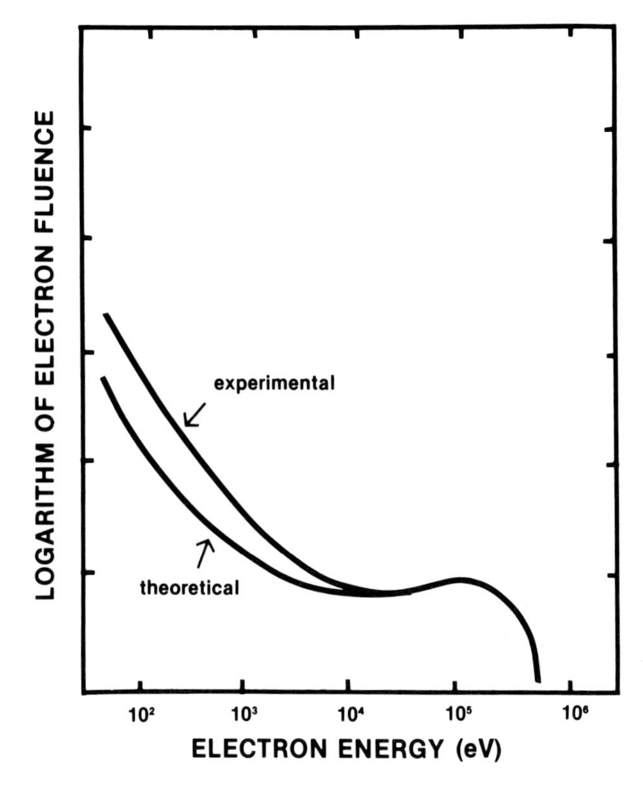

Figure 11.3 The electron spectrum for ^{198}Au beta particles under uniform degradation in aluminum is shown (6). Experimental results show an excess of low-energy electrons. Energy is referenced to the bottom of the conduction band on the abscissa.

trons occurred below 100 keV, as indicated by the minimum on the figure. Narrow-beam equilibrium electron spectra produced by photon beams from radionuclides other than ^{60}Co have been measured (10) and have a gross shape similar to that on Figure 11.3.

The equilibrium electron spectrum calculated with Equation 11.7 by assuming Compton interactions of ^{60}Co photons in aluminum is also shown on Figure 11.4 for comparison. Notice that the average energy of the calculated uniform fluence spectrum is considerably smaller than the average energy of the narrow-beam emission spectrum. A deficit of low-energy electrons scattered at large angles is apparent in the spectrum from the narrow unidirectional beam. This difference is related to the fact that low-energy Compton electrons are ejected at large angles with respect to the incident photon direction. The narrow-beam result is unlikely to include many of these electrons.

Although forward-directed electrons are the most prevalent at high energies, some electrons are ejected from a surface at angles between 90° and 180° from the incident beam direction. At small photon energies, these would include photoelectrons emitted at large angles; at larger energies, where Compton interactions dominate, backward emission occurs because of multiple scattering of the newly ejected electrons. The ratio of the backward electron emission current to the forward emission current is about 0.04 for aluminum and carbon targets

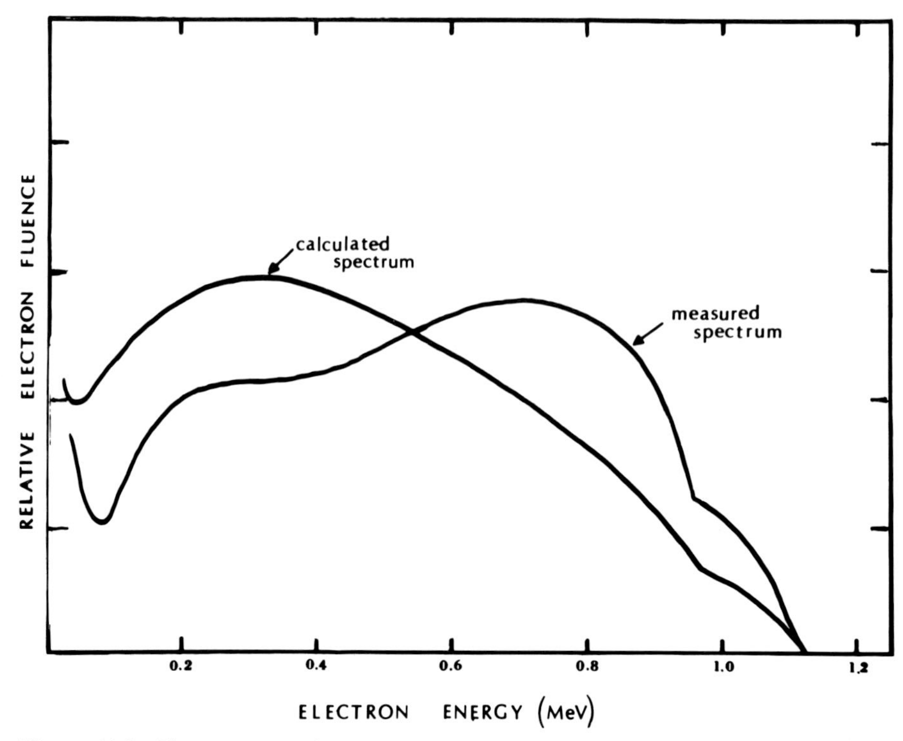

Figure 11.4 The measured electron emission spectrum from aluminum, induced by a narrow beam of ^{60}Co gamma rays, has relatively more high-energy electrons than the spectrum calculated for uniform degradation (9). The curves have been normalized so that the areas are equal.

when ^{60}Co photons are used (11). It is much larger when high atomic number materials are used (12). The average energy of the backward-directed electrons is considerably smaller than the average energy of the forward electrons.

Involved calculation techniques are often needed to calculate spectra for nonuniform and nonequilibrium irradiations. Figures 11.5 and 11.6 give calculated spectra for electron beams in water. Figure 11.5 shows the energy spectrum at depth 3.3 cm from a 10-MeV monoenergetic beam (13); notice that the distribution has a substantial degraded tail partially due to ejected atomic electrons. Figure 11.6 shows the electron fluence rate for a 20-MeV beam in a water absorber. Notice the substantial numbers of ejected electrons that contribute in depth (14).

11.3 Ion Production in Gaseous Absorbers

In this section, equations will be developed to describe the amount of charge liberated in a gaseous absorber. Simple expressions will be given for charge

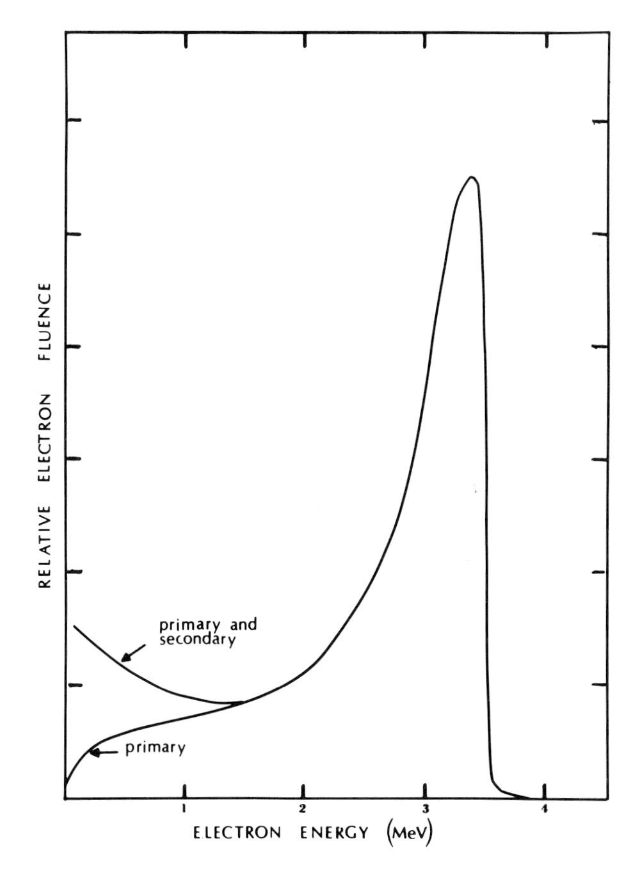

Figure 11.5 The electron spectrum generated by a 10-MeV electron beam at depth 3.3 cm in water is shown (13).

densities and current densities. The aim of the section is to provide background information necessary for understanding charge production phenomena in ion chambers (15, 16).

Recall that an atom or molecule is said to be ionized when one or more of the orbital electrons has been removed. If an atom or molecule is excited, the energy of the system is increased above the ground-state value, but no charge is removed. The energy equivalent of the ionization potential, E_{IP}, is the minimum energy required to remove an electron. Experimental values of E_{IP} were given for various molecules in Table 1.1.

W, the average energy expended per ion pair produced, is the quotient of the kinetic energy T of a directly ionizing particle by the mean number of ion pairs, N_{ip}, formed when the particle is completely stopped in a gas (17):

$$W = T/N_{ip} \tag{11.8}$$

It is assumed that N_{ip} is the ion pair number prior to any charge recombination.

If the molecular targets are subjected to a charged-particle fluence with spectrum function $\phi(T)dT$, the value of N_{ip} is given by

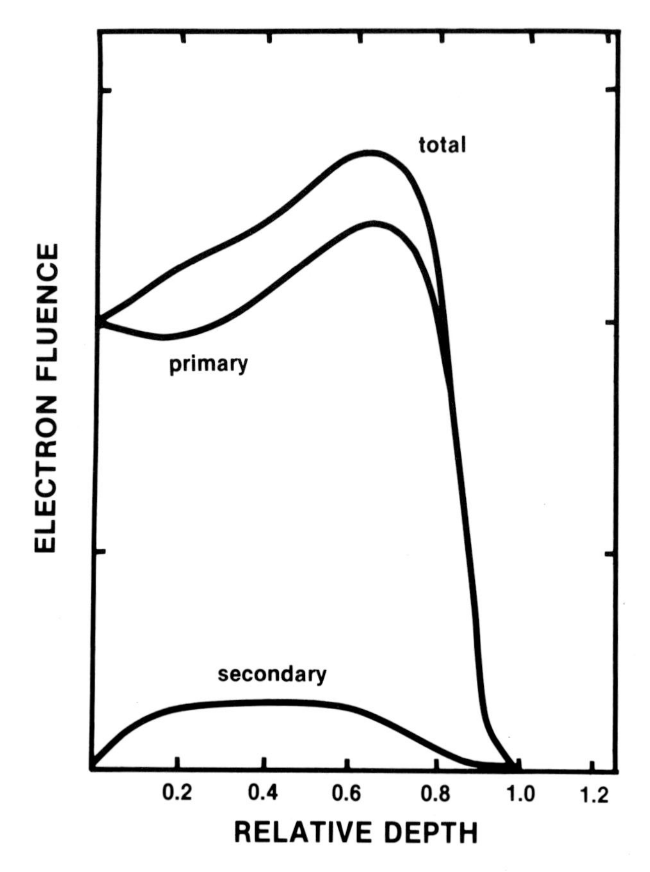

Figure 11.6 The electron fluence for a 20-MeV beam in water shows a considerable contribution from secondaries (14). Adapted from J. S. Laughlin, "Electron Beams," in *Radiation Dosimetry,* Vol. III, edited by F. H. Attix and E. Tochilin (Academic Press, New York, 1969).

$$N_{\rm ip} = n_v V \int_{E_{\rm IP}}^{T_{\max}} \sigma_i(T')\phi(T')dT' = n_m \rho V \int_{E_{\rm IP}}^{T_{\max}} \sigma_i(T')\phi(T')dT' \quad (11.9)$$

Here $n_v = n_m \rho$ is the number of molecules per unit volume, where n_m is the number of molecules per unit mass, and ρ is the density. The volume of gas irradiated is V, and $\sigma_i(T)$ is the ionization cross section per atom of gas expressed as a function of kinetic energy of the incident particles. The assumption that the fluence is uniform throughout the volume is implicit in Equation 11.9. If E is the total energy absorbed from the charged-particle beam, then

$$W = \frac{E}{N_{\rm ip}} = \frac{E}{n_v V \int_{E_{\rm IP}}^{T_{\max}} \sigma_i(T')\phi(T')dT'} \quad (11.10)$$

The value of W is independent of charged particle kinetic energy for a given material, except when the particle speed approaches the speed of the electrons in the target atoms. At that point, W rises sharply (18). Figure 11.7 shows a plot of $N_{\rm ip}$ versus T for an electron (19, 20). The line on that figure has the form

$$N_{ip} = (T - T_0)/C_s = T/W, \quad W = TC_s/(T - T_0) = C_s/(1 - T_0/T) \tag{11.11}$$

where C_s is the reciprocal of the slope and T_0 is the intercept on the energy axis. T_0 is slightly smaller than the energy equivalent of the ionization potential, as is shown on the figure.

Table 11.1 shows a list of experimental values for W (18, 20). The values for air and the tissue-equivalent material are especially important. Generally, all values are in the range from 25 to 40 eV/ion pair. One should notice that experimental W values for protons and for alpha particles are generally greater than those for electrons and photons. This difference may be partially due to recombination along the particle track, since the ionization density is large along a heavy ion track.

The phrase "average energy per ion pair" can be misleading since energy lost in excitation processes is included, although only ion pairs are explicitly counted. Suppose N_{ex} is the average number of excitations produced and ϵ_{ex} is the average excitation energy. If ϵ_{pi} is the average energy expended in production of a positive ion and ϵ_{se} is the average kinetic energy of the subexcitation electrons, then

$$T = N_{ip}(\epsilon_{pi} + \epsilon_{se}) + N_{ex}\epsilon_{ex},$$

$$W = T/N_{ip} = \epsilon_{pi} + \epsilon_{se} + (N_{ex}/N_{ip})\epsilon_{ex} \tag{11.12}$$

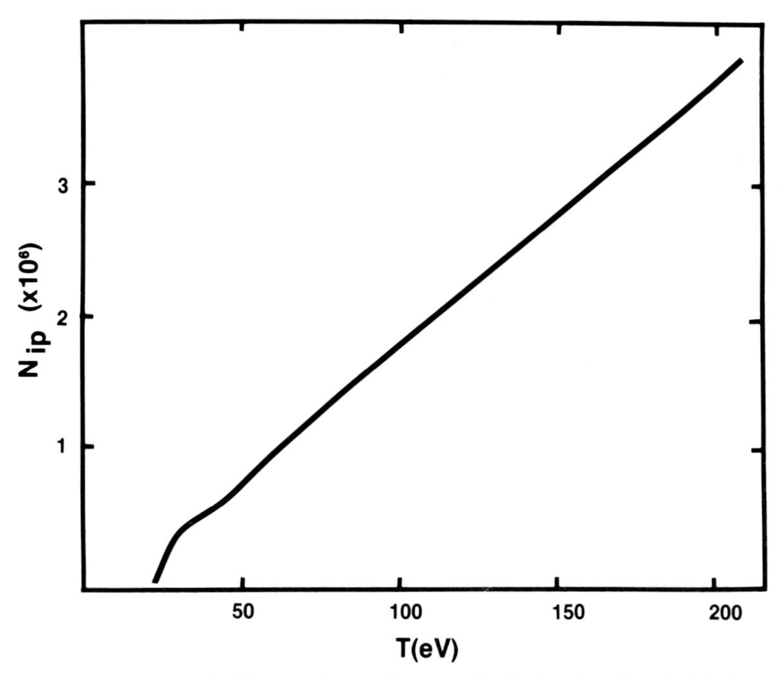

Figure 11.7 For kinetic energies well above the ionization threshold, the number of ion pairs, N_{ip}, produced is proportional to the kinetic energy dissipated (19).

Table 11.1 Average Energy per Ion Pair

Gas	*Electrons and Photons*	*Neutrons 1 → 14 MeV*	*Protons*	*α Particles*
Air	33.8		35.2	35.1
Argon	26.4		26.7	26.4
N_2	34.8		36.7	36.4
TE[a]	28.1	31.0	30.0	30.0
Water	29.6			37.6

Data from Refs. 18 and 20.
[a] Tissue-equivalent material.

Of course, electrons liberated in an ionization process with substantial kinetic energy will themselves induce further ionization and excitation until they are degraded to the subexcitation level. Subexcitation electrons cannot ionize or cause electronic excitation but may lose energy in other processes during collisions (21).

Generally, ϵ_{pi} is larger than the ionization potential; $\epsilon_{pi} = E_{IP} + \Delta$. The reason is that ionization processes do not always occur from valence states. More tightly bound electrons may be expelled, and the resulting ions are left in excited configurations. In this circumstance, Δ is the average excitation energy of the residual positive ions. In addition, multiple ionizations are possible, so doubly charged ions may be produced in a single process. Estimates of the double-to-single ratio (22) are usually less than 0.1, however.

In general, the average excitation energy ϵ_{ex} includes contributions from several different excited states. Table 11.2 gives the relative frequency of various collision processes for electrons in atomic hydrogen (23). Notice that excitation events occur more often than ionizations but that the excitation probability decreases with increasing excitation energy.

The average energy per ion pair is found to be directly related to the energy equivalent of the ionization potential E_{IP}. This dependence is suggested from Equations 11.12 because of the influence of ϵ_{ip}, a substantial part of W. In fact, one can use the approximations (18, 20)

Table 11.2 Percentage of Various Collision Processes in Atomic Hydrogen

Collision Type	*% of Collisions*	
	T = 1 keV	*T = 1 MeV*
Excitation to first excited state	43	50
Excitation to second excited state	6	8
Excitation to higher excited state	6	6
Ionization	36	32
Elastic process	9	4

Data from Ref. 23.
T, Electron energy.

$$\text{noble gases: } W \simeq 1.7E_{IP}, \quad W/E_{IP} \simeq 1.7,$$

$$\text{molecular gases: } W \simeq 1.7E_{IP} + 10, \quad W/E_{IP} \simeq 2.3 \tag{11.13}$$

The increase in the W/E_{IP} ratio for molecular gases can be identified with an increased frequency of excitations. A possible reason for the different W values for molecular gases and noble gases is that noble gases generally have fewer low-energy excited states available for transitions than the molecular gases. Transitions in the closed-shell systems require substantial energy when compared to the energy transitions of valence electrons.

Another possible reason for different W values involves the favored formation of excimers in noble gases. Excimers are excited dimeric species with a repulsive ground state. A reaction for excimers is (21)

$$R + R^* \rightarrow R_2^* \rightarrow R_2^+ + e^- \tag{11.14}$$

where the asterisk indicates an excited configuration of the noble gas atom R and the dimer R_2. The reaction of Equation 11.14 tends to increase the total ionization at the expense of excitation energy and to lower the ratio W/E_{IP}.

Platzman has made numerical estimates (24) of the quantities in Equation 11.12 for the case of helium gas:

$$W = \epsilon_{pi} + \epsilon_{se} + (N_{ex}/N_{ip})E_{ex} = 1.06E_{IP} + 0.31E_{IP} + (0.40)(0.85)E_{IP},$$

$$1.37E_{IP} + 0.34E_{IP} = 1.71E_{IP} = 42.1 \text{ eV} \tag{11.15}$$

Thus, for helium at least, the positive ion energy is expected to be slightly greater than the ionization potential. Furthermore, subexcitation electron energy is substantial, and the average excitation energy is large: $E_{ex} = 0.85E_{IP}$. Excitation events occur much less frequently than ionization events for noble gases since $N_{ex}/N_i = 0.40$. For a molecular gas, the excitation-to-ionization ratio might be one or larger (see Table 11.2).

If a volume of gas, V, of density ρ absorbs energy E to produce a uniform dose D throughout, then

$$W = \frac{E}{N_{ip}} = \frac{D\rho V}{N_{ip}}, \quad N_{ip} = \frac{D\rho V}{W}, \quad Q/V = \frac{eN_{ip}}{V} = \frac{D\rho}{W/e} \tag{11.16}$$

where Q is the total charge of one sign on ions produced by the radiation. Equations 11.16 shows that the number of ion pairs per unit volume produced in a gaseous absorber is proportional to the dose. Recombination of any of the ions is neglected in this equation. Here N_{ip} refers to the number of ion pairs produced rather than the number that can be collected at an electrode.

The last equation in Equations 11.16 indicates that the charge produced per unit volume is proportional to the density of the gas. Assuming that the system behaves like a perfect gas (25),

$$pV = NkT^\circ, \quad \rho = \frac{NM}{V} = \frac{M}{k}\left[\frac{p}{T^\circ}\right] \tag{11.17}$$

where N is the number of molecules contained in volume V at pressure p and absolute temperature $T°$. The value k is the Boltzmann constant, and M is the mass per gas molecule. From Equations 11.16 and 11.17 it is evident that the charge produced in a chamber open to the atmosphere is directly proportional to the pressure and inversely proportional to the temperature of the air in the chamber.

Suppose electrodes are put inside the gas-filled volume, an electric field is applied, and the charge liberated by the radiation is collected. In this case the positive ions and electrons collected at the electrodes and the positive ions and electrons that recombine both serve to decrease the equilibrium charged-particle density in the gaseous volume. Suppose P_v is the ion pair production rate per unit volume due to radiation, ${}_e\tau_c$ is the mean electron lifetime in the gas because of collection at the electrodes, and ${}_e\tau_r$ is the mean electron lifetime because of recombination. From Equations 11.16

$$P_v = \frac{\rho}{W}\left[\frac{dD}{dt}\right], \quad \frac{d({}_en_v)}{dt} = P_v - \frac{{}_en_v}{{}_e\tau_r} - \frac{{}_en_v}{{}_e\tau_c} \tag{11.18}$$

Here ${}_en_v$ is the instantaneous electron density in the gas. A similar equation involving ${}_in_v$, the number of positive ions per unit volume, could also be written, but a different value of τ_c would be required.

In fact, the equation describing the liberated electron density will be further complicated when some of the electrons attach to neutral gaseous molecules prior to charge collection. The negative ions formed in this manner have their own characteristic collection times, which are generally longer than ${}_e\tau_c$. Furthermore, the possibility that charge multiplication might occur has been ignored in Equation 11.18. High-energy transfer collisions between the electrons and neutral molecules can cause release of additional electrons and an increase in charge density. Equation 11.18 holds only when the applied electric field is below the value necessary for charge multiplication.

With the help of Figure 11.8, a simple expression for ${}_e\tau_r$ can be obtained. Suppose that initial recombination along a charged-particle track can be neglected and assume that further recombination events are random processes that occur whenever an electron passes the neighborhood of a positive ion. If the effective ion pair recombination cross section is σ_{ip}, an electron with speed

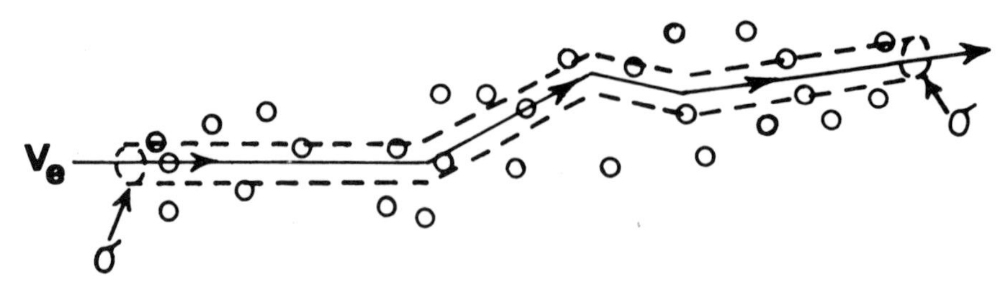

Figure 11.8 The path of an electron with speed v_e through a gaseous absorber is depicted. The magnitude of the cross section for ion pair recombination is labeled σ; the circles represent positive ions distributed randomly around the electron path.

v_e can be assumed to sweep out an effective recombination volume $(v_e)\sigma_{\text{ip}}$ in one second. The product ${}_in_v(v_e)\sigma_{\text{ip}}$ gives the number of positive ions that are in the sweeping volume in a second. If the electron speed is much greater than the positive ion speed, $v_e \gg v_i$, and if each collision produces recombination, this product is equivalent to the recombination probability per second. The mean lifetime for recombination is the reciprocal of the recombination probability per second:

$$\frac{1}{{}_e\tau_r} = {}_in_v(v_e)\sigma_{\text{ip}} \tag{11.19}$$

Substituting Equation 11.19 in the right-hand side of Equations 11.18 yields

$$\frac{d({}_en_v)}{dt} = P_v - \frac{{}_en_v}{{}_e\tau_c} - ({}_en_v){}_in_v(v_e)\sigma_{\text{ip}} = P_v - \frac{{}_en_v}{{}_e\tau_c} - \alpha_r({}_en_v)({}_in_v) \tag{11.20}$$

where α_r can be called the ion pair recombination coefficient. The last term in Equation 11.20 shows that recombination loss is related to the product of the electron density and the ion density. Thus, recombination becomes more important when dose rates are very high. For air, α_r is about 10^{-13} to 10^{-16} m^3/s when the electrons themselves are charge carriers. When the electrons attach to neutral atoms to form negative ions, $\alpha \simeq 10^{-12}$ m^3/s (16). The slower negative ions recombine more readily.

Equation 11.20 can be simplified by neglecting the recombination term when ${}_e\tau_r \gg {}_e\tau_c$. Under this assumption, for an equilibrium electron density and a steady beam,

$$\frac{d({}_en_v)}{dt} = P_v - \frac{({}_en_v)}{{}_e\tau_c} = 0, \quad ({}_en_v) = P_v({}_e\tau_c) = \frac{\rho({}_e\tau_c)}{W}\left[\frac{dD}{dt}\right] \tag{11.21}$$

When recombination is negligible, the equilibrium value of the electron density is proportional to the first power of the dose rate. If recombination can not be neglected, the equilibrium electron density is proportional to a fractional power (less than one) of the dose rate.

The basic equation describing the movement of charged particles in a gaseous medium under the influence of an electric field involves the drift velocity $\boldsymbol{\omega}$. If $\mathbf{J}$ is the current density in amperes per square meter and $\pm en_v$ is the charge density in coulombs per cubic meter,

$$\mathbf{J} = \mathbf{J}_i + \mathbf{J}_e = e({}_in_v)\boldsymbol{\omega}_i - e({}_en_v)\boldsymbol{\omega}_e \tag{11.22}$$

where subscript i or e stands for positive ions or electrons, respectively. Diffusion contributions and electron attachment have been ignored in Equation 11.22.

The drift velocity for positive and negative ions can be linearly related to the ionic mobility μ, the electric field E and inversely related to the gas pressure p:

$$\boldsymbol{\omega}_i = (\mu)\frac{E}{p} \tag{11.23}$$

Table 11.3 Drift Mobilities of Positive and Negative Ions

	$(m/sec)\cdot(mmHg)\cdot(V/m)^{-1}$		
Gas	*Air*	*Argon*	*Carbon Dioxide*
$\mu+$	0.107	0.104	0.060
$\mu-$	0.135	0.129	0.072

Some values for drift mobilities (16) are given in Table 11.3 for positive and negative ions. The drift velocity for electrons is a function of E/p, although it is not always linearly related to that quantity. Figure 11.9 shows the functional relationship (26) for electrons in argon. It is not unusual for the electron drift speed to approach a limiting value of about 10^4 to 10^5 m/s (16), even when the electric field is increased further.

Equations 11.21 and 11.22 can be used to provide an expression for the total current density in terms of the dose rate:

$$\mathbf{J} = eP_v[\omega_i({}_i\tau_c) + \omega_e({}_e\tau_c)]$$

$$= \frac{\rho}{(W/e)}[\omega_i({}_i\tau_c) + \omega_e({}_e\tau_c)](dD/dt) \qquad (11.24)$$

where ${}_i\tau_c$ and ${}_e\tau_c$ stand for the mean collection times for the positive ions and electrons, respectively. Equation 11.24 holds for the steady-state situation, when diffusion and recombination are negligible.

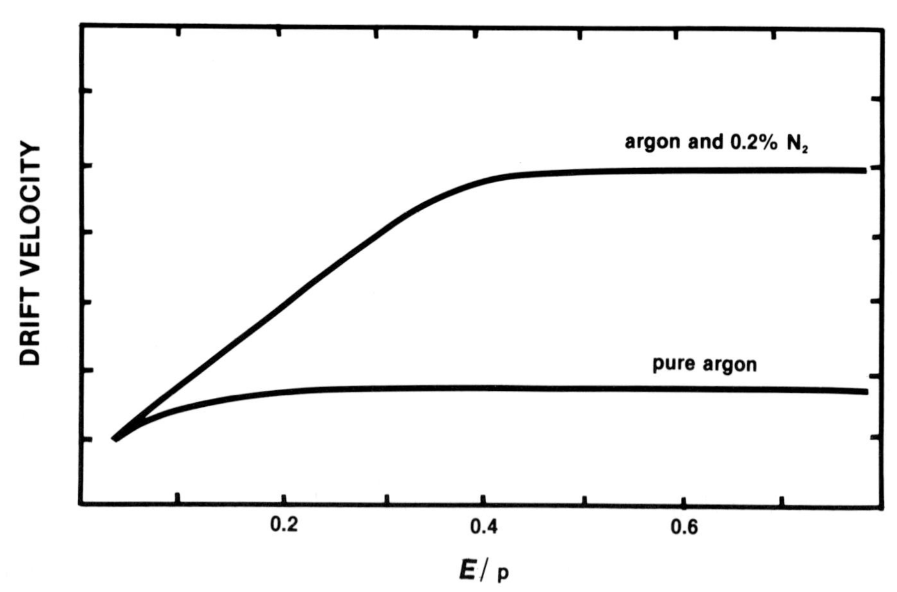

Figure 11.9 The magnitude of the electron drift velocity is shown as a function of the electric field per unit pressure (E/p) for argon (26).

11.4 Cavity-Dose Relationships

As was mentioned previously, one usually measures the dose in an irradiated medium using the response of a device inserted in the medium. Generally, the device has somewhat different properties for absorption of radiation than the medium itself. Thus, the actual measured quantity must be corrected to give the dose that would have been present if the dosimeter had not perturbed the system.

If the measuring device is filled with gas and is placed in a solid or liquid medium, it is commonly called a cavity. Whether the device is gaseous, liquid, or solid, it represents a discontinuity in the physical properties of the medium under irradiation and can be considered a cavity in that sense. Examples of cavity dosimeters for the gaseous, liquid, and solid categories are ion chambers, Fricke dosimeters, and thermoluminescent dosimeters, respectively. The theory dealing with the relationship between dose measurements made with gaseous, liquid, or solid devices and the dose in the surrounding medium is generally called cavity theory (3, 27, 28).

In the following paragraphs, a simplified cavity theory will be presented. Irradiation with photons and electrons will be considered. The development leads to analytical expressions involving ratios of mass energy absorption coefficients for photons in the cavity and in the medium, as well as ratios of mass stopping powers for electrons in the cavity and in the medium. The expressions that result could be applied to measurements with neutron beams if appropriate energy absorption coefficients and stopping powers for released ions were utilized.

The theory is most easily understood if all possible cavities are divided into three categories according to relative size: large, intermediate, and small. The cavity size is of great importance for measurements in indirectly ionizing radiation beams, where the liberated charged secondaries distribute nearly all of the dose. In this case one can compare the effective cavity diameter l with the average range $\langle R \rangle$ of the charged particles produced in the cavity material. The categories are defined by:

1. large cavity, $l \gg \langle R \rangle$;
2. intermediate cavity, $l \sim \langle R \rangle$;
3. small cavity, $l \ll \langle R \rangle$.

The situation is illustrated in Figure 11.10. In a large cavity, the dose is deposited locally by secondary charged particles that are released in the cavity. Comparatively little energy is deposited by secondaries liberated in the medium that pass into the cavity. For a small cavity, the bulk of the dose is from secondaries generated in the medium that cross the cavity. Relatively few charged secondaries are generated in the cavity itself, and they deposit little energy.

In Table 11.4 the dimensions of some practical dosimeters can be compared to the ranges of electrons (29) produced by photon beams. Appropriate size categories can be identified. For example, an air-filled ionization chamber with a diameter of approximately one centimeter is a small cavity for photon beams with an average energy above 300 keV, since the electron range in air is much

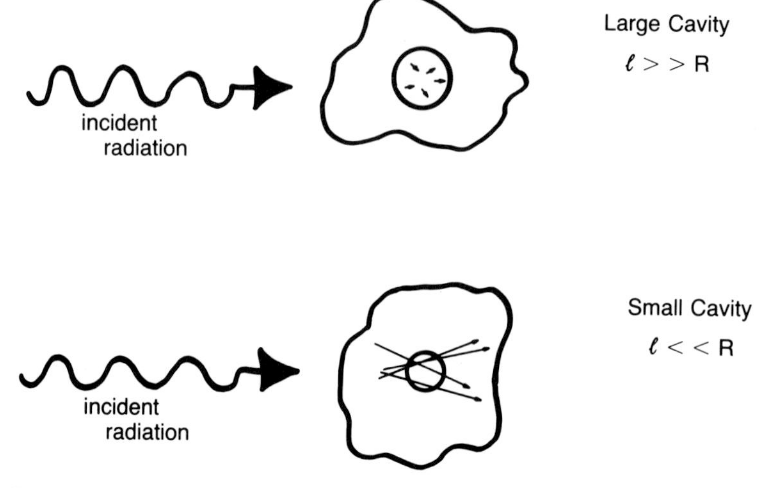

Figure 11.10 A large cavity and a small cavity are illustrated. The electron range, indicated by the arrows, should be compared to the cavity diameter.

larger than the air-cavity dimensions. The relatively low density of air is the determining factor in this classification. On the other hand, a solid dosimeter with effective diameter of about 1 cm must be considered a large cavity for all low-energy photon beams because of the relatively large mass density and resulting small charged-particle range.

Consider first the case of a large cavity with dimensions much greater than the average range of the electrons in the cavity material. Because of the large size, interchange in energy between the cavity and the medium is neglected, and the cavity dose is ascribed to photons interacting in the cavity. Using Equations 5.8 and 5.17,

$$K = E\phi(\mu_{tr}/\rho) = E\phi \frac{(\mu_{en}/\rho)}{(1-g)} \tag{11.25}$$

where K is the kerma, E is the energy of the indirectly ionizing particle, and ϕ is the monoenergetic fluence of such particles. The factor g is the fraction of the energy of the secondary electrons that is radiated.

Table 11.4 Ranges of Electrons Produced by Photons

Photon Energy (MeV)	*Maximum Compton Electron Energy (MeV)*	*Maximum Range in Air (m)*	*Maximum Range in Water (cm)*
0.3	0.16	0.3	0.03
1.0	0.80	2.9	0.33
3.0	2.76	11.8	1.32
10.0	9.75	39.3	4.76

Data from Ref. 29.

Suppose that the kerma-to-dose conversion ratio in the cavity is b_c. Then

$$D_c = b_c E \phi_c \frac{(\mu_{en}/\rho)_c}{(1-g)_c} \tag{11.26}$$

If the cavity were removed, the dose received by the medium in the same position would be given by

$$D_m = b_m \mathrm{E} \phi_m \frac{(\mu_{en}/\rho)_m}{(1-g)_m} \tag{11.27}$$

where b_m is the kerma-to-dose conversion ratio for the medium m. Equation 11.27 can be justified whether or not the electron range *in the medium* is smaller than the cavity dimensions. A sufficient condition is that an equilibrium in energy exchange occurs, so that as much energy is carried into the region of interest by radiation as is carried out.

Dividing Equation 11.26 by Equation 11.27 gives

$$D_c = \frac{b_c \phi_c/(1-g)_c}{b_m \phi_m/(1-g)_m} \frac{(\mu_{en}/\rho)_c}{(\mu_{en}/\rho)_m} D_m = A_{cm} \frac{(\mu_{en}/\rho)_c}{(\mu_{en}/\rho)_m} \mathrm{D}_m \tag{11.28}$$

A_{cm} the cavity-over-medium perturbation factor, is introduced to simplify the notation. If A_{cm} is different from one, the effects of beam attenuation, bremsstrahlung emission, or charged-particle buildup are different in the two materials considered.

Equation 11.28 involves a ratio of mass energy absorption coefficients, and is well defined for monoenergetic beams. In most cases continuous spectrum beams are employed, so an average mass energy absorption coefficient ratio must be utilized. Then

$$(R_\mu)_{cm} = \frac{\int \frac{b_c}{(1-g)_c} E(\mu_{en}/\rho)_c \; \phi_c(E)dE}{\int \frac{b_m}{(1-g)_m} E(\mu_{en}/\rho)_m \; \phi_m(E)dE,}$$

$$D_c = (\mathrm{R}_\mu)_{cm} D_m \tag{11.29}$$

In this equation $\phi(E)dE$ is the spectrum function for indirectly ionizing radiation and $(R_\mu)_{cm}$ is the cavity-over-medium ratio for spectrum averaged dose values.

For the case of a small cavity, the dimensions involved are much less than the average range of the secondary charged particles in the cavity material. Deposited energy is attributable to charged particles traversing the cavity material and is nearly independent of indirectly ionizing components absorbed in the cavity. As mentioned previously, ion chambers often qualify as small cavities.

The response of ion chambers to photon beams has been considered by Bragg (30) and Gray (31, 32). The principle of equivalence as set forth by Gray was an important step in understanding dose in a cavity. According to Gray, the energy lost per unit volume by electrons in a small gas-filled cavity is equal to the energy absorbed per unit volume from the radiation in the surrounding medium multiplied by a constant factor. The factor was shown to be the ratio of the collisional loss stopping power in the cavity, $(S_c)_c$, to the collisional loss stopping power in the medium, $(S_c)_m$. Then

$$(-\Delta T/\Delta V)_c = \frac{(S_c)_c}{(S_c)_m} (\Delta E/\Delta V)_m \tag{11.30}$$

where $-\Delta T$ stands for the kinetic energy loss and ΔE for the energy absorbed.

An important assumption implicit in Equation 11.30 is that the photon fluence is uniform in the region around the cavity. If the cavity were removed, an equilibrium in charged-particle energy exchange in the region of interest is presupposed. Under these conditions it is not necessary that the charged-particle range *in the medium* be greater than the cavity dimensions.

Incorporating mass density factors ρ_c and ρ_m in Equation 11.30 gives

$$(-\Delta T/\rho\Delta V)_c = \frac{(S_c/\rho)_c}{(S_c/\rho)_m} (\Delta E/\rho\Delta V)_m \tag{11.31}$$

Suppose we assume that the energy loss $-\Delta T$ is equal to the energy absorbed ΔE in the cavity. A relationship between the doses follows immediately. However, the assumption is implicit that the net effect of the exchange of secondary radiations such as bremsstrahlung and delta rays between the cavity and the medium is negligible (33). If the cavity and the medium produce similar secondaries in comparable amounts, their effects on energy deposition will be identical. However, atomic number differences might affect this balance, particularly as regards bremsstrahlung. Matching the two absorbers for effective atomic number is useful in an experimental situation. In general, the possibility of radiation fluence perturbations can be acknowledged with the use of a cavity perturbation factor p_{cm}. Then for monoenergetic charged particles, Equation 11.31 can be written

$$D_c = p_{cm} \frac{(S_c/\rho)_c}{(S_c/\rho)_m} D_m \tag{11.32}$$

When indirectly ionizing beams are employed, the secondary charged particles are never monoenergetic and the mass stopping power values must be averaged over the electron spectrum (34). The ratio of spectrum-averaged mass stopping powers is given by

$$(R_s)_{cm} = \frac{\int (S_c/\rho)_c \, \phi_c(T)dT}{\int (S_c/\rho)_m \, \phi_m(T)dT}, \quad D_c = (R_s)_{cm} D_m \tag{11.33}$$

In this equation $\phi(T)dT$ is the charged-particle spectrum function, and $(R_s)_{cm}$ is the cavity-over-medium ratio for spectrum-averaged mass stopping powers.

An intermediate-size cavity is one with dimensions comparable to the charged secondary average range. In this case photon absorption in the cavity is appreciable and electron interchange between the cavity and the medium cannot be neglected. Neither Equations 11.29 nor Equations 11.33 are exactly applicable. The actual cavity dose can be considered to be a weighted sum of doses given by the two limiting conditions:

$$D_c = [d(R_s)_{cm} + (1 - d)(R_\mu)_{cm}]D_m \tag{11.34}$$

where d is the weighting factor such that $d = 0$ indicates the large-cavity limit, and $d = 1$ indicates the small-cavity limit. The factor d is directly related to the average transmission of electrons through the cavity material.

If the electron spectrum is similar to that from a negative beta source, the factor d can be estimated. Figure 4.4 shows that the transmission of a beta spectrum of electrons is approximately exponential (35) except at small values. Using Equations 4.8,

$$\phi/\phi_0 = \exp(-\beta\rho x), \quad \beta = 1.7/(T_{\max})^{1.14},$$

$$d = \frac{\int_0^l \exp(-\beta\rho x)dx}{\int_0^l dx} \tag{11.35}$$

where ϕ_0 and ϕ are the incident and transmitted beta-particle fluence; β is the effective mass attenuation coefficient in square meters per kilogram; $T_{\max}$ is the beta end-point energy in megaelectron volts; and the parameter l is the effective cavity diameter. Equation 11.34 has been applied to situations involving electron fluences generated by photon beams. However, one should recognize that the actual electron spectrum may differ appreciably from a beta spectrum (compare Figures 11.2 and 11.4) in a practical situation.

11.5 Ion Chamber Response

Some algebraic relationships involving the dose at a point in a medium and the exposure at that point were given in Chapter 5. In this section, equations will be given that can be used to calculate water dose from the response of an ion chamber irradiated in water. The development will utilize some of the relationships given in the preceding section.

In the following paragraphs, two extreme cases will be discussed in detail: in the first case the walls (usually made of plastic) of the ion chamber will be treated as a large cavity in the water; in the second case, the wall material will be treated as a small cavity in the water. In both cases the central air chamber encompassed by the walls will be represented as a small cavity in the plastic, and the electrode composition will be ignored. In the result the response to an arbitrary beam is related to its response in a ^{60}Co photon beam as a reference. This subject have been discussed previously (34, 36–38).

For the first case, the ion chamber measurement is assumed to be made in water at some depth using a photon beam characterized by $h\nu$. The beam may be monoenergetic or have a continuous spectrum. In the latter case $h\nu$ would be an average energy. Suppose the maximum range of the electrons set free by photon interactions is much smaller than the thickness of the plastic wall. D_w is the dose that would have been deposited in the water at the chamber position if the chamber had been removed. D_p is the dose in the plastic cavity. Applying Equation 11.29, it is evident that

$$D_p = (R_\mu)_{pw} D_w \tag{11.36}$$

The subscript pw indicates that the ratio R_μ involves the averaged mass energy absorption coefficient for plastic divided by the averaged coefficient for water.

The dose to the air cavity, D_a, confined within the plastic walls of the ion chamber is given by

$$D_a = (R_s)_{ap} D_p = b_a \frac{33.8}{(1-g)_a} X_a \tag{11.37}$$

Equation 11.33 and Equation 5.31 were used to obtain this relationship. The subscript ap on $(R_s)_{ap}$ indicates that the ratio involves the averaged mass stopping power in air divided by the value in plastic. Since the subscript a stands for a value measured in air, X_a is the exposure in coulombs per kilogram and b_a is the kerma-to-dose conversion ratio for the air cavity. Combining Equations 11.36 and 11.37 to eliminate D_p gives

$$D_w = \frac{D_a}{[(R_s)_{ap}][(R_\mu)_{pw}]}$$

$$= \frac{b_a 33.8/(1-g)_a}{[(R_s)_{ap}][(R_\mu)_{pw}]} X_a = C_\lambda X_a \tag{11.38}$$

The value C_λ defined in Equation 11.38 is the factor needed to calculate dose when measurements are made in water with ion chambers calibrated for exposure in air.

As a second illustration, consider measurements in a high-energy beam (subscript $h\nu$) generating electrons with average range much greater than the thickness of the plastic walls of the ion chamber. In this case the dose in the air cavity can be related directly to the dose in the water, D_w:

$$D_a = [(R_s)_{aw}]_{h\nu}(D_w)_{h\nu} \tag{11.39}$$

Suppose that a dose equal to D_a is deposited in the same air cavity by a ^{60}Co beam and that the thickness of the ion chamber wall is sufficient to use the large cavity limit for ^{60}Co photons:

$$(D_w)_{Co} = (C_\lambda)_{Co} X_a = \frac{D_a}{[(R_s)_{ap}]_{Co}[(R_\mu)_{pw}]_{Co}},$$

$$(D_w)_{h\nu} = (C_\lambda)_{h\nu} X_a = \frac{D_a}{[(R_s)_{aw}]_{h\nu}} \tag{11.40}$$

where the subscript Co stands for the ^{60}Co photon beam. Eliminating D_a in Equations 11.40 allows identification of

$$(C_\lambda)_{h\nu} = \frac{[(R_s)_{ap}]_{Co}[(R_\mu)_{pw}]_{Co}(C_\lambda)_{Co}}{[(R_s)_{aw}]_{h\nu}} \tag{11.41}$$

Equation 11.41 can be used to calculate the C_λ factor to convert ion chamber response to the dose in water when an ion chamber calibrated for use with ^{60}Co is available. However, Equation 11.41 is only approximately correct in many practical circumstances. Many ion chambers commonly available have walls with thickness characteristic of an intermediate cavity when used in bremsstrahlung beams of moderate energy. In this case a relationship utilizing Equation 11.34 would be more appropriate in converting between the doses in the air chamber and the water. Several workers have evaluated C_λ for ion chambers calibrated with ^{60}Co or equivalent beams (34, 37, 39, 40). Table 11.5 shows recommended values (34).

The calibration of electron beams in water by ion chambers is also of considerable significance. Fortunately, the small-cavity assumptions are often valid for high-energy electron beams. In this situation, one can relabel Equations 11.40 for electron beams of energy T:

$$(D_w)_{Co} = (C_\lambda)_{Co}X_a = \frac{D_a}{[(R_s)_{ap}]_{Co}[(R_\mu)_{pw}]_{Co}},$$

$$(D_w)_T = (C_E)_T X_a = \frac{D_a}{[(R_s)_{aw}]_T} \tag{11.42}$$

The constant C_E is the correction factor used when electron beam doses are measured with an ion chamber calibrated in a ^{60}Co beam. One should note that the energy T used as the subscript refers to the degraded electron energy at the depth of the chamber and not the energy of the beam incident on the surface. Equations 11.42 can be used to show that

$$(C_E)_T = \frac{[(R_s)_{ap}]_{Co}[(R_\mu)_{pw}]_{Co}}{[(R_s)_{aw}]_T}(C_\lambda)_{Co} \tag{11.43}$$

Tables of C_E values at various depths are available (36).

Table 11.5 Conversion Factor for Water C_λ

Radiation	*Gy·(C/kg)⁻¹*	*rad/R*
^{60}Co	36.8	0.95
5 MV	36.4	0.94
10 MV	36.0	0.93
15 MV	35.7	0.92
25 MV	34.9	0.90
35 MV	34.1	0.88

Data from Ref. 34.

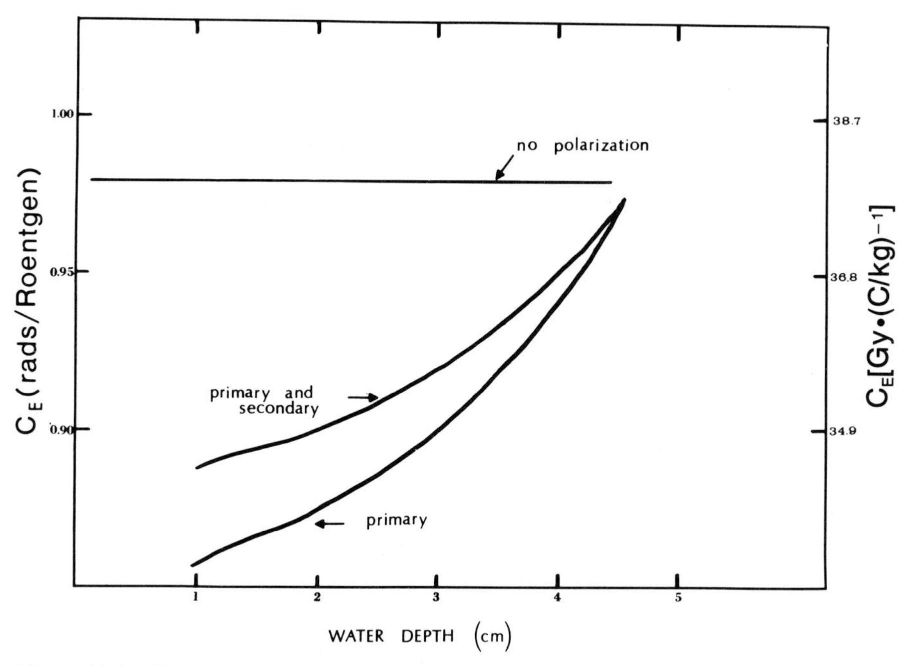

Figure 11.11 The conversion factor C_E for ion chamber calibration of a 10-MeV electron beam incident on a water surface is shown as a function of depth (14). Adapted from N. D. Kessaris, Radiat. Res. **43**, 288 (1970).

Figure 11.11 shows some values of C_E that have been calculated using the spectra like those shown on Figure 11.5. Notice that the major feature in the energy dependence of C_E is that its value decreases with increasing energy T. This is because the mass stopping power ratio indicated in the denominator of Equation 11.43 is changed because of the density effect. The mass stopping power for electrons in water is substantially reduced at high energies because of the density effect. The density effect is negligible for the air in the chamber. The line labeled *no polarization* on the figure shows that the mass stopping power ratio is nearly constant if the density effect is neglected. Note also that the dependence of C_E on depth is changed substantially when secondary electrons are included.

11.6 Problems

1. Suppose that the electron emission spectrum from a source has a constant value of $1/(\text{cm}^3 \cdot \text{keV})$ from 0 to 1 MeV and drops to zero immediately at higher energies. Plot the primary electron degradation spectrum from 10 keV to 1 MeV for a spatially uniform source distribution in water. Use stopping power values at representative points from tables in Appendix 7. Ignore secondaries.

2. Suppose that the electron emission spectrum from a source is linear with an intercept on the number axis at 1/cm³ and an intercept on the energy axis at 1 MeV. Plot the primary electron degradation spectrum from 10 keV to 1 MeV for a spatially uniform source distribution in water. Use stopping power values at representative points from tables in Appendix 7. Ignore secondaries.
3. In Problem 1, what is the maximum secondary electron energy? Why?
4. Suppose that the theoretical estimates of

$$\epsilon_{pi} = 1.06\ E_{IP}$$
$$\epsilon_{se} = 0.31\ E_{IP}$$
$$\epsilon_{ex} = 0.85\ E_{IP}$$

for helium gas are also good approximations for argon for electron bombardment. What is the ratio of N_{ex}/N_{ip} for electrons in argon and what fraction of the total energy would be excitation energy under these assumptions?
5. An ion chamber with a ^{60}Co calibration factor in air of 1.045 at 22°C and 760 mm Hg gave a reading of 150 roentgens at 5 cm depth in water after a 1.5-min exposure using a 25-MV photon beam. The water temperature was 23°C and the chamber air pressure was 725 mm Hg. What was the dose rate in water at 5 cm depth in grays per minute?
6. An ion chamber with a ^{60}Co calibration factor in air of 0.965 at 22°C and 760 mm Hg gave a reading of 250 R at 3 cm depth in water after a 2-min exposure using 10-MeV electrons incident on the water surface. The water temperature was 21°C and the chamber air pressure was 735 mm Hg. What was the dose rate in water at 3 cm depth in grays per minute?
7. An ion chamber with a ^{60}Co calibration factor in air of 0.983 at 22°C and 760 mm Hg measured 285 R at 4 cm depth in water after a 3-min exposure when ^{60}Co photons were incident on the water surface. The water temperature was 20°C and the chamber air pressure was 695 mm Hg. What was the dose rate in water at 4 cm depth in grays per minute?
8. Calculate the cavity theory weighting factor d for an electron spectrum similar to a ^{32}P beta-emission spectrum. Assume that the cavity is a spherical ion chamber with effective diameter of 5 mm sealed at 1 atm with a plastic wall.
9. Calculate the cavity theory weighting factor d for an electron spectrum similar to a ^{32}P beta-emission spectrum. Assume that the cavity is a lithium fluoride thermoluminescent dosimeter with 1-mm effective diameter embedded in a plexiglass phantom.
10. Calculate the cavity theory weighting factor d for the electrons produced by photons from an x-ray tube with a molybdenum target. Use the beta-spectrum approximation with maximum electron energy of 19 keV and assume that the cavity is a spherical ion chamber with effective diameter of 5 mm sealed at 1 atm with a plastic wall.
11. Calculate the ratio of the dose in water to the dose in a lithium fluoride chip of 1-mm effective diameter placed in the water in a beam of ^{60}Co photons. Assume the electron spectrum generated is similar to a beta spec-

trum with end-point energy 1.1 MeV. Assume that the average mass stopping power ratio can be approximated by the stopping power ratio at 300 keV. Ignore cavity perturbation factors.

12. Calculate the value of C_λ for ^{60}Co photons for an ion chamber with air-equivalent plastic walls that are thicker than the range of the electrons produced by ^{60}Co photons. Assume that the kerma-to-dose ratio is one and that the cavity perturbation factor for ^{60}Co photons for air cavities in water is 1.015.
13. Calculate the value of C_E for 25-MeV electrons for an ion chamber with air-equivalent plastic walls that are thicker than the range of the electrons produced by ^{60}Co photons but not thick enough to stop 25-MeV electrons. Assume that all cavity perturbation factors are one except the one for ^{60}Co photons for air cavities in water, which is 1.015. Use the C_λ value for ^{60}Co from Table 11.5.

11.7 References

1. U. Fano, Phys. Rev. **92**, 328 (1953).
2. L. V. Spencer and U. Fano, Phys. Rev. **93**, 1172 (1954).
3. L. V. Spencer and F. H. Attix, Radiat. Res. **3**, 239 (1955).
4. R. D. Birkhoff, H. H. Hubbell, Jr., J. S. Cheka, and R. H. Ritchie, Health Phys. **1**, 27 (1958).
5. R. D. Birkhoff, "Dose from electrons and β-rays," in *Principles of Radiation Protection,* edited by K. Z. Morgan and J. E. Turner (John Wiley and Sons, New York, (1967), p. 259.
6. W. J. McConnell, R. D. Birhoff, R. N. Hamm, and R. H. Ritchie, Radiat. Res. **33**, 216 (1968).
7. L. C. Emerson, R. D. Birkhoff, V. E. Anderson, R. H. Ritchie, Phys. Rev. Sect. B **7**, 1798 (1973).
8. S. C. Soong, Radiat. Res. **67**, 187 (1976).
9. F. St. George, D. W. Anderson, and L. McDonald, Radiat. Res. **75**, 453 (1978).
10. K. K. Aglintsev, V. P. Kasatkin, V. V. Mitrofanov, and V. V. Smirnov, "Application of nuclear spectroscopy methods to beta and gamma ray dosimetry," in *Proceedings of the Second United Nations International Conference on Peaceful Uses of Atomic Energy* (United Nations, Geneva, 1958), p. 165.
11. P. J. Ebert and A. F. Lauzon, IEEE Trans. Nucl. Sci. **13**, 735 (1966).
12. M. Mladjenovic, *Radioisotope and Radiation Physics* (Academic, New York, 1973), p. 136.
13. N. D. Kessaris, Radiat. Res. **43**, 281 (1970).
14. N. D. Kessaris, Phys. Rev. **145**, 164 (1966).
15. J. W. Boag, "Ionization chambers," in *Radiation Dosimetry,* vol. 2, edited by F. H. Attix and W. C. Roesch (Academic, New York, 1966), pp. 1–72.
16. W. J. Price, *Nuclear Radiation Detection* (McGraw-Hill, New York, 1964), pp. 70–114.
17. *Radiation Quantities and Units, ICRU Report 19* (International Commission on Radiation Units and Measurements, Washington, 1971).
18. *Average Energy Required to Produce an Ion Pair, ICRU Report 31* (International Commission on Radiation Units and Measurements, Washington, 1979).

19. M. Inokuti, Radiat. Res. **64**, 6 (1975).
20. I. T. Myers, "Ionization," in *Radiation Dosimetry,* vol. 1, edited by F. H. Attix and W. C. Roesch (Academic, New York, 1968), pp. 317–325.
21. L. G. Christophorou, *Atomic and Molecular Radiation Physics* (Wiley-Interscience, New York, 1971).
22. A. Ore, "The role of multiple ionization in radiation action," in *Radiation Research 1966,* edited by G. Silini (North-Holland, Amsterdam, 1967), p. 55.
23. F. Mott and J. B. Massey, *Theory of Atomic Collisions* (Oxford University, London, 1949).
24. R. L. Platzman, Int. J. Appl. Radiat. Isot. **10**, 116 (1961).
25. M. W. Zemansky, *Heat and Thermodynamics,* 5th ed. (McGraw-Hill, New York, 1968), pp. 111–115.
26. T. E. Bortner, G. S. Hurst, and W. G. Stone, Rev. Sci. Instrum. **28**, 103 (1957).
27. T. E. Burlin, Br. J. Radiol. **39**, 727 (1966).
28. T. E. Burlin, "Cavity chamber theory," in *Radiation Dosimetry,* vol. 1, edited by F. H. Attix and W. C. Roesch (Academic, New York, 1968), p. 331.
29. M. J. Berger and S. M. Seltzer, "Tables of energy losses and ranges of electrons and positrons," in *Studies in Penetration of Charged Particles in Matter, Publication 1133* (National Academy of Sciences-National Research Council, Washington, 1964), p. 205.
30. W. H. Bragg, *Studies in Radioactivity* (MacMillan, London, 1912).
31. L. H. Gray, Proc, R. Soc. Ser. A **156**, 578 (1936).
32. L. H. Gray, Br. J. Radiol. **22**, 677 (1949).
33. U. Fano, Radiat. Res. **1**, 237 (1954).
34. *Radiation Dosimetry: X Rays and Gamma Rays with Maximum Photon Energies between 0.6 and 50 MeV, ICRU Report 14* (International Commission on Radiation Units and Measurements, Washington, 1969).
35. B. R. Paliwal and P. R. Almond, Phys. Med. Biol. **20**, 547 (1975).
36. *Radiation Dosimetry: Electrons with Initial Energies between 1 and 50 MeV* (International Commission on Radiation Units and Measurements, Washington, 1972), p. 43.
37. H. E. Johns and J. R. Cunningham, *The Physics of Radiology,* 3rd ed. (Charles Thomas, Springfield, MA, 1969), p. 286.
38. R. Loevinger, Med. Phys. **8**, 1 (1981).
39. A. E. Nahun and J. R. Greening, Phys. Med. Biol. **21**, 862 (1976).
40. G. P. Bernard, Phys. Med. Biol. **9**, 321 (1964).

12

Excitation and De-excitation in Crystalline Solids

12.1 Crystalline Solids and Energy Bands

The solid state of matter is characterized by the ability to resist a deforming force so that an object tends to maintain its shape (1). On an atomic scale, this property can be related to small interatomic distances with substantial binding forces and a stable arrangement.

Amorphous materials are a subgroup of the solids. They show definite order in their atomic arrangement over a short range, including a few of the nearest neighbors. Thus, all identical atoms in the amorphous material may have a similar array of nearest neighbors. However, the complete array extending many interatomic distances will not be repetitive. The ordering in an amorphous solid is similar to that of a liquid except that the configuration of the solid is stable over long periods of time (2).

Crystalline solids are of special interest because they exhibit long-range order; that is, they occur in an extended periodic array on an atomic scale. The basis group in a crystal is the simplest group of atoms, often just a molecule. Since the geometrical pattern of basis groups is largely uninterrupted, a three-dimensional regular array of lattice points can be assigned, one point per basis group. Lines constructed through these lattice points would divide the crystal into identical volume elements, called unit cells. The lengths of the edges of a minimum volume cell are the lattice constants. A single unit cell can be made to describe the entire crystal by a series of translations, along the directions of the unit cell edges. Figure 12.1 shows a simple lattice and a unit cell with lattice constants a, b, and c. Table 12.1 gives some representative lattice constant values for cubic crystals ($a = b = c$) of interest. Several ionic radii are also given for comparison (3).

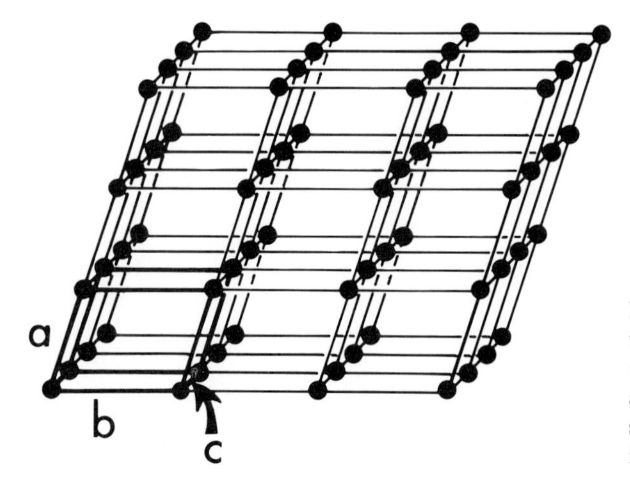

Figure 12.1 A unit cell with lattice constants *a*, *b*, and *c* is illustrated for a simple crystal. The small dark spheres represent atomic cores.

The periodic arrangement of atoms in a crystal gives rise to allowed and forbidden energy bands for electrons. Radiation-induced phenomena in crystalline solids are often described with terms relating to these bands. An introduction to the origins of the bands and to band terminology is appropriate at this point, prior to the discussion of specific absorption phenomena.

Imagine a three-dimensional array of identical atoms with uniform separation between nearest neighbors. Initially assume a separation so large that interatomic forces are entirely negligible. As is illustrated on Figure 11.2, the electron energy levels for this case are the same as the levels for the individual atoms. However, for the array, each level has many allowed states. In fact, the allowed level population is the product of the number of atoms and the allowed population of the individual atomic level.

Consider the effect on the electron energy levels when the atoms in the array are uniformly brought together. As the interatomic spacing decreases, charge distributions from adjacent atoms begin to interact and perturb the energy of the degenerate levels. For an array of N atoms, a given atomic level will be split by the perturbing forces into a number of separate levels, and these will diverge over a range of energies (see Figure 12.2). For macroscopic arrays,

Table 12.1 Ionic Radii and Lattice Constants

Material	*Lattice Constant* (*Å*)	*Ion*	*Ionic Radius* (*Å*)
Silicon (Si)	5.4	Li^+	0.7
Copper (Cu)	3.6	Na^+	1.0
Germanium (Ge)	5.7	Ag^+	1.3
Lead (Pb)	4.9	F^-	1.4
Lithium fluoride (LiF)	4.0	Cl^-	1.8
Sodium iodide (NaI)	6.5	Br^-	2.0
Silver bromide (AgBr)	5.8	I^-	2.2

Data from Ref. 3.

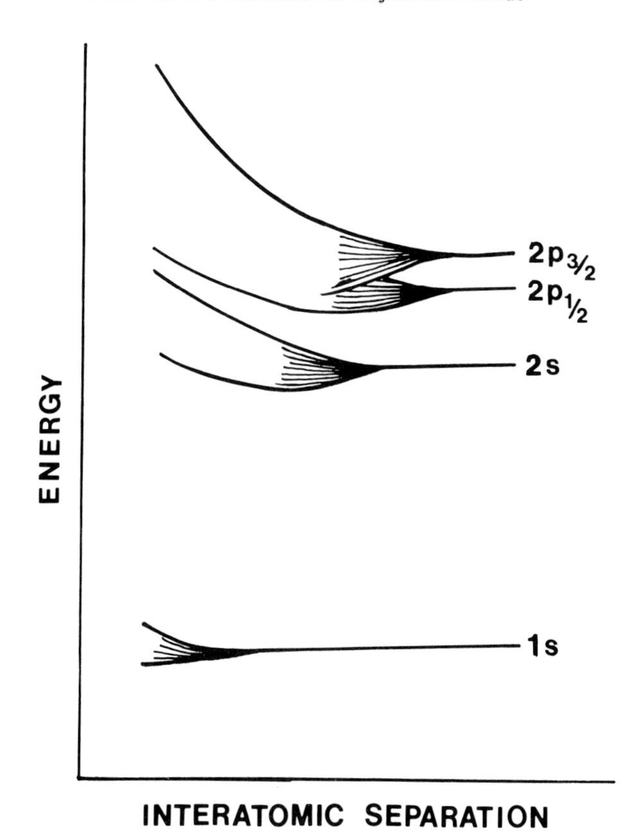

Figure 12.2 The electron energy levels for a periodic system of atoms are shown as a function of separation between the atoms. Energy levels broaden into bands when adjacent atoms perturb each other appreciably.

the number of atoms involved is sufficient so that the spacing between levels is less than the level width, and a continuous band of allowed energies exists. Note that the width of the energy band is related to the spacing between atoms and to the amount of overlap between atomic electron clouds. At an appropriate interatomic separation, well localized K-shell electron levels are not affected much by splitting, but the levels for external valence electrons are widely split. Calculations of energy band parameters have been published for many specific crystals (4, 5).

A simplified general argument for energy bands will be outlined in the following paragraphs. Only one dimension of a three-dimensional periodic array will be considered. Although incomplete, this will be sufficient to illustrate some major effects of the periodicity (6). Three-dimensional general solutions to the problem of the periodic potential are available (3, 7) but are much more advanced.

The time-dependent one-dimensional Schrödinger equation for an electron in a repetitive potential can be written as

$$\frac{-\hbar^2}{2m_e}\frac{\partial^2\psi(x,t)}{\partial x^2} + V(x)\psi(x,t) = i\hbar\frac{\partial\psi(x,t)}{\partial t}, \qquad V(x) = V(x+a) \tag{12.1}$$

The function $V(x)$ is the electrostatic potential energy of the electron under consideration. It must include the effect of attraction of the nuclear cores and repulsion of all other electrons averaged together. The equality to the right in Equations 12.1 are an expression of the periodicity, since a is the atomic spacing. The electron wave function Ψ is separable. $\psi(x,t) = \phi(x)f(t)$. Dividing through by $\psi(x,t)$ yields

$$\left\{\frac{-\hbar^2}{2m_e}\left[\frac{d^2\phi(x)}{dx^2}\right] + V(x)\phi(x)\right\}\frac{1}{\phi(x)} = i\hbar\left[\frac{df(t)}{dt}\right]\frac{1}{f(t)} = E \tag{12.2}$$

where the separation constant E represents the total electron energy. The equation on the right side of Equation 12.2 can easily be integrated:

$$\psi(x,t) = \phi(x)\exp\left(\frac{-iEt}{\hbar}\right) = \phi(x)\exp(-i\omega t)$$

$$H(x)\phi(x) = \frac{-\hbar^2}{2m_e}\left[\frac{d^2\phi(x)}{dx^2}\right] + V(x)\phi(x) = E\phi(x), \quad V(x) = V(x+a) \tag{12.3}$$

where $H(x)$ stands for the Hamiltonian operator defined by Equations 12.3.

Choose a subgroup of the infinite one-dimensional array of atoms. The subgroup has N atoms arranged linearly along the x axis with spacing between each atom of magnitude a. The boundary condition,

$$\phi(x) = \phi(x + Na) \tag{12.4}$$

expresses the fact that the wave function will be periodic for a subgroup of the array since the potential is periodic as given in Equation 12.1.

Suppose $\tilde{T}$ is the operator that translates the argument of a function by distance a. Since it is evident that $H(x) = H(x + a)$, it is easy to show that $\tilde{T}$ commutes with $H(x)$:

$$\tilde{T}H(x)\phi(x) = H(x+a)\tilde{T}\phi(x) = H(x)\tilde{T}\phi(x),$$

$$[\tilde{T}H(x) - H(x)\tilde{T}]\phi(x) = 0 \tag{12.5}$$

One can show that nondegenerate eigenfunctions are simultaneous to operators that commute (7). Thus, there exist solutions $\phi(x)$ of the equation $\tilde{T}\phi(x) = T_0\phi(x)$ with eigenvalue T_0 that are also solutions of Equations 12.3. The plan is to apply $\tilde{T}$ many times to find such solutions:

$$\tilde{T}\phi(x) = T_0\phi(x) = \phi(x+a)$$

$$(\tilde{T})^2\phi(x) = \tilde{T}[T_0\phi(x)] = T_0\tilde{T}\phi(x) = (T_0)^2\phi(x) = \tilde{T}\phi(x+a) = \phi(x+2a)$$

$$(\tilde{T})^N\phi(x) = (T_0)^N\phi(x) = \phi(x+Na) \tag{12.6}$$

Combining the boundary condition from Equation 12.4 with the final result in Equations 12.6 yields

$$(T_0)^N\phi(x) = \phi(x+Na) = \phi(x),$$

$$(T_0)^N = 1, \quad T_0 = \exp\frac{i2\pi l}{N} = \exp(ik_l a),$$

$$k_l = \frac{2\pi l}{Na}, \; l = 0, \pm 1, \pm 2, \ldots \tag{12.7}$$

The eigenvalues T_0 are the complex roots of unity. Note that the range of l for unique solutions is N in this case. The solutions obtained from larger l values contain no new phase factors, since $\exp(i2\pi l/N) = \exp[i2\pi(l + N)/N]$. If both plus and minus l values are chosen, unique solutions are confined to $|l| \leq N/2$. The wave number k_l was introduced for ease of reference in the foregoing discussion.

One can obtain a solution for the wave function by applying Equations 12.6 and 12.7:

$$\tilde{T}\phi(x) = T_0\phi(x),$$
$$\phi(x + a) = \exp(ik_l a)\phi(x) = u_l(x + a)\exp[ik_l(x + a)],$$
$$\phi(x) = u_l(x)\exp(ik_l x), \quad u_l(x) = u_l(x + a) \tag{12.8}$$

The function $u_l(x)$ is used to indicate the part of $\phi(x)$ that is not a phase factor. These last equations for $\phi(x)$ can be identified as solutions to Equations 12.3 under appropriate conditions (6). Combining Equations 12.3 and 12.8 gives

$$\psi(x,t) = u_l(x)\exp[i(k_l x - \omega t)],$$
$$k_l = \frac{2\pi}{\lambda} = \frac{2\pi l}{Na} \quad l = 0, \quad \pm 1, \quad \pm 2, \; \ldots, \; |l| \leq N/2 \tag{12.9}$$

The exponential part of $\psi(x,t)$ represents a wave traveling in the $+x$ direction when $l > 0$ or the $-x$ direction when $l < 0$. A traveling wave is normally associated with particles free to range over the coordinate space. Since the function $u_l(x)$ has period a, it is identified with localized configurations near nuclear cores in the array. Thus, the functions $\psi(x,t)$ are a hybrid of localized wave function modulated by free-particle plane waves.

The total energy values associated with the solutions of Equations 12.9 are of great interest. The plane wave kinetic energy expressions T_l are

$$T_l = \frac{p_l^2}{2m_e} = \frac{\hbar^2 k_l^2}{2m_e} = \frac{h^2 l^2}{2m_e(Na)^2}, \quad |l| \leq N/2$$

$$\langle E \rangle_l = \frac{\hbar^2 k_l^2}{2m_e} + \langle T_u \rangle_l + \langle V \rangle_l \tag{12.10}$$

The quantity $\langle E \rangle_l$ is the expectation value of the total energy. The energy $\langle V \rangle_l$ is an expectation value of the electron potential energy due to the periodic potential. The energy $\langle T_u \rangle_l$ is an expectation value of the kinetic energy attributable to the core functions.

The valence electron wave functions overlap from atom to atom, so these electrons are relatively free to move over many interatomic distances. The range of allowed free-particle energies is approximately $2[h^2(N/2)^2/2m_e(Na)^2] = h^2/8m_e a^2$. When N is large, many possible free electron levels are available and the spacing between adjacent levels must be small. In fact, a continuum or band of allowed energies is available. The band edge occurs when l approaches $N/2$. At this energy the electron wave number is given by $|k| = \frac{2\pi l}{Na} = \pi/a$ and the wavelength $\lambda = \frac{2\pi}{k} = 2a$. This condition corresponds to the situation for first-order Bragg reflection of waves at normal incidence from crystal planes separated by the lattice spacing a (6).

The discussion given in the preceding paragraphs was simplified, but some of the major properties discussed apply to actual three-dimensional crystalline materials. For example, three-dimensional crystal wave functions are called the Bloch functions and have the form (3, 7, 8):

$$\phi(\mathbf{r}) = u_k(\mathbf{r})\exp(i\mathbf{k}\cdot\mathbf{r}),$$

$$\mathbf{k} = 2\pi\left[\frac{l}{N_x a}\,\hat{\imath} + \frac{m}{N_y b}\,\hat{\jmath} + \frac{n}{N_z c}\,\hat{k}\right],$$

$$l, m, n = 0, \pm 1, \pm 2, \ldots \tag{12.11}$$

where $\hat{\imath}$, $\hat{\jmath}$, and $\hat{k}$ are unit vectors in the x, y, and z directions, respectively. The actual kinetic energy values depend on the wave number product $\mathbf{k}\cdot\mathbf{k}$. Energy bands occur for valence electrons, and discontinuities or gaps in the energy bands are apparent when $\mathbf{k}$ satisfies the Bragg condition. However, the three-unit cell dimensions are not necessarily equal, so the magnitude of $\mathbf{k}$ that

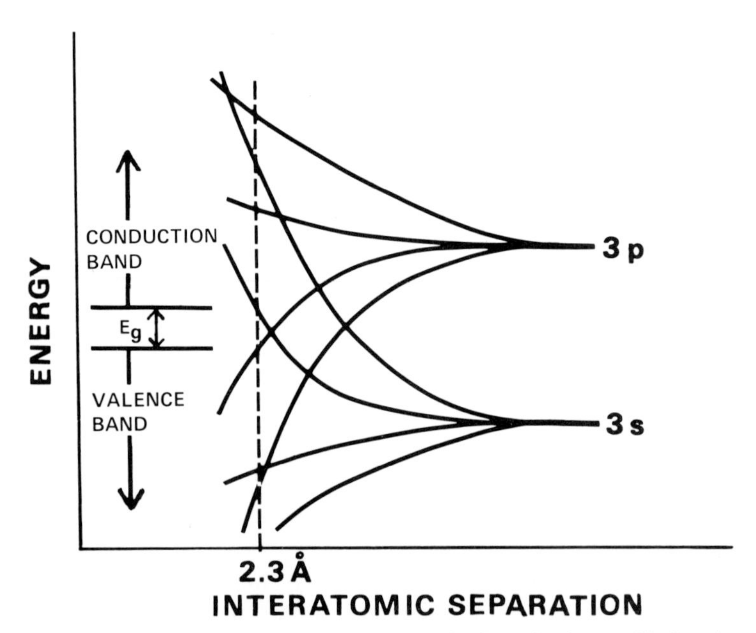

Figure 12.3 The energy band structure of silicon is shown. Notice that the calculations indicate band splitting and crossing. E_g is the band gap width. Adapted from Ref. 9.

produces an energy gap is not necessarily the same in all directions. In this case, forbidden gaps related to different directions in the crystal do not always have the same energy.

Many other complications are found for real crystals. Figure 12.3 is an illustration of the energy band structure for the covalent crystalline state of silicon. Notice that the splitting of the 3*s* and 3*p* states of atomic silicon are a function of equilibrium separation, as might be expected (9). However, parts of the 3*s* and 3*p* bands overlap, so the equilibrium silicon band gap is actually quite small.

12.2 Excitations in Metals, Semiconductors, and Insulators

In the preceding section some general arguments were presented for the existence of bands of allowed electron energies in crystalline solids. With this is an introduction, band terminology and excitation processes common to crystalline solids can be discussed. Discussions of several related phenomena with widespread applications will be presented in the later sections.

The intervals with forbidden energy values between the allowed bands of energy are called gaps, as was mentioned in Section 12.1. Lattice structure, the atomic constituents, and their bonding determine variations in the gap width in different materials. Solids are often classified as electrical conductors (metals), semiconductors, or insulators on the basis of the relative population of the allowed electron levels within the last occupied band and the size of the energy gap between bands. Figure 12.4 shows three representative situations for crystalline solids at a temperature of absolute zero (3).

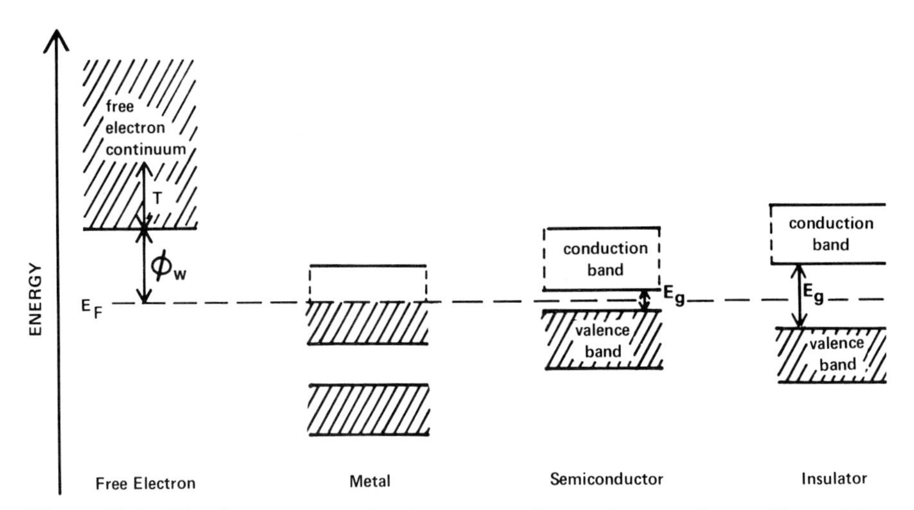

Figure 12.4 The electron energy band structure of several types of crystalline solids is shown. E_g is the band gap width; E_F, the Fermi energy.

The energy of the topmost filled electron level at absolute zero is the Fermi energy, E_F, as shown on the figure. This energy serves as a reference for electron binding energies in the crystal and for electrons in free space. If E is the total electron energy in the crystal and T_{max} is the maximum value of the free-space electron kinetic energy (10), then

$$E = E_F + \phi_W + T_{max}, \quad E - E_F = \phi_W + T_{max} \tag{12.12}$$

The work function ϕ_W is the minimum energy in excess of the Fermi level that must be added to free an electron from the crystal surface. The onset of the free-electron continuum always occurs at energies larger than E_F because escaping electrons are attracted back to the crystal surface by the positive charge induced on the surface.

In the first case shown in Figure 12.4, that of a metal, the Fermi level occurs inside a band of allowed energies. In such a situation, there exists a great many loosely bound electrons in the crystal, in random motion. Upon application of an electric field, some will obtain a net momentum antiparallel to the field, concomitant with an increase in energy, and become carriers of current. A small increase in energy is possible because many unfilled levels are available above the Fermi level. In this case the material is classified as an electrical conductor.

In the second and third cases shown in Figure 12.4, semiconductor energy bands and insulator energy bands are illustrated, respectively. In each of these cases the topmost filled electron energy band is called the valence band and the band above it is called the conduction band. At absolute zero the conduction band is not populated for these crystalline solids. For perfect crystals, states with energies within the gap are forbidden and there are no empty states available in the valence band. Thus, the electron momentum distribution can not be altered appreciably by the applied electric field. The materials do not conduct electrical current well.

An important difference between semiconductor crystals and insulator crystals concerns the band gap width E_g. If E_g is small enough, at room temperature many valence-band electrons will be thermally exicted into states with energies in the conduction band. Since these electrons are available as carriers of current, a crystalline material with small energy gap is called a semiconductor. If E_g is so large that thermal excitation does not produce an appreciable conduction band population at room temperature, the material is called an insulator. Semiconductors are sometimes defined as substances with $E_g \leq 2$ eV (10). Table 12.2 shows the valence-to-conduction-band gap energy for some useful materials (3, 10, 11).

In the last several paragraphs we have considered the effect of conduction-band electrons on electrical conductivity. For semiconductors and insulators, the vacancy left in the valence band can also contribute to the conductivity. A vacancy in an otherwise filled band is called a hole, and holes can move through a crystal under the influence of an electric field. One can picture hole motion as a series of displacements from one atomic site to an adjacent one. This can occur when an electron jumps to a vacancy in an orbit of a neighboring

Table 12.2 Band Gap Energies

Crystal	E_g (*eV*)
Carbon (C)[a]	5.4
Silicon (Si)	1.1
Germanium (Ge)	0.7
Lithium fluoride (LiF)	12.1
Lithium iodide (LiI)	5.9
Sodium iodide (NaI)	5.8
Zinc sulfide (ZnS)	3.7
Cesium antimonide (Cs_3Sb)	1.6
Silver bromide (AgBr)	2.5

Data from Refs. 3, 10, and 11.
[a] Diamond.

atom. In semiconductors or insulators, hole motion can contribute significantly to the transport of electrical current.

The existence of perfect crystals assumed in the foregoing descriptions does not occur. Imperfections or defects in the regular long-range ordering in crystalline solids are common. Point defects can be grouped under four categories: vacant lattice sites, interstitial atoms or ions, impurity atoms or ions, and combinations thereof. These categories apply to a localized site within the crystal. Defects involving large numbers of atoms (planes) can also be identified.

Two common types of vacancy defects are illustrated in Figure 12.5. The Frenkel defect is a lattice vacancy with an associated interstitial atom or ion, although the two may be separated by several lattice distances. The Schottky defect is a lattice vacancy wherein the missing atom or ion has been completely removed from the neighborhood of the vacancy (3). These defects occur in all crystals. They can be produced thermally and with ionizing radiation.

Vacant lattice sites can serve to trap charge carriers and prevent their further movement. An important type of electron trap in ionic crystals is the

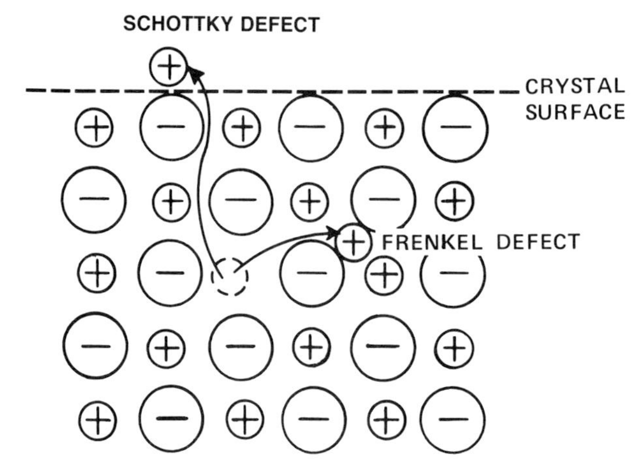

Figure 12.5 Frenkel and Schottky defects are illustrated.

negative-ion vacancy. A negative-ion vacancy that contains a trapped electron is called an F center (see Figure 12.6). The filled-hole trap analogous to the F center occurs only rarely. The most common hole trap in ionic crystals is the V_K center (3), shown on Figure 12.6. The V_K center does not involve a vacancy, rather two adjacent anions combine with a hole so that the local interatomic distance is changed and the energy is reduced.

Impurity atoms or ions can be located interstitially or substituted into lattice sites in place of the usual atom or ion. The electron energy levels for impurity atoms are perturbed by the local electric fields in the crystal. Local fields can split degenerate levels. They serve to displace allowed energy levels with respect to the free-particle continuum (3). Impurity atom energy levels that fall in the band gap play a part in excitation and de-excitation processes in the crystal.

Impurity atoms which donate electrons to the conduction band are called donors; if the atoms capture electrons from the valence band, they are called acceptors. Examples of donor and acceptor atoms can be found in the substitutional impurities in covalently bonded crystals. Germanium and silicon are tetravalent, thus impurity atoms with more than four valence-shell electrons tend to act as electron donors in these crystals. If the number of impurity valence electrons is less than four, the substituted atoms tend to act as acceptors.

Radiation-induced excitations in metals, semiconductors, and insulators

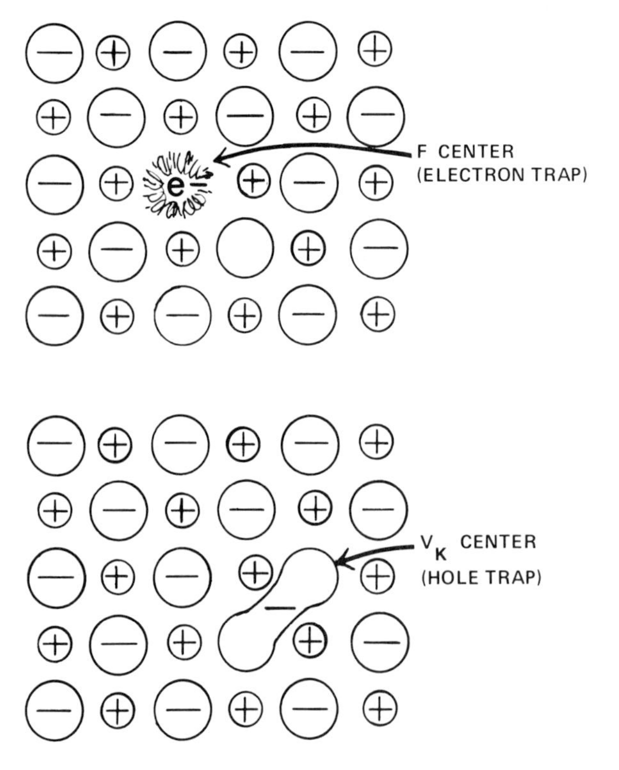

Figure 12.6 An F center is shown at top. A V_K center is shown at bottom.

can involve electron transitions that belong to at least four major categories: intraband, interband, collective, and discrete-state electron transitions. Two important categories for lattice energy loss processes are the collective and local processes.

Intraband transitions involve electron energy changes within a single continuous energy band, whereas interband transitions occur between two different bands (7). Collective electron transitions refer to changes in energy of an electron plasma as a whole rather than to individual electrons. Discrete-state electron transitions involve one or more levels that do not fall within a band. They are associated with impurity atoms or with the inner shells in lattice atoms. Collective lattice processes refer to oscillations of the lattice atoms or ions as a whole, whereas local processes refer to lattice excitations at a particular site in the crystal.

Any of these excitations can be induced by a coulomb collision in a crystalline solid. These collisions can be initiated by a fast charged primary particle traveling through the crystal or by charged secondaries set free after interactions of photons or neutrons in the crystal. Excitation of the electrons in the absorbing crystal will be described first; discussion of lattice excitations will follow.

At lowest energies, in a metal, charged particles can induce intraband transitions with a restricted number of crystal electrons with energies near the Fermi energy. Participation of the vast majority of the electrons is prohibited since their neighboring levels are filled. However, as the energy available for excitation is increased, an increasing number of crystal electrons become available for participation. Thus, the intraband excitation probability increases. When sufficient energy is available, collective, discrete-state, and interband transitions also become possible (12).

The situation in semiconductors and insulators is different than that for a metal. The electron energy bands are either nearly full or nearly empty at room temperature. Thus, the population available for intraband transitions is restricted in any case. However, when the excitation energy exceeds the band gap energy, the probability of excitation increases rapidly because of electron-hole pair formation (13). An electron-hole pair is produced when an electron is promoted from a state in a filled band to an empty state in another band. The hole is identified with the empty state left behind, as is shown on Figure 12.7. If subscripts e and h stand for the electron and hole, respectively, a necessary condition for pair formation is that the energy transferred $\Delta E = E_e - E_h \geq E_g$.

In many cases the newly formed electron and hole drift apart as free charge carriers and are never again correlated in space. However, they can associate after formation into a weakly bound pair called an exciton, which moves through the crystal as a single entity. The exciton system has discrete energy states, which are illustrated on Figure 12.7.

Collective electron excitations in a crystal involve vast numbers of charges that respond to stimuli much as a plasma would respond. Such excitations, when induced by energetic charged particles, take the form of longitudinal collective oscillations. The negative electron plasma vibrates with respect to the positive

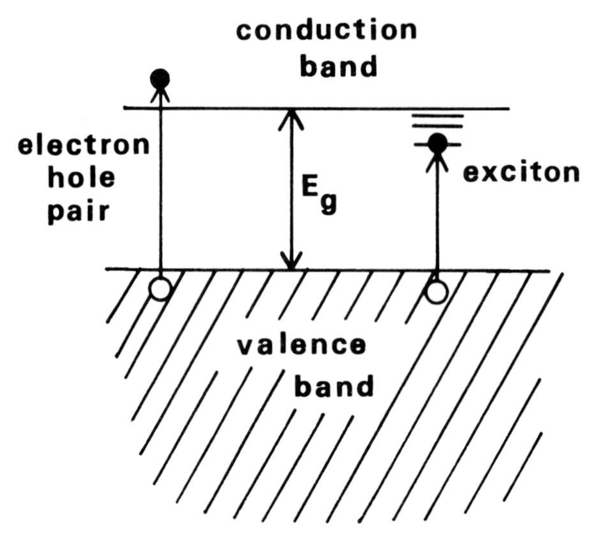

Figure 12.7 The formation of an electron-hole pair is illustrated at left; the excited states of an exciton are shown at right.

ion cores. The oscillations are harmonic, with frequency $f = \omega/2\pi$, and the energy of the oscillation is quantized with value $\hbar\omega$. The quantum is called the plasmon.

The allowed number of plasmons is limited only by the energy delivered (3). Because the plasma oscillation is quantized, the excitation energy must exceed a threshold between 3 and 15 eV in most common metals (3). Beyond the threshold energy, plasma oscillations become an important mode for energy transfer from fast charged particles to absorbers. Plasmons are frequently intermediaries that can lead to interband and intraband electron excitations (12). Figure 12.8 shows the relative cross section for various electron excitation processes in metallic aluminum (14). Note that intraband processes are most probable at lowest energies, but volume plasmon excitation dominates at somewhat higher energies. The *K*- and *L*-shell discrete-state excitations are relatively improbable.

The most important lattice excitation process is the induction of collective vibrations of the atoms about their lattice sites. Lattice atoms are arranged with well defined equilibrium positions but they can oscillate with limited amplitude about these positions. Since adjacent atoms interact through relatively strong forces, such oscillations can be coupled through an entire crystal. The situation is most easily analyzed in terms of collective harmonic motion in normal modes with well defined phase relationships (7). Figure 12.9 is a schematic representation of two such oscillation modes for a simple one-dimensional array of atoms. Longitudinal (L) and transverse (T) motions occur in acoustical (A) and optical (O) modes (3); thus, the oscillations may be designated by LA, LO, TA, and TO. Normal modes in three-dimensional crystals may be more complicated, but a similar terminology is used.

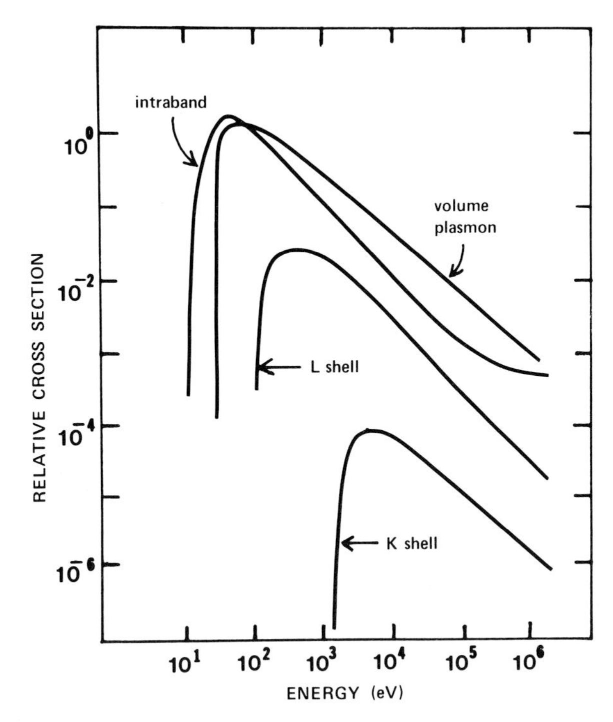

Figure 12.8 The relative cross section for several excitation processes in aluminum is shown. The zero value of the energy scale is taken at the bottom of the conduction band. Adapted from Ref. 14.

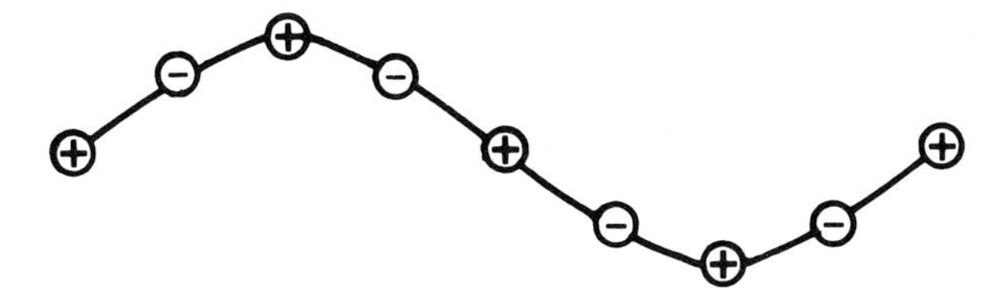

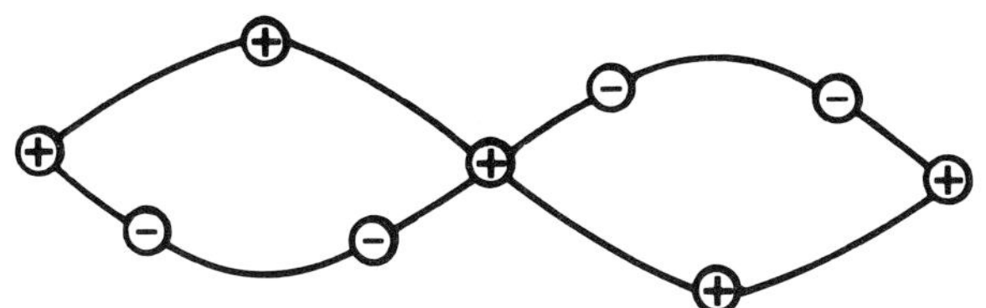

Figure 12.9 A transverse acoustical (TA) mode of oscillation is shown (*top*) for a string of ions. A transverse optical (TO) mode is shown at bottom.

The energy of a collective lattice oscillation is quantized, and the quantum is called a phonon, with energy $\hbar\omega$. Individual phonon energies are usually small (0.06 → 1.10 eV), but unlimited numbers of phonons can accrue in a crystal at one time (3). Phonon losses are important when energetic charged particles are degraded in crystalline materials. They are often produced concomitantly with electron-hole transitions, particularly when the initial and final electron states have different wave number **k**; the phonons provide for momentum conservation in such circumstances. Indirect transitions are interband transitions requiring phonons for momentum conservation.

12.3 Production of Mobile Charge Carriers

In this section, changes in the electrical conductivity of a semiconducting or insulating crystal produced by ionizing radiation will be discussed. Basic processes will be mentioned and simple expressions for the charge-carrier densities will be given. The aim of the section is to provide background information necessary for understanding conductivity dosimeters (15) and semiconductor detectors (16, 17).

Definitions for basic electrical parameters are given in many texts (3, 18). For example, the electrical conductivity σ is the ratio of the current density J in amperes per square meter divided by the impressed electric field E in volts per meter. The expression can be regarded as a form of Ohm's law:

$$J = \sigma E, \qquad \sigma = J/E \tag{12.13}$$

Conductivity has units amperes per volt meter, which is equivalent to reciprocal ohm-meters [(ohm-m)$^{-1}$] in SI units.

The conductivity is also related to the mobility μ of the charge carriers. If v is the drift speed of mobile charges q due to an electric field and J is the current density caused by these carriers, then

$$\mu = \frac{v}{E}, \qquad \frac{J}{E} = \frac{n_v q\, v}{E} = n_v q\, \mu \tag{12.14}$$

where n_v is the number of mobile carriers (with charge q) per unit volume. For semiconductors and insulators, it is best to consider the contribution of mobile holes (subindex h) separately from those of mobile electrons (subindex e):

$$\sigma = {}_e n_v\, e\mu_e + {}_h n_v\, e\mu_h \tag{12.15}$$

Table 12.3 shows measured values of σ and μ for several important crystalline semiconductors and insulators (15, 17). In some cases values occur within a wide range when different samples of a given element or compound are used. The tabulated values shown are intermediate for the measured range. The conductivity of metallic copper, $\sigma = (6)10^7$ (ohm-m)$^{-1}$ at 300 K (3) is useful for comparison with the values in the table.

If some of the valence-band electrons are promoted into the conduction

Table 12.3 Charge Conduction Properties at 300 K

Crystal	E_g (eV)	$\sigma(ohm \cdot m)^{-1}$	$[m^2/(V \cdot s)]$ μ_e	μ_h	τ (s)
Carbon (C)[a]	5.4	$(5)10^{-11}$	0.180	0.120	
Cadmium sulfide (CdS)	2.4	$(5)10^{-9}$	0.030	0.001	$(1)10^{-3}$
Gallium arsenide (GaAs)	1.4	$(1)10^{-4}$	0.850	0.040	$(1)10^{-7}$
Silicon (Si)	1.1	$(4)10^{-4}$	0.150	0.050	$(3)10^{-3}$
Germanium (Ge)	0.7	$(3)10^{0}$	0.380	0.180	$(1)10^{-3}$

Data from Refs. 15 and 17.
[a] Diamond.

band, the electrical properties of a nonconducting crystal will be changed. Extra conduction-band electrons and valence-band holes add to the mobile carrier concentration and therefore increase the electrical conductivity. When electron-hole pair formation occurs because of thermal excitation, the charge-carrier density can be related to kT°, where k is the Boltzmann constant and T° is the absolute temperature. If the carrier population is completely intrinsic and effects of impurities and traps can be neglected, the charge-carrier density for nonconductors is given by

$$ {}_e n_v = {}_h n_v = C_n (kT^\circ)^{3/2} \exp(-E_g/2kT^\circ) \tag{12.16} $$

where C_n is a constant for a given crystalline material (3). Equation 12.16 shows that when $kT^\circ \ll E_g$, the intrinsic charge-carrier concentration is small, but it increases with increasing temperature. For silicon at 300 K under intrinsic conditions, ${}_e n_v \simeq {}_h n_v \simeq (1)10^{16}/m^3$. For comparison, the mobile electron concentration for metallic copper is about $(1)10^{29}/m^3$.

When thermal electron-hole production is negligible, ionizing radiation can have an appreciable effect on the electrical properties of nonconducting crystals. Under such circumstances it may be important to be able to estimate the numbers of mobile electrons and holes produced by the radiation.

Figure 12.10 illustrates the charge-carrier population distribution after valence-band electrons have been excited by ionizing radiation. The liberated secondary electrons with excess energy induce additional interband transitions via collisional processes and produce more electrons and holes. Energy in the form of phonons is invariably coupled to the lattice in the entire process. A short time after excitation, a pool of mobile electrons and holes is present in the crystal.

Assume that the initial kinetic energy T of a fast charged particle was dissipated in a crystal in producing N_{eh} electron-hole pairs and N_p phonons (19, 20). If W_{eh} is the average energy dissipated per pair,

$$ W_{eh} = T/N_{eh} \tag{12.17} $$

when $T \gg E_g$. The concept of the average energy per electron-hole pair is analogous in many ways to the average energy per ion pair discussed in Chapter 11 for gaseous absorbers.

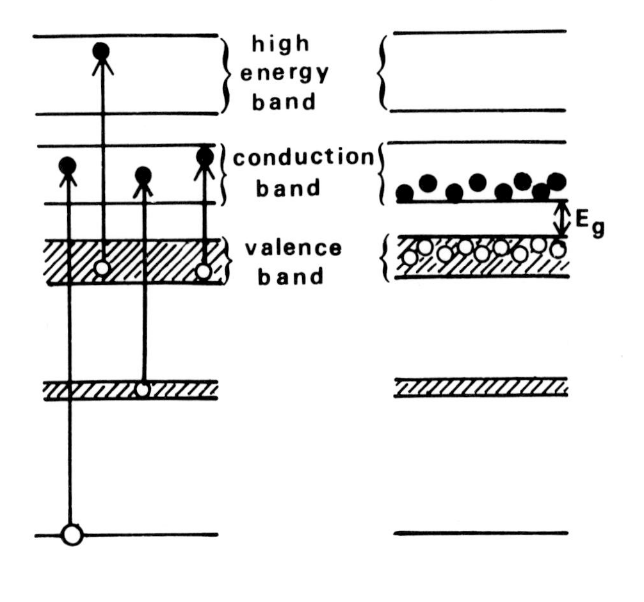

Figure 12.10 Several interband transitions to highly excited states are shown at left. The residual electron hole excitations are shown at right.

Suppose ϵ_p is the average energy of the phonons produced, ϵ_e is the average energy above the conduction-band edge for the mobile electrons, and ϵ_h is the average energy of the mobile holes prior to thermalization. Then,

$$T = N_{eh}(E_g + \epsilon_e + \epsilon_h) + N_p \epsilon_p,$$

$$W_{eh} = T/N_{eh} = E_g + \epsilon_h + \epsilon_e + (N_p/N_{eh})\epsilon_p \tag{12.18}$$

Since electrons and holes with sufficient initial energy will produce more electron-hole pairs in the crystal, the values of ϵ_e and ϵ_h are always less than E_g.

Numerical estimates have been made for the quantities in Equations 12.18 for silicon (21). If $\epsilon_h \simeq \epsilon_e \simeq 0.6E_g$ and about 18 phonons with average energy $\epsilon_p = 0.06$ eV are produced for each pair,

$$W_{eh} \simeq E_g + (0.6 + 0.6)E_g + 18\epsilon_p$$

$$\simeq 1.1 \text{ eV} + 1.3 \text{ eV} + 1.1 \text{ eV} = 3.5 \text{ eV} \tag{12.19}$$

For silicon, the gap energy, the residual energy of the charge carriers, and lattice pronons contribute about equally to W_{eh}. Table 12.4 gives some experimental values of W_{eh} (19, 20) for important materials. Notice that an increase in E_g generally correlates with an increase in W_{eh}. However, no single numerical ratio between E_g and W_{eh} applies to all crystals.

If a crystal of density ρ absorbs energy E resulting in a uniform dose D throughout a volume V, one can write

$$W_{eh} = \frac{E}{N_{eh}} = \frac{D\rho V}{N_{eh}}, \quad N_{eh} = \frac{D\rho V}{W_{eh}}, \quad \frac{Q}{V} = \frac{D\rho}{W_{eh}/e} \tag{12.20}$$

Table 12.4 Average Energy per Electron-Hole Pair

Crystal	E_g (*eV*)	W_{eh} (*eV*)
Silver bromide (AgBr)	2.5	5.8
Carbon (C)[a]	5.4	10.0
Cadmium sulfide (CdS)	2.4	5.2
Gallium arsenide (GaAs)	1.4	6.3
Silicon (Si)	1.1	3.6
Germanium (Ge)	0.7	2.9

Data from Refs. 19 and 20.
[a] Diamond.

where Q is the magnitude of liberated charge of either sign. Values of W_{eh} are much less than the values for the average energy per ion pair for gases. At usual pressures, ρ is about a thousand times larger for a solid than for a gas. Thus, there are a great many more charge carriers produced per unit volume in a crystalline solid than in a gas for the same dose of ionizing radiation.

The middle equation in Equations 12.20 relates the total number of electron-hole pairs produced to the absorbed dose. From Equation 12.15 it is evident that the conductivity is related to ${}_en_v$ and ${}_hn_v$, the numbers of mobile charge carriers per unit volume. A discussion of the instantaneous mobile carrier densities will be given in the following paragraphs. An expression relating them to the dose rate in the crystal will be developed (15, 22).

Suppose a bias is applied to electrodes on opposite faces of a nonconducting crystal. If the thermally generated carriers occur in insignificant numbers, the electron-hole pairs produced by ionizing radiation can be important. After irradiation they will disappear from the mobile population because of collection at the electrodes, recombination, and capture in traps. In a steady-state situation, the rate of mobile electron production per unit volume, P_v, is related to the dose rate and electron density by the following:

$$P_v = \frac{\rho}{W_{eh}}\left[\frac{dD}{dt}\right], \quad \frac{d({}_en_v)}{dt} = P_v - \frac{{}_en_v}{{}_e\tau_c} - \frac{{}_en_v}{{}_e\tau_t} - \frac{{}_en_v}{{}_e\tau_r} \tag{12.21}$$

In this case, ${}_e\tau_c$ is the mean electron lifetime for collection, ${}_e\tau_t$ is the mean electron lifetime for trapping, and ${}_e\tau_r$ is the mean electron lifetime for recombination with holes. It is assumed that trapping and recombination are independent processes. Equation 12.21 is written for ${}_en_v$, but a similar equation can be obtained for ${}_hn_v$ using different values for lifetimes.

Consider ${}_e\tau_r$ first. Assume that recombination occurs whenever an electron comes into the neighborhood of a hole, i.e., no recombination transitions are forbidden and there is no hindrance. If the effective electron-hole recombination cross section is σ_{eh}, the electron sweeps out an effective volume for recombination of $v_e\sigma_{eh}$ in one second. The electron speed v_e is assumed to be much larger than the hole speed in this case. The product ${}_hn_v(v_e)\sigma_{eh}$ gives the number of holes that are in the sweeping volume in a second. This is equivalent to the recombination probability per second for the electron under the assumptions listed. The mean lifetime for recombination is the reciprocal of the recombination probability:

$$\frac{1}{{}_e\tau_r} = {}_h n_v(v_e)\sigma_{eh} \tag{12.22}$$

An analogous expression for ${}_e\tau_t$ can be written using σ_{et} for the effective cross section for electron trapping and ${}_t n_v$ for the electron trap density. Combining the two gives

$$\frac{1}{\tau} = \frac{1}{{}_e\tau_r} + \frac{1}{{}_e\tau_t} = {}_h n_v(v_e)\sigma_{eh} + {}_t n_v(v_e)\sigma_{et} \tag{12.23}$$

Using Equation 12.23, the right-hand side of Equation 12.21 can then be written

$$\frac{d({}_e n_v)}{dt} = P_v - \frac{{}_e n_v}{{}_e\tau_c} - ({}_e n_v){}_t n_v(v_e)\sigma_{et} - ({}_e n_v){}_h n_v(v_e)\sigma_{eh}$$

$$= P_v - \frac{{}_e n_v}{{}_e\tau_c} - \alpha_t({}_t n_v){}_e n_v - \alpha_r({}_h n_v){}_e n_v \tag{12.24}$$

where α_r is called the recombination coefficient and α_t is called the electron trapping coefficient.

In real crystals a variety of trapping and recombination modes can occur. In some important cases, notably for germanium and silicon of detector quality, collection at electrodes is the dominant means for loss of mobile carriers when the production rate is low enough (17). When trapping and recombination can be neglected, Equation 12.24 can be simplified by neglecting the last two terms; then for a low-level steady beam at equilibrium,

$$\frac{d({}_e n_v)}{dt} = P_v - \frac{{}_e n_v}{{}_e\tau_c} = 0, \quad {}_e n_v = P_v({}_e\tau_c) = \frac{\rho({}_e\tau_c)}{W_{eh}}(dD/dt) \tag{12.25}$$

Equation 12.25 shows that the equilibrium value of the mobile electron density is proportional to the first power of the dose rate when trapping and recombination are negligible.

When an electric field E is applied to the crystal electrodes during the time of irradiation, a current will be obtained. If J_e is the electron current density, Equations 12.13, 12.15, and 12.25 can be combined to give

$$J_e = \sigma_e E = e\,\mu_e E\,P_v\,({}_e\tau_c) = \frac{\mu_e E\rho({}_e\tau_c)}{W_{eh}/e}\left[\frac{dD}{dt}\right] \tag{12.26}$$

A similar equation can be obtained for the current density attributable to holes. Under the circumstances assumed, the currents are proportional to the first power of the dose rate. When direct recombination is an important mode for loss of carriers, the instantaneous electron density is proportional to a fractional power of the dose rate less than one (15).

The effects of trapping are undesirable in a crystal used to detect radiation, since charge is lost from the mobile carrier population and the relevant radiation-induced signal is reduced. In addition, trapping produces a space charge, which generally opposes the impressed electric field and thereby impedes charge collection. Thus, for a detector crystal, high-mobility carriers and low trap density are needed.

12.4 Luminescence from Inorganic Solids

Luminescence is the emission of electromagnetic radiation in excess of thermal radiation after some form of excitation (23, 24). Usually it is assumed that visible light is emitted during luminescence, but this need not be the case exclusively. Ultraviolet light is often included in the emissions. The reference to thermal radiation in the definition serves to distinguish luminescence from incandescence. In the following section, luminescence induced by ionizing radiation will be discussed. Three categories will be included: fluorescence, phosphorescence, and thermoluminescence.

Fluorescence, phosphorescence, and thermoluminescence differ from each other in part because of the duration of the emission process after the excitation is removed. Suppose luminescent energy is emitted from excited configurations of independent atomic systems. If the number of atomic states that decay in time interval dt is called dN, and if λ is the decay probability per second for any state, then

$$-\frac{dN}{dt} = \lambda N, \quad \int \frac{dN}{N} = -\lambda \int dt, \quad \ln N = -\lambda t + C,$$

$$N = \exp(C)\exp(-\lambda t), \quad N = N_0 \exp(-\lambda t) \tag{12.27}$$

The number of excited atoms, N, was taken to be N_0 at $t = 0$ in order to evaluate the integration constant C. Since the emitted photon fluence ϕ is proportional to λN,

$$\phi = \phi_0 \exp(-\lambda t) = \phi_0 \exp(-t/\tau) \tag{12.28}$$

where $\tau = 1/\lambda$ is the mean lifetime of the luminescence.

Some authors refer to fluorescence as luminescence that occurs during the energy stimulus, whereas phosphorescence persists after removal of the source of excitation. A common dividing line for mean life values is $\tau = 10^{-8}$ s. Fluorescence is said to occur when $\tau \leq 10^{-8}$ s; phosphorescence occurs when $\tau > 10^{-8}$ s (3). Ten nanoseconds is appropriate because it is somewhat longer than the lifetime of ordinary allowed atomic electron transitions. Thus, phosphorescence is associated with metastable atomic or molecular states or with shallow traps in a crystal. The name "thermoluminescence" comes from thermally stimulated luminescence (1), which occurs when the mean lifetime at room temperature is very long and heating is required to release the electromagnetic energy. Thermoluminescence involves relatively deep traps in a crystal (25).

For crystalline solids, phosphorescence involving shallow traps and thermoluminescence attributable to deep traps can both be discussed using the trap depth ϵ_t in energy units. The effect of the absolute temperature $T°$ on the escape probability for a carrier in a trap must be considered (26). Suppose a charge carrier is in oscillatory thermal motion inside a trap such that f is the striking frequency at the confining boundary. Under this circumstance the escape probability per second, p, is given by

$$p = f\exp(-\epsilon_t/kT°), \quad \epsilon_t > kT° \tag{12.29}$$

where k is the Boltzmann constant. The exponential factor in Equation 12.29 is called the Boltzmann factor. As long as it is small, it accurately represents the probability a system will be found with excess energy ϵ_t due to thermal excitation (6).

Call N_t the number of trapped carriers and substitute p from Equations 12.29 for λ in Equations 12.27, and the result is $N_t = N_{0t}\exp(-pt)$ when N_{0t} is the number of carriers trapped at $t = 0$. Suppose the mean time for escape of a charge carrier from a trap is much larger than the mean time elapsed thereafter for photon emission. Then the mean life of the luminescence can be identified with $1/p$:

$$\tau = \frac{1}{p} = \frac{\exp(\epsilon_t/kT°)}{f} \tag{12.30}$$

From this form the condition for thermoluminescence can be written as $\{\exp[\epsilon_t/kT°]/f\} \gg 1$ when $T°$ is the room temperature. Trap-related phosphorescence is associated with much reduced values of ϵ_t, leading to smaller values of τ at room temperature.

Consider electron excitation and de-excitation energies in luminescent crystals. Figure 12.11 is an illustration of energy functions for the ground state and excited states of an electron bound in an atom or ion in a crystal. Since

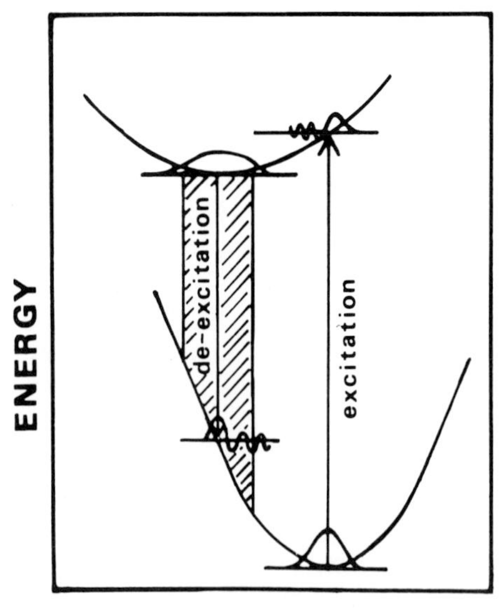

Figure 12.11 Electron energy curves are illustrated for the case of an impurity atom in a crystal. The allowed energy values are a function of displacement of the nucleus from the equilibrium position.

the nuclear centers are vibrating in their lattice positions, the curves are displayed with parent nuclear displacement from the equilibrium lattice position as the abscissa. The parabolic minimum for the ground state defines the equilibrium position for that state. The increase in energy required to move away from equilibrium occurs because of perturbing coulomb forces from neighboring lattice atoms. Since charge distributions are different, ground-state and excited-state equilibrium positions are somewhat different.

The lifetime of most excited electronic states is sufficient to allow the atomic system to relax to a new equilibrium displacement prior to de-excitation. This is accompanied by release of phonons in crystal systems. Luminescent de-excitation most often occurs from the new position, determined by the excited-state charge distribution. It is evident from the figure that the luminescent energy is less than the energy required to excite the system originally. This phenomenon is called the Stokes shift (27). Practically, it means that luminescent photons are likely to escape from a small emitting crystal, since their energy is insufficient for immediate reabsorption.

The Franck-Condon principle says that the actual time required for electronic motion in atoms is negligible compared to times necessary for substantial movement of nuclear centers (28–30). This is necessarily true since electron periods are much shorter than the periods for molecular or lattice vibrations. This means that electronic excitation or de-excitation transitions must be shown as vertical lines on Figure 12.11. They occur at essentially constant nuclear displacement.

The excited-state energy function is closely related to the luminescent spectrum. The minimum in the energy function means that a range of displacement values of excited atoms produces a range of possible photon energies. When many luminescent sites are considered, the energy spectrum of the emitted photons is broadened into bands (23). Anisotropic local charge distributions located close to emitting sites produced a similar broadening of the energy spectrum (31). Thus, one often finds a continuous emission spectrum from luminescent materials, even when a multiplet of supposedly discrete states are involved. The luminescence from ZnS:Cl (32) and thermoluminescence from LiF:Mn,Ti (33) shown on Figure 12.12 have this character.

The discrete spectrum from Gd_2O_2S:Tb (34) is an exception because the active 4*f* electrons in the emitting ion (Tb^{+++}) are confined inside the parent atom and are well shielded from external charge distributions. Thus, they are not greatly affected by changes in environmental conditions. Emissions from other rare earth impurities show narrow-line emission patterns for the same reason (35).

Interatomic forces often couple excitation energy away before luminescence can occur (36). In fact, nonradiative de-excitation dominates in most crystals. Sometimes impurities can be added to the crystal in small amounts (1%), which increase the luminescence. They are called activators if they help determine the emission spectrum. In this section, activator impurity atoms are indicated by a colon and atomic symbol following the chemical formula for the crystalline material.

Table 12.5 summarizes important properties of several useful luminescent

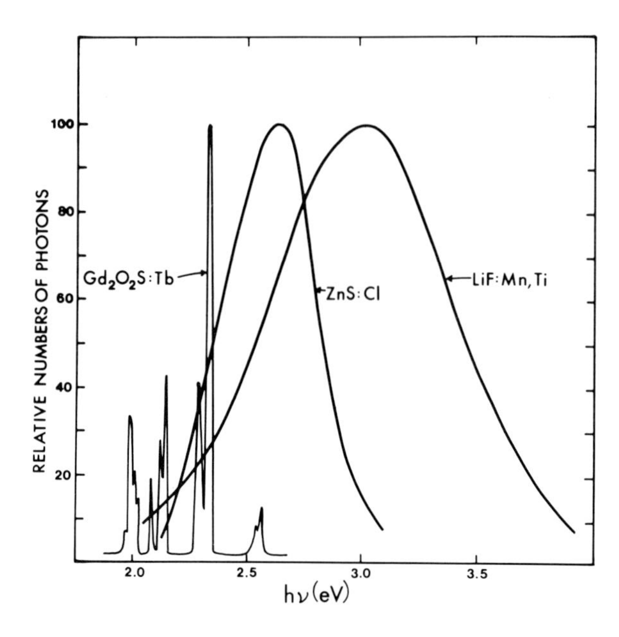

Figure 12.12 Emission spectra are shown for several luminescent materials (32–34). The curve for LiF:Mn,Ti is for thermoluminescence. The spectra are individually normalized to 100 at their peaks.

materials at 300 K (25, 34, 37); most are activated. The efficiency values given in the table are the quotient of the photon energy emitted, divided by the exciting energy and multiplied by 100. Notice that nonradiative de-excitation predominates. Even the most efficient luminescent materials in the list emit less than 25% of the available energy.

In the paragraphs to follow, a discussion of several types of luminescent transitions in crystals will be given. We assume that the crystal contains a large number of radiation-produced electrons and holes, some associated in pairs as excitons. We suppose that the pair recombination process supplies the energy for the luminescence, although the photon emission process may occur in a variety of ways.

Figure 12.13 shows some important recombination processes (23). Band-

Table 12.5 Properties of Common Phosphors

Chemical Formula	*Mean Lifetime* (μs)	*Luminescent Efficiency* (%)	*Spectral Maximum* hν (*eV*)	λ (*Å*)
$CaWO_4$	3	4	2.9	4300
CsI:Tl	1	6	2.2 → 3.0	4100 → 5800
Gd_2O_2S:Tb		18	2.3	5400
LiF:Mg,Ti	(>1 yr)		3.1	4000
LiI:Eu	0.2	4	2.6	4700
NaI:Tl	0.2	13	3.0	4100
ZnS:Ag	10	23	2.6	4700

Data from Refs. 25, 34, and 37.

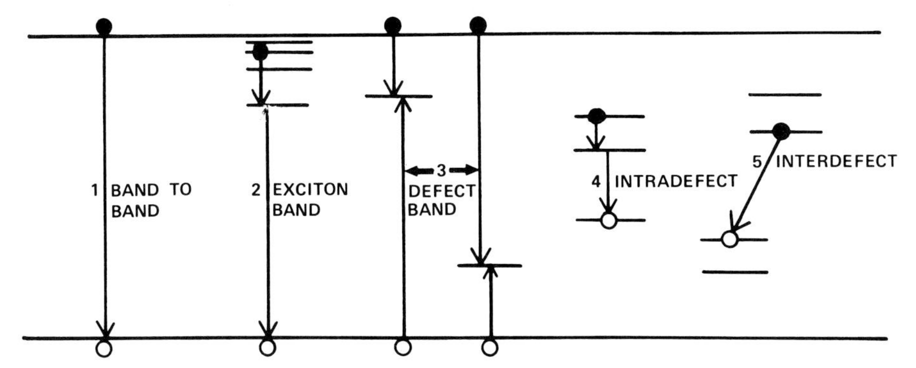

Figure 12.13 Several types of de-excitation transitions are illustrated.

to-band direct recombination and exciton-mediated recombination are important in some very pure crystals (24). If excitons are involved, the low-energy emissions will be characteristic of the spacing between the exciton energy levels. Three other processes shown on the figure are defect band recombination, intradefect recombination for a single site, and interdefect recombination when two separate sites are involved.

If the crystals are very pure, defect-related recombination processes often involve interstitials, vacancies, and the like, which act as traps. In pure alkali halide crystals, a sequential process is initiated by hole trapping at V_K centers, followed by electron-initiated recombination (38). When a conduction-band electron is involved, the process can be called defect band recombination. If the electron is itself trapped in a defect near the trapped hole, recombination after tunneling through the barrier is a possible interdefect process.

When impurity atoms are present in the crystal, they can function as luminescent sites. If activator impurity concentrations are small, the emission is probably host sensitized. This means that excited species are produced at random in the crystal. Capture of an exciton or charge carrier at the impurity, or nearby, follows. For an intradefect process, some of the energy upon recombination is coupled through an excited state of the impurity system and one or more luminescent photons are emitted (38).

As an example, consider the thallous ion Tl^+, a widely used activator impurity in the alkali halide phosphors. Examples of thallium-activated phosphors include NaI:Tl, KI:Tl, and CsI:Tl. The optical transitions in the Tl^+ ion involve the two electrons in the $n = 6$ orbit, the $6s^2$ configuration (3). Figure 12.14 shows that energy levels for these highly exposed electrons are drastically modified by the ionic lattice potential. It has been proposed (38) that Tl^+ luminescence at room temperature occurs by the transition from the triplet $^3P_0$ electron state to the singlet $^1S_0$ ground state. Presumably many electronic states initially excited in the ion de-excite thermally to populate the $^3P_0$ state at room temperature. The $^3P_0 \rightarrow {}^1S_0$ transition energy is about 2.9 eV for Tl^+ in NaI. This energy is near the maximum of the emission spectrum of NaI:Tl.

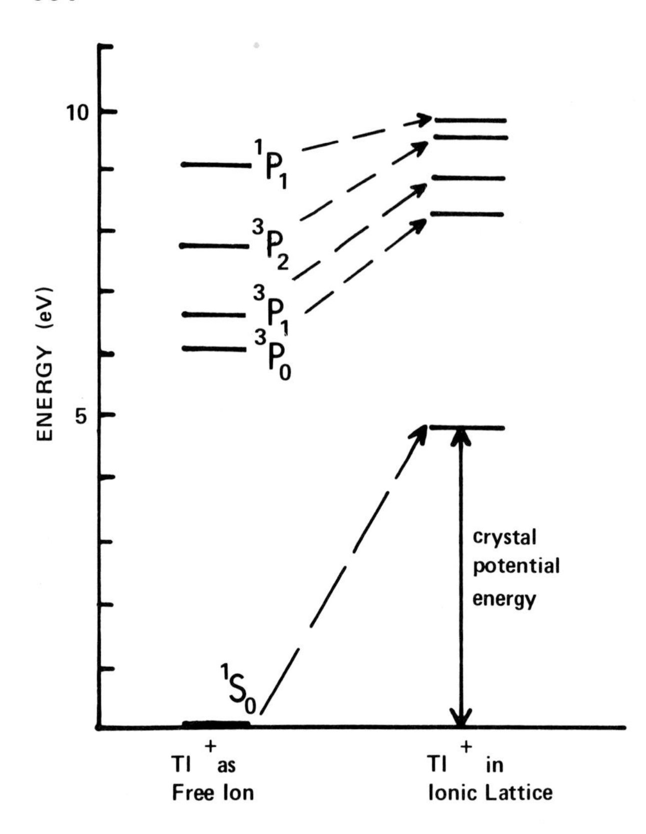

Figure 12.14 Energy levels of the thallous ion are shown both for a free ion and as a substitutional impurity in an ionic lattice (38).

The thermoluminescence of LiF:Mg,Ti is another example of impurity-activated luminescence. Various mechanisms have been proposed for lithium fluoride thermoluminescence (25). Electron traps are thought to be associated with the Mg^{++} impurity or F centers. V_3 centers, the double-hole traps, are important since V_K traps are unstable in lithium fluoride. The luminescent site is associated with the Ti^{++} impurity. Christy and Mayhugh (39) proposed that the emission process is initiated when an electron is thermally released from a Mg^{++} trap site. The newly mobile electron recombines with one of the holes trapped at a V_3 center and initiates release of the second hole. This mobile hole is captured for recombination near a Ti^{++} impurity. It is annihilated when a nearby F center electron tunnels through a separating barrier. Thus, an interdefect process is proposed as the direct cause of the photon emission.

Since charge carriers are trapped in lithium fluoride for very long times under ordinary circumstances, the phosphor can be used as a thermoluminescent dosimeter. Suppose N_{0t} is the total number of active traps in a lithium fluoride crystal and N_t is the number filled because of radiation. Then if dD is the dose increment producing dN_t newly trapped carriers,

$$dN_t = C_D(N_{0t} - N_t)dD, \qquad \frac{dN_t}{N_{0t} - N_t} = C_D dD,$$

$$\ln(N_{0t} - N_t) = -C_D D + C,$$

$$N_{0t} - N_t = \exp C \exp(-C_D D), \quad N_{0t} - N_t = N_{0t} \exp(-C_D D),$$

$$N_t = N_{0t}[1 - \exp(-C_D D)] \tag{12.31}$$

Equation 12.31 shows that the number of trapped carriers is proportional to the absorbed dose when $D \ll 1/C_D$. Saturation of the trapping sites should be expected at high doses. Actually, Equation 12.31 is approximate only; supralinearity (25) and other phenomena related to lithium fluoride thermoluminescence have been ignored in the foregoing development.

12.5 Silver Bromide Radiolysis in Emulsions

Consideration will be given to decomposition by ionizing radiation (radiolysis) of the ionic silver bromide (AgBr) molecule in this section. This process is fundamental to the formation of a latent image in a radiographic emulsion. The discussion will include an analytical relationship between the optical density of a processed film and the dose in the original emulsion.

Emulsions usually contain between 40% and 80% (by weight) AgBr crystals, suspended in a gelatin (40). The gelatin provides a support matrix for the crystals during handling and exposure, yet allows access for developing chemicals later. In radiographic film emulsion layers are usually made to adhere to both sides of a nearly transparent base (polyester), which provides mechanical support. In this stable configuration, the latent image is the spatial pattern of AgBr crystals sensitized by radiation prior to development. Sensitized crystals are subject to the action of a chemical developing solution. Nonsensitized crystals are more difficult to develop.

The process of latent image formation in an emulsion requires several steps. Some phenomena involved are unusual if not entirely unique to AgBr crystals. The size and shape of AgBr crystals used in x-ray emulsion are illustrated in Figure 12.15. The electronic properties of AgBr are best described with reference to the electron energy bands. The band gap in AgBr is about 2.5 eV (3); thus the crystals would ordinarily be classed as insulators. In many cases iodine and sulfur impurities can be identified in the AgBr used in emulsions. Electron traps are quite numerous, and some are associated with the impurities.

AgBr crystals are unusual since they support ionic conductivity very well (41); i.e. electrical charge can be transported in these crystals by the motion of positive ions. This property is a characteristic of several types of alkali halide and silver halide crystals. Ionic conductivity can occur by several mechanisms (3). We discuss only interstitial migration and vacancy position exchange.

In Table 12.1 the lattice spacing in some ionic crystals was compared with the ionic radii. From the values given for AgBr, it is not surprising that there is sufficient space for interstitial Ag^+ to move easily between lattice sites. At the same time, a silver ion vacancy can move through the crystal by a succession of ion-transfer processes from lattice sites into vacancies.

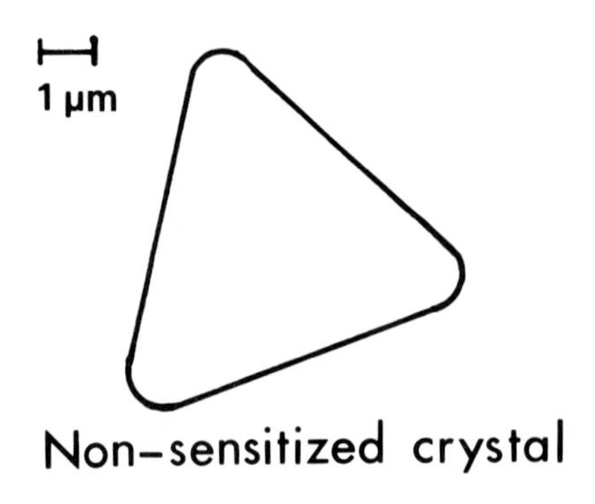

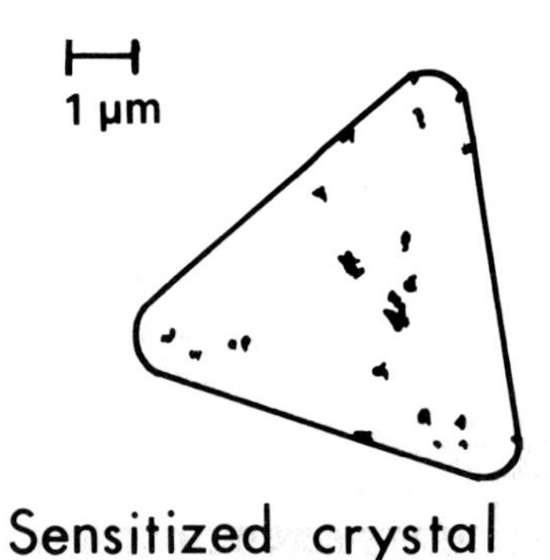

Figure 12.15 Two silver bromide crystals are illustrated. The sensitized crystal has several free silver (black) centers.

By analogy to Equation 12.29, the expression for the probability per second, p, for thermal excitation of Ag^+ lattice ions over an energy barrier of height ϵ_a is

$$p = f\exp(-\epsilon_a/kT°) \tag{12.32}$$

where k is the Boltzmann constant, $T°$ is the absolute temperature, and f is the striking frequency on the barrier. For this case, ϵ_a is called the activation energy. In ionic crystals, it is attributable to the charges on the surrounding ions. Table 12.6 shows some activation energies for interstitial motion and vacancy exchange in alkali and silver halide crystals (3). The low energies for AgBr indicate that the transport of positive charge in AgBr crystals is relatively easy. Also shown on the table are values for the energy of formation of Frenkel defects. This is also relatively low for AgBr. Thus, AgBr crystals are likely to be relatively rich in mobile interstitial Ag^+ ions.

A detailed explanation for the latent image in AgBr emulsion was presented by Gurney and Mott in 1938 (42). Their basic hypothesis is widely accepted although not uncontested (43). In the Gurney and Mott theory, latent image

Table 12.6 Activation Energies for Positive Ions

Crystal	Vacancy Exchange (eV)	Interstitial Motion (eV)	Frenkel Defect (eV)[a]
Sodium chloride (NaCl)	0.86		2.02
Lithium fluoride (LiF)	0.65		2.68
Lithium iodide (LiI)	0.38		1.34
Silver chloride (AgCl)	0.39	0.14	1.43
Silver bromide (AgBr)	0.25	0.11	1.12

Data from Ref. 3.
[a] Divide these values by 2 to apply them to Equation 12.32.

formation can be divided into the following steps after exposure of a crystal to ionizing radiation:

1. The radiation produces electron-hole pairs in the crystal.
2. Some electrons are trapped at impurity sites or defects.
3. Silver ions migrate to sites holding trapped electrons and react, yielding atomic silver Ag^0.
4. Holes migrate to the crystal surface and release bromine.
5. Further trapping and migration occurs until clusters of Ag^0 atoms exist, some of which are large enough to catalyze later chemical development.

The electron-hole pairs produced in the AgBr by the ionizing radiation are the result of collisional energy losses of the fast charged particles in the crystal. (Photoelectric effect may be the dominant contributor at optical energies.) The charged particles may be from a primary electron beam or they may be secondaries from photon or neutron beams. For relatively high energy beams, a substantial number of pairs are produced in a single pass of the charged particle. Some of the valence-band electrons excited to conduction-band energies are eventually trapped. When silver sulfide is present as an impurity, the sulfide ion positions are often identified as trapping centers and called sensitivity specks. Traps occurring near the surface are of most significance since they lead most directly to sensitized crystals.

The most important immediate consequence of an electron in a trap is that the surrounding electric field can attract a mobile silver ion. Gurney and Mott assumed that the ions were located interstitially and associated with Frenkel defects (42). Whatever the source, the Ag^+ ions migrate to the trapped electrons (labeled $\textcircled{e^-}$) and the reaction $Ag^+ + \textcircled{e^-} \rightarrow Ag^0$ follows. Simultaneously, many of the residual holes migrate to the crystal surface; they can be trapped by a Br^- ion near an imperfection. If the hole is shown as $\textcircled{+}$, the reaction that occurs may be indicated by $Br^- + \textcircled{+} \rightarrow Br$. The bromine liberated in this way may diffuse into the gelatin in the emulsion.

One Ag^0 atom on the surface of a crystal is usually not sufficient to sensitize it; a cluster of silver atoms, the minimum number between 3 and 6 (43), is usually necessary to effectively catalyze development. In the process of accumulating a cluster, it may happen that a single Ag^0 atom acts as a trap for a second electron. The resulting negatively charged site can attract another mobile Ag^+ ion and free silver formation can occur again. Thus, cluster growth can take place atom by atom. This simple stepwise process may be over simplified when ionizing radiation is utilized. By some process several Ag^0 atoms must accumulate locally. Figure 12.15 shows a crystal with a collection of very large, black latent image centers.

In the paragraph to follow it will be useful to relate n_A, the number of AgBr crystals per unit area sensitized by ionizing radiation, to n_{A0}, the total number of crystals per unit area in an emulsion. Suppose the radiation dose D applied to the emulsion is the primary physical parameter governing the n_A/n_{A0} ratio and suppose the boundary condition is $n_A = 0$ when $D = 0$. If C_D is the sensitizing probability per unit dose and if traversal by a single charged particle is sufficient to sensitize a crystal, C_D is independent of D. The equations below follow from a development similar to that given in Equations 12.31:

$$dn_A = C_D(n_{A0} - n_A)dD, \quad n_A = n_{A0}[1 - \exp(-C_D D)] \tag{12.33}$$

The form of this relationship was obtained by different means by Silberstein in 1922 (44). It is usually appropiate for ionizing radiation but not appropriate for exposures utilizing visible light.

Many sources of information are available on the chemistry of the development of photographic film (45). Suffice it to say that developing solutions preferentially reduce the Ag^+ ions in AgBr crystals possessing latent image centers on the surface. Chemical or direct development involves reduction of the Ag^+ ions from the sensitized crystal directly into Ag^0 at or near the site.

Organic molecules containing aromatic groups are often used as developing agents. Hydroquinone and phenidone are commonly used in solutions for radiographs. Solution temperature, concentration, and time of development affect the amount of AgBr reduced. In any case, substantially more black atomic silver is deposited on the film after development than was present in the latent image. The development process produces a very large gain in blackness, perhaps 10^8.

After fixing with a thiosulfate solution, washing and drying, a quantitative measure of the amount of silver deposited on various areas of the film may be of interest. The optical density is the relevant physical variable in this situation and is obtained by measuring the film's light transmission. The optical density (OD) is the logarithm to the base 10 of the ratio of the visible light fluence incident on the film, ϕ_i, to that fluence transmitted through the film, ϕ_t:

$$\mathrm{OD} = \log \frac{\phi_i}{\phi_t} = \log \frac{1}{Tm} \tag{12.34}$$

where Tm is the light transmission. Small-diameter light beams are necessary when OD measurements are made because they produce minimum scatter.

In 1913 Nutting developed an equation (46) for the OD of film in terms of the number of crystals per unit area reduced and deposited, and the average cross-sectional area of the developed silver globules. A similar result can be obtained by a different method if we suppose that the number of developed crystals per unit volume is n_v and the average cross-sectional area of the reduced silver globule is $\langle\sigma\rangle$. Consider first an emulsion with average thickness $\langle t\rangle$. For a light fluence ϕ from a viewbox or light source,

$$\frac{-d\phi}{\phi} = n_v\langle\sigma\rangle dx,$$

$$\int_{\phi_i}^{\phi_t} \frac{d\phi}{\phi} = -n_v\langle\sigma\rangle \int_0^{\langle t\rangle} dx, \qquad \ln(\phi_t/\phi_i) = -n_v\langle\sigma\rangle\langle t\rangle,$$

$$\phi_t/\phi_i = \exp(-n_v\langle t\rangle\langle\sigma\rangle), \qquad \mathrm{OD} = \log[\exp(n_v\langle\sigma\rangle\langle t\rangle)] = 0.434 n_A\langle\sigma\rangle \qquad (12.35)$$

The number of globules per unit area is $n_a = n_v\langle t\rangle$. The algebraic derivation used for Equation 12.35 assumes that every light photon striking a developed silver globule is absorbed and that there is no overlap between developed globules in different layers. Under these limitations the OD is directly proportional to the area density of the deposited silver.

A more comprehensive result can be obtained from Equations 12.33 and 12.35 if the number of sensitized crystals per unit area can be equated to the number of developed silver globules per unit area:

$$\mathrm{OD} = 0.434\langle\sigma\rangle n_{A0}[1 - \exp(-C_D D)] \qquad (12.36)$$

This result is plotted on Figure 12.16, with a linear abscissa (*left*) and a logarithmic abscissa (*right*). The linear plot is often called a sensitivity curve or sensitometric curve; the semilogarithmic plot is often called the characteristic curve. When visible light is involved, even the sensitivity curves have a pro-

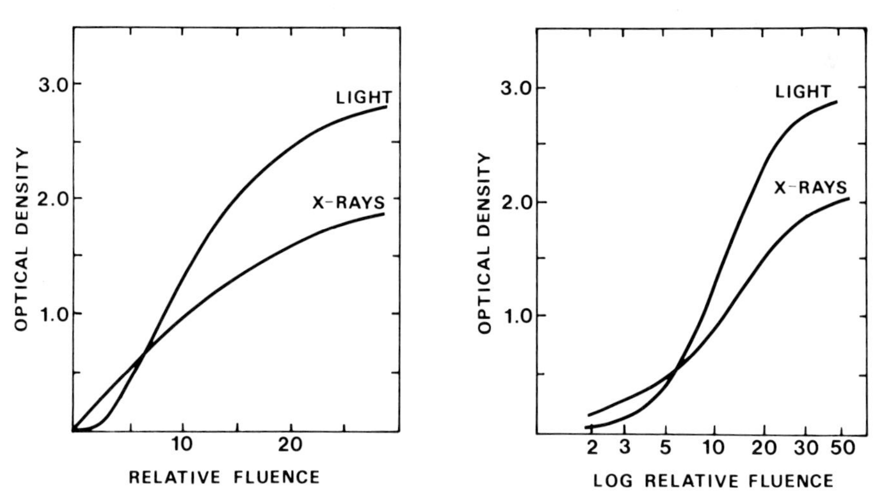

Figure 12.16 Sensitivity curves for direct exposure by x-rays and visible light are shown at left. The semilogarithmic plots of the same data shown at right are often called the characteristic curves. No comparison of relative film sensitivity to light versus x-rays is intended. Adapted from Ref. 47.

nounced sigmoid shape (47). If the radiographic system is to be used for dosimetric purposes, the sigmoid shape causes an unnecessary complication; thus film packs without screens are often preferred.

12.6 Problems

1. Suppose atoms are equally spaced along the x axis with separation $a =$ 3Å. Calculate the difference in kinetic energy between a free electron state with $l = N/2$ and one with $l = 0$.
2. A perfect diamond appears colorless, cadmium sulfide is yellow-orange, and crystalline silicon has a metallic gray color. Relate these facts to the band gaps for each crystalline substance.
3. The conversion efficiency of energy to light for crystalline NaI:Tl is about 13%. If all of the energy from a 1-MeV photon were absorbed in the crystal, estimate the number of light photons produced. Use the energy corresponding to the maximum of the spectral emission curve.
4. The light conversion efficiency for crystalline Gd_2O_2S:Tb is about 18%. If all of the energy from a 30-keV x ray were absorbed in the crystal, estimate the number of light photons produced. Use the energy corresponding to the dominant peak in the emission spectrum.
5. Why does the cross section for volume plasmon production change so rapidly at about 20 eV in aluminum? Justify your conclusions using harmonic oscillator calculations with $\hbar\omega = \hbar\sqrt{k/m}$. The restoring force constant is $k = n_v e^2/\epsilon_0$ and the electron density is $n_v = (1.8)10^{29}$ electrons/m^3.
6. Silicon electronic devices often show more stability to high temperature environments than equivalent germanium devices. Calculate the ratio of the intrinsic carrier density for these two semiconductor materials at room temperature (22°C) and at 38°C. Relate your results to the temperature stability of electronic components.
7. Calculate the electron density for silicon under intrinsic conditions at 300 K using Equation 12.15 and values from Table 12.3.
8. The resolution of a charged-particle detector is defined as the full width half maximum (FWHM) in energy of the spectral peak divided by the energy of the charged-particle deposited in the detector. Suppose 1-MeV electrons are completely stopped in the active volume of a germanium detector. What is theoretically the best resolution the detector could have for 100% collection efficiency? Use the standard deviation from the Poisson distribution and relate it to the FWHM.
9. Suppose the depth of a trap in a lithium fluoride thermoluminescent dosimeter is 1.2 eV and the frequency factor is 8×10^{11} Hz. What is the half-life of the trap at room temperature (22°C)? What is the half-life at a readout temperature of 250°C?
10. Suppose it takes 0.65 eV to produce a positive-ion Frenkel defect in lithium fluoride. Is it possible for a 1-keV electron to produce such a defect in a head-on collision with the Li^+ lattice ion?

11. Suppose a silver bromide crystal has a mean diameter of 1.0 μm. How many electron-hole pairs are produced in the crystal by the passage of a single 1-MeV electron?
12. Calculate the fraction of the Ag^+ ions associated with Frenkel defects in silver bromide at room temperature (22°C). Calculate the fraction of the Ag^+ interstitials that are mobile at room temperature in silver bromide.
13. Assume a single crystal with only 6 Ag^0 atoms is developed. Calculate the maximum blackness gain due to development of this crystal. Assume that the crystal is equivalent in size to a sphere with diameter 1 μm.
14. Silver bromide film shows signoid sensitivity curves for exposures with visible light. Why is this?

12.7 References

1. J. Thewlis, *Concise Dictionary of Physics* (Pergamon, New York (1973).
2. H. N. V. Temperley and D. H. Trevena, *Liquids and Their Properties* (Ellis Horwood, Chicester, England, 1978).
3. C. Kittel, *Introduction to Solid State Physics,* 3rd ed. (John Wiley and Sons, New York, 1966).
4. J. C. Slater, Phys. Rev. **45**, 794 (1934).
5. W. Shockley, *Electrons and Holes in Semiconductors* (D. Van Nostrand, Princeton, NJ, 1950).
6. R. B. Leighton, *Principles of Modern Physics* (McGraw-Hill, New York, 1959).
7. M. Morrison, T. Estle, and N. Lane, *Quantum States of Atoms, Molecules and Solids* (Prentice-Hall, Englewood Cliffs, NJ, 1976).
8. F. Block, Z. Phys. **52**, 555 (1928).
9. H. D. Young, *Fundamentals of Optics and Modern Physics* (McGraw-Hill, New York, 1963).
10. A. H. Sommer, *Photoemissive Materials* (John Wiley and Sons, New York, 1968), p. 95.
11. G. F. Garlick, "Cathodo and radioluminescence," in *Luminescence of Inorganic Solids,* edited by P. Goldberg (Academic, New York, 1966).
12. R. H. Ritchie, F. W. Garber, M. Y. Nakai, and R. D. Birkhoff, "Low energy electron mean free paths in solids, in *Advances in Radiation Biology,* vol. 3, edited by L. G. Augenstein, R. Mason, and M. Zelle (Academic, New York, 1969).
13. L. C. Emerson, R. D. Birkhoff, V. E. Anderson, and R. H. Ritchie, Phys. Rev. Sect. B **7**, 1798 (1973).
14. W. J. McConnell, R. D. Birkhoff, R. N. Hamm, and R. H. Ritchie, Radiat. Res. **33**, 216 (1968).
15. J. F. Fowler, "Solid state electrical conductivity dosimeters," in *Radiation Dosimetry,* vol. 2, edited by F. H. Attix and W. C. Roesch (Academic, New York, 1966), pp. 291–325.
16. W. J. Price, *Nuclear Radiation Detection* (McGraw-Hill, New York, 1958).
17. F. Goulding and Y. Stone, "Semiconductor radiation detectors," in *Semiconductor Detectors in the Future of Nuclear Medicine,* edited by P. B. Hoffer, R. N. Beck, and A. Gottschalk (Society of Nuclear Medicine, New York, 1971), pp. 1–91.
18. W. K. H. Panofsky and M. Phillips. *Classical Electricity and Magnetism* (Addison-Wesley, Reading, MA, 1956).

19. I. T. Myers, "Ionization," in *Radiation Dosimetry,* vol. 1, edited by F. H. Attix and W. C. Roesch (Academic, New York, 1968), pp. 317–331.
20. *Average Energy Required to Produce an Ion Pair, ICRU Report 31* (International Commission on Radiation Units and Measurements, Washington, 1979).
21. W. Shockley, Czech. J. Phys. Sect. B **11**, 81 (1961).
22. G. Sadasiv, "Photoconductivity," in *Photoelectric Imaging Devices,* vol. 1, edited by L. M. Biberman and S. Nudelman (Plenum, New York, 1971), pp. 111–133.
23. F. Williams, "Theoretical basis for solid state luminescence," in *Luminescence of Inorganic Solids,* edited by P. Goldberg (Academic, New York, 1966), pp. 1–52.
24. P. J. Dean, "Lattices of the diamond type," in *Luminescence of Inorganic Solids,* edited by P. Goldberg (Academic, New York, 1966), p. 120.
25. K. Becker, *Solid State Dosimetry* (Chemical Rubber Co., Cleveland, 1973), pp. 27–109.
26. J. T. Randall and M. H. F. Wilkins, Proc. R. Soc. Ser. A **184**, 366 (1945).
27. G. Stokes, Philos. Trans. R. Soc. London Ser. A **142**, 463 (1852).
28. J. B. Birks, *Photophysics of Aromatic Molecules* (Wiley-Interscience, London 1970), p. 52.
29. J. Franck, Trans. Faraday Soc. **21**, 536 (1925).
30. E. U. Condon, Phys. Rev. **32**, 858 (1928).
31. L. G. Van Uitert, "Luminescence of insulating solids for optical masers," in *Luminescence of Inorganic Solids,* edited by P. Goldberg (Academic, New York, 1966), p. 471.
32. S. Shionoya, T. Koda, K. Era, and H. Fujiwara, Acta Phys. Pol. **26**, 801 (1965).
33. B. G. Oltman, J. Kastner, and C. Paden, "Spectral analysis of thermoluminescence curves," in *Proceedings of the Second International Symposium on Luminescence Dosimetry, CONF-680920* (1968), p. 23.
34. R. A. Buchanan, S. I. Finkelstein, and K. A. Wickersheim, Radiology **105**, 185 (1972).
35. P. D. Johnson, "Oxygen dominated lattices," in *Luminescence of Inorganic Solids,* edited by P. Goldberg (Academic, New York, 1966), p. 315.
36. W. J. Ramm, "Scintillation detectors," in *Radiation Dosimetry,* vol. 2, edited by F. H. Attix and W. C. Roesch (Academic, New York, 1966), pp. 124–125.
37. J. B. Birks, *The Theory and Practice of Scintillation Counting* (Pergamon, New York, 1964).
38. K. Teagarden, "Halide lattices," in *Luminescence of Inorganic Solids,* edited by P. Goldberg (Academic, New York, 1966), pp. 53–119.
39. R. W. Christy and M. R. Mayhugh, J. Appl. Phys. **43**, 3216 (1972).
40. R. A. Dudley, "Dosimetry with photographic emulsions," in *Radiation Dosimetry,* vol. 2, edited by F. H. Attix and W. C. Roesch (Academic, New York, 1966), pp. 325–388.
41. C. R. Berry, "Structural defects and ionic conductivity," in *The Theory of the Photographic Process,* 3rd ed., edited by C. E. Mees and T. H. James (Macmillan, London, 1966), pp. 12–19.
42. R. W. Gurney and N. F. Mott, Proc. R. Soc. London Ser. A **164**, 151 (1938).
43. J. F. Hamilton and F. Urback, "The mechanism of the formation of the latent image," in *The Theory of the Photographic Process,* 3rd ed., edited by C. E. Mees and T. H. James (Macmillan, London, 1966), pp. 87–119.
44. L. Silberstein, Philos. Mag. **44**, 257 (1922).
45. W. E. Lee, "The developing agents and their reactions," in *The Theory of the Photo-*

graphic Process, 3rd ed., edited by C. E. Mees and T. H. James (Macmillan, London, 1966), pp. 278–311.

46. P. G. Nutting, Philos. Mag. **26**, 423 (1913).
47. J. C. Dainty and R. Shaw, *Image Science, Principles Analysis and Evaluation of Photographic-Type Imaging Processes* (Academic, New York, 1974), p. 50.

13

Ionization, Excitation, and Dissociation in Molecular Solutions

13.1 The Liquid State

The geometrical arrangement of molecules in a liquid is intermediate between the random arrays characterizing gases and the rigid ordering of a crystalline solid. Intermolecular forces in the liquid state are sufficient to produce a contiguous macroscopic mass in a containing vessel but insufficient for rigidity. On the atomic scale, there is a short-range or local order in the molecular arrangements of liquids. Although there is no periodicity, each molecule has a set of nearest neighbors whose positions fall within a narrow range of distances. On the other hand, thermal energies are sufficient to produce constant changes in the geometrical configurations, and intermolecular movement is unceasing (1–3).

Considerable variety exists in the intermolecular forces occurring in different liquids (2). Nevertheless, it is generally true that liquid-state forces are strongly repulsive at very short distances (hard core). When chemical bonding is not involved, the forces are only weakly attractive in a restricted range of intermediate distances. They diminish in strength very rapidly thereafter. Part of the short-range repulsive component is due to interpenetration of molecular electron clouds. In addition to the electron-electron repulsion, the positive nuclear cores repel each other strongly when interpenetration is substantial.

The intermolecular attractive force is often called the Van der Waals force (2, 4). It is electrostatic in nature, but operates for electrically neutral molecules. An electric dipole moment is central to the description of Van der Waals forces; in some cases a permanent moment is involved, but more often the moment is transient. Suppose one end of a neutral molecule has an excess of positive charge while the other end remains negative. If several such molecules are in proximity, a cohesive energy will result from the attraction of the positive end of one for the negative end of a neighbor.

Assume that an electric dipole moment **d** exists on a given molecule at some instant of time. The magnitude of **d** is equal to the product of the excess in positive charge on one end of the molecule and the distance between the charge distributions at opposite ends. The vector direction points from negative charge to positive charge. The electric field E at distance r from the dipole center can be written (4, 5)

$$E(r,\theta) = \frac{1}{4\pi\epsilon_0}(\mathbf{d} - \hat{\mathbf{r}}\, 3d\cos\theta)\frac{1}{r^3} \tag{13.1}$$

where θ is the angle between vector **d** and the radial unit vector $\hat{\mathbf{r}}$.

A dipole moment $\mathbf{d}'$ can be induced on an adjacent molecule in the field of the first dipole. If the polarizability is α, the dipole moment is $\mathbf{d}' = \alpha E$. The potential energy of the dipole-dipole configuration is

$$\begin{aligned} V(R) &= -E\cdot\mathbf{d}' = -E\cdot\alpha E \\ &= \frac{-\alpha}{(4\pi\epsilon_0)^2}\left[\frac{d^2}{R^6} - \frac{6d^2\cos^2\theta}{R^6} + \frac{9d^2\cos^2\theta}{R^6}\right] \\ &= \frac{-\alpha}{(4\pi\epsilon_0)^2}(1 + 3\cos^2\theta)\frac{d^2}{R^6} \end{aligned} \tag{13.2}$$

Here R is the distance between the centers of the dipoles.

Equation 13.2 shows that the dipole-dipole cohesive potential energy varies as the inverse sixth power of the distance and the square of the electric dipole moment. The force producing such a potential is always attractive since d^2 and $\cos^2\theta$ are always positive. The force diminishes very rapidly with increasing dipole separation.

A polar molecule has a permanent electric dipole moment. Although most molecules are not highly polar, they are usually polarizable. Transient molecular dipole moments can be expected since atomic electrons are in constant orbital motion. Transient polar configurations can induce transient antiparallel moments in neighbors. On the average an attractive dipole-dipole force can be expected in this situation.

The total potential energy between two molecules in a liquid can be represented by (2, 6)

$$V(R) = -\frac{A}{R^6} + \frac{B}{R^{12}} \tag{13.3}$$

The coefficients A and B depend on the liquids involved. Equation 13.3 is the Lennard-Jones potential (7), suitable for nonpolar or weakly polar liquids. The term $-A/R^6$ represents the dipole-dipole attractive component similar to that of Equation 13.2. The term B/R^{12} represents a repulsive part that dominates at small intermolecular separation. The total molecular cohesive energy at equilibrium separation due to the Van der Waals force is usually between 0.01 and 0.1 eV/molecule (2, 4).

Liquid-state forces maintain cohesion but are insufficient to resist shearing stresses. They produce short-range order, but can not establish periodicity. Liquid-state ordering can be described (2, 6) by a radial distribution function $g(r)$. If a particular molecule is taken to be at the origin of a coordinate system and dN is the number of liquid-state molecules in a spherical shell with volume $4\pi r^2 dr$,

$$g(r) = \frac{dN(r)}{4\pi r^2 n_v dr} \tag{13.4}$$

where n_v is the average number of molecules per unit volume for the liquid.

Figure 13.1 shows a radial distribution function for the gaseous, liquid, and solid (crystalline) forms of a metal (8, 9). The gaseous state is represented by the horizontal line at $g(r) = 1$; density fluctuations are ignored on the figure. For the crystalline solid, only thermal vibrations and minor defects are assumed to perturb the fixed molecular positions in the periodic array. Thus, a set of narrow discrete lines are shown. For the liquid state, maxima are apparent in a continuous functional form. For liquids, when r is small, $g(r) = 0$ but increases to a peak thereafter. Beyond the peak the distribution function oscillates about an average, with $g(r) = 1$. At large distances the oscillation damps out.

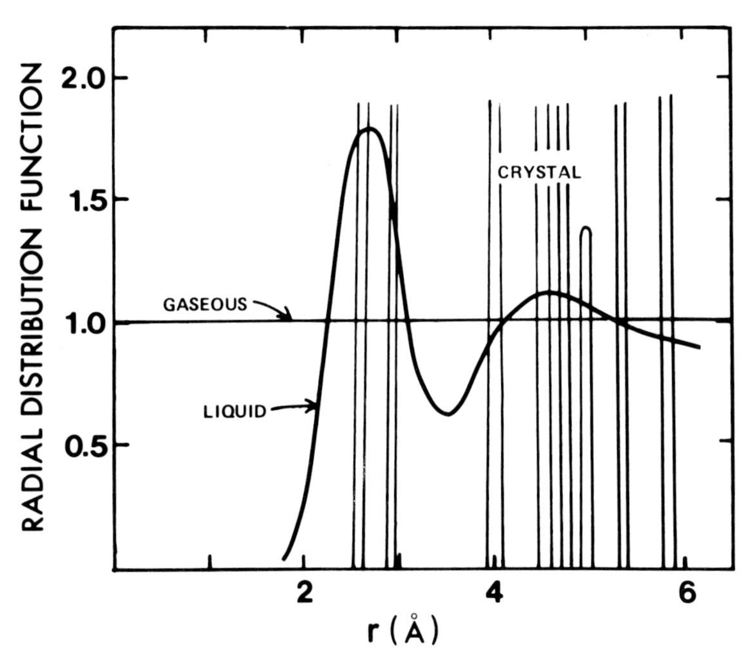

Figure 13.1 The radial distribution functions for the gaseous, liquid, and solid (crystalline) states of a metal are illustrated schematically (8, 9).

The region where $g(r)$ is small includes a volume surrounding the central molecule through which it is relatively free to move. The first peak in the distribution function represents a shell of nearest neighbors (2), which is quite well defined. Secondary peaks represent secondary shells of neighbors that are not as well defined.

The effect of temperature on liquid-state ordering is illustrated on Figure 13.2 with water as the example (10, 11). At 4°C there is a well defined shell of nearest neighbors at about 3Å from the central molecule; a secondary shell, less well defined, exists at about 5Å. The distribution function at 100°C shows maxima that are considerably less distinct. Thus, increased thermal energy tends to destroy the order and randomize the array.

The region of relatively free movement at small r and the shell of nearest neighbors is common to most liquids. The situation suggests a model wherein a given molecule in a liquid is confined to a cage of neighbors during short time periods (2). Figure 13.3 illustrates such a model. The cage is constantly being deformed, broken apart, and then reformed because of thermal motion. The central molecule in each cage moves back and forth across the region of freedom in a relatively unrestricted manner and collides with nearest neighbors at the wall.

The frequency of collisions, f_c, with the neighboring molecules and mean time between collisions, τ_c, can be crudely estimated for a liquid using the cage model. If the molecular speeds v have a Maxwellian distribution, the

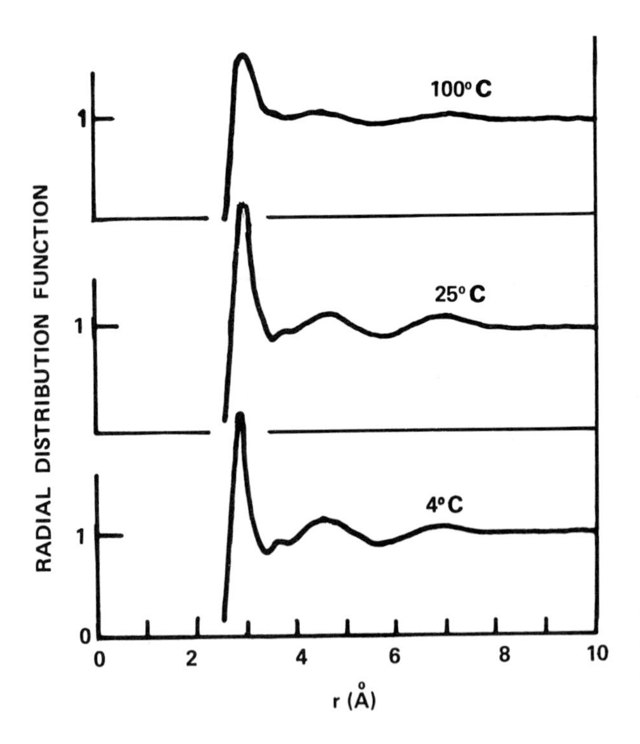

Figure 13.2 The radial distribution functions (10, 11) for liquid water are shown at 4°, 25°, and 100°C.

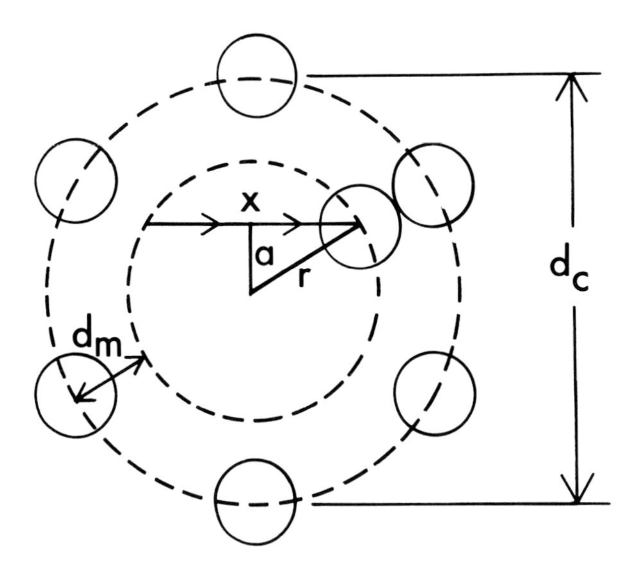

Figure 13.3 A two-dimensional cage of nearest neighbors is illustrated for a liquid.

average collision frequency can be estimated from the average distance $\langle d \rangle$ between collisions:

$$\tau_c = \frac{\langle d \rangle}{\langle v \rangle} = \frac{\langle x \rangle}{(8k\text{T}^\circ/\pi M)^{1/2}} = \frac{1}{f_c} \tag{13.5}$$

where $\langle v \rangle = (8k\text{T}^\circ/\pi M)^{1/2}$ is the Maxwellian average speed, k is Boltzmann's constant, T° the absolute temperature, and M the molecular mass (12). The variable $\langle x \rangle$ is the average chord length for a spherical cage of radius r.

The random probability $P(x)dx$ of occurrence of a chord of length between x and $x + dx$ can be determined by considering a hypothetical cylindrical shell with length between x and $x + dx$ and radius between a and $a + da$ that is inscribed within the spherical cage (see Figure 13.3). The ratio of the area of the end of the cylindrical shell divided by the cross-sectional area of the sphere equals the random probability desired. Thus,

$$P(x)dx = \frac{|2\pi a da|}{\pi r^2}, \quad (x/2)^2 + a^2 = r^2,$$

$$\frac{2xdx}{4} = -2ada, \quad P(x)dx = \frac{xdx}{2r^2} \tag{13.6}$$

The average chord length is obtained from

$$\langle x \rangle = \frac{\int xP(x)dx}{\int P(x)dx} = \frac{\int x^2\, dx}{\int x\, dx} = \frac{2d_c}{3} \tag{13.7}$$

where d_c is the diameter of the spherical cage.

Two identical molecules make contact when their centers are separated by d_m, the molecular diameter. Thus, the cage diameter must be reduced by twice the molecular diameter to obtain the average distance between collisions. Then

$$\tau_c = \frac{\langle d \rangle}{\langle v \rangle} = \frac{(2/3)(d_c - 2d_m)}{(8kT°/\pi M)^{1/2}} = \frac{1}{f_c} \tag{13.8}$$

Equation 13.8 can be used to estimate the mean time between collisions for water molecules using the information on Figure 13.2. If $d_c = 2(3\text{Å}) = 6\text{Å}$ and the molecular diameter is taken as 1.4Å (13), $\tau_c = (2)10^{-13}$ s at 293 K. This value is an order of magnitude estimate for the period between liquid-state collisions in water (14).

13.2 Water Structure

Water is an appropriate example for further discussion because it is commonplace and occupies a special prominence in living systems. Human cells are composed of about 80% water (15). Absorption processes for radiation in water, the excited states produced, and the chemical reactions which take place in organic solutions are of basic importance in radiation science.

The water molecule is unique in a physical sense because it is polar and tends to combine in tetrahedral arrays (16). Both of these properties can be related to the electronic properties of oxygen, especially to the configurations available for bonding with other atoms. A review of atomic orbitals and introduction to hybridization in bonding is helpful in understanding the molecular structure.

Recall that atomic electrons can be specified by the principal quantum number n and the orbital angular momentum quantum number l. The orbitals corresponding to $l = 0$, 1, 2, 3, and so forth, are given letter designations s, p, d, f, and so forth, respectively. The geometrical shapes of s- and p-electron clouds are illustrated on Figure 13.4, where surfaces of constant charge density are shown. Notice the spherical symmetry of the s-electron cloud and the lobular configurations for each of the p electrons, oriented about the x, y, and z axes. Configurations for d and f electrons are generally more complex, involving spherical, lobular, and ring shapes (4).

The electronic configuration of an atom can be specified by the configurations of all the individual electrons. Electronic configurations for the first few elements are shown on Table 13.1. The table is constructed according to Hund's rule (4), which dictates that electrons in a subshell do not pair their spins until all orbitals of equal energy are populated.

Hybrid orbitals are combinations of atomic orbitals with specific geometrical configurations. Hybrid orbitals between s and p electrons sometimes form covalent systems with reduced total energy. Prior to bonding they represent an excited configuration of the electrons in the atom. Elements with $2s$ and $2p$ electrons often show hybridization in bonding (9, 17).

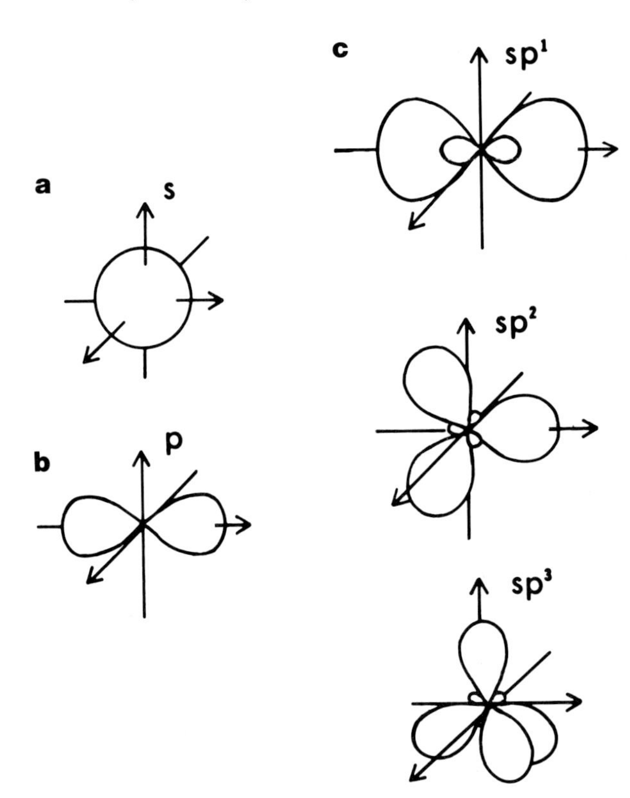

Figure 13.4 Geometrical shapes of surfaces of constant electron density are shown. a) *s* electrons; b) p_x, p_y, and p_z electrons; c) sp^1, sp^2, and sp^3 hybrid orbitals.

Suppose a 2*s* state is mixed with a single 2*p* state to form sp^1 hybrids. In this case the hybridized electron orbitals project from the parent at 180° as illustrated in Figure 13.4. If a 2*s* state is mixed with two 2*p* states, the resulting configurations are called sp^2 hybrids. These orbitals project at 120° from the parent in a planar array. If a 2*s* state is mixed with three 2*p* states, the configuration has electron density lobes forming the corners of a tetrahedron

Table 13.1 Atomic Structures of Some Elements

Element	*Atomic Number (z)*	*Atomic Structure*	*Occupancy of Orbitals*[a]				
			1s	*2s*	$2p_x$	$2p_y$	$2p_z$
Hydrogen (H)	1	$1s$	↑				
Helium (He)	2	$1s^2$	↑↓				
Lithium (Li)	3	$1s^2 2s$	↑↓	↑			
Beryllium (Be)	4	$1s^2 2s^2$	↑↓	↑↓			
Boron (B)	5	$1s^2 2s^2 2p$	↑↓	↑↓	↑		
Carbon (C)	6	$1s^2 2s^2 2p^2$	↑↓	↑↓	↑	↑	
Nitrogen (N)	7	$1s^2 2s^2 2p^3$	↑↓	↑↓	↑	↑	↑
Oxygen (O)	8	$1s^2 2s^2 2p^4$	↑↓	↑↓	↑↓	↑	↑

[a] ↑, Spin up; ↓, Spin down.

with included angles of 109.5°. This hybrid configuration is labeled sp^3 and is also illustrated on Figure 13.4. For a carbon parent atom, the molecules acetylene, ethylene, and methane are examples of molecules with sp^1, sp^2, and sp^3 covalent bonds, respectively. The geometrical configurations (4) relate directly to the hybridized carbon configurations.

The oxygen atom in the water molecule provides an excellent example of the effects of hybridization. The oxygen $2s$ and $2p$ electrons normally are part of the configuration $(1s^2 2s^2 2p_x{}^2 2p_y 2p_z)$. When oxygen $n = 2$ states are fully hybridized, six sp^3 configurations result. Two contain unpaired electrons available for covalent bonding and the others are paired sets, the so-called "lone pairs." The tetrahedral configuration of the water molecule is related to this hybridization, since the two lone-pair lobes point toward adjacent corners of a tetrahedron and two unpaired electron lobes point toward the other corners.

For the water molecule, each of the oxygen unpaired electron hybrid states form a covalent bond with a hydrogen atom. The bond length is 1.0Å and the bonding energy is 4.7 eV in this case. Actually, the angle included between the covalent bonds with the hydrogen atoms is 105° for water, rather than 109.5° for the perfect tetrahedral configuration, because of electrostatic distortion (16).

The polarity of the water molecule is a result of the configuration of the bonded hydrogen atoms and the sets of lone-pair electrons. In the covalent bonding configuration, part of the electronic charge normally surrounding a hydrogen atom is preferentially maintained between the hydrogen and oxygen atomic centers. Thus, the extended hydrogen atoms are each partially electron deficient and when taken together they form a region with a net positive charge. The two sets of lone-pair electrons on the oxygen atom opposite the hydrogen atoms produce a region where negative charge dominates; thus, the neutral water molecule has a strong electric dipole moment.

This polar situation is ideal for hydrogen bonding between adjacent water molecules. Each of the positive hydrogen atoms seeks a lone pair of electrons on a neighboring oxygen atom. The hydrogen bonds formed are intermediate in strength between covalent bonds and Van der Waals forces. When the hydrogen atoms are on a line between two oxygen atoms, the hydrogen-bond energy is about 0.2 eV (16).

In the liquid state, an array of hydrogen-bonded water molecules forms with oxygen atoms about 3Å apart (16), located at the four corners of a tetrahedron. The configuration is illustrated in Figure 13.5. It is a direct result of the geometrical shape of the sp^3 hybridized oxygen orbitals. Thus, in the ideal case, each molecule participates in four hydrogen bonds, one for each hydrogen atom and one for each set of lone-pair electrons. However, some distortion and bond bending occurs since the included angle is 105°. Indeed, at any instant of time in liquid water, many hydrogen bonds are strained and perhaps 20% to 40% are broken (16). There is continual intermolecular movement, so a given configuration may last only about 10^{-12} s (18) before alteration and reformation takes place.

When ions are dissolved in water, there is some reformation of the local

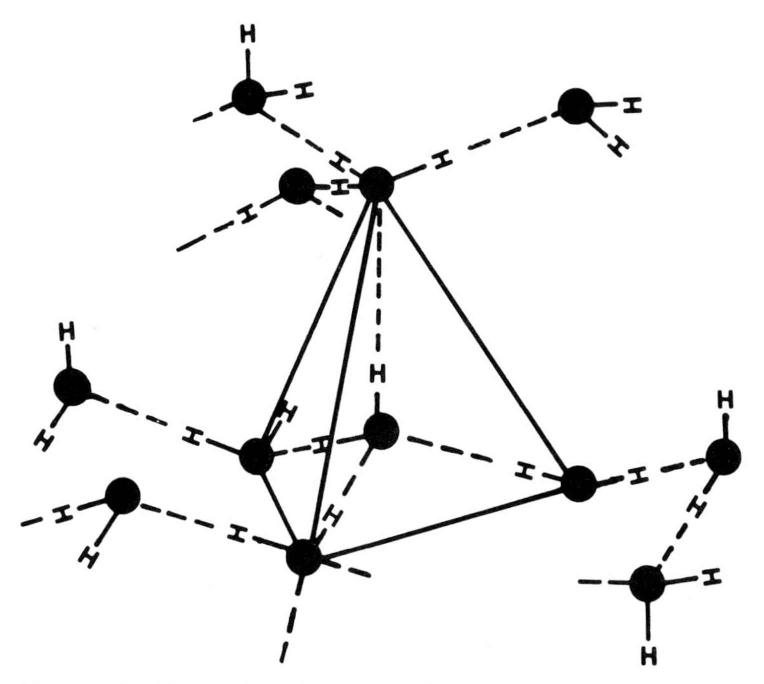

Figure 13.5 The tetrahedral structure of liquid water is shown for the ideal case with complete hydrogen bonding.

water structures around the charged species. The water dipoles are expected to be oriented by the ionic charge, at least in the first shell of nearest neighbors. The hydrated electron, e^-_{aq}, as shown on Figure 13.6, is a good example of such a situation (19). This species forms after thermalization of electrons liberated from their parent structures by radiation in water. In general, an electron is associated with four to six water molecules. Since the dielectric relaxation time of water is about 10^{-11} s (18), at least this much time is necessary for the water molecules to break hydrogen bonds and align themselves with the ion. The situation for positive ions in a water solution is similar to that for thermalized electrons; hydration is expected, with the water dipoles pointing away from a positive ion.

13.3 Molecular Energy Levels

Energy level diagrams for atomic systems are insufficient when the atoms are bound together into molecules. Molecular energy level schemes are generally more complex than atomic schemes because of the extra degrees of freedom associated with spatial configurations and because valence electrons involved in chemical bonding can participate in several energy states.

The total energy E for a molecule can be described by (20)

$$E = E_t + E_r + E_v + E_e \tag{13.9}$$

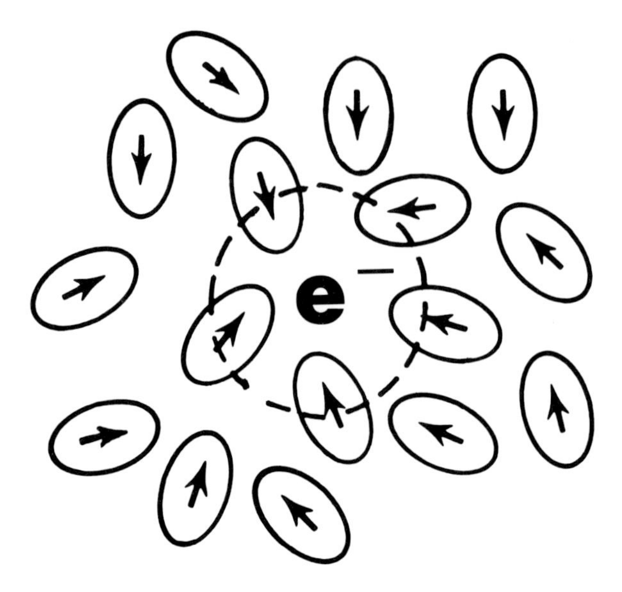

Figure 13.6 Hydration of a free electron in water solution is illustrated. Arrows represent water molecules.

where E_t is translational energy, E_r is rotational energy, E_v is vibrational energy, and E_e is electronic energy.

There are three degrees of translational freedom corresponding to the three directions of space. According to the equipartition of energy principle (12), each degree of freedom has kinetic energy $kT°/2$. The average translational kinetic energy $\langle T \rangle$ is related to the absolute temperature $T°$ by $\langle T \rangle = 3kT°/2$, where k is the Boltzmann constant. For molecules free of confinement, the allowed translational energies form a continuous distribution. Translational kinetic energy is an eventual respository for excess rotational, vibrational, and electronic energies.

Rotational energies must be considered when molecules are free to turn about an axis through the center of mass. Rotational motion is quantized: in the gaseous state, discrete transitions are readily observable: in the more dense states, restrictions on movement may occur. In liquids, rotations are possible (21) when the cage of nearest neighbors is sufficiently large and well defined. Rotational level spacing is of the order of a few hundredths of an electron volt in the gaseous state. In many cases individual levels can not be resolved in more dense matter.

Molecular vibrational energy levels occur when the atomic nuclei oscillate about the molecular center of mass. The level spacing is usually a few tenths of an electron volt. The level positions can be perturbed by fields from the surrounding molecules in high-density physical states. Molecular electronic levels resemble the levels of the constituent atom except when valence electrons are involved in chemical bonding. In any case the electronic level spacing varies over a wide range, from a few electron volts for valence electrons to many kiloelectron volts for tightly bound K electrons in high Z atoms.

The overall energy level configuration for a molecule is a superposition of the levels resulting from different effects, as illustrated on Figure 13.7. Vibrational levels are effectively superposed on electronic levels, and rotational levels are superposed on the vibrational levels. Thus, the resulting array of energies can be very complex.

Consider the simple case of rotation of a diatomic molecule (4, 22). The rotational angular momentum L_r is quantized:

$$L_r = I\omega = \sqrt{J(J+1)}\,\hbar, \quad J = 0, 1, 2, 3 \ldots \tag{13.10}$$

where I is the molecular moment of inertia about a rotational axis through the center of mass, ω is the angular speed, and J is the rotational quantum number. The rotational energy E_r is given by

$$E_r = \frac{1}{2} I\omega^2 = \frac{L_r^2}{2I} = \frac{J(J+1)\hbar^2}{2I} \tag{13.11}$$

According to Equation 13.11 the spacing between rotational levels depends inversely on the moment of inertia about the center of mass. In general, successive levels occur at 0, 2, 6, 12, and the like, units of $\hbar^2/2I$, so the level spacing occurs in the sequence 2, 4, 6, 8, and so forth, respectively, as shown on Figure 13.7. In many actual diatomic molecules, the configuration is stretched during

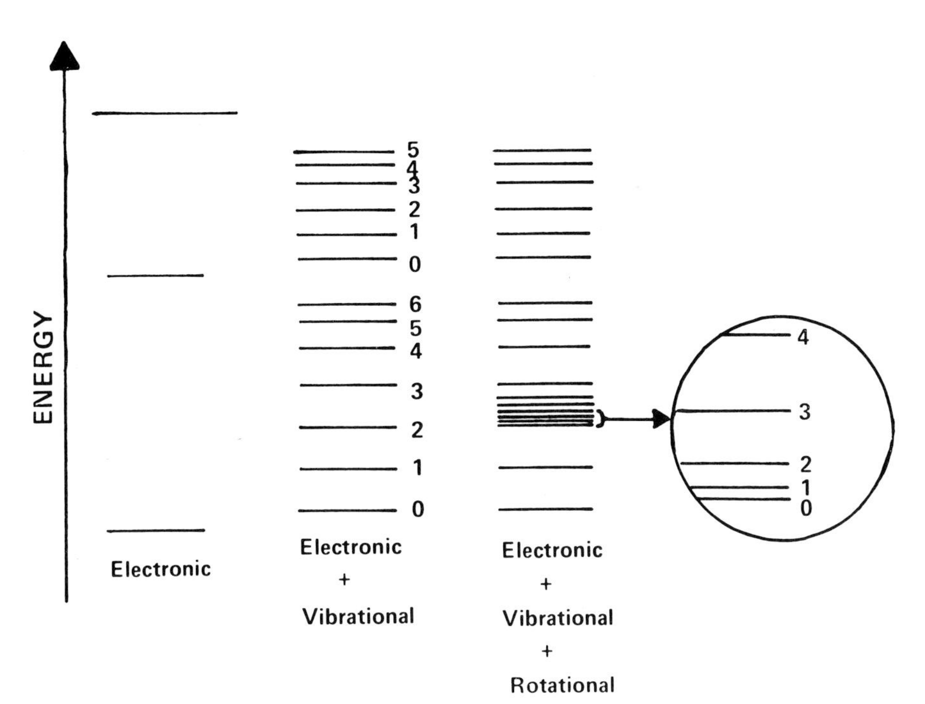

Figure 13.7 The energy levels of a molecule free to rotate and vibrate are illustrated. The rotational levels are clearly shown only in the magnified circular region because of the narrow spacing. Adapted from A. Beiser, *Concepts of Modern Physics* (McGraw-Hill, New York, 1967).

rotation and the moment of inertia increases. This results in lowering the discrete rotational energies (22).

In general, for a polyatomic molecule, rotations can occur about three perpendicular axes passing through the center of mass. Then

$$L^2 = (L_x{}^2 + L_y{}^2 + L_z{}^2) = J(J+1)\hbar,$$

$$E_r = \frac{L_x{}^2}{2I_x} + \frac{L_y{}^2}{2I_y} + \frac{L_z{}^2}{2I_z} \qquad (13.12)$$

If there exists a unique axis of symmetry for the molecule, such that $I_x = I_y$, the energy level structure can be further analyzed in terms of a rotational quantum number K where $L_z = \hbar K$, $K = 0, \pm 1, \pm 2, \ldots$. If there is no axis of symmetry, a general form for energy can not be written so easily, and each case must be considered separately.

Molecular internuclear vibrations depend on the resiliency of the chemical bonds involved, particularly covalent bonds. A short digression on the origins of the covalent bond is in order at this time (9). Consider constituent atoms A and B with atomic wave functions Ψ_A and Ψ_B representing the respective electronic configuration, as shown on Figure 13.8. Atoms A and B can be

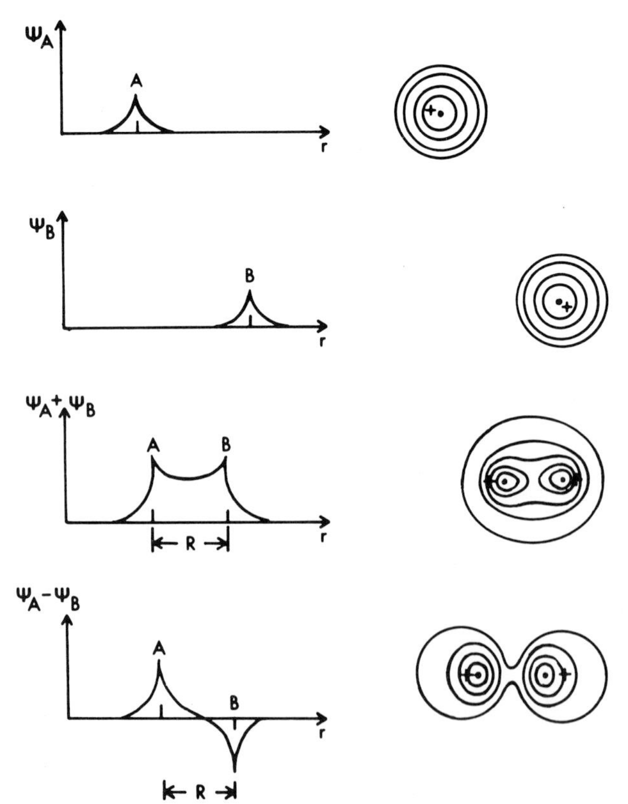

Figure 13.8 Wave functions for electronic charges are shown at left for the case $\psi_A + \psi_B$ (bonding) and $\psi_A - \psi_B$ (antibonding). Contours of electron probability density are shown at right.

brought together with two different wave-function configurations: If the two-atom system wave function is represented by Ψ_A and Ψ_B, at the equilibrium separation, the wave functions overlap and a buildup of electronic charge between the positive atomic cores is expected. This is a bonding configuration, since the positive cores are each attracted to the residual internuclear negative charge. In this situation the energy of the atomic combination is lower than the energy of the separated atoms. In the antibonding configuration $\Psi_B - \Psi_B$, the electron charge density vanishes between the positive nuclei, and the repulsive force between the cores dominates.

The potential energy diagrams for the bonding and antibonding cases are shown on Figure 13.9 as a function of internuclear separation R. The antibonding potential energy has a $1/R$ dependence characteristic of a repulsive coulomb force. The potential energy curve for the bonding configuration includes a well with a minimum at separation distance R_e. The well occurs because of the attraction of the nuclear centers for the confined electronic charge. A short-range repulsive part attributable to interpenetration of atomic electron charges is also involved. The covalent bonding potential energy curve is well represented by a form from Morse (23):

$$V(R) = V_B\{1 - \exp[-a(R - R_e)]\}^2 - V_B \tag{13.13}$$

with width parameter a and depth parameter V_B.

Internuclear vibration in molecules involves an oscillation in internuclear separation distance R. Thus, pairs of nuclei are assumed to move apart to a maximum value of R (amplitude) and then to come back together again. The chemical bonding force provides the impetus for restoration. For small amplitudes this motion reduces to a harmonic oscillation (14). If $R - R_e \ll 1/a$,

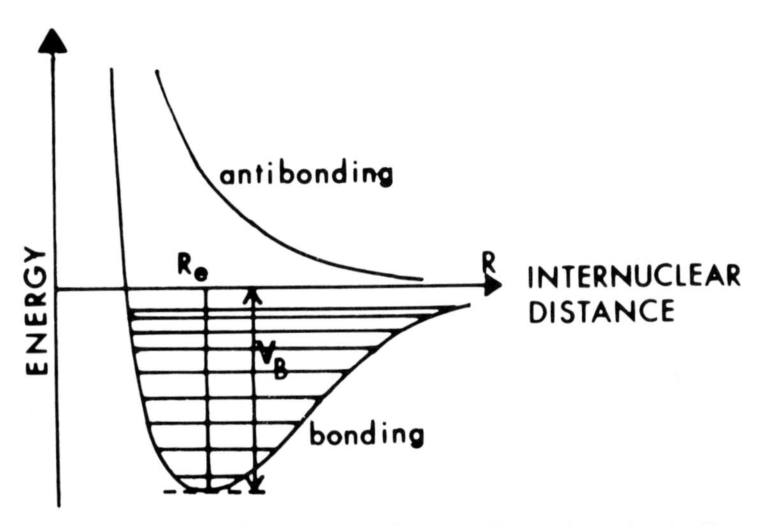

Figure 13.9 Potential energy curves for a bonding configuration (well) and an antibonding configuration are shown.

$$\exp[-a(R-R_e)] \simeq 1 - a(R-R_e)$$
$$\{1-\exp[-a(R-R_e)]\}^2 = a^2(R-R_e)^2 \tag{13.14}$$

For this case a parabolic potential form $V(R)$ results. The restoring force F is proportional to the displacement from the equilibrium separation:

$$V(R) \simeq V_B a^2(R-R_e)^2 - V_B, \quad F = -\frac{dV(R)}{dR} = -2V_B a^2(R-R_e),$$
$$k = 2V_B a^2 \tag{13.15}$$

The potential energy expression in Equations 13.15 is analogous to a harmonic oscillator potential if k is the restoring force constant.

The oscillation frequency ν is determined by the force constant k and the reduced mass M_r of the center of mass. Since the vibrational energy E_v is quantized,

$$E_v = (v+\tfrac{1}{2})\hbar\omega, \quad v = 0, 1, 2, 3, \ldots,$$
$$\nu = \frac{\omega}{2\pi} = \frac{1}{2\pi}\sqrt{k/M_r} = 1/\tau \tag{13.16}$$

where v is the vibrational quantum number and τ is the period of the oscillation. A zero-point vibration exists with energy $E_0 = \frac{1}{2}\hbar\omega$ when $v = 0$. Energy levels are equally spaced ($\hbar\omega$) when oscillations have small amplitude, as indicated on Figure 13.7. Equations 13.16 can be used to estimate vibrational periods for atoms bonded together; they are of the order of 10^{-14} to 10^{-13} s (14, 20).

For large-amplitude, high-energy oscillations, anharmonic motion components are included and the molecular potential does not have a parabolic form. In this situation the potential well becomes wider at the top (14), and the level spacing decreases near the energy zero. Other terms that are proportional to higher powers of $(v + \frac{1}{2})\hbar\omega$ can be added to the energy expression in Equation 13.13 to describe this situation more closely. Vibrational motions with kinetic energy greater than V_B will have sufficiently high amplitude to produce dissociation of the involved chemical bond. This is discussed in Section 13.5.

Molecules composed of several atoms can vibrate in several different modes. A normal vibrational mode is one in which the oscillations of the individual molecules are in phase. Generally all possible oscillations can be analyzed into combinations of normal-mode oscillations. Figure 13.10 shows the three normal modes for the water molecule together with associated energies in the vapor phase (4). In a solution the energies and frequencies listed will be perturbed by the surrounding media (21). For very large molecules with many atomic constituents, some of the normal modes of vibration will involve groups of atoms vibrating more or less independently of the rest of the molecule. Thus, the —OH group has a characteristic vibrational frequency of 1.1×10^{14} Hz, and the $—NH_2$ group has a frequency of 1.0×10^{14} Hz. The characteristic vibrational frequency of a carbon-carbon unit depends upon the number of

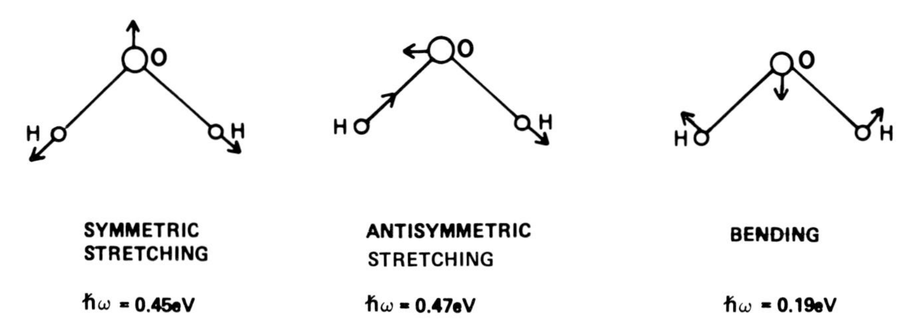

Figure 13.10 The normal modes of vibration of the water molecule are shown, together with vapor-state values for the separation between energy levels (4).

bonds between the C atoms: the —C—C— unit vibrates at about 3.3×10^{13} Hz; >C=C< vibrates at about 5.0×10^{13} Hz; and the —C≡C— group vibrates at about 6.7×10^{13} Hz (4). As one would expect, the greater the number of carbon-carbon bonds, the larger the value of the force constant k and the higher the frequency. However, the vibrational frequencies and energies for these groups are not entirely independent of the rest of the molecule. Group frequencies vary from molecule to molecule.

Nonvalence electron energy levels for a molecule are essentially the same as the levels of the constituent atoms, although the energies may be perturbed by the surrounding electrostatic fields. Bonding and antibonding configurations for valence electrons were mentioned in Section 13.2. In the following we will discuss σ and π molecular orbitals and give some typical electron configurations.

A molecular orbital is a particular electronic configuration that is characteristic of the molecule taken as a whole. Molecular orbitals are usually classified according to the symmetry of the electron configuration about the internuclear axis (taken as z axis). Thus, a σ orbital has cylindrical symmetry about the z axis. π, δ, and ϕ orbitals are invariant to rotation about that axis by angle $2\pi/n$, with π orbitals indicated by $n = 1$, δ orbitals by $n = 2$, and so forth (24). Since σ and π orbitals are most common for low atomic number molecules, we discuss them in some detail.

Table 13.2 indicates that σ orbitals can be formed from paired combina-

Table 13.2 Atomic Orbital Constituents of Molecular Orbitals

σ *Orbitals*	π *Orbitals*
s	p_x
p_z	p_y
sp^n	

$n = 1, 2, 3$

tions of *s*, p_z, and *sp* hybrid atomic orbitals. The π orbitals are formed from p_x and p_y atomic orbitals. Figure 13.11 is an illustration of the electron density configurations for *s*-*s*, p_z-p_z, p_x-p_x, and *sp*-*sp* bonded pairs (14, 17). Other mixed combinations such as *s*-sp^n, *s*-p_z, and p_z-sp^n, and p_z-sp^n, and the like are also possible. The figure also shows the antibonding electron density configurations. Notice that antibonding configurations show minimal electron charge density in the plane separating the atomic centers. The bonding σ orbitals have a substantial localization of charge density on a line between the atomic centers. For the bonding π orbitals, the charge concentration is removed from the line of centers but is nonetheless substantial in the plane separating the atomic centers.

Because of the concentrations of charge shown on Figure 13.11, one might expect the bonding σ orbitals to be located at lower potential energy than the bonding π orbitals. Figure 13.12 shows that this is the case for molecular orbitals in a homogeneous diatomic bonding system. Notice that the antibonding levels are at higher energy than bonding levels. However, pairs of levels are equally spaced above and below the levels for the isolated atoms. Lone-pair electron levels occur at intermediate energies since they do not participate in bonding. Because of the energy level spacing, when corresponding bonding and antibonding states are both filled, there is no net energy change and therefore no net bonding effect (14).

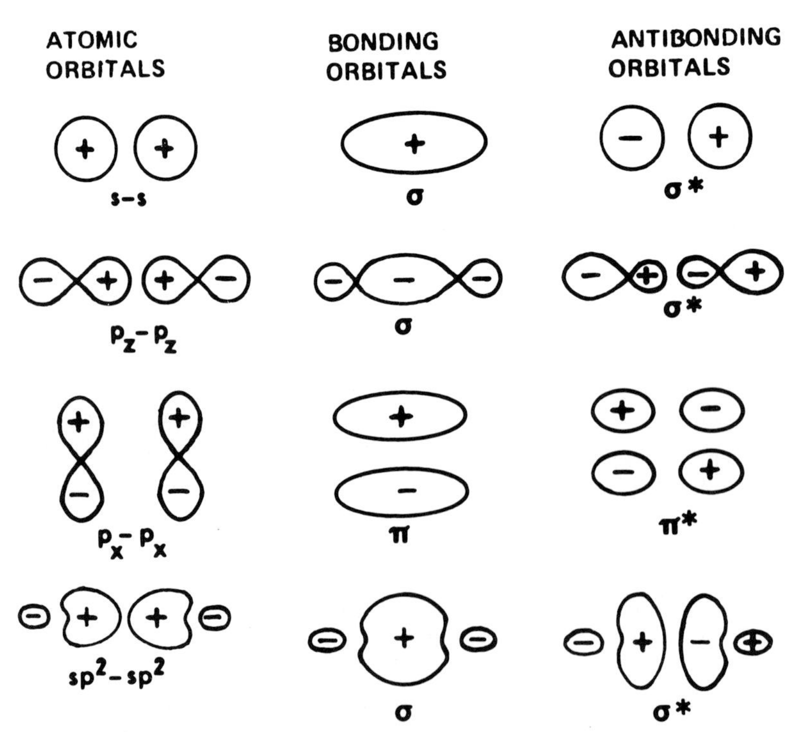

Figure 13.11 Bonding and antibonding orbital charge distributions are illustrated for various configurations. Asterisks indicate antibonding orbitals.

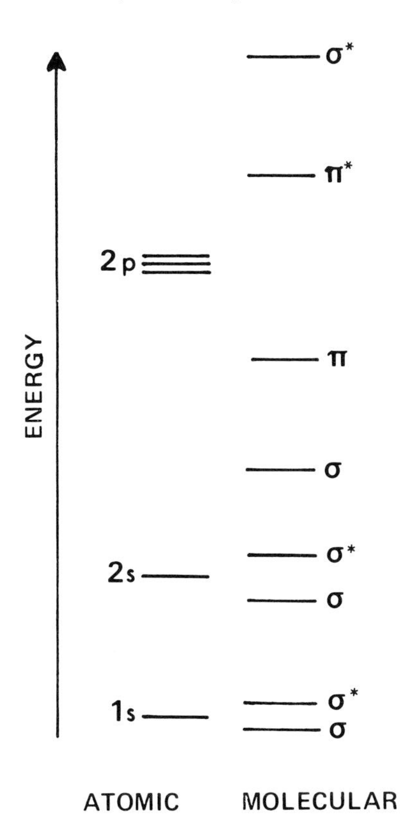

Figure 13.12 The general energy level scheme for σ and π molecular orbitals is illustrated. Asterisks indicate antibonding orbitals.

Figure 13.13 shows the molecular orbital levels for the heterogeneous molecule carbon monoxide. Note that the level sequence is similar to that of Figure 13.12. However, the atomic orbitals are hybridized, and these levels do not occur at the same energies for the dissimilar atoms. We have chosen to indicate antibonding configurations by the asterisk as with σ^* and π^*. Figure 13.14 shows the complete energy level scheme for cytosine, a pyrimidine base in DNA (25). The structural form of the compound is also shown in the figure. The *K*-electron levels are not involved in bonding and are shown at energies of from 300 to 600 eV below the zero of energy. The *L*-shell atomic electrons for oxygen, carbon, and nitrogen participate in π and σ bonds in a variety of configurations and are represented as groups of levels on the figure. The arrows indicate groups of electron transitions that have been identified experimentally. Apparently, $\pi \rightarrow \pi^*$ transitions occur from 5 to 20 eV and above; the $\sigma \rightarrow \sigma^*$ processes have higher energies.

Each of the molecular levels illustrated on Figures 13.12, 13.13, and 13.14 has a capacity for two electrons, one with spin up and another with spin down. For the usual case in molecules, covalent levels are either filled or empty, so the spin quantum number $s = 0$. In excited states the spins may be paired ($s = 0$) with the multiplicity ($2s + 1 = 1$); this state is called a singlet. If the spins are not all paired, the values of ½ add and $s = 1$. This state is called a

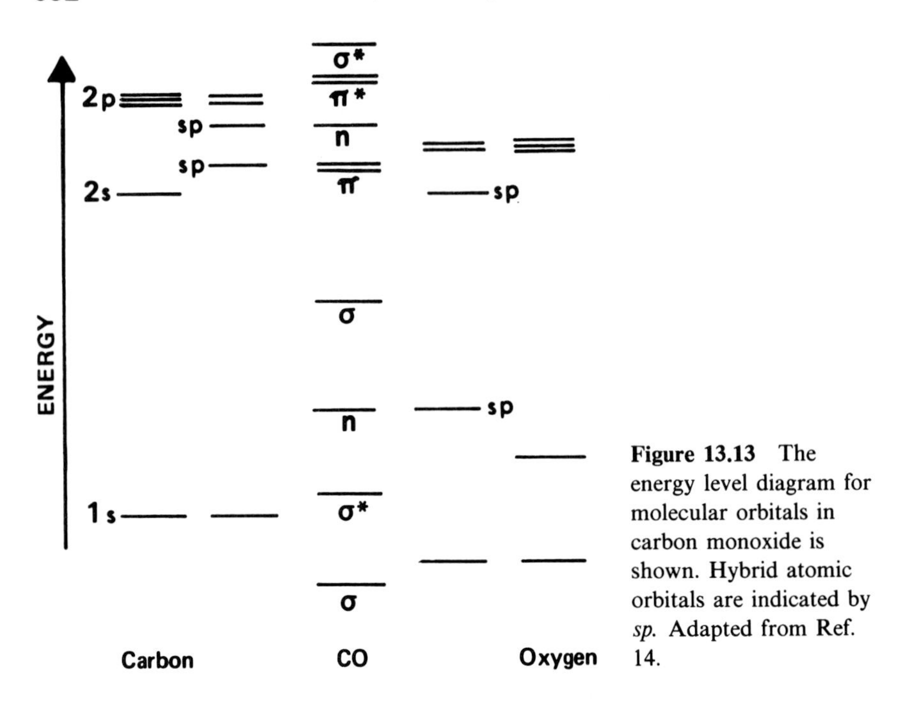

Figure 13.13 The energy level diagram for molecular orbitals in carbon monoxide is shown. Hybrid atomic orbitals are indicated by *sp*. Adapted from Ref. 14.

triplet because of the three possible total spin orientations ($2s + 1 = 3$). Singlet- and triplet-state properties are especially important in the discussion of liquid-state luminescence (26).

13.4 De-excitation Processes

Several interesting processes can be included in the category of molecular de-excitation phenomena following absorption of radiation. Liquid-state luminescence is of considerable importance because of the wide-spread use of liquid scintillation detectors and liquid state fluorescence in research. Bond dissociation processes are of interest because free radicals produced in water solutions are precursors to chemical and biological changes induced by ionizing radiation in living systems. In this section an overview of several common molecular excited-state processes is given, including liquid-state luminescence. In the next section, dissociation and free radical formation are discussed.

Molecular processes that may be involved in a de-excitation sequence are grouped in categories (20):

1. Energy transfer (intermolecular),
2. Luminescence (intramolecular),
3. Nonradiative processes (intramolecular).

Ionic and free radical chemical reactions are not included in this list since they are assigned to a later stage in the sequence of radiation-generated processes.

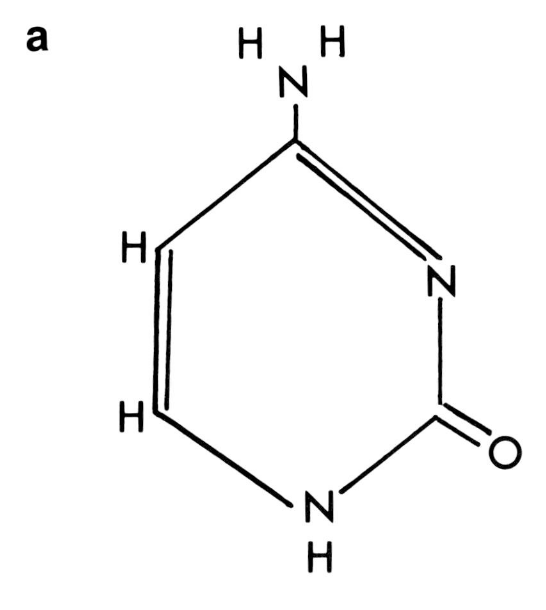

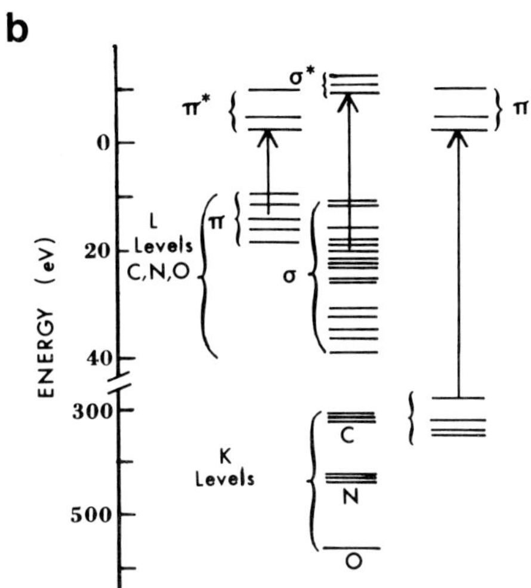

Figure 13.14 The molecular configuration (a) and energy level sequence (b) for cytosine are shown (25). Typical groups of excitations are indicated by vertical arrows.

Energy transfer between neighboring molecules occurs most often in collisions or encounters in the solution where the molecular wave functions overlap. In Section 13.1 it was estimated that this sort of collision may occur as often as 10^{13} times/s in water solutions. These collisions constitute an important mechanism for vibrational relaxation and eventual thermalization of energy. For excess vibrational energy, processes for transfer to translational ($V \rightarrow T$), transfer to rotational ($V \rightarrow R$), and transfer to vibrational ($V \rightarrow V$) states of adjacent molecules have been identified (27).

Several other intermolecular energy transfer processes which do not involve collisions can also be identified. Resonance transfer of a photon from an excited molecule (donor) to an absorbing molecule (acceptor) can occur (20). Energy transfer by virtue of a dipole-dipole interaction with no photon involved is also known (20).

In Chapter 12, luminescence or radiative de-excitation was defined as the emission of electromagnetic energy in excess of that from thermal sources after energy absorption. In liquids, luminescence is an intramolecular process involving an excited state of the emitting molecule. Fluorescence and phosphorescence are the common luminescent subcategories. Actual photon emission occurs in times that are short compared with the time necessary for substantial internuclear motion. This is in agreement with the Franck-Condon principle (20). Of course, the lifetime of the excited states involved can be much longer than internuclear vibrational periods.

The scheme for fluorescence (short-lived luminescence) in a liquid system is illustrated in Figure 13.15 (*left*). The molecular potential energy functions for the electronic ground state and an electronic excited state are shown on the figure. The wells indicate that both are bonding configurations, although the equilibrium internuclear separation distances are different for the two states. The first few vibrational levels are shown in each of the wells and the corresponding singlet states are indicated by S_{nv}, where the quantum number n is shown with values 0 and 1 for the electronic states, and v is the vibrational quantum number shown with values 0, 1, 2, 3 on the figure. An excitation induced by radiation is shown as an arrow pointing upward.

The most common occurrence is an excitation into some state S_{nv}, with $v > 0$. Vibrational relaxation brought on by liquid-state collisions ensues until

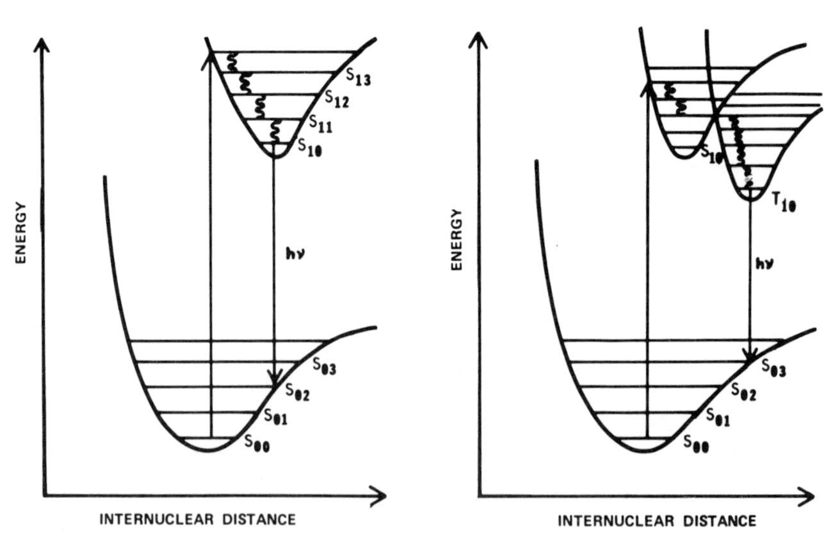

Figure 13.15 The liquid-state transitions appropriate for fluorescence (*left*) and phosphorescence (*right*) are illustrated. Wavy lines indicate vibrational relaxation.

another state S_{n0} is reached. If the transition $S_{n0} \rightarrow S_{0v}$ follows, the photon energy is determined by the energies of the singlet states involved:

$$E_{n0} - E_{0v} = h\nu \quad \text{(singlet-singlet)} \tag{13.17}$$

This sort of singlet-singlet transition is allowed, and the process is called fluorescence (26).

The scheme for phosphorescence or delayed luminescence is also shown in Figure 13.15 (*right*). In this case the excitation is indicated as before, but a system crossing to some triplet state T_{nv} occurs prior to complete vibrational relaxation. Relaxation in the triplet system to state T_{n0} follows. A photon is released in the transition $T_{n0} \rightarrow S_{0v}$ with energy

$$E_{n0} - E_{0v} = h\nu \quad \text{(triplet-singlet)} \tag{13.18}$$

Since triplet-to-singlet spin-flip transition is forbidden, the triplet state is termed metastable. The transition can occur, but the lifetime of the emission is relatively long.

Scintillators are materials that emit electromagnetic radiation in short bursts (1). Liquids can be used as scintillators if the luminescent yield ω_n for excited state n is sufficiently large:

$$\omega_n = \frac{\text{number of characteristic photons emitted}}{\text{number of excited states formed}} \tag{13.19}$$

This quantity is commonly called the fluorescent yield, although phosphorescent processes are not excluded. ω is considerably less than one for the best of scintillators because of competing nonradiative processes. Organic liquids with substantial numbers of π molecular bonds are often used as scintillators. Conjugated carbon-carbon double-bond arrays around a ring structure often de-excite via $\pi^* \rightarrow \pi$ transitions and produce characteristic ultraviolet- or blue-light emissions (26).

Three general types of intramolecular nonradiative processes will be included in the discussions to follow:

1. Internal conversion and intersystem crossing,
2. Pre-ionization,
3. Dissociation and predissociation.

This list is not exhaustive. The third category will be discussed in Section 13.5.

Internal conversion and intersystem crossing involve the exchange of excitation energy from one electron site to another in a molecule (20). These processes can lead to delocalization of the excitation energy in a complex molecule. Both processes involve an isoenergetic crossover from one state to another. Such a process can occur with high probability when electronic wave functions for the states overlap sufficiently (20). Figure 13.15 includes an illustration of a singlet-to-triplet intersystem crossing (*right*). Phosphorescence is regularly associated with intersystem crossing. Internal conversion is an isoenergetic crossover between two singlet states or two triplet states (26).

The theoretical probabilities (28, 29) for ionization and excitation of a water molecule are shown in Figure 13.16. The pronounced peaks are indicative of particular processes and specific states. The region of ionization is clearly indicated in the figure, and excitations are shown below the energy corresponding to ionization potential $E_{IP} = 12.6$ eV (19). Another category of transitions called superexcitations is also indicated. For molecule AB, superexcitations are defined by

$$\begin{aligned} \text{AB} + \text{energy} \rightarrow (\text{AB})^*;\ E_{ex} &\leq E_{IP} \quad \text{excitations,} \\ E_{ex} &> E_{IP} \quad \text{superexcitations} \end{aligned} \tag{13.20}$$

where E_{ex} is excitation energy and the asterisk indicates the excited state. Superexcitations can occur if excitation energy is localized on one atom in the molecule when other easily ionized systems exist at distant sites on the same molecule. They can occur when excitations involve tightly bound inner-shell electrons rather than valence electrons. Pre-ionization is the decay of a superexcited state by radiationless transition into a state in the ionization continuum (20). Thus, when $(AB)^*$ is a superexcited configuration, $(AB)^* \rightarrow (AB)^+ + e^-$. If a relatively modest excitation energy is involved, singly charged ions may be produced. This process is often called the Auger effect if an inner-shell vacancy in the atom is involved.

The Auger effect (30) was discussed in Chapter 10. Since it is in competition with luminescent emission, the Auger yield Y_n for an excited state n can be related to the luminescent yield ω_n by

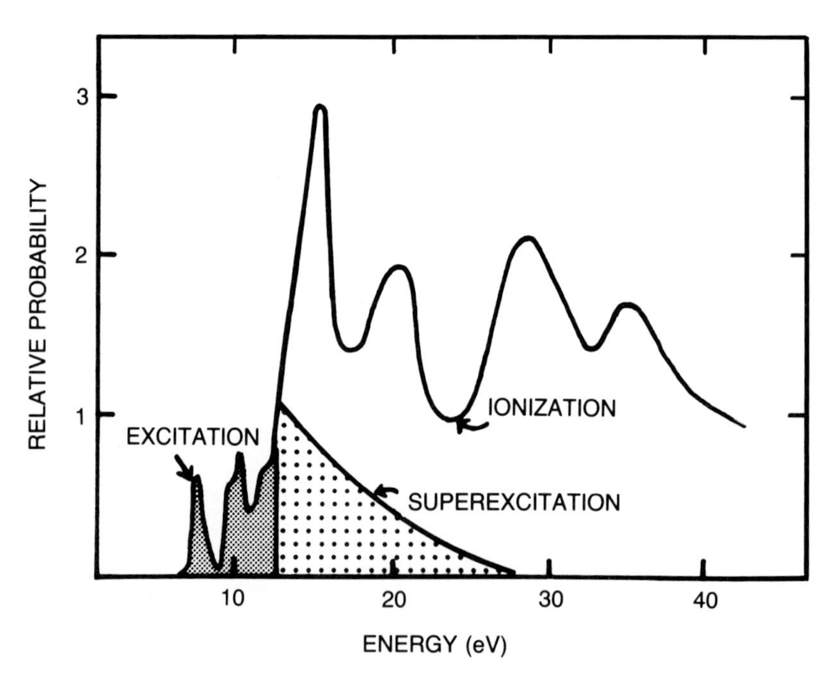

Figure 13.16 The excitation spectrum for water is shown as a function of energy (28).

$$Y_n = 1 - \omega_n \approx \frac{a_n}{a_n + Z^4} \tag{13.21}$$

where a_n is a constant. The analytical form involving the atomic number Z is an empirical approximation (31). For a K-shell vacancy, $a_K = (1.1)10^6$. Because of the Z dependence, the Auger process is expected to dominate for inner-shell vacancies in biological molecules. Since each Auger process leaves two vacancies behind in place of the one present originally, the potential exists for considerable disruption of the bonds in a complex molecule (32).

13.5 Dissociation and Radical Yields

Dissociation is the severing of a chemical bond and the subsequent separation of a molecule into constituent fragments (1). Dissociation is one of several de-excitation processes that can occur after radiation is absorbed. It is often associated with an internuclear vibration of very large amplitude. In this case the restoring force associated with the chemical bond is unable to bring the nuclei back together. Because of this, dissociation may require a time (20) comparable with the period of vibrational oscillations, (10^{-14}–10^{-13} s). The dissociation energy is equal to the depth of the bonding potential well minus the zero-point energy (Figure 13.9). Dissociation energies for representative chemical bonds are shown in Table 1.2. They usually range from 2 to 5 eV/bonding electron pair in organic molecules.

Figure 13.17 shows three types of excitations often followed by molecular dissociation. They are:

a) Excitation into nonbonding or antibonding electronic states;
b) Excitation into a vibrational state with energy above the dissociation limit;

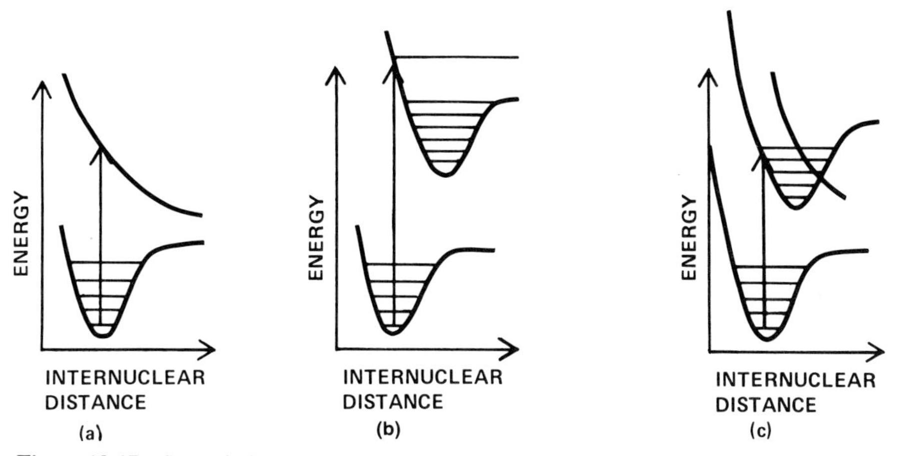

Figure 13.17 Several dissociation processes are shown. a) Excitation into an antibonding state is shown. b) Excitation into a vibrational state with energy above the dissociation limit is shown. c) Excitation followed by crossover into an antibonding state is shown.

c) Excitation into a bound state followed by crossover into a state that allows dissociation.

The first process includes either ionization by removal of a bonding electron or excitation of bonding electrons, e.g. $\sigma \rightarrow \sigma^*$, $\pi \rightarrow \pi^*$, and so forth. If the resulting potential energy diagram is repulsive and no other bonds compensate, separation must result. The second process occurs whenever the vibrational energy exceeds the well depth; such an excitation might be the result of energy redistribution in a molecule or the result of a collision with an excited neighbor. In any case the vibrational amplitude is not bounded, and the internuclear distance will increase without limit if no other dissipative mechanism intervenes.

The third process is called predissociation since the initial system must cross over into a nonbonding or antibonding state or one with a vibrational amplitude without bound. The time for predissociation is larger than the vibration period since the time for the internal conversion or intersystem crossing may be significant.

Free radicals are atomic species or molecular fragments with an unpaired electron (9). They are often indicated by the appropriate chemical symbol but include a single dot on the side. Free radicals are the first products of dissociation and are often involved in chemical reactions induced by radiation. Since they can be very reactive species, free radical reactions in aqueous solution are precursors of many radiation effects.

The molecular effects of radiation can be due to either direct action or indirect action (33). These names do not refer to the type of radiation (photons, electrons, and so forth); rather, they refer to the manner in which the damage was produced. If a molecule in a solution has been damaged by direct action, the molecule was itself ionized or excited by the passing radiation. If the action was indirect, the molecular damage was the result of a chemical transfer of energy to the site via radicals or ions produced in the surrounding solvent.

Since living cells are about 80% water (15), irradiation of cellular systems must result in interactions with water molecules. Thus, the indirect effect of radiation in living systems often involves the ions and free radicals produced in water.

The basic ionization and excitation processes in the water molecule can be written as

$$
\begin{aligned}
&H_2O + \text{energy} \rightarrow H_2O^+ + e^- \qquad \text{(ionization)},\\
&H_2O + \text{energy} \rightarrow H_2O^* \qquad \text{(excitation)}
\end{aligned}
\tag{13.22}
$$

Calculations for the ratio of the number of excitation events divided by the number of ionization events give values of about 0.4 (19).

After an ionization process, the ejected electron is thermalized and captured in a molecular cage with N water molecules:

$$
e^- + N(H_2O) \rightarrow e^-_{aq} \tag{13.23}
$$

The hydrated electron e^-_{aq}, was discussed in Section 13.2. In principle the excited water molecule can dissociate in several ways. The most important for our purposes is free radical dissociation:

$$H_2O^* \rightarrow H\cdot + \cdot OH \tag{13.24}$$

$H\cdot$, $\cdot OH$, and e_{aq}^- are usually considered to be the primary radical species produced in water radiolysis (19). They may recombine on the spot or diffuse through the solution and engage in reactions with molecules and radicals in the vicinity. The average lifetime and diffusion distance for the hydroxyl radical $\cdot OH$ are estimated at about 5×10^{-9} s and 75Å, respectively. The lifetimes of the hydrogen atom $H\cdot$ and the hydrated electron e_{aq}^- are as much as 5×10^{-6} s (19). The numerical values are variable depending on the conditions involved. Some of the free radicals may recombine. Examples of recombination reactions are

$$\cdot OH + \cdot OH \rightarrow H_2O_2,$$

$$H\cdot + H\cdot \rightarrow H_2 \tag{13.25}$$

Of course many other reactions are possible, but the species H_2 and H_2O_2 are considered (19) the primary molecular products of water radiolysis.

An important quantity used to indicate the yield of radiation products is the G value. The quantity G is the number of species N_s produced per 100 eV of energy absorbed. Thus, for a charged particle with energy T dissipated in a molecular absorber,

$$G = \frac{N_s}{T/100} = \frac{100 N_s}{T} \tag{13.26}$$

where T is in electron volts. Some G values for the primary radical and molecular products of water radiolysis are given in Table 13.3 (19). The values shown are averages for ^{60}Co gamma irradiation or irradiation with fast electrons. They are functions of the ionization density of the radiation because of recombination and radical-radical reactions along the track.

Table 13.3 *G* Values for the Primary Radical and Molecular Products of Water

Primary Radical and Molecular Products	*G Value*
e_{eq}^-	2.7
$H\cdot$	0.6
$\cdot OH$	2.7
H_2	0.45
H_2O_2	0.7

13.6 Problems

1. Assume that the Van der Waals potential energy is 0.04 eV for two polar molecules separated by 3Å. If the dipoles are aligned antiparallel, calculate the electric dipole moment d.

2. Calculate the ratio of the Van der Waals force for antiparallel dipoles divided by the force for dipoles oriented at 45°. What is the effect on the magnitude of the attractive force when the separation distance is doubled?
3. Calculate the collision frequency for liquid water molecules at 4°C if the cage diameter $d_c = 6\text{Å}$.
4. Using the law of cosines, calculate the angle between O—H bonds for a perfect tetrahedral configuration.
5. Assume that $-0.5e$ of charge is held on a line halfway between two atoms separated by distance R. Each atom has a net excess charge of $+0.25e$. Is this charge configuration sufficient to bind the atoms together?
6. The carbon monoxide molecule has a length $R = 1.1\text{Å}$. Calculate the reduced mass and the moment of inertia about an axis perpendicular to the line between the atoms and through the center of mass. Calculate the energy in electron volts for the rotational level, with $J = 1$ for this configuration.
7. Assume $L_z = K\hbar (K = 0, \pm 1, \pm 2)$ for a symmetric-top molecule with $I_x = I_y$. Develop an expression for the rotational energy E in terms of J, K, I_x, and I_z.
8. Assume that the H_2 molecule behaves like a harmonic oscillator with $k = 573$ N/m. Find the value of v for the lowest state having sufficient energy for dissociation. The dissociation energy is 4.5 eV. Neglect anharmonic effects and the change in level spacing they produce. Estimate the time elapsed between excitation and dissociation for this molecule.
9. Calculate the photon energy in electron volts for 10-cm^{-1}, 1 000-cm^{-1} and 30 000-cm^{-1} transitions characteristic of rotational, vibrational, and electronic levels, respectively, in molecules.
10. The vibrational period for many molecular modes is about 10^{-13} s. The mean life for a phosphorescent emission may be as large as 10^{-2} s. Does the Franck-Condon principle hold when the luminescent lifetime is much greater than the vibrational period? How is this possible?
11. An energy of 7.4 eV is required to dissociate H_2O into $H\cdot$ and $\cdot OH$. Calculate this value in kilocalories per mole.
12. A ferrous sulfate dosimeter has a Fe^{++} concentration that is 10^{-3} M. Calculate the radiation dose required to deplete the Fe^{++} concentration to $(4)10^{-4}$ M if the G value is 15.5.
13. Using Table 13.3 calculate the number of hydrogen atoms, hydroxyl radicals, and hydrated electrons produced when a 1-MeV electron passes through a cell nucleus of 1-μm diameter.
14. Supposing a radiochemical radical is produced with G value of 4 in a water solution. What is the concentration in moles per liter of the solution produced when 10 Gy are absorbed in the solution? If 10^{-5} moles per liter is the lower limit for identification of the radical chemically, what is the smallest dose one can use in an identification experiment?

13.7 References

1. J. Thewlis, *Concise Dictionary of Physics* (Pergamon, New York, 1973).
2. H. N. V. Temperly, and D. H. Trevena, *Liquids and Their Properties* (John Wiley and Sons, New York, 1978).
3. A. J. Watton and A. G. Woodruff, Contemp. Phys. **10**, 59 (1969).
4. A. Beiser, *Perspectives in Modern Physics* (McGraw-Hill, New York, 1969).
5. R. L. Armstrong and J. D. King, *The Electromagnetic Interaction* (Prentice-Hall, Englewood Cliffs, NJ, 1973), pp. 25–37.
6. N. H. March and M. P. Tosi, *Atomic Dynamics in Liquids* (John Wiley and Sons, New York, 1976).
7. J. E. Lennard-Jones and A. F. Devonshire, Proc. R. Soc. Ser. A **163**, 53 (1937).
8. L. S. Darken and R. W. Gurry, *Physical Chemistry of Metals* (McGraw-Hill, New York, 1953), p. 112.
9. F. Daniels and R. A. Alberty, *Physical Chemistry* (John Wiley and Sons, New York, 1967).
10. A. H. Narten, M. D. Danford, and H. A. Levy, Discuss. Faraday Soc. **43**, 97 (1967).
11. F. Vaslow, "Thermodynamics of solutions of electrolytes," in *Water and Aqueous Solutions: Structure, Thermodynamics and Transport Processes,* edited by R. A. Horne (Wiley-Interscience, New York, 1972), pp. 465–518.
12. M. W. Zemansky, *Heat and Thermodynamics,* 5th ed. (McGraw-Hill, New York, 1957).
13. A. Ben-Naim and F. H. Stillinger, Jr., "Aspects of the statistical mechanical theory of water," in *Water and Aqueous Solutions: Structure, Thermodynamics and Transport Processes,* edited by R. A. Horne (Wiley-Interscience, New York, 1972), pp. 295–330.
14. M. Orchin and H. H. Jaffé, *Symmetry Orbitals and Spectra* (Wiley-Interscience, New York, 1971), p. 298.
15. G. N. Ling, C. Miller, and M. M. Ochsenfeld, "The physical state of solutes and water in living cells according to the association-induction hypothesis," in *Physicochemical State of Ions and Water in Living Tissues and Model Systems,* edited by C. J. Hazlewood, Ann. NY Acad. Sci. **204** (1973), pp. 6–50.
16. F. H. Stillinger, Science **209**, 451 (1980).
17. J. P. Lowe, *Quantum Chemistry* (Academic, New York, 1978).
18. C. M. Davis and J. Jarzynski, "Mixture models of water," in *Water and Aqueous Solutions: Structure, Thermodynamics and Transport Processes,* edited by R. A. Horne (Wiley-Interscience, New York, 1972), pp. 377–424.
19. I. G. Draganic and Z. D. Draganic, *The Radiation Chemistry of Water* (Academic, New York, 1971).
20. L. G. Christophorou, *Atomic and Molecular Radiation Physics* (Wiley-Interscience, New York, 1971).
21. G. S. Kell, "Continuum theories of liquid water," in *Water and Aqueous Solutions: Structure, Thermodynamics and Transport Processes,* edited by R. A. Horne (Wiley-Interscience, New York, 1972), pp. 331–376.
22. P.O.'D. Affenhartz, *Atomic and Molecular Orbital Theory* (McGraw-Hill, New York, 1970), pp. 75–85.
23. P. M. Morse, Phys. Rev. **34**, 57 (1929).
24. M. A. Morrison, T. L. Estle, and N. F. Lane, *Quantum States of Atoms, Molecules and Solids* (Prentice-Hall, Englewood Cliffs, NY, 1976).
25. D. E. Johnson, Radiat. Res. **49**, 63 (1972).

26. J. B. Birks, *Photophysics of Aromatic Molecules* (Wiley-Interscience, New York, 1970).
27. I. W. M. Smith, "Vibrational relaxation in small molecules," in *Molecular Energy Transfer,* edited by R. Levine and J. Jortner (John Wiley and Sons, New York, 1976), pp. 1–15.
28. R. L. Platzman, "Energy spectrum of primary activations in the action of ionizing radiation," in *Radiation Research, 1966,* edited by G. Silini (North-Holland, Amsterdam, 1967), p. 20.
29. A. Skerbele, M. A. Dillion, and E. N. Lassettre, J. Chem. Phys. **49**, 5042 (1968).
30. D. Chattorji, *The Theory of Auger Transitions* (Academic, New York, 1976).
31. W. Bambynek, B. Crasena, R. W. Fink, H. U. Freund, H. Mork, C. D. Surft, R. E. Price, and P. V. Rao, Rev. Mod. Phys. **44**, 714 (1972).
32. T. A. Carlson and R. M. White, J. Chem. Phys. **44**, 4510 (1966).
33. H. L. Andrews, *Radiation Biophysics* (Prentice-Hall, Englewood Cliffs, NJ, 1961), pp. 224–227.

14

Microscopic Energy Distribution and Biological Radiation Damage

14.1 Introduction

In the preceding chapter, excited molecules were mentioned as an immediate result of the absorption of energy from ionizing radiation. Dissociation of chemical bonds was discussed as an important process in de-excitation. In addition, because indirect radiation action is important, the physical properties of water were discussed and G values for many of the free radicals produced in water were given. These topics are forerunners of the consideration of radiation damage in biological molecules. They provide background for the discussion of radiation-induced lesions in the macromolecular targets existing in solution in living cells.

In Section 14.2 some relationships between important macroscopic physical quantities will be established that are helpful in describing radiation effects on biological systems. In brief statements in Chapters 1 and 3, the importance of the absorbed dose and linear energy transfer was mentioned. The dose is often assumed to be continuously variable in a homogeneous medium (1) and can be used as a distribution function. For example, the distribution in absorbed dose as a function of linear energy transfer can be used in an integral definition of the average quality factor. The quality factor and the dose equivalent are useful quantities when the effects of different types of radiation are compared for humans.

On a much diminished scale, a distinguishing feature of ionizing radiation is the concentration of deposited energy along charged-particle tracks (2). Since relatively small amounts of energy are transported to regions between well separated tracks, the energy density is not uniform on a molecular scale unless the track density is very large. Consideration of distributions and regional energy density fluctuations is appropriate in this situation. In Section 14.3, microscopic analogs of the absorbed dose and the linear energy transfer will be defined that can be used conveniently in describing local distributions in energy. The magnitude of radiation damage in localized regions is more easily appreciated with the aid of these distributions.

Since the earliest effects of radiation occur in molecular systems, consideration of radiation damage in living systems usually involves discussions of effects at the molecular level. Damage in the nucleus of the individual cell, particularly damage to the double-stranded DNA, is important for mammalian cells. Cytoplasmic damage may also be of consequence, but the cytoplasm usually participates most directly in repair of nuclear constituents (3). Thus, targets for energy deposition that are the size of the mammalian cell nucleus, about 1–10 μm in diameter (4), should be considered. Even smaller targets the size of the DNA double strand, about 20Å in diameter (5), are also appropriate.

In Sections 14.4 and 14.5, some mathematical models will be presented that describe molecular radiation damage and cellular damage in terms of chemical bonds that have been severed in sensitive target molecules. The models are presented because they are useful for introducing concepts. They serve as background for more complete theories of cell survival and related topics in radiation biology. They are not necessarily the most realistic descriptions available.

14.2 Biological Effects and the Dose Equivalent

In Chapter 1 the absorbed dose was identified as the physical parameter most widely used for relating the amount of radiation delivered to its effect on a system. It should be emphasized that the dose is not entirely sufficient for this purpose. In particular, it is well known that one type of radiation may produce more pronounced effects than another, although the absorbed dose and time interval for delivery are the same for both (3, 6). This fact has led to the use of other quantities which relate closely to radiation effects, particularly for biological systems.

The relative biological effectiveness (RBE) is often applied in experimental situations when living systems or biologically active molecules are exposed to different types of radiation (3). The standard radiation used for comparison is usually taken to be x rays from a generator operating with a peak of 200 kV of potential (200 kVp) between the filament and target. In an experimental determination, the dose of test radiation, D_t, must be found that will produce the same biological end point as the dose for 200-kVp x rays, D_x. Under these circumstances,

$$\mathrm{RBE} = \frac{D_x}{D_t}, \quad D_x = \mathrm{RBE}(D_t) \tag{14.1}$$

The definition of RBE is quite general since the only constraint is that both doses produce the same end point, and the particular end point is not specified. In fact, the RBE may take on several different values in a given experiment if several different end points are considered. For example, the RBE for an *in vitro* cellular irradiation might be quite different if the end point is changed from death for 50% of the sample cells to death for 95% of the sample cells.

For mammalian cells in culture, the RBE is usually greater than one for densely ionizing radiation. This indicates that the effective damage per gray is greater for densely ionizing radiation than for the sparsely ionizing secondary electrons from 200-kVp x rays. Figure 14.1 is a plot of RBE versus linear energy transfer L for several systems (7–9). The usual shape of the RBE-versus-L curve for mammalian cells rises from unity at low linear energy transfer, peaks somewhere in the neighborhood of 100 keV/μm, and falls at higher L values.

A rationale given to explain the increase in RBE with increasing linear energy transfer involves the energy transferred when directly ionizing radiation penetrates through a sensitive site in a cell (10). For sparsely ionizing electrons, the energy deposited in the local site may not be sufficient to cause severe damage; several traversals by several electrons may be required to cause substantive biological effects. A densely ionizing particle striking the same sensitive volume is more likely to cause severe damage in a single traversal; thus, the RBE may be increased in the latter case. Of course, when the linear energy transfer is sufficiently large, each traversal leads to severe damage and limited function. A further increase in energy deposition in the site may only lower the overall efficiency of the radiation for production of damage. The extra deposited energy will have no noticeable effect and the RBE will decrease.

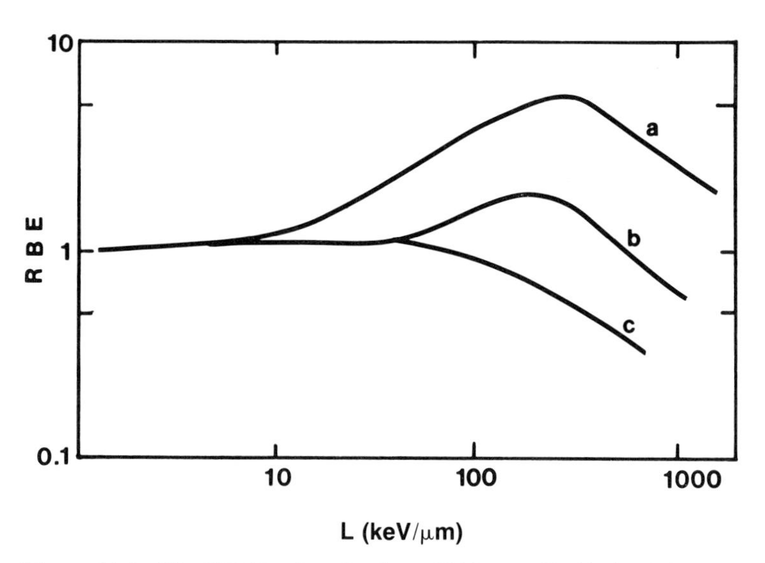

Figure 14.1 The RBE for inactivation of kidney cells (a), haploid yeast cells (b), and bacteria (c) is shown as a function of the linear energy transfer (7–9). Adapted from Ref. 6.

Table 14.1 RBE for Inhibition of Cell Cloning

Type of Neutrons	$\langle T_n \rangle$(*MeV*)	*S = 0.80*[a]	*S = 0.01*[b]
Tritium [T(*d,n*)]	15	2.9	1.6
Beryllium [Be(*d,n*)]	8	3.6	1.8
Deuterium [D(*d,n*)]	3	5.0	2.0
Uranium fission	1	6.1	2.3

Data from Ref. 11.
$\langle T_n \rangle$, average neutron kinetic energy.
[a] Surviving fraction, 80%.
[b] Surviving fraction, 1%.

The RBE curve for bacteria (7) shown on Figure 14.1 does not show a peak and should serve as a reminder that the effect of radiation depends on the system tested. Table 14.1 shows RBE results for prevention of cell cloning, which was studied as a function of the absorbed dose (11). Note that the RBE taken when 80% of the cells retain their ability to form clones in culture (S = 0.8) is quite different than the RBE for only 1% cloning (S = 0.01). In addition to dose, the RBE for a given test system may depend on whether

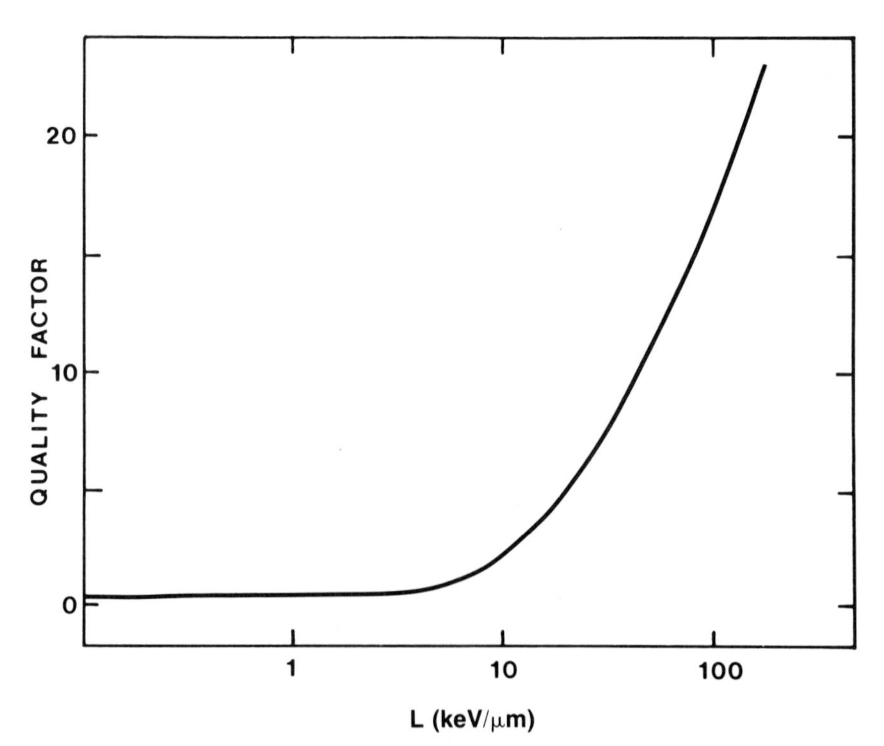

Figure 14.2 The quality factor is shown as a function of the linear energy transfer (6).

the irradiation is fractionated, allowing for repair of the damage accumulated, or whether it is given in a single treatment. It may depend on whether oxygenated or hypoxic conditions prevail, and so forth.

The quality factor Q is a quantity similar in some respects to RBE but much less general. It is used in the context of radiation protection (1). The quality factor is a weighting factor for biological effectiveness that depends on the type of radiation used to deliver the dose. It has been evaluated with the aid of experimental RBE results, but is independent of a definite biological end point. It is not directly related to any experiment but has been chosen (6) to be a smooth function of linear energy transfer by the International Commission on Radiological Protection (ICRP). The quality factor is appropriate for low-level irradiation of humans.

The dependence of the quality factor on linear energy transfer is illustrated on Figure 14.2. Similarities with curve *a* in Figure 14.1 are evident except at large values of linear energy transfer. Values of the quality factor recommended for some general radiation categories (12) are shown on Table 14.2. They can be used in radiation protection problems where more detailed linear energy transfer information is not readily available.

The quality factor is a modifying factor used in calculation of the dose equivalent. The dose equivalent H is the product of D, Q, and N at a point of interest in tissue, where D is the absorbed dose, Q is the quality factor, and N is the product of all other dose-modifying factors (1):

$$H = DQN \tag{14.2}$$

For example, N might depend on spatial dose distribution in an organ or rate of delivery of the dose.

Since Q and N are dimensionless, the appropriate units for dose equivalent are joules per kilogram; the name "Sievert" is used with symbol Sv. A special unit of dose equivalent with historical significance is the rem (roentgen equivalent man):

$$1\ \text{Sv} = 1\ \text{J/kg}, \qquad 1\ \text{rem} = (1.0)10^{-2}\ \text{J/kg} \tag{14.3}$$

Generally speaking, the charged-particle energy spectrum is inhomogeneous in an irradiated medium. This is assured because of accumulation of secondary charged particles and degradation of both primaries and secondaries. The

Table 14.2 Practical Quality Factors

Type of Radiation	*Suggested* Q
X rays, gamma rays	1
Electrons, positrons	1
Neutrons: $T_n < 10$ keV	3
Neutrons: $T_n \geq 10$ keV	10
Fission fragments	20

Data from Ref. 12.
T_n, kinetic energy of neutrons.

mathematical formalism presented in Chapter 11 can be used to calculate electron spectra under equilibrium conditions.

For a given type of charged particle and a given absorbing medium, the particle kinetic energy is sufficient to determine its linear energy transfer for a given cutoff energy and vice versa. Because of this one-for-one correspondence, the linear energy transfer can be used as an independent variable instead of the kinetic energy. The linear energy transfer in water is a convenient variable for discussions involving the biological effects of radiation.

Figure 14.3 shows several examples of the distribution in absorbed dose expressed as a function of linear energy transfer (13). Notice that the dose deposited by ^{60}Co photon beams in water occurs in the linear energy transfer range between 0.2 and 0.4 keV/μm for the most part. The range of linear energy transfer for electron secondaries from 200-kVp x rays is quite broad (0.5–30 keV/μm) and encompasses higher L values than the ^{60}Co distribution. The neutron dose is deposited at the highest L values shown (0.8–90 keV/μm), since the secondaries are densely ionizing heavy particles.

Since absorbed dose is deposited over a range of linear energy transfer values, and since the relative biological effectiveness and the quality factor both depend on linear energy transfer, appropriate averages must be defined for use with Equations 14.1 and 14.2. Suppose $[dD(L)/dL]dL$ represents the dose delivered with linear energy transfer between L and $L + dL$. The quantity $(1/D)[dD(L)/dL]$ is a probability distribution for absorbed dose since

$$\frac{1}{D}\int_0^\infty \frac{dD(L)}{dL}\, dL = 1.0 \tag{14.4}$$

Dose-averaged values of linear energy transfer, $\langle L\rangle$, and of relative biological effectiveness, $\langle \text{RBE}\rangle$, are given by

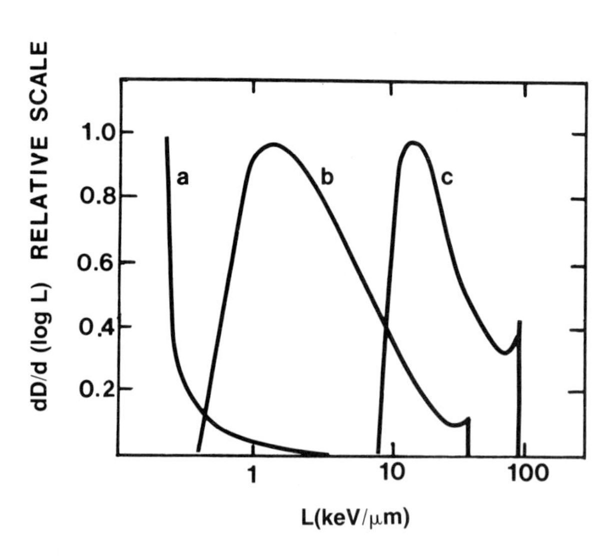

Figure 14.3 Several dose distributions as functions of linear energy transfer are shown (13). The curves are for ^{60}Co gamma rays (a), 200-kVp x rays (b), and 5-MeV neutrons (c). Adapted from Ref. 10.

$$\langle L\rangle = \frac{1}{D}\int_0^\infty L\,\frac{dD(L)}{dL}\,dL, \quad \langle \text{RBE}\rangle = \frac{1}{D}\int_0^\infty \text{RBE}(L)\,\frac{dD(L)}{dL}\,dL \tag{14.5}$$

The average quality factor $\langle Q\rangle$ and the dose equivalent H can be found from

$$\langle Q\rangle = \frac{1}{D}\int_0^\infty Q(L)\,\frac{dD(L)}{dL}\,dL, \quad H = D\langle Q\rangle N \tag{14.6}$$

When the distributions are known, these calculated quantities can be used rather than estimates such as those in Table 14.2.

14.3 Specific Energy and Lineal Energy

The macroscopic variables absorbed dose and linear energy transfer, as commonly used, are averages over spatial regions. As such they are adequate when radiation effects are considered on human organs or on living species consisting of large numbers of cells. They are less relevant when radiation damage in individual mammalian cells or in macromolecules is considered. The averaging process may very well camouflage the most important information. Fluctuations in energy density between neighboring cells are relevant because during irradiation some cells sustain lethal damage while others survive the effects of the radiation unscathed. For a similar reason, fortuitous spatial distributions of ionization and excitation processes are important in the dissociation of chemical bonds in macromolecules. The specific energy z and the lineal energy y are microscopic quantities that are analogs of the dose and linear energy transfer, respectively, and relate more easily to energy deposition on this smaller scale (1).

The specific energy z is the quotient of ΔE by Δm, where ΔE is the energy imparted to matter of mass Δm by ionizing radiation:

$$z = \Delta E/\Delta m \tag{14.7}$$

The unit for specific energy is the joule per kilogram, also called the gray (Gy). The rad, which is equal to $\frac{1}{100}$ gray, also can be used in this context, since the specific energy has the same units as absorbed dose (1). However the specific energy is a stochastic quantity, and fluctuations are important, while the absorbed dose is an average value, and is usually assumed to vary continuously from place to place in a medium.

The relationship between the specific energy and the absorbed dose is illustrated in Figure 14.4. The energy density $\Delta E/\Delta m$ for a collection of volumes with equal mass Δm is shown as a function of the magnitude of Δm (14). The radiation field utilized is assumed to be uniform on a macroscopic scale.

When Δm is sufficiently large, many charged-particle tracks pass through each indicated volume, and the fluctuation level from volume to volume is small. All volumes in a homogeneous radiation field sustain an energy density very close to the average energy density. In this region of large Δm, the absorbed dose D is appropriate.

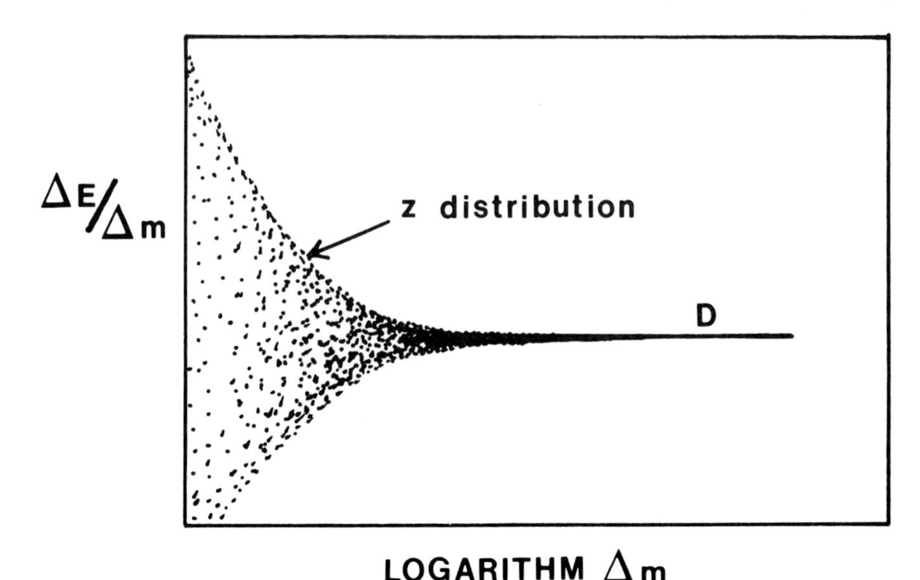

Figure 14.4 The energy density for a large number of masses Δm is shown as a function of the logarithm of Δm. Adapted from Ref. 14.

As the mass Δm decreases, the fluctuations become more significant. When the enclosed volumes are sufficiently small, some absorb no energy at all, whereas the energy density for others is large. When the distribution in deposited energy is important, the specific energy z is useful.

The relationship between z and D involves an averaging process (1). If $P(z)dz$ is the probability distribution for occurrence of specific energy between z and $z + dz$, and if $\langle z \rangle$ is the average specific energy,

$$\langle z \rangle = \int_0^\infty zP(z)dz, \quad D = \lim_{\Delta m \to 0} \langle z \rangle = \frac{dE}{dm} \tag{14.8}$$

The lineal energy is the microscopic quantity analogous to the linear energy transfer. The lineal energy y is the quotient of ΔE by $\langle x \rangle$ where ΔE is the energy imparted to the matter in a specified volume by an energy deposition event and $\langle x \rangle$ is the mean chord length for that volume:

$$y = \frac{\Delta E}{\langle x \rangle} \tag{14.9}$$

The unit used for this quantity is the joule per meter, although in some cases electron volts per centimeter or kiloelectron volts per micrometer are convenient:

$$1 \text{ eV/cm} = (1.6)10^{-17} \text{ J/m}, \quad 1 \text{ keV}/\mu\text{m} = (1.6)10^{-10} \text{ J/m} \tag{14.10}$$

The diagrams shown in Figure 14.5 can be used to develop an expression for the probability distribution in z or y for a spherical volume (14). Suppose a charged-particle track with chord length between x and $x + dx$ cuts through

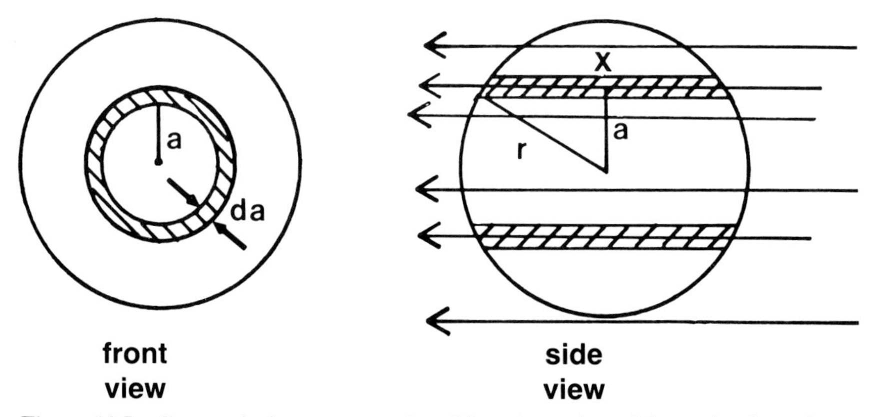

Figure 14.5 Geometrical parameters describing charged-particle tracks through a spherical target are illustrated. Adapted from Ref. 14.

a sphere with radius r. The path of the track is within the open-ended cylindrical shell of radius a (shaded on the figure). Assume that

1. The charged particle tracks are always straight;
2. All charged particles incident on the sphere pass entirely through the sphere;
3. The rate of loss of energy is constant in the sphere;
4. All energy liberated along the track is absorbed on the spot.

The probability $P(x)dx$ of charged particles striking the sphere and passing through the cylindrical shell with radius between a and $a + da$ is equal to the ratio of the area of the end of the shell to the cross-sectional area of the sphere (14):

$$P(x)dx = \frac{|2\pi a da|}{\pi r^2} \tag{14.11}$$

The probability is related to the chord length x by:

$$P(x)dx = \frac{|2ada|}{r^2} = \frac{xdx}{2r^2} \tag{14.12}$$

This result was obtained for a different purpose in Section 13.2.

Suppose N is the total number of particles incident on the sphere with linear energy transfer L. Then the expression for the energy lost in the sphere for tracks with lengths between x and $x + dx$ is

$$dE_x = N\,P(x)dx(xL) = \frac{NL}{2r^2}\,x^2dx \tag{14.13}$$

The average value of the total energy lost for all N particles is given by

$$E = \int_0^{2r} dE_x = \left[\frac{NLx^3}{6r^2}\right]_0^{2r} = \frac{4NLr}{3} \tag{14.14}$$

The probability distribution $E(x)dx$ for the energy loss when the path length is between x and $x + dx$ can be found using Equations 14.13 and 14.14:

$$E(x)dx = \frac{dE_x}{E} = \frac{3x^2dx}{8r^3} \tag{14.15}$$

The probability distribution $P(z)dz$ for specific energy between z and $z + dz$ is equal to $E(x)dx$ since x determines z for the sphere:

$$z = \Delta E/\Delta m = xL/\Delta m, \quad dz = (L/\Delta m)dx, \quad z_{max} = 2rL/\Delta m \tag{14.16}$$

and thus

$$P(z)dz = \frac{3x^2dx}{8r^3} = \frac{3z^2dz}{8r^3(L/\Delta m)^3} = \frac{3z^2dz}{(z_{max})^3} \tag{14.17}$$

The distribution $P(z)dz$ for specific energy in a spherical absorber is shown on Figure 14.6. The parabolic function shown as the solid line is plotted from Equation 14.17 and applies only to monoenergetic particles when conditions 1 through 4 hold, as stated earlier. The value of z_{max} shown on the figure is the theoretical upper limit of z obtained for track lengths equal to the spherical diameter. The distribution shown as the dashed line on the figure is also for monoenergetic particles but simulates results from nonideal systems. The edges and end points are rounded because the particle tracks do not necessarily follow

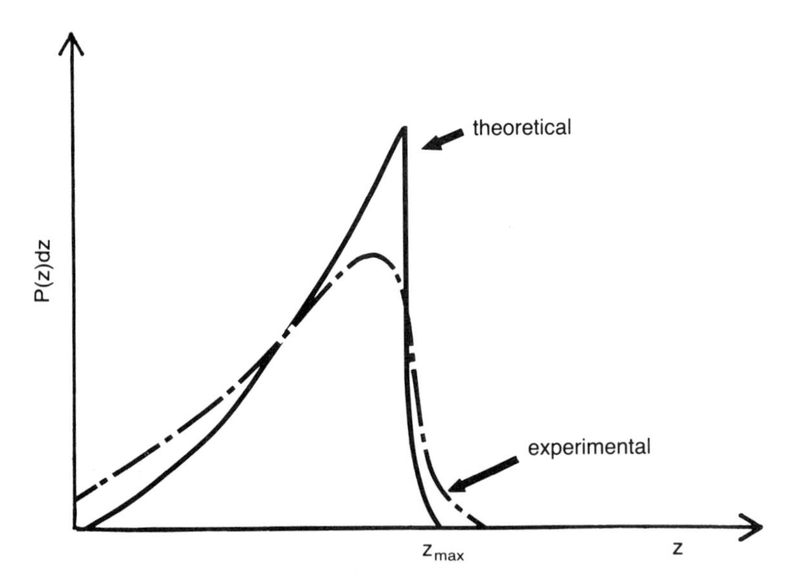

Figure 14.6 The distribution in specific energy is shown for passage of a single charged particle through a spherical absorber. The solid curve is the theoretical result of Equation 14.17. The dashed curve indicates the additional effects of delta rays, energy loss fluctuations, and path-length fluctuations.

straight lines. Furthermore, delta rays and energy loss fluctuations may produce spurious contributions for small spheres.

Actual z distributions from indirectly ionizing sources are composites with contributions from charged particles with different stopping power values. Of course, the high L particles produce high z events and vice versa.

The curves shown on Figure 14.7 are illustrative of specific energy probability distributions produced by the passage of charged particles through tissue spheres with 1-μm diameter (14, 15). In each case the dose was only a small fraction of a gray, so effects of traversals by single particles are indicated. For the distribution labeled ^{60}Co, the charged particles are secondary electrons; for the distribution labeled 1-MeV neutrons, liberated protons are very important. Both distributions show the effect of a spectrum of linear energy transfer values for the secondary particles. They are both rather broad and skewed (when compared to curves on Figure 14.6). It is of primary importance that the distribution produced by neutrons occurs at values of z over 10^2 larger than the distribution produced by ^{60}Co gamma rays. This is because proton secondaries are much more densely ionizing than electron secondaries.

Figure 14.8 shows four number spectra for 1-μm diameter spheres (14).

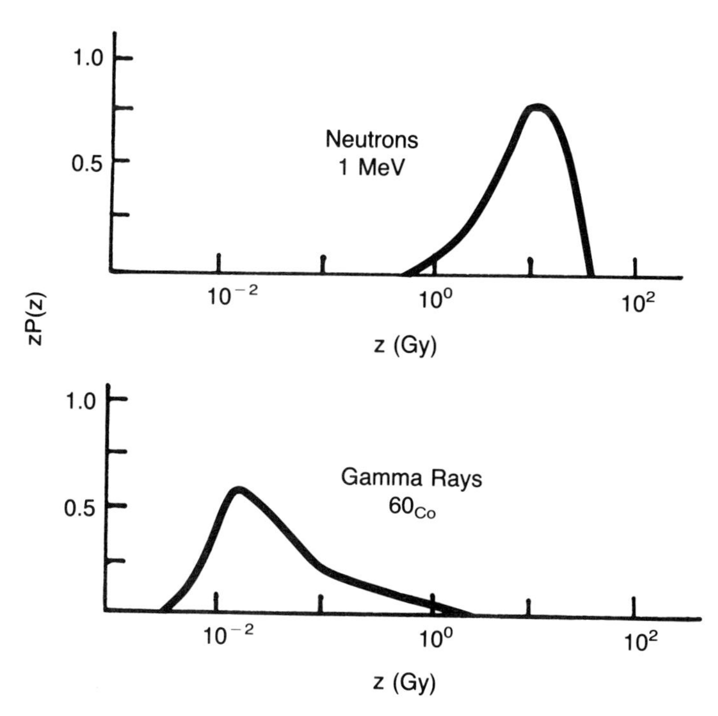

Figure 14.7 The probability distributions for specific energy per unit logarithm interval [$P(z)dz/d(\log z) = zP(z)$] are shown for ^{60}Co x rays and 1-MeV neutrons for single particle traversals through a 1-μm tissue sphere (14).

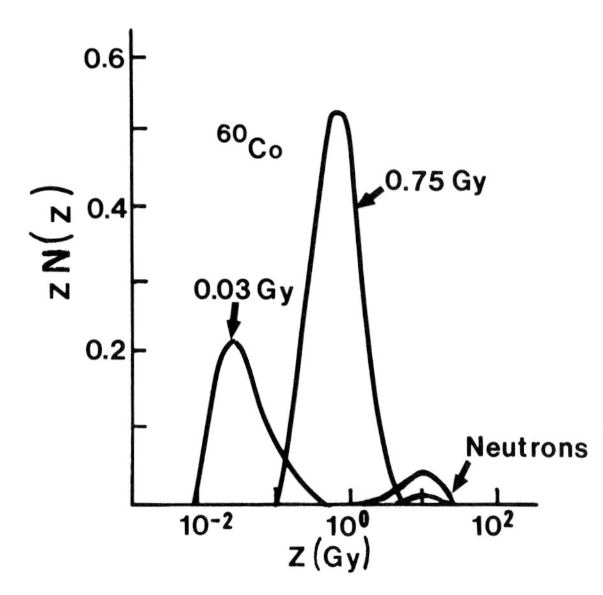

Figure 14.8 The relative number of spheres with specific energy z per unit logarithmic interval $[N(z)dz/d(\log z) = zN(z)]$ is shown for 1-μm tissue spheres. ^{60}Co gamma rays and 1-MeV neutrons are the primary radiations. Distributions are shown for doses of 0.03 and 0.75 Gy for each type of radiation. Adapted from Ref. 14.

Two curves for ^{60}Co gamma rays and two for 1-MeV neutrons are shown (0.03 Gy and 0.75 Gy for each radiation type). The peak areas are not normalized on this figure, as was the case on Figure 14.7. At the doses chosen, it is evident that the number of spheres affected by neutron irradiation is far less than the number affected by ^{60}Co irradiation at the same dose level. The heavily ionizing radiation produces fewer tracks per unit of dose.

Notice that the curve labeled ^{60}Co shifts to larger values of z and is more nearly symmetrical at the larger value of absorbed dose. This is due to multiple hits on the same 1-μm sphere by different secondary electrons. The 0.03-Gy (smaller) curve labeled 1-MeV neutrons is not shifted much to larger z when the dose is increased to 0.75 Gy; rather the height of the spectrum is increased uniformly. This indicates that multiple hits for the 1-μm targets are relatively infrequent for proton secondaries at the dose levels indicated.

14.4 Target Theory

In the preceding section, the specific energy and the lineal energy were defined and some representative distributions were shown for microscopic volumes. At modest values of absorbed dose, a high level of fluctuation in specific energy exists for volumes the size of a cell nucleus and smaller. Since traversal of a volume by a charged particle is necessary for energy deposition, the entire process is loosely analogous to a random hit being scored on a target. The terms "target" and "hit" are often used in discussions of the microscopic effects of radiation. In particular the volume occupied by a specific molecule can be considered to be a target. A hit deposits energy and thereby affects the subsequent activity or function of the molecule. Statistical theories involving hits by charged particles have been applied to explain experimental results on radiation inactivation of enzymes and antigens (16, 17). They also have been evoked to aid in the understanding of the action of radiation on cellular systems (17–19).

Ionizing radiation can act directly or indirectly at the microscopic level (3). These terms were discussed in Section 13.5. In the case of direct action, chemical changes are elicited in a target because a charged particle has passed directly through the sensitive molecular volume and deposited energy there. Indirect action occurs when energy is initially deposited outside the sensitive volume, often in a group of solvent molecules, for example. Free radicals from the deposition site then migrate to the sensitive volume and perturb the chemical structure of the molecule. Target theory is most easily understood when direct action dominates. In that case, hits outside of the actual sensitive target volume can be considered ineffective. Dry samples are often used in experimental tests of target theory.

In this section, charged-particle energy deposition will be treated as a random event. Elementary statistical principles will be applied to obtain the probability of a hit at a particular molecular site. A similar mathematical development was first applied to estimate the volumes of macromolecules in the dry state (16, 17).

Suppose a fluence Φ of charged particles is incident on a uniform dry sample consisting of N active macromolecules. Suppose that a single effective hit in a molecule of cross-sectional area σ is sufficient to destroy its potential for subsequent biological activity. However, some charged-particle traversals through the molecule are not effective because they fail to deposit sufficient energy. Call P the probability of effectiveness per traversal. The infinitesimal increase in damaged targets, $-dN$, is related directly to the infinitesimal fluence increment $d\Phi$:

$$-dN = -NP\sigma\, d\Phi, \quad \ln N = -P\sigma\Phi + C,$$

$$N = \exp(-P\sigma\Phi)\exp(C), \quad N = N_0 \exp(-P\sigma\Phi) \tag{14.18}$$

The constant of integration, C, was evaluated in Equations 14.18 using the boundary condition that N_0 was the number of active targets present when $\Phi = 0$. An expression for P must be obtained to develop the exponential in Equations 14.18 further.

The binomial distribution (20, 21) is generally applicable to random events. It can be used to obtain an expression for the probability of effectiveness when a charged-particle traverses a molecule. Suppose that

1. p is the probability that an event occurs in a single trial;
2. $q = 1 - p$ is the nonoccurrence probability per trial;
3. k is the total number of trials involved;
4. j is a running index representing the number of events that occur in k trials;
5. P_j is the probability that some number of events j occur in k trials.

The binomial expansion of $(p + q)^k$ yields

$$(p+q)^k = p^k + \frac{kp^{k-1}}{1!}q + \frac{k(k-1)p^{k-2}}{2!}q^2 + \ldots,$$

$$1 = P_k + P_{k-1} + P_{k-2} + \ldots \tag{14.19}$$

where each term in the expansion is identified with a particular P_j. In the first term to the right of the equal sign, the running index $j = k$; in the second term, $j = k - 1$, and so forth. Of course, the total probability that some number of events occur must be equal to one.

The general expression for the probability of j events occurring in k random trials can be identified from Equations 14.19:

$$P_j = \frac{k!}{(k-j)!}\left[\frac{p^j}{j!}\right] q^{k-j} = \frac{k!}{(k-j)!}\left[\frac{p^j}{j!}\right](1-p)^{k-j} \tag{14.20}$$

This expression is applicable without specific conditions on j and k except that they be positive integers or zero.

The Poisson distribution can be obtained as a special case of Equation 14.20 when $p \ll 1$ and $k \gg 1$. It is relatively easy to use since it involves the mean value m of the number of events. Since $p \ll 1$ and $j \ll k$,

$$(1-p)^{k-j} \simeq [\exp(-p)]^{k-j} \simeq \exp(-kp),$$

$$\frac{k!}{(k-j)!j!} = \frac{k(k-1)\ldots(k-j)\ldots(2)(1)}{j!(k-j)(k-j-1)\ldots(2)(1)}$$

$$= \frac{k(k-1)\ldots(k-j+1)}{j!} \simeq \frac{k^j}{j!} \tag{14.21}$$

Substituting results from Equations 14.21 into Equation 14.20 yields

$$\frac{k!}{(k-j)!}\left[\frac{p^j}{j!}\right](1-p)^{k-j} \simeq \frac{k^j}{j!}p^j\exp(-kp) = \frac{(kp)^j}{j!}\exp(-kp) \tag{14.22}$$

The mean value of the number of events is given by $m = kp$. With this substitution Equation 14.22 becomes the Poisson distribution for the probability that j events occur when the average is m:

$$P_j = \frac{m^j}{j!}\exp(-m), \quad k \gg 1, \quad p \ll 1 \tag{14.23}$$

The Poisson distribution as given in Equation 14.23 can be substituted into Equations 14.18 to give the number of targets that have not been damaged by radiation. Take $\langle x \rangle$ to be the mean path length of a charged particle in a target molecule and E_e to be the average energy expended for an effective deactivation event. The probability for no effective hits from a single traversal is found by setting $j = 0$ in Equation 14.23: $P_0 = \frac{m^0}{0!}\exp(-m) = \exp(-m)$. The probability of one or more effective hits is equal to one minus the probability of no effective hits:

$$P_{1\to\infty} = 1 - P_0 = 1 - \exp(-m) = [1 - \exp(L\langle x\rangle/E_e)] \tag{14.24}$$

In this case $m = L\langle x\rangle/E_e$, where L is the linear energy transfer.

We assume targets are very small relative to the distance between ionization and excitation events: $L\langle x\rangle/E_e \ll 1$. Then Equations 14.18 and 14.24 yield

$$1 - \exp(-L\langle x\rangle/E_e) \simeq L\langle x\rangle/E_e = P$$

$$N/N_0 = \exp(-P\sigma\Phi) = \exp[-(L\langle x\rangle/E_e)\sigma\Phi]$$

$$= \exp[-(L/E_e)V_e\Phi] \tag{14.25}$$

where the effective target volume is taken as $V_e = \langle x\rangle\sigma$. When a large number of charged particles are incident on the sample, the average absorbed dose is given by $D = L\Phi/\rho$:

$$\exp[-(L/E_e)V_e\Phi] = \exp[-(\rho V_e/E_e)D] = \exp(-D/D_0) \tag{14.26}$$

$D_0 = E_e/\rho V_e$ is the mean inactivation dose. The fraction of targets not inactivated is called the surviving fraction (S). From Equations 14.25 and 14.26,

$$S = N/N_0 = \exp(-D/D_0) \tag{14.27}$$

Figure 14.9 shows several curves of surviving fraction versus absorbed dose. The exponential of Equation 14.27 is represented by the curve labeled H1T1 (single hit, single target) on the semilogarithmic plot. In some cases, the fractions of enzyme and antigen molecules that escape inactivation after irradiation are seen to fall along exponential curves (16). These results reinforce the assumption that a single hit by radiation can inactivate a macromolecule.

Single-hit target theory has been used to estimate molecular weights for biologically active macromolecules (16) through equations similar to Equations 14.26 and 14.27. The value of D_0 can be found from the experimental surviving fraction data: If E_e can be estimated, the product ρV_e can be calcualted; if ρ is measured, the molecular weight can be obtained from V_e. Results for macromolecular weights obtained by applying this theory often agree with accepted values within an order of magnitude.

The curves on Figure 14.9 labeled H2T1 (double hit, single target) and H1T2 (single hit, double target) both show a low-dose shoulder but become more nearly exponential at high-dose values. In the following paragraphs, we will obtain analytical expressions that describe both of these curves.

Suppose n effective random hits are required on a single sensitive target to cause sufficient damage. The probability of less than n hits occurring is the sum of the n probability terms representing all cases from 0 hits to $n - 1$ hits. From Equation 14.23 this sum can be written

$$P_{0\to n-1} = \left[\frac{m^0}{0!} + \frac{m^1}{1!} + \ldots \frac{m^{n-1}}{(n-1)!}\right]\exp(-m)$$

$$= \sum_{j=0}^{n-1} \frac{m^j}{j!}\exp(-m) \tag{14.28}$$

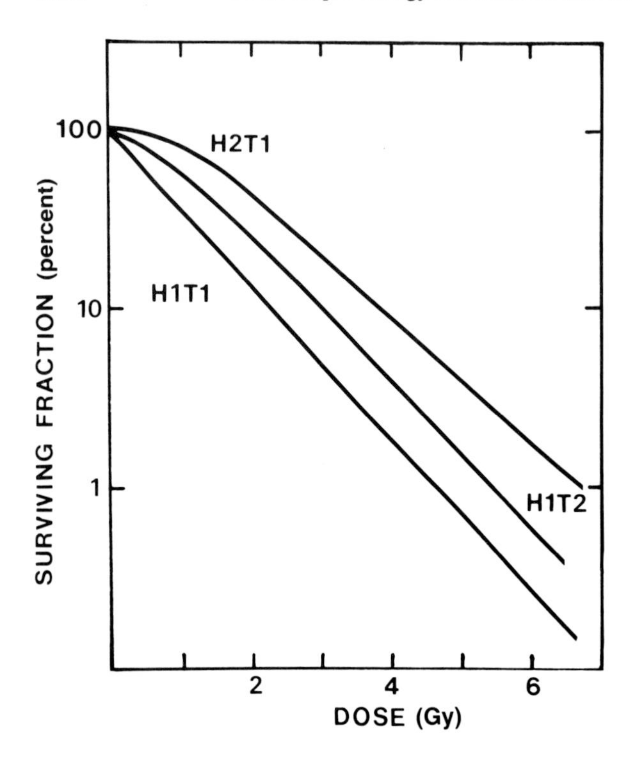

Figure 14.9 Curves of surviving fractions for the cases of single-hit-single-target (H1T1), double-hit-single-target (H2T1), and single-hit-double-target (H1T2) inactivation are illustrated.

where m represents the mean value of the cumulative number of hits per target. It is given by the ratio of the absorbed dose to the mean inactivation dose, $m = D/D_0$. For this situation the surviving fraction is equal to $P_{0 \to n-1}$:

$$S = \sum_{j=0}^{n-1} \frac{(D/D_0)^j}{j!} \exp\left(-\frac{D}{D_0}\right) \tag{14.29}$$

Substituting $n = 1$ into Equation 14.29 produces a result identical to Equation 14.27, the single-hit-single-target result.

An analytical expression for the single-hit-multiple-target surviving fraction (22) can also be obtained within the limits of the Poisson distribution. Suppose n separate but identical targets within a given structure must each sustain one or more effective hits for sufficient damage. Since the probability for no hits per target is $\exp(-m)$, the probability for one or more effective hits on a single target is $[1 - \exp(-m)]$, where m is the mean value of the cumulative number of hits in a single target. Furthermore, the probability for at least one random hit on n identical targets taken in any order is the product of n individual effective hit probabilities. The surviving fraction must be equal to one minus this product:

$$S = 1 - [1 - \exp(-m)]^n = 1 - [1 - \exp(-D/D_0)]^n \tag{14.30}$$

Notice that Equation 14.30 reduces to the exponential single-hit-single-target result when $n = 1$.

Generally speaking, survival curves for mammalian cells after radiation-induced injury are approximately exponential when high linear energy transfer radiation is utilized. When sparsely ionizing radiation is used, the curves show a shoulder more similar to the multiple-hit or multiple-target curves on Figure 14.9. Target theory is sometimes utilized to interpret cellular survival in terms of single- or multiple-event damage; single-event damage is associated with high linear energy transfer radiation; multiple-event damage is associated with low linear energy transfer radiation. One should examine the premises carefully before applying the simple expressions developed in this section to surviving-fraction data for cellular systems.

14.5 Molecular Lesions and Cell Survival

The target theory outlined in the previous section is a description of the accumulation of effective radiation damage in large samples of macromolecules. As mentioned, some of the concepts introduced can also be used in a description of the effects of radiation-induced damage to cellular systems. Suppose that chemical bonds in critical molecules, notably those of DNA, are severed by radiation action. Serious damage can often be related to a site in which both of the associated DNA strands are broken (2, 19). The bond can be severed by indirect action or direct action, since free radicals produced in nearby water molecules can migrate to a sensitive site and effect a molecular structure change at that point (10).

If the damage at a molecular site is potentially lethal, it may be termed a lesion. Some evidence exists to support the hypothesis that lesion formation often involves formation of dual sublesions in molecules (23, 24). Thus, the molecular damage sites that cause cell death are presumed to be paired. A simple statistical development for the formation of pairs of sublesions will be given in this section (25). The descriptions given are not complete, since well known phenomena, such as repair of lesions by the cells, will not be included (26). Furthermore, the nature of the lesions and sublesions and their particular modes of formation will not be discussed within the confines of this section (2).

Dual sublesions can be produced either by a single-step or a two-step process. In Figure 14.10 state 1 represents the undamaged cell nucleus, state 2 represents the partially damaged configuration, and state 3 represents a nucleus with critical damage. P_{13} represents the probability of producing critical damage in a single step. P_{12} represents the probability that the radiation will produce a sublesion by inducing a change from state 1 to state 2. When state 2 is populated, the step to the critically damaged state 3 occurs with probability P_{23}.

High linear energy transfer radiations are presumed to act via the single-step process on many occasions. Thus, P_{13} is expected to be relatively large for these radiations. Lower linear energy transfer radiations cause damage by

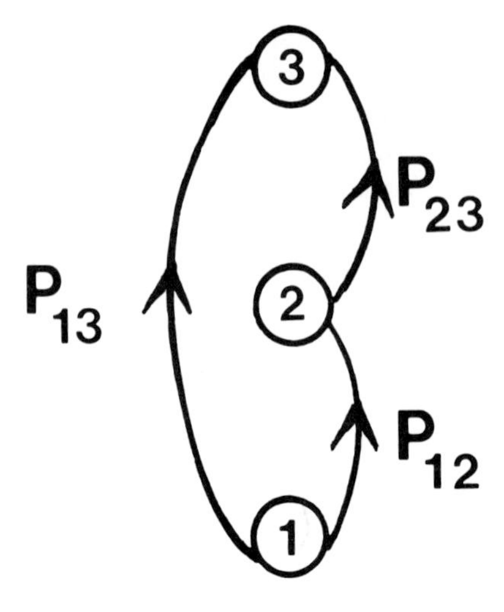

Figure 14.10 The schematic for single-step and two-step inactivation is shown (23).

two-step processes in many cases; thus, the product $P_{12}P_{23}$ is expected to be relatively large (24) in this case.

In the paragraphs immediately following, a simplified molecular theory of cell survival (25) involving paired damage sites is presented. Some assumptions are:

1. The immediate expression of radiation-induced damage is the dissociation of chemical bonds in critical molecules to produce sublesions.
2. Potentially lethal lesions consist of at least two associated sublesions.
3. Potentially lethal lesions can be produced in a single-step process or in two random independent steps.
4. The single-step production of sublesions or lesions is governed by an exponential law.

The following parameters are necessary:

N_1 represents the number of appropriate sites available for single-step production of lesions;
N_2 represents the number of appropriate sites available for production of a sublesion;
c_1 is the probability per unit dose for producing a lesion at a particular site;
c_2 is the probability per unit dose for producing a sublesion at a particular site.

From Equation 14.27 the expression for $\langle \Delta N_{13} \rangle$, the average number of sites where dual lesions are produced in a single step, can be written

$$\langle \Delta N_{13} \rangle = N_1[1 - \exp(-c_1 D)] \simeq N_1 c_1 D, \quad c_1 D \ll 1 \tag{14.31}$$

In this case c_1 is assumed to be sufficiently small to allow one to approximate the exponential by the first two terms, $\exp(-c_1 D) \simeq 1 - c_1 D$.

The average probability for production of a lone sublesion is given by

$$\frac{\langle \Delta N_{12} \rangle}{N_2} = [1 - \exp(-c_2 D)] \simeq c_2 D, \quad c_2 D \ll 1 \tag{14.32}$$

The probability of the production of a lone sublesion and subsequent production of another associated sublesion is proportional to the product of the individual formation probabilities. We assume the processes are independent (21). Thus, the mean number of dual sublesions, $\langle \Delta N_{123} \rangle$, formed by two-step processes, is given by

$$\langle \Delta N_{123} \rangle = \epsilon (N_2 c_2 D)^2 \tag{14.33}$$

The constant ϵ is the probability the two sublesions are associated so that a potentially lethal lesion results.

The mean number of lesions produced by single-step and two-step processes is given as $\langle \Delta N_T \rangle$, where

$$\langle \Delta N_T \rangle = \langle \Delta N_{13} \rangle + \langle N_{123} \rangle = N_1 c_1 D + \epsilon (N_2 c_2 D)^2 \tag{14.34}$$

Suppose p is the probability that a dual sublesion is lethal. The surviving fraction must be equal to the probability that no lethal lesions are produced. From the Poisson distribution of Equation 14.23, this probability is expressed by $\exp(-m) = \exp(-p\langle \Delta N_T \rangle)$.

$$\begin{aligned} S &= \exp(-p\langle \Delta N_T \rangle) \\ &= \exp[-pN_1 c_1 D - p\epsilon (N_2 c_2 D)^2] \\ &= \exp(-aD - bD^2) \\ a &= pN_1 c_1; \quad b = p\epsilon (N_2 c_2)^2 \end{aligned} \tag{14.35}$$

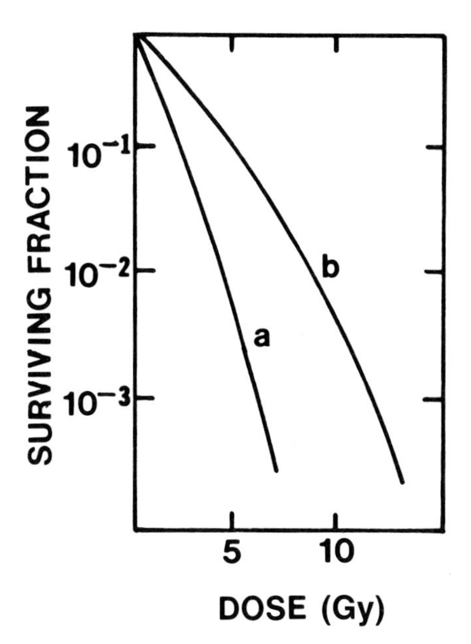

Figure 14.11 Survival curves for cultured cells of human origin are shown (26): curve a is for 15-MeV neutrons; curve b is for 250-kVp x rays.

Of course the development of Equations 14.36 is not completely realistic since the repair of sublesions and lesions and the effects of oxygen pressure, among other phenomena, have not been included (26). Nevertheless, the form of the final result can be shown to reproduce the shape of some experimental data for survival of certain cells in culture (27).

Figure 14.11 shows two curves generated using the exponential form in Equation 14.35. When high linear energy transfer radiation is used, the specific energy is large and single-step lesion formation is important. Thus, for the first few grays, $aD \gg bD^2$ and a nearly exponential survival curve results (curve a). For photons and electrons, the specific energy is moderate, so two-step lesion formation would be more important and a noticeable curvature would be observed throughout the survival curve (curve b).

The general form of the exponential in Equation 14.35 can be related to the relative biological effectivensss. Suppose $a_n D_n \gg b_n D_n^2$ for some high linear energy transfer radiation indicated by index n at some dose D_n. For the same cellular system, suppose $a_x D_x \ll b_x D_x^2$ for irradiation with 200 kVp x rays at an appropriate dose D_x. When the surviving fractions are equal,

$$S_n \simeq \exp(-a_n D_n) = \exp(-b_x D_x^2) \simeq S_x,$$

$$a_n D_n = b_x D_x^2, \quad D_x = \left[\left(\frac{a_n}{b_x}\right) D_n\right]^{1/2} \tag{14.36}$$

The RBE can be expressed as the ratio of the doses for equal surviving fractions:

$$\text{RBE} = \frac{D_x}{D_n} = \frac{(a_n/b_x)^{1/2}}{D_n} D_n^{1/2} = \left(\frac{a_n}{b_x}\right)^{1/2} D_n^{-1/2},$$

$$\log \text{RBE} = \log\left(\frac{a_n}{b_x}\right)^{1/2} - \frac{1}{2} \log D_n \tag{14.37}$$

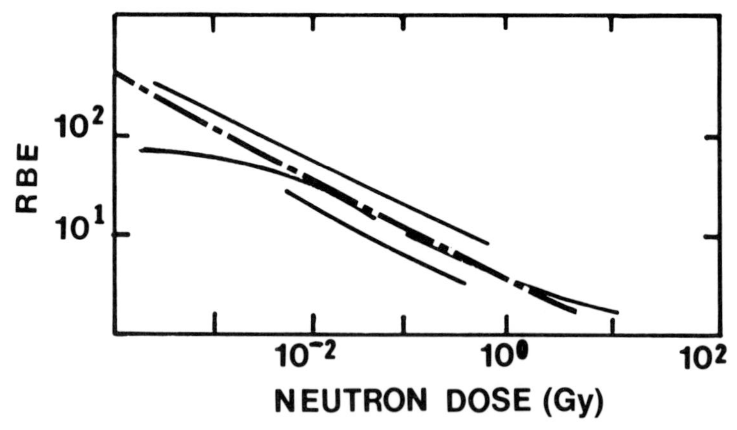

Figure 14.12 The logarithm of the RBE is shown as a function of the logarithm of the neutron dose. The dashed line is the theoretical result from Equation 14.38; other lines are experimental results. Adapted from Ref. 23.

The logarithmic function in Equation 14.37 has the form of the dashed curve shown on Figure 14.12. Notice that the slope is $-\frac{1}{2}$ on this log-log plot. Some experimental RBE curves are also shown on the figure for several different systems (24). Many of the experimental curves also have slopes near $-\frac{1}{2}$. If the formation of dual sublesions is important in the action of radiation on cellular systems, these sorts of RBE curves would be expected.

14.6 Problems

1. Assume that a DNA double strand can be represented by a cylinder with diameter of 20Å and that a cell nucleus is spherical and 2 μm in diameter. Assume that an ionizing particle passes directly through the structures on a diameter and that the medium is water equivalent. Use stopping power tables from the Appendixes to calculate how many ionizations would be produced on the average in the
   a. DNA cylinder by a 1-MeV electron passing through the diameter;
   b. DNA cylinder by a 10-MeV alpha particle passing through the diameter;
   c. Cell nucleus by a 1-MeV electron through the diameter;
   d. Cell nucleus by a 10-MeV alpha particle through the diameter.
   What conclusions can you draw about the effects of the disruptions caused?
2. Calculate the RBE for 15-MeV neutrons at 10% survival and 1% survival using the expressions

$$S = \exp(-8 \times 10^{-1}\, D)\exp(-5 \times 10^{-2}\, D^2), \text{ for neutrons;}$$

$$S = \exp(-2 \times 10^{-1}\, D)\exp(-3 \times 10^{-2}\, D^2), \text{ for 200-kVp x rays.}$$

3. Suppose radioactive ^{125}I is incorporated into nuclear DNA in a mammlian cell. Assume the ^{125}I atom is situated at the center of the spherical cell nucleus of diameter 1 μm and assume the nucleus is very similar to water. Calculate the specific energy for the cell nucleus due to decay of a single ^{125}I nucleus. Ignore contributions due to x, gamma, and delta rays. Use range tables for the electrons to estimate the energy deposited in the cell nucleus. The electrons produced upon decay of ^{125}I are:

Type	*Mean no./ Disintegration*	*Average Electron Energy (keV)*
K int. conv.	0.746	3.7
L int. conv.	0.107	30.9
M int. conv.	0.080	34.7
KLL Auger	0.137	22.7
KLX Auger	0.058	26.4
KXY Auger	0.009	30.2
LMM Auger	1.490	2.9
MXY Auger	3.590	0.8

4. Calculate the maximum lineal energy in kiloelectron volts per micrometer for passage of a 1.0-MeV electron through a spherical cell nucleus with diameter 1 μm. Use stopping power information from Chapter 3.
5. Suppose 0.1 Gy is deposited by 1-MeV electrons normally incident on a 1-mm-thick water-equivalent slab. From the number of tracks per unit area required, calculate the average separation between tracks. Assume each track is isolated at the center of a square surface element. Perform the same calculation for 1-MeV protons instead of the electrons.
6. For a spherical volume show that the average track length $\langle x \rangle = 4r/3$. Then show that $\langle y \rangle$, the average lineal energy, is independent of the sphere diameter.
7. If $\langle x \rangle = 4r/3$ for a spherical volume, show that the lineal energy $y = \left[\frac{\rho A}{4}\right] z$, where ρ is the density, A the spherical surface area, and z the specific energy.
8. If $\langle x \rangle = 4\ r/3$ for a spherical volume, show that

$$\exp[-(L\langle x \rangle/E_e)\sigma\Phi] = \exp[-(L/E_e)V_e\Phi]$$

9. A target 20Å thick is hit by two charged particles, one with $L = 100$ keV/μm and another with $L = 1$ keV/μm. Calculate the probability that an average deposition event with expenditure of 110 eV occurs in the target in each case.
10. A monomolecular layer of enzyme preparation is bombarded with ^{60}Co gamma rays, and the dose for 63% deactivation is 5.8×10^5 Gy. Calculate the sensitive volume of the enzyme and the molecular weight if the density is 1.3 g/cm.
11. Find an expression for the slope of the survival curve at $D = 0$ for a double-hit-single-target theory calculation using Equation 14.29 and the single-hit-double-target theory given by Equation 14.30. Which is more realistic? Why?
12. Calculate the surviving fraction for cells in culture after receiving a dose of 2.0 Gy. Assume that D_0 for the cells is 1.2 Gy. In one case, assume that the single-hit-single-target theory is sufficient. In a second case, assume that two hits are required in a single target for lethal effect. In a third case, assume that a single hit is required in each of two targets for lethal effects.
13. Assume a tumor contains 10^{11} cells. Find the dose of neutrons (single-hit theory) required to ensure that there be only a 1-in-10 chance that a single cell in the tumor survives. The mean inactivation dose D_0 is 1.2 Gy for these cells.

14.7 References

1. *Radiation Quantities and Units, ICRU Report 33* (International Commission on Radiation Units and Measurements, Washington, 1980).

2. A. M. Kellerer, Int. J. Radiat. Oncol. Biol. Phys. **5**, 1041 (1979).
3. D. J. Pizzarello and R. L. Witcofski, *Basic Radiation Biology* (Lea and Febiger, Philadelphia, 1967).
4. S. W. Jacob and C. A. Francone, *Structure and Function in Man* (W. B. Saunders, Philadelphia, 1965), pp. 20–36.
5. G. V. Dalrymple and M. L. Baker, "Molecular biology," in *Medical Radiation Biology,* edited by G. V. Dalrymple, M. E. Gaulden, G. M. Kollmorgen, and H. H. Vogel (W. B. Saunders, Philadelphia, 1973), pp. 39–43.
6. *Linear Energy Transfer, ICRU Report 16* (International Commission on Radiation Units and Measurements, Washington, 1970).
7. T. Brustad, Radiat. Res. **15**, 139 (1961).
8. T. R. Manney, T. Brustad, and C. A. Tobias, Radiat. Res. **18**, 374 (1963).
9. P. Todd, Med. Coll. Va. Q. **1**, 2 (1965).
10. H. E. Johns and J. R. Cunningham, *The Physics of Radiology* (Charles C. Thomas, Springfield, IL, 1969), pp. 674–728.
11. G. W. Barendsen, "Radiobiological dose-effect relations for radiation characterized by a wide spectrum of LET: implications in radiotherapy," in *Particle Accelerators in Radiation Therapy, Report LA5180-C* (Los Alamos Scientific Laboratory, Los Alamos, NM, 1973), pp. 120–125.
12. *Basic Radiation Protection Criteria, NCRP Report 39* (National Council on Radiation Protection and Measurements, Washington, 1971), pp. 81–84.
13. *Report of the International Commission on Radiological Units and Measurements, NBS Handbook 62* (National Bureau of Standards, Washington, 1957).
14. H. H. Rossi, "Microscopic energy distribution in irradiated matter," in *Radiation Dosimetry,* vol 1, edited by F. H. Attix and W. C. Roesch (Academic, New York, 1968), pp. 43–91.
15. A. M. Kellerer and H. H. Rossi, Br. J. Radiol. **45**, 656 (1972).
16. E. C. Pollard, W. R. Guild, F. Hutchinson, and R. B. Setlow, Prog. Biophysics. Biophys. Chem. **5**, 72 (1955).
17. H. L. Andrews, *Radiation Biophysics* (Prentice-Hall, Englewood Cliffs, NJ, 1961), p. 12.
18. R. Katz, Phys. Med. Biol. **23**, 909 (1978).
19. P. R. J. Burch and M. S. Chesters, Phys. Med. Biol. **26**, 997 (1981).
20. R. D. Evans, *The Atomic Nucleus* (McGraw-Hill, New York, 1955), pp. 746–754.
21. B. R. Martin, *Statistics for Physicists* (Academic, London, 1971), pp. 36–40.
22. *Quantitative Concepts and Dosimetry in Radiobiology, ICRU Report 30* (International Commission on Radiation Units and Measurements, Washington, 1979), p. 5.
23. A. M. Kellerer and H. H. Rossi, Radiat. Res. **47**, 15 (1971).
24. A. M. Kellerer and H. H. Rossi, Curr Top. Radiat. Res. Q. **8**, 85 (1972).
25. K. H. Chadwick and H. P. Leenhouts, Phys. Med. Biol. **18**, 78 (1973).
26. M. M. Elkind, Int. J. Radiat. Oncol. Biol. Phys. **5**, 1089 (1979).
27. G. W. Barendsen, Curr. Top. Radiat. Res. **4**, 293 (1968).

Appendixes

Appendix 1 The Greek Alphabet

A	α	Alpha	N	ν	Nu
B	β	Beta	Ξ	ξ	Xi
Γ	γ	Gamma	O	o	Omicron
Δ	δ	Delta	Π	π	Pi
E	ϵ	Epsilon	P	ρ	Rho
Z	ζ	Zeta	Σ	σ	Sigma
H	η	Eta	T	τ	Tau
Θ	θ	Theta	Y	υ	Upsilon
I	ι	Iota	Φ	ϕ	Phi
K	κ	Kappa	Ξ	χ	Chi
Λ	λ	Lambda	Ψ	ψ	Psi
M	μ	Mu	Ω	ω	Omega

Appendix 2 International System of Units (SI)

Some Basic and Supplementary Units

Quantity	*Name*	*Symbol*
Length	meter	m
Mass	kilogram	kg
Time	second	s
Electric current	ampere	A
Temperature	kelvin	K
Amount of substance	mole	mol
Luminous intensity	candela	cd
Plane angle	radian	rad
Solid angle	steradian	sr

Some Derived Units

Quantity	*Name*	*Symbol*	*Equivalent Basic Units*
Frequency	hertz	Hz	s^{-1}
Energy, work	joule	J	m^2 kg s^{-2}
Power	watt	W	m^2 kg s^{-3}
Force	newton	N	m kg s^{-2}
Pressure	pascal	Pa	m^{-1} kg s^{-2}
Activity	becquerel	Bq	s^{-1}
Absorbed dose	gray	Gy	m^2 s^{-2}
Dose equivalent	sievert	Sv	m^2 s^{-2}
Electric charge	coulomb	C	s A
Electric potential	volt	V	m^2 kg s^{-3} A^{-1}
Electric resistance	ohm	Ω	m^2 kg s^{-3} A^{-2}
Conductance	siemens	S	m^{-2} kg^{-1} s^3 A^2
Inductance	henry	H	m^2 kg s^{-2} A^{-2}
Capacitance	farad	F	m^{-2} kg^{-1} s^4 A^2
Magnetic flux	weber	Wb	m^2 kg s^{-2} A^{-1}
Magnetic induction	tesla	T	kg s^{-2} A^{-1}

Some Prefixes

Prefix	*Factor*	*Symbol*
10^{12}	tera	T
10^{9}	giga	G
10^{6}	mega	M
10^{3}	kilo	k
10^{1}	deca	da
10^{-1}	deci	d
10^{-2}	centi	c
10^{-3}	milli	m
10^{-6}	micro	μ
10^{-9}	nano	n
10^{-12}	pico	p

Some Conversion Factors for Other Units

Name	*Symbol*	*Value in Basic Units*
angstrom	Å	1.00×10^{-10} m
barn	b	1.00×10^{-28} m^2
calorie	cal	4.184 J
degree	°	1.7453×10^{-2} rad
electron volt	eV	1.6022×10^{-19} J
erg	erg	1.00×10^{-7} J
gauss	gs	1.00×10^{-4} T
liter	l	1.00×10^{-3} m^3
roentgen	R	2.58×10^{-4} s A kg^{-1}

Appendix 3 Table of Constants

Physical Constants

Constant	*Symbol*	*Value*
Atomic mass unit	u	1.6606×10^{-27} kg
Avogadro's constant	N_A	6.0220×10^{26} kmol^{-1}
Bohr magneton	μ_B	9.2741×10^{-24} J T^{-1}
Bohr radius	a_0	5.2918×10^{-11} m
Boltzmann constant	k	1.3807×10^{-23} J K^{-1}
Electron charge	e	1.6602×10^{-19} C
Electron radius	r_0	2.8179×10^{-15} m
Electron rest mass	m_e	9.1095×10^{-31} kg
Fine structure constant	α	7.2974×10^{-3}
Proton rest mass	m_p	1.6726×10^{-27} kg
Neutron rest mass	m_n	1.6749×10^{-27} kg
Planck constant	h	6.6262×10^{-34} J·s
Vacuum speed of light	c	2.9979×10^{8} m s^{-1}
Vacuum permeability	μ_0	1.2566×10^{-6} H m^{-1}
Vacuum permittivity	ϵ_0	8.8542×10^{-12} C^2 J^{-1} m^{-1}

Rest Mass Energies

Electron	$m_e c^2 = 0.511$ MeV
Proton	$m_p c^2 = 938.3$ MeV
Neutron	$m_n c^2 = 939.6$ MeV
Mass unit	$uc^2 = 931.5$ MeV

Appendix 4 Electron Binding Energies

Atomic Energy Levels (Shells K through M_V) in Kiloelectron Volts

Z	K	L_I	L_{II}	L_{III}	M_I	M_{II}	M_{III}	M_{IV}	M_V
1	0.014								
2	0.025	0.001							
3	0.055	0.003	0.001	0.001					
4	0.111	0.006	0.002	0.002					
5	0.188	0.009	0.004	0.004					
6	0.284	0.013	0.005	0.005					
7	0.400	0.018	0.007	0.007					
8	0.533	0.024	0.009	0.009					
9	0.687	0.032	0.012	0.012					
10	0.867	0.045	0.018	0.18	0.001				

Data selected from E. Storm and H. I. Israel, "Photon cross sections from 1 keV to 100 Mev for elements $Z = 1$ to $Z = 100$," Nucl. Data Tables **A7**, 641 (1970).

Z	K	L_I	L_{II}	L_{III}	M_I	M_{II}	M_{III}	M_{IV}	M_V
11	1.073	0.063	0.032	0.032	0.002				
12	1.305	0.088	0.050	0.050	0.003				
13	1.560	0.118	0.073	0.073	0.005				
14	1.839	0.151	0.099	0.100	0.007	0.001	0.001		
15	2.144	0.188	0.130	0.130	0.010	0.002	0.002		
16	2.472	0.227	0.165	0.165	0.014	0.004	0.004		
17	2.824	0.270	0.203	0.202	0.018	0.007	0.007		
18	3.203	0.320	0.247	0.245	0.025	0.012	0.012		
19	3.607	0.377	0.296	0.294	0.034	0.018	0.018		
20	4.037	0.438	0.350	0.346	0.044	0.025	0.025		
21	4.491	0.500	0.406	0.401	0.053	0.032	0.032		
22	4.966	0.563	0.462	0.456	0.060	0.035	0.035		
23	5.465	0.628	0.521	0.513	0.066	0.038	0.038		
24	5.989	0.696	0.584	0.575	0.074	0.042	0.042	0.001	0.001
25	6.539	0.769	0.651	0.640	0.084	0.047	0.047	0.002	0.002
26	7.112	0.846	0.721	0.708	0.093	0.053	0.053	0.003	0.003
27	7.709	0.926	0.794	0.779	0.101	0.060	0.060	0.004	0.004
28	8.332	1.008	0.871	0.854	0.111	0.067	0.067	0.005	0.005
29	8.981	1.096	0.953	0.933	0.122	0.074	0.074	0.007	0.007
30	9.659	1.193	1.043	1.020	0.138	0.088	0.087	0.010	0.010
31	10.367	1.300	1.142	1.115	0.158	0.106	0.103	0.017	0.017
32	11.104	1.413	1.248	1.217	0.180	0.126	0.121	0.028	0.028
33	11.867	1.530	1.359	1.323	0.204	0.146	0.140	0.041	0.041
34	12.658	1.652	1.475	1.434	0.230	0.168	0.161	0.055	0.055
35	13.474	1.782	1.597	1.551	0.257	0.191	0.184	0.072	0.071
36	14.323	1.921	1.727	1.675	0.288	0.219	0.210	0.091	0.090
37	15.200	2.065	1.863	1.805	0.322	0.248	0.239	0.112	0.110
38	16.105	2.216	2.007	1.940	0.358	0.280	0.269	0.135	0.133
39	17.038	2.373	2.155	2.079	0.394	0.312	0.299	0.158	0.156
40	17.998	2.533	2.307	2.223	0.430	0.344	0.330	0.182	0.180
41	18.986	2.698	2.464	2.370	0.467	0.377	0.361	0.206	0.204
42	20.000	2.867	2.625	2.521	0.505	0.410	0.392	0.230	0.228
43	21.044	3.043	2.793	2.677	0.545	0.445	0.426	0.256	0.253
44	22.117	3.224	2.967	2.838	0.585	0.483	0.461	0.284	0.280
45	23.220	3.412	3.146	3.004	0.627	0.521	0.496	0.312	0.307
46	24.350	3.605	3.330	3.174	0.670	0.559	0.532	0.340	0.335
47	25.514	3.806	3.524	3.351	0.718	0.602	0.571	0.373	0.367
48	26.711	4.018	3.727	3.537	0.770	0.651	0.616	0.410	0.404
49	27.940	4.238	3.938	3.730	0.826	0.702	0.664	0.451	0.443
50	29.200	4.465	4.156	3.929	0.884	0.756	0.714	0.493	0.485
51	30.491	4.698	4.381	4.132	0.944	0.812	0.766	0.537	0.528
52	31.814	4.939	4.612	4.341	1.006	0.870	0.819	0.583	0.572
53	33.170	5.188	4.852	4.557	1.072	0.931	0.876	0.633	0.619
54	34.561	5.445	5.102	4.782	1.143	0.997	0.936	0.686	0.672
55	35.985	5.713	5.360	5.012	1.217	1.065	0.998	0.740	0.726
56	37.441	5.987	5.623	5.247	1.291	1.135	1.061	0.794	0.780
57	38.925	6.266	5.891	5.484	1.363	1.204	1.124	0.848	0.832
58	40.443	6.549	6.164	5.723	1.435	1.273	1.185	0.901	0.883
59	41.991	6.835	6.440	5.964	1.505	1.337	1.242	0.951	0.931

Appendix 4 Electron Binding Energies—Continued

Z	K	L_I	L_{II}	L_{III}	M_I	M_{II}	M_{III}	M_{IV}	M_V
60	43.569	7.128	6.722	6.208	1.575	1.403	1.298	1.001	0.978
61	45.184	7.428	7.013	6.459	1.648	1.471	1.357	1.052	1.027
62	46.834	7.736	7.312	6.716	1.723	1.541	1.419	1.106	1.078
63	48.519	8.052	7.618	6.977	1.800	1.614	1.481	1.161	1.131
64	50.239	8.375	7.930	7.243	1.881	1.688	1.544	1.217	1.185
65	51.996	8.708	8.252	7.514	1.963	1.765	1.610	1.274	1.240
66	53.788	9.046	8.580	7.790	2.046	1.842	1.676	1.332	1.295
67	55.618	9.394	8.918	8.072	2.130	1.923	1.743	1.392	1.351
68	57.486	9.752	9.264	8.358	2.217	2.006	1.812	1.453	1.409
69	59.390	10.116	9.617	8.648	2.306	2.090	1.881	1.515	1.468
70	61.332	10.489	9.978	8.943	2.398	2.175	1.950	1.577	1.528
71	63.316	10.874	10.349	9.245	2.494	2.264	2.024	1.641	1.591
72	65.345	11.272	10.739	9.560	2.600	2.364	2.108	1.716	1.662
73	67.416	11.680	11.136	9.880	2.709	2.469	2.194	1.793	1.735
74	69.525	12.098	11.541	10.204	2.820	2.575	2.281	1.871	1.809
75	71.676	12.528	11.957	10.534	2.934	2.682	2.368	1.950	1.883
76	73.871	12.969	12.385	10.871	3.052	2.792	2.457	2.031	1.960
77	76.111	13.419	12.824	11.215	3.173	2.908	2.551	2.116	2.040
78	78.395	13.880	13.273	11.564	3.297	3.027	2.645	2.202	2.122
79	80.725	14.353	13.734	11.919	3.425	3.150	2.743	2.291	2.206
80	83.102	14.842	14.209	12.283	3.562	3.280	2.847	2.385	2.295
81	85.530	15.346	14.697	12.656	3.704	3.416	2.956	2.485	2.389
82	88.004	15.861	15.200	13.035	3.851	3.554	3.066	2.586	2.484
83	90.526	16.391	15.714	13.420	4.000	3.696	3.177	2.689	2.581
84	93.105	16.936	16.244	13.814	4.156	3.849	3.295	2.798	2.683
85	95.730	17.491	16.785	14.214	4.317	4.006	3.416	2.909	2.787
86	98.404	18.055	17.337	14.619	4.482	4.164	3.538	3.022	2.892
87	101.137	18.639	17.904	15.030	4.652	4.325	3.664	3.136	3.000
88	103.922	19.237	18.484	15.446	4.824	4.490	3.791	3.253	3.109
89	106.759	19.845	19.083	15.870	5.002	4.658	3.918	3.371	3.219
90	109.651	20.466	19.693	16.300	5.182	4.830	4.046	3.490	3.332
91	112.601	21.105	20.314	16.733	5.364	5.003	4.174	3.609	3.442
92	115.606	21.759	20.948	17.170	5.548	5.181	4.304	3.728	3.552
93	118.670	22.427	21.600	17.613	5.735	5.366	4.435	3.850	3.664
94	121.797	23.109	22.270	18.063	5.927	5.555	4.568	3.973	3.778
95	124.990	23.812	22.958	18.519	6.122	5.748	4.703	4.100	3.894
96	128.253	24.535	23.663	18.982	6.322	5.945	4.839	4.230	4.012
97	131.590	25.275	24.385	19.452	6.526	6.147	4.977	4.364	4.132
98	135.005	26.030	25.125	19.929	6.735	6.353	5.117	4.502	4.254
99	138.502	26.803	25.883	20.414	6.949	6.564	5.259	4.644	4.378
100	142.085	27.594	26.659	20.907	7.168	6.780	5.403	4.790	4.504

Appendix 5 Atomic Mass Tables

Z		A	M-A(keV)	Z		A	M-A(keV)	Z		A	M-A(keV)	Z		A	M-A(keV)	Z		A	M-A(keV)
0	N	1	8071.44	4	Be	11	20181.	9	F	18	872.4	13	Al	24	−15.	16	S	31	−18992.
1	H	1	7288.99	5	B	11	8667.68	10	Ne	18	5319.3	11	Na	25	−9356.	14	Si	32	−24089.
1	H	2	13135.91	6	C	11	10648.4	8	O	19	3332.7	12	Mg	25	−13190.7	15	P	32	−24302.7
1	H	3	14949.95	5	B	12	13370.2	9	F	19	−1486.0	13	Al	25	−8931.	16	S	32	−26012.7
2	He	3	14931.34	6	C	12	0	10	Ne	19	1752.0	11	Na	26	−7690.	17	Cl	32	−13537.
2	He	4	2424.75	7	N	12	17342.	8	O	20	3799.	12	Mg	26	−16214.2	15	P	33	−26334.6
2	He	5	11454.	5	B	13	16561.6	9	F	20	−11.9	13	Al	26	−12210.8	16	S	33	−26582.6
3	Li	5	11679.	6	C	13	3124.6	10	Ne	20	−7041.5	14	Si	26	−7141.	17	Cl	33	−21014.
2	He	6	17598.2	7	N	13	5345.2	11	Na	20	6870.	12	Mg	27	−14582.6	15	P	34	−24830.
3	Li	6	14088.4	6	C	14	3019.82	9	F	21	−46.	13	Al	27	−17196.1	16	S	34	−29933.5
4	Be	6	18376.	7	N	14	2863.73	10	Ne	21	−5729.9	14	Si	27	−12386.0	17	Cl	34	−24449.
3	Li	7	14907.3	8	C	14	8008.00	11	Na	21	−2185.	12	Mg	28	−15020.	18	Ar	34	−18394.
4	Be	7	15768.9	6	C	15	9873.2	12	Mg	21	10620.	13	Al	28	−16855.4	16	S	35	−28847.1
2	He	8	31650.	7	N	15	100.4	9	F	22	2828.	14	Si	28	−21489.9	17	Cl	35	−29014.5
3	Li	8	20946.2	8	O	15	2859.9	10	Ne	22	−8024.9	15	P	28	−7152.	18	Ar	35	−23051.
4	Be	8	4944.2	6	C	16	13695.	11	Na	22	−5182.2	13	Al	29	−18218.	16	S	36	−30655.
5	B	8	22923.1	7	N	16	5685.1	12	Mg	22	−380.	14	Si	29	−21893.6	17	Cl	36	−29519.6
3	Li	9	24965.	8	O	16	−4736.55	10	Ne	23	−5148.3	15	P	29	−16945.	18	Ar	36	−30231.6
4	Be	9	11350.5	9	F	16	10693.	11	Na	23	−9528.3	13	Al	30	−15900.	16	S	37	−36890.
5	B	9	12418.6	7	N	17	7871.	12	Mg	23	−5472.	14	Si	30	−24439.4	17	Cl	37	−31764.8
6	C	9	28990.	8	O	17	−807.7	13	Al	23	6770.	15	P	30	−20197.	18	Ar	37	−30950.9
4	Be	10	12607.0	9	F	17	1951.9	10	Ne	24	−5949.	16	S	30	−14063.	19	K	37	−24799.6
5	B	10	12052.2	10	Ne	17	16479.	11	Na	24	−8418.4	14	Si	31	−22962.	20	Ca	37	−13240.
6	C	10	15702.6	8	O	18	−782.43	12	Mg	24	−13933.3	15	P	31	−24437.6	16	S	38	−26800.

Data prepared by S. Morse and F. Ajzenberg-Selove from J. E. Mattauch, L. Thiele, and A. H. Wapstra, Nucl. Phys. **67,** 1 (1965); T. Lauritsen and F. Ajzenberg-Selove, Nucl. Phys. **78,** 1 (1966); and P. M. Endt and J. Van der Leun, Nucl. Phys. Sect. A. **105,** 1 (1967).

Appendix 5 Atomic Mass Tables—Continued

Z		A	M-A(keV)	Z		A	M-A(keV)	Z		A	M-A(keV)	Z		A	M-A(keV)	Z		A	M-A(keV)
17	Cl	38	−29803.	20	Ca	44	−41459.6	20	Ca	50	−39572.	26	Fe	55	−57473.5	29	Cu	61	−61984.
18	Ar	38	−34718.2	21	Sc	44	−37813.	21	Sc	50	−44540.	27	Cu	55	−54014.	30	Zn	61	−56580.
19	K	38	−28786.	22	Ti	44	−37658.	22	Ti	50	−51430.7	24	Cr	56	−55290.	27	Co	62	−61528.
20	Ca	38	−22050.	19	K	45	−36730.	23	V	50	−49215.8	25	Mn	56	−56903.8	28	Ni	62	−66748.
17	Cl	39	−29800.	20	Ca	45	−40808.5	24	Cr	50	−50249.	26	Fe	56	−60605.4	29	Cu	62	−62813.
18	Ar	39	−33238.	21	Sc	45	−41060.6	25	Mn	50	−42648.	27	Co	56	−56031.	30	Zn	62	−61123.
19	K	39	−33803.3	22	Ti	45	−39002.0	21	Sc	51	−43220.	28	Ni	56	−53918.	27	Co	63	−61920.
20	Ca	39	−27283.	19	K	46	−35420.	22	Ti	51	−49738.	25	Mn	57	−57480.	28	Ni	63	−65516.
17	Cl	40	−27500.	20	Ca	46	−43140.	23	V	51	−52198.9	26	Fe	57	−60175.5	29	Cu	63	−65583.1
18	Ar	40	−35038.3	21	Sc	46	−41755.7	24	Cr	51	−51447.2	27	Co	57	−59338.9	30	Zn	63	−62217.
19	K	40	−33533.3	22	Ti	46	−44122.6	25	Mn	51	−48260.	28	Ni	57	−56104.	31	Ga	63	−56720.
20	Ca	40	−34847.6	23	V	46	−37060.	22	Ti	52	−49468.	25	Mn	58	−55650.	28	Ni	64	−67106.
21	Sc	40	−20523.	19	K	47	−35703.	23	V	52	−51436.	26	Fe	58	−62146.5	29	Cu	64	−65427.6
18	Ar	41	−33067.4	20	Ca	47	−42347.	24	Cr	52	−55410.7	27	Co	58	−59838.	30	Zn	64	−66000.3
19	K	41	−35552.4	21	Sc	47	−44326.3	25	Mn	52	−50702.	28	Ni	58	−60228.	31	Ga	64	−58928.
20	Ca	41	−35140.	22	Ti	47	−44926.6	26	Fe	52	−48328.	29	Co	58	−51659.	28	Ni	65	−65137.
21	Sc	41	−28643.	23	V	47	−42010.	23	V	53	−52180.	26	Fe	59	−60659.9	29	Cu	65	−67266.
18	Ar	42	−34420.	20	Ca	48	−44216.	24	Cr	53	−55280.7	27	Co	59	−62232.7	30	Zn	65	−65917.
19	K	42	−35018.	21	Sc	48	−44505.	25	Mn	53	−54682.	28	Ni	59	−61158.7	31	Ga	65	−62658.
20	Ca	42	−38539.7	22	Ti	48	−48483.1	26	Fe	53	−50698.	29	Cu	59	−56359.	32	Ge	65	−56260.
21	Sc	42	−32106.4	23	V	48	−44470.	23	V	54	−49630.	26	Fe	60	−61511.	28	Ni	66	−66055.
22	Ti	42	−25123.	24	Cr	48	−43070.	24	Cr	54	−56930.5	27	Co	60	−61651.3	29	Cu	66	−66255.
19	K	43	−36579.	20	Ca	49	−41288.	25	Mn	54	−55552.	28	Ni	60	−64470.7	30	Zn	66	−68881.
20	Ca	43	−38395.9	21	Sc	49	−46549.	26	Fe	54	−56245.5	29	Cu	60	−58346.	31	Ga	66	−63706.
21	Sc	43	−36174.	22	Ti	49	−48557.7	27	Co	54	−47994.	26	Fe	61	−59130.	32	Ge	66	−60740.
22	Ti	43	−29324.	23	V	49	−47950.2	24	Cr	55	−55113.	27	Co	61	−62930.	29	Cu	67	−67291.
19	K	44	−35780.	24	Cr	49	−45390.	25	Mn	55	−57704.8	28	Ni	61	−64220.	30	Zn	67	−67863.

31	Ga	67	−66865.
32	Ge	67	−62460.
29	Cu	68	−65410.
30	Zn	68	−69994.
31	Ga	68	−67074.
32	Ge	68	−66570.
30	Zn	69	−68425.
31	Ga	69	−69326.2
32	Ge	69	−67100.7
33	As	69	−63200.
30	Zn	70	−69550.
31	Ga	70	−68897.
32	Ge	70	−70558.
33	As	70	−64322.
30	Zn	71	−67520.
31	Ga	71	−70134.7
32	Ge	71	−69902.
33	As	71	−67893.
34	Se	71	−63490.
30	Zn	72	−68144.
31	Ga	72	−68583.
32	Ge	72	−72579.1
33	As	72	−68219.
34	Se	72	−67620.
31	Ga	73	−69743.
32	Ge	73	−71293.
33	As	73	−70921.
34	Se	73	−68171.
31	Ga	74	−67820.
32	Ge	74	−73418.5
33	As	74	−70855.0
34	Se	74	−72212.2
35	Br	74	−65410.
36	Kr	74	−62310.
32	Ge	75	−71833.
33	Se	75	−72166.3
35	Br	75	−69444.
36	Kr	75	−64340.
32	Ge	76	−73209.3
33	As	76	−72286.
34	Se	76	−75257.
35	Br	76	−70630.
36	Kr	76	−69430.
32	Ge	77	−71170.
33	As	77	−73917.
34	Se	77	−74601.
35	Hr	77	−73236.
36	Kr	77	−70350.
33	As	78	−72750.
34	Se	78	−77020.5
35	Br	78	−73447.
36	Kr	78	−74143.
33	As	79	−73690.
34	Se	79	−75920.8
35	Br	79	−76074.7
36	Kr	79	−74455.
33	As	80	−71750.
34	Se	80	−77753.0
35	Br	80	−75882.2
36	Kr	80	−77891.
37	Rb	80	−72800.
34	Se	81	−76396.
35	Br	81	−77972.
36	Kr	81	−77670.
37	Rb	81	−75430.
34	Se	82	−77586.
35	Br	82	−77497.
36	Kr	82	−80589.4
37	Rb	82	−76419.
38	Sr	82	−76020.
35	Br	83	−79019.
36	Kr	83	−79984.7
37	Rb	83	−79420.
38	Sr	83	−77120.
35	Br	84	−77730.
36	Kr	84	−82432.6
37	Rb	84	−79752.5
38	Sr	84	−80638.0
39	Y	84	−74340.
35	Br	85	−78680.
36	Kr	85	−81483.
37	Rb	85	−82156.
38	Sr	85	−81049.
39	Y	85	−77789.
35	Br	86	−76200.
36	Kr	86	−83259.3
37	Rb	86	−82722.
38	Sr	86	−84499.1
39	Y	86	−79226.
40	Zr	86	−78030.
36	Kr	87	−80698.
37	Rb	87	−84590.8
38	Sr	87	−84864.9
39	Y	87	−83150.
40	Zr	87	−79650.
35	Br	88	
36	Kr	88	−79850.
37	Kr	88	−82650.
38	Sr	88	−87874.
39	Y	88	−84273.
40	Zr	88	−83770.
41	Nb	88	−76570.
36	Kr	89	−77700.
37	Rb	89	−82300.
38	Sr	89	−86215.
39	Y	89	−87678.3
40	Zr	89	−84845.
41	Nb	89	−80960.
36	Kr	90	−74780.
37	Rb	90	−79340.
38	Sr	90	−85932.
39	Y	90	−86476.
40	Zr	90	−88770.2
41	Nb	90	−82660.
42	Mo	90	−80160.
37	Rb	91	−78180.
38	Sr	91	−83683.
39	Y	91	−86353.
40	Zr	91	−87892.8
41	Nb	91	−86750.
42	Mo	91	−82290.
37	Rb	92	−75320.
38	Sr	92	−82920.
39	Y	92	−84834.
40	Zr	92	−88461.7
41	Nb	92	−86431.
42	Mo	92	−86804.3
43	Tc	92	−78750.
38	Sr	93	−79450.
39	Y	93	−84250.
40	Zr	93	−87140.
41	Nb	93	−87203.5
42	Mo	93	−86785.
43	Tc	93	−83599.
38	Sr	94	−78820.
39	Y	94	−82270.
40	Zr	94	−87267.0
41	Nb	94	−86346.
42	Mo	94	−88406.5
43	Tc	94	−84146.
39	Y	95	−81460.
40	Zr	95	−85663.1
41	Nb	95	−86784.1
42	Mo	95	−87708.9
43	Tc	95	−86050.
44	Ru	95	−84018.
39	Y	96	−78530.
40	Zr	96	−85429.8
41	Nb	96	−85644.
42	Mo	96	−88794.2
43	Tc	96	−85860.
44	Ru	96	−86071.
40	Zr	97	−82934.
41	Nb	97	−85606.
42	Mo	97	−87538.9
43	Tc	97	−87240.
44	Ru	97	−86040.
45	Rh	97	−82550.
40	Zr	98	−82010.
41	Nb	98	−83510.

Appendix 5 Atomic Mass Tables—Continued

Z		A	M-A(keV)
42	Md	98	−88109.7
43	Tc	98	−86520.
44	Ru	98	−88221.5
45	Rh	98	−84020.
41	Nb	99	−82860.
42	Mo	99	−85957.
43	Tc	99	−87327.
44	Ru	99	−87619.0
45	Rh	99	−85519.
46	Pd	99	−81720.
41	Nb	100	−80090.
42	Mo	100	−86185.3
43	Tc	100	−85850.
44	Ru	100	−89218.7
45	Rh	100	−85579.
46	Pd	100	−84980.
42	Mo	101	−83504.
43	Tc	101	−86324.
44	Ru	101	−87953.2
45	Rh	101	−87393.
46	Pd	101	−85630.
42	Mo	102	−83600.
43	Tc	102	−84600.
44	Ru	102	−89097.9
45	Rh	102	−86774.
46	Pd	102	−87923.
47	Ag	102	−82620.
43	Tc	103	−84920.

Z		A	M-A(keV)
44	Ru	103	−87274.
45	Rh	103	−88014.4
46	Pd	103	−87460.
47	Ag	103	−84870.
43	Tc	104	−82240.
44	Ru	104	−88089.9
45	Rh	104	−86945.
46	Pd	104	−89411.
47	Ag	104	−85141.
48	Cd	104	−83940.
43	Tc	105	−82590.
44	Ru	105	−85995.
45	Rh	105	−87866.
46	Pd	105	−88431.
47	Ag	105	−87130.
48	Cd	105	−84330.
44	Ru	106	−86328.
45	Rh	106	−86367.
46	Pd	106	−89907.
47	Ag	106	−86943.
48	Cd	106	−87128.1
49	In	106	−80630.
44	Ru	107	−83710.
45	Rh	107	−86858.
46	Pd	107	−88367.9
47	Ag	107	−88402.8
48	Cd	107	−86986.
49	In	107	−83500.

Z		A	M-A(keV)
44	Ru	108	−83700.
45	Rh	108	−85000.
46	Pd	108	−89524.
47	Ag	108	−87607.
48	Cd	108	−89248.1
49	In	108	−84100.
50	Sn	108	
45	Rh	109	−85100.
46	Po	109	−87602.
47	Ag	109	−88717.4
48	Cd	109	−88558.
49	In	109	−86538.
45	Rh	110	−82800.
46	Pd	110	−88338.
47	Ag	110	−87470.
48	Cd	110	−90342.4
49	In	110	−86412.
46	Pd	111	−86010.
47	Ag	111	−88196.
48	Cd	111	−89246.4
49	In	111	−88160.
50	Sn	111	−85640.
46	Pd	112	−86268.
47	Ag	112	−86568.
48	Cd	112	−90574.6
49	In	112	−87984.
50	Sn	112	−88644.

Z		A	M-A(keV)
47	Ag	113	−87041.
48	Cd	113	−89041.3
49	In	113	−89339.
50	Sn	113	−88316.
51	Sb	113	−83846.
47	Ag	114	−85420.
48	Cd	114	−90017.8
49	In	114	−88579.
50	Sn	114	−90565.
51	Sb	114	−84290.
47	Ag	115	−84830.
48	Cd	115	−88089.
49	In	115	−89542.
50	Sn	115	−90031.
51	Sb	115	−87001.
47	Ag	116	−82610.
48	Cd	116	−88712.3
49	In	116	−88195.
50	Sn	116	−91522.7
51	Sb	116	−86970.
52	Te	116	−85410.
48	Cd	117	−86405.
49	In	117	−88925.
50	Sn	117	−90392.4
51	Sb	117	−88572.
52	Ie	117	−85070.
48	Cd	118	−86650.
49	In	118	−87450.

Z		A	M-A(keV)
50	Sn	118	−91652.0
51	Sb	118	−87956.
52	Te	118	−87660.
48	Cd	119	−84070.
49	In	119	−87570.
50	Sn	119	−90061.6
51	Sb	119	−89483.
52	Te	119	−87189.
49	In	120	−85700.
50	Sn	120	−91100.2
51	Sb	120	−88415.
52	Te	120	−89400.
53	I	120	−84000.
49	In	121	−85610.
50	Sn	121	−89210.
51	Sb	121	−89593.2
52	Te	121	−88305.
53	I	121	−85950.
54	Xe	121	−82160.
49	In	122	−83200.
50	Sn	122	−89942.5
51	Sb	122	−88320.
52	Te	122	−90291.
53	I	122	−86151.
49	In	123	−83300.
50	Sn	123	−87803.
51	Sb	123	−89223.8
52	Te	123	−89163.

53	I	123	−87810.
54	Xe	123	−85010.
49	In	124	−80900.
50	Sn	124	−88237.0
51	Sb	124	−87584.
52	Te	124	−90500.
53	I	124	−87330.
54	Xe	124	−87450.
50	Sn	125	−85933.
51	Sb	125	−88275.
52	Te	125	−89032.
53	I	125	−88883.
54	Xe	125	−86980.
55	Cs	125	−83910.
50	Sn	126	−86030.
51	Sb	126	−86330.
52	Te	126	−90053.
53	I	126	−87903.
54	Xe	126	−89154.
55	Cs	126	−84350.
50	Sn	127	−83600.
51	Sb	127	−86695.
52	Te	127	−88295.
53	I	127	−88984.3
54	Xe	127	−88280.
55	Cs	127	−86180.
56	Ba	127	−82580.
50	Sn	128	−83400.
51	Sb	128	−84700.
52	Te	128	−88978.
53	I	128	−87710.
54	Xe	128	−89850.
55	Cs	128	−85920.
56	Ba	128	−85220.
51	Sb	129	−84520.
52	Te	129	−87023.
53	I	129	−88503.
54	Xe	129	−88692.0
55	Cs	129	−87590.
56	Ba	129	−85140.
57	La	129	−81140.
51	Sb	130	−81940.
52	Te	130	−87337.
53	I	130	−86930.
54	Xe	130	−89880.
55	Cs	130	−86889.
56	Ba	130	−87331.
57	La	130	−81730.
52	Te	131	−85161.
53	I	131	−87440.6
54	Xe	131	−88411.0
55	Cs	131	−88056.
56	Ba	131	−86892.
57	La	131	−83930.
58	Ce	131	−78710.
52	Te	132	−85209.
53	I	132	−85714.
54	Xe	132	−89271.9
55	Cs	132	−87193.
56	Ba	132	−88380.
57	La	132	−83560.
58	Ce	132	−82360.
53	I	133	−85930.
54	Xe	133	−87732.
55	Cs	133	−88160.
56	Ba	133	−87672.
57	La	133	−85470.
58	Ce	133	−82670.
53	I	134	−83970.
54	Xe	134	−88120.5
55	Cs	134	−86793.
56	Ba	134	−88852.
57	La	134	−85080.
58	Ce	134	−84940.
53	I	135	−83810.
54	Xe	135	−86610.
55	Cs	135	−87770.
56	Ba	135	−87980.
57	La	135	−86730.
58	Ce	135	−84630.
53	I	136	−79420.
54	Xe	136	−86422.
55	Cs	136	−86310.
56	Ba	136	−89140.
57	La	136	−86270.
58	Ce	136	−86550.
54	Xe	137	−82810.
55	Cs	137	−86850.
56	Ba	137	−88020.
57	La	137	−87520.
58	Ce	137	−86320.
59	Pr	137	−83500.
54	Xe	138	−80290.
55	Cs	138	−83090.
56	Ba	138	−88490.
57	La	138	−86710.
58	Ce	138	−87720.
59	Pr	138	−83400.
54	Xe	139	−76530.
55	Cs	139	−81130.
56	Ba	139	−85130.
57	La	139	−87428.
58	Ce	139	−87158.
59	Pr	139	−85160.
60	Nd	139	−82360.
55	Cs	140	−77210.
56	Ba	140	−83307.
57	La	140	−84357.
58	Ce	140	−88125.
59	Pr	140	−84758.
60	Nd	140	−84460.
56	Ba	141	−80060.
57	La	141	−83062.
58	Ce	141	−85492.
59	Pr	141	−86072.
60	Nd	141	−84272.
61	Pm	141	−80650.
56	Ba	142	−77920.
57	La	142	−80120.
58	Ce	142	−84631.
59	Pr	142	−83854.
60	Nd	142	−86010.
61	Pm	142	−81210.
57	La	143	−78370.
58	Ce	143	−81665.
59	Pr	143	−83106.
60	Nd	143	−84039.
61	Pm	143	−82910.
62	Sm	143	−79590.
58	Ce	144	−80488.
59	Pr	144	−80808.
60	Nd	144	−83797.
61	Pm	144	−81500.
62	Sm	144	−81980.
58	Ce	145	−77060.
59	Pr	145	−79664.
60	Nd	145	−81469.
61	Pm	145	−81326.
62	Sm	145	−80672.
63	Eu	145	−77880.
58	Ce	146	−75760.
59	Pr	146	−76760.
60	Nd	146	−80959.
61	Pm	146	−79518.
62	Sm	146	−81046.
63	Eu	146	−77184.
64	Gd	146	−76080.
59	Pr	147	−75630.
60	Nd	147	−78175.
61	Pm	147	−79075.
62	Sm	147	−79300.
63	Eu	147	−77500.
64	Gd	147	−75300.
59	Pr	148	−72730.
60	Nd	148	−77435.
61	Pm	148	−76921.
62	Sm	148	−79371.
63	Eu	148	−76280.
64	Gd	148	−76287.
65	Tb	148	−70670.
60	Nd	149	−74405.

Appendix 5 Atomic Mass Tables—Continued

Z		A	M-A(keV)	Z		A	M-A(keV)	Z		A	M-A(keV)	Z		A	M-A(keV)	Z		A	M-A(keV)
61	Pm	149	−76074.	63	Eu	153	−73361.	65	Tb	158	−69428.	65	Tb	164	−62150.	70	Yb	169	−60050.
62	Sm	149	−77145.	64	Gd	153	−73118.	66	Dy	158	−70374.	66	Dy	164	−65949.	71	Lu	169	−57790.
63	Eu	149	−76390.	65	Tb	153	−71270.	67	Ho	158	−66330.	67	Ho	164	−64840.	68	Er	170	−60020.
64	Gd	149	−75170.	66	Dy	153	−69170.	63	Eu	159	−66290.	68	Er	164	−65867.	69	Tm	170	−59560.
65	Tb	149	−71400.	67	Ho	153	−64950.	64	Gd	159	−68586.	69	Tm	164	−61905.	70	Yb	170	−60530.
60	Nd	150	−73666.	62	Sm	154	−72393.	65	Tb	159	−69534.	66	Dy	165	−63512.	71	Lu	170	−56980.
61	Pm	150	−73630.	63	Eu	154	−71675.	66	Dy	159	−69154.	67	Ho	165	−64811.	68	Er	171	−57630.
62	Sm	150	−77056.	64	Gd	154	−73653.	67	Ho	159	−67350.	68	Er	165	−64440.	69	Tm	171	−59120.
63	Eu	150	−74807.	65	Tb	154	−70250.	64	Gd	160	−67891.	69	Tm	165	−62840.	70	Yb	171	−59220.
64	Gd	150	−75818.	66	Dy	154	−70460.	65	Tb	160	−67862.	70	Yb	165	−60140.	71	Lu	171	−57620.
65	Tb	150	−71027.	67	Ho	154	−64960.	66	Dy	160	−69673.	66	Dy	166	−62589.	68	Er	172	−56510.
66	Dy	150	−69310.	68	Er	154	−62630.	67	Ho	160	−66370.	67	Ho	166	−63071.	69	Tm	172	−57400.
60	Nd	151	−71000.	62	Sm	155	−70140.	64	Gd	161	−65460.	68	Er	166	−64918.	70	Yb	172	−59280.
61	Pm	151	−73403.	63	Eu	155	−71789.	65	Tb	161	−67465.	69	Tm	166	−61940.	71	Lu	172	−56580.
62	Sm	151	−74594.	64	Gd	155	−72037.	66	Dy	161	−68049.	70	Yb	166	−61620.	69	Tm	173	−56370.
63	Eu	151	−74670.	65	Tb	155	−71140.	67	Ho	161	−67250.	67	Ho	167	−62280.	70	Yb	173	−57690.
64	Gd	151	−74270.	66	Dy	155	−69040.	68	Er	161	−65250.	68	Er	167	−63285.	71	Lu	173	−57000.
65	Tb	151	−71580.	62	Sm	156	−69331.	69	Tm	161	−61730.	69	Tm	167	−62380.	69	Tm	174	−54060.
66	Dy	151	−68700.	63	Eu	156	−70046.	65	Tb	162	−65380.	70	Yb	167	−60420.	70	Yb	174	−57060.
61	Pm	152	−71250.	64	Gd	156	−72493.	66	Dy	162	−68182.	71	Lu	167	−57380.	71	Lu	174	−55560.
62	Sm	152	−74746.	65	Tb	156	−70090.	67	Ho	162	−66022.	67	Ho	168	−59680.	72	Hf	174	−55550.
63	Eu	152	−72889.	66	Dy	156	−70860.	68	Er	162	−66370.	68	Er	168	−62983.	69	Tm	175	−52320.
64	Gd	152	−74710.	63	Eu	157	−69500.	69	Tm	162	−61480.	69	Tm	168	−61266.	70	Yb	175	−54820.
65	Tb	152	−70530.	64	Gd	157	−70769.	65	Tb	163	−64680.	70	Yb	168	−61330.	71	Lu	175	−55290.
66	Dy	152	−70113.	65	Tb	157	−70709.	66	Dy	163	−66363.	71	Lu	168	−56730.	72	Hf	175	−54390.
67	Ho	152	−63750.	66	Dy	157	−69610.	67	Ho	163	−66353.	67	Ho	169	−58810.	69	Tm	176	−49190.
61	Pm	153	−70760.	63	Eu	158	−67130.	68	Er	163	−65143.	68	Er	169	−60909.	70	Yb	176	−53390.
62	Sm	153	−72560.	64	Gd	158	−70627.	69	Tm	163	−62873.	69	Tm	169	−61249.	71	Lu	176	−53410.

72	Hf	176	−54430.
70	Yb	177	−50850.
71	Lu	177	−52230.
72	Hf	177	−52720.
73	Ta	177	−51560.
71	Lu	178	−50020.
72	Hf	178	−52270.
73	Ta	178	−50360.
71	Lu	179	−48930.
72	Hf	179	−50270.
73	Ta	179	−50150.
71	Lu	180	−46230.
72	Hf	180	−49530.
73	Ta	180	−48862.
74	W	180	−49365.
72	Hf	181	−47407.
73	Ta	181	−48430.
74	W	181	−48240.
72	Hf	182	−45920.
73	Ta	182	−46418.
74	W	182	−48156.
75	Re	182	−45296.
72	Hf	183	−43000.
73	Ta	183	−45204.
74	W	183	−46272.
75	Re	183	−45400.
73	Ta	184	−42870.
74	W	184	−45619.
75	Re	184	−43990.
76	Os	184	−44010.
73	Ta	185	−41400.
74	W	185	−43296.
75	Re	185	−43725.
76	Os	185	−42743.
73	Ta	186	−38740.
74	W	186	−42438.
75	Re	186	−41900.
76	Os	186	−42970.
77	Ir	186	−39140.
74	W	187	−39827.
75	Re	187	−41140.
76	Os	187	−41141.
77	Ir	187	−39530.
74	W	188	−38362.
75	Re	188	−38793.
76	Os	188	−40909.
77	Ir	188	−38077.
78	Pt	188	−37570.
75	Re	189	−37840.
76	Os	189	−38840.
77	Ir	189	−38270.
78	Pt	189	−36690.
75	Re	190	−35440.
76	Os	190	−38840.
77	Ir	190	−36490.
78	Pt	190	−37300.
79	Au	190	−32870.
76	Os	191	−36360.
77	Ir	191	−36670.
78	Pt	191	−35910.
79	Au	191	−33960.
76	Os	192	−35910.
77	Ir	192	−34740.
78	Pt	192	−36190.
79	Au	192	−32950.
80	Hg	192	−31520.
76	Os	193	−33322.
77	Ir	193	−34454.
78	Pt	193	−34409.
79	Au	193	−33310.
80	Hg	193	−30970.
76	Os	194	−32388.
77	Ir	194	−32485.
78	Pt	194	−34721.
79	Au	194	−32212.
80	Hg	194	−31860.
81	Tl	194	−26480.
77	Ir	195	−31780.
78	Pt	195	−32776.
79	Au	195	−32555.
80	Hg	195	−31090.
81	Tl	195	−28090.
77	Ir	196	−29570.
78	Pt	196	−32633.
79	Au	196	−31154.
80	Hg	196	−31838.
81	Tl	196	−27240.
82	Pb	196	−24410.
77	Ir	197	−28420.
78	Pt	197	−30415.
79	Au	197	−31166.
80	Hg	197	−30403.
81	Tl	197	−28200.
82	Pb	197	−24130.
78	Pt	198	−29905.
79	Au	198	−29592.
80	Hg	198	−30966.
81	Tl	198	−27510.
82	Pb	198	−25700.
83	Bi	198	−18290.
78	Pt	199	−27404.
79	Au	199	−29088.
80	Hg	199	−29547.
81	Tl	199	−28450.
82	Pb	199	−25280.
83	Bi	199	−20080.
78	Pt	200	−26610.
79	Au	200	−27290.
80	Hg	200	−29503.
81	Tl	200	−27049.
82	Pb	200	−26110.
83	Bi	200	−19620.
84	Po	200	−16000.
79	Au	201	−26160.
80	Hg	201	−27658.
81	Tl	201	−27250.
82	Pb	201	−25280.
83	Bi	201	−21080.
84	Po	201	−15820.
79	Au	202	−24110.
80	Hg	202	−27346.
81	Tl	202	−26128.
82	Pb	202	−26078.
83	Bi	202	−20610.
84	Po	202	−17580.
79	Au	203	−23160.
80	Hg	203	−25262.
81	Tl	203	−25753.
82	Pb	203	−24936.
83	Bi	203	−21750.
84	Po	203	−17260.
80	Hg	204	−24689.
81	Tl	204	−24344.
82	Pb	204	−25109.
83	Bi	204	−20670.
84	Po	204	−18200.
80	Hg	205	−22160.
81	Tl	205	−23807.
82	Pb	205	−23772.
83	Bi	205	−21068.
84	Po	205	−17510.
81	Tl	206	−22259.
82	Pb	206	−23783.
83	Bi	206	−20131.
84	Po	206	−18328.
85	At	206	−12300.
81	Tl	207	−21005.
82	Pb	207	−22446.
83	Bi	207	−20085.
84	Po	207	−17178.
85	At	207	−13450.
81	Tl	208	−16754.
82	Pb	208	−21750.
83	Bi	208	−18880.
84	Po	208	−17472.
85	At	208	−12470.
81	Tl	209	−13697.
82	Pb	209	−17622.
83	Bi	209	−18262.
84	Po	209	−16370.
85	At	209	−12885.

Appendix 5 Atomic Mass Tables—Continued

Z		A	M-A(keV)
81	Tl	210	−9264.
82	Pb	210	−14730.
83	Bi	210	−14791.
84	Po	210	−15950.
85	At	210	−12075.
86	Rn	210	−9743.
82	Pb	211	−10486.
83	Bi	211	−11830.
84	Po	211	−12429.
85	At	211	−11679.
86	Rn	211	−8787.
82	Pb	212	−7541.
83	Bi	212	−8124.
84	Po	212	−10371.
85	At	212	−8641.
86	Rn	212	−8656.
82	Pb	213	−3450.
83	Bi	213	−5294.
84	Po	213	−6683.
85	At	213	−6460.
86	Rn	213	−5652.
82	Pb	214	−218.
83	Bi	214	−1224.
84	Po	214	−4470.
85	At	214	−3410.
86	Rn	214	−4310.
87	Fr	214	−930.
83	Bi	215	1700.
84	Po	215	−537.
85	At	215	−1245.
86	Rn	215	−1220.
84	Po	216	1790.
85	At	216	2246.
86	Rn	216	254.
87	Fr	216	3080.
85	At	217	4329.
86	Rn	217	3629.
87	Fr	217	4430.
84	Po	218	8318.
85	At	218	8017.
86	Rn	218	5219.
87	Fr	218	7020.
85	At	219	10520.
86	Rn	219	8832.
87	Fr	219	8622.
88	Ra	219	9360.
86	Rn	220	10620.
87	Fr	220	11492.
88	Ra	220	10274.
86	Rn	221	14190.
87	Fr	221	13211.
88	Ra	221	12940.
89	Ac	221	14600.
86	Rn	222	16329.
87	Fr	222	16420.
88	Ra	222	14322.
89	Ac	222	16540.
87	Fr	223	18383.
88	Ra	223	17233.
89	Ac	223	17832.
87	Fr	224	21970.
88	Ra	224	18832.
89	Ac	224	20200.
90	Th	224	20005.
88	Ra	225	21916.
89	Ac	225	21566.
90	Th	225	22288.
88	Ra	226	23623.
89	Ac	226	24370.
90	Th	226	23195.
91	Pa	226	25900.
88	Ra	227	27161.
89	Ac	227	25851.
90	Th	227	25807.
91	Pa	227	26837.
88	Ra	228	29005.
89	Ac	228	28950.
90	Th	228	26780.
91	Pa	228	28880.
92	U	228	29236.
90	Th	229	29483.
91	Pa	229	29828.
92	U	229	31187.
90	Th	230	30820.
91	Pa	230	32074.
92	U	230	31612.
90	Th	231	33804.
91	Pa	231	33419.
92	U	231	33780.
90	Th	232	35512.
91	Pa	232	35966.
92	U	232	34621.
90	Th	233	38628.
91	Pa	233	37383.
92	U	233	36814.
93	Np	233	37880.
90	Th	234	40596.
91	Pa	234	40332.
92	U	234	38102.
93	Np	234	39920.
91	Pa	235	42310.
92	U	235	40906.
93	Np	235	41031.
91	Pa	236	45860.
92	U	236	42510.
93	Np	236	43429.
94	Pu	236	42914.
91	Pa	237	47580.
92	U	237	45277.
93	Np	237	44763.
94	Pu	237	44989.
92	U	238	47291.
93	Np	238	47408.
94	Pu	238	46118.
95	Am	238	48380.
92	U	239	50579.
93	Np	239	49297.
94	Pu	239	48573.
95	Am	239	49383.
92	U	240	52716.
93	Np	240	52240.
94	Pu	240	50190.
95	Am	240	51490.
96	Cm	240	51739.
93	Np	241	54210.
94	Pu	241	52849.
95	Am	241	52838.
96	Cm	241	53599.

Appendix 6 Mass Stopping Powers and Ranges, Protons

	Beryllium, Z = 4		*Carbon, Z = 6*		*Aluminum, Z = 13*	
Energy (MeV)	*Stopping power (MeV · m²/kg)*	*Range (kg/m²)*	*Stopping power (MeV · m²/kg)*	*Range (kg/m²)*	*Stopping power (Mev · m²/kg)*	*Range (kg/m²)*
1	22.3	0.029	22.9	0.039	17.4	0.042
2	13.4	0.089	14.1	0.097	11.0	0.116
3	9.85	0.177	10.5	0.180	8.31	0.222
4	7.87	0.291	8.40	0.287	6.75	0.356
5	6.60	0.431	7.07	0.418	5.72	0.518
6	5.70	0.594	6.13	0.570	5.00	0.705
7	5.04	0.781	5.43	0.744	4.45	0.918
8	4.52	0.991	4.83	0.938	4.02	1.16
9	4.11	1.22	4.45	1.15	3.67	1.42
10	3.77	1.48	4.09	1.39	3.39	1.70
15	2.71	3.07	2.95	2.85	2.47	3.45
20	2.14	5.16	2.33	4.77	1.94	5.73
25	1.78	7.74	1.95	7.14	1.66	8.5
30	1.54	10.8	1.68	9.91	1.43	11.8
35	1.36	14.2	1.48	13.1	1.27	15.5
40	1.22	18.1	1.33	16.7	1.14	19.6
45	1.11	22.5	1.21	20.6	1.04	24.2
50	1.02	27.2	1.11	24.9	0.961	29.2
55	0.944	32.3	1.03	29.6	0.893	34.6
60	0.880	37.8	0.965	34.6	0.835	40.4
65	0.827	43.6	0.906	39.9	0.785	46.6
70	0.781	49.8	0.856	45.6	0.742	53.1
75	0.740	56.4	0.811	51.6	0.705	60.1
80	0.704	63.3	0.772	57.9	0.671	67.3
85	0.672	70.6	0.737	64.6	0.641	75.0
90	0.643	78.2	0.706	71.5	0.615	82.9
95	0.618	86.2	0.676	78.7	0.591	91.2
100	0.595	94.4	0.653	86.2	0.569	99.9

Appendix 6 Mass Stopping Powers and Ranges, Protons—Continued

	Copper, Z = 29		*Lead, Z = 82*		*Water, H_2O*	
Energy (MeV)	*Stopping power (Mev · m²/kg)*	*Range (kg/m²)*	*Stopping power (MeV · m²/kg)*	*Range (kg/m²)*	*Stopping power (MeV · m²/kg)*	*Range (kg/m²)*
1	12.1	0.061	6.29	0.116	27.1	0.039
2	8.05	0.165	4.48	0.308	16.4	0.088
3	6.22	0.308	3.59	0.560	12.1	0.160
4	5.15	0.486	3.06	0.863	9.67	0.253
5	4.43	0.696	2.69	1.21	8.13	0.367
6	3.90	0.937	2.41	1.61	7.04	0.499
7	3.50	1.21	2.20	2.04	6.23	0.650
8	3.19	1.51	2.02	2.52	5.60	0.820
9	2.93	1.83	1.88	3.03	5.09	1.01
10	2.71	2.19	1.75	3.58	4.68	1.21
15	2.01	4.36	1.34	6.88	3.37	2.49
20	1.62	7.14	1.10	11.0	2.66	4.18
25	1.37	10.5	0.944	16.0	2.22	6.25
30	1.19	14.4	0.830	21.6	1.91	8.68
35	1.06	18.9	0.744	28.0	1.69	11.5
40	0.959	23.9	0.677	35.0	1.52	14.6
45	0.877	29.3	0.622	42.8	1.38	18.1
50	0.810	35.2	0.578	51.1	1.27	21.8
55	0.754	41.7	0.540	60.1	1.18	25.9
60	0.707	48.5	0.507	69.6	1.10	30.4
65	0.666	55.8	0.480	79.8	1.03	35.1
70	0.630	63.5	0.455	90.5	0.973	40.1
75	0.599	71.7	0.434	102.	0.922	45.3
80	0.571	80.2	0.415	114.	0.877	50.9
85	0.546	89.2	0.397	126.	0.838	56.7
90	0.524	98.5	0.382	139.	0.802	62.8
95	0.504	108.	0.368	152.	0.770	69.2
100	0.486	118.	0.355	166.	0.741	75.8

Data selected from H. Bischel, "Charged particle interactions," in *Radiation Dosimetry,* vol. 1, edited by F. H. Attix and W. C. Roesch (Academic, New York, 1968), pp. 177 and 205.

Appendix 7 Mass Stopping Powers and Ranges, Electrons

Hydrogen, $Z = 1$

	Stopping Power (MeV · m²/kg)			
Energy (MeV)	*Collision*	*Radiation*	*Total*	*Range (kg/m²)*
0.010	5.147E 00	1.970E−04	5.147E 00	1.071E−03
0.020	2.928E 00	1.969E−04	2.928E 00	3.767E−03
0.030	2.118E 00	1.983E−04	2.118E 00	7.846E−03
0.040	1.693E 00	2.003E−04	1.693E 00	1.317E−02
0.050	1.429E 00	2.026E−04	1.429E 00	1.963E−02
0.060	1.249E 00	2.050E−04	1.249E 00	2.713E−02
0.070	1.118E 00	2.076E−04	1.118E 00	3.562E−02
0.080	1.018E 00	2.103E−04	1.018E 00	4.500E−02
0.090	9.398E−01	2.122E−04	9.400E−01	5.523E−02
0.100	8.766E−01	2.152E−04	8.768E−01	6.626E−02
0.200	5.869E−01	2.480E−04	5.871E−01	2.111E−01
0.300	4.912E−01	2.874E−04	4.915E−01	3.993E−01
0.400	4.458E−01	3.305E−04	4.461E−01	6.139E−01
0.500	4.205E−01	3.779E−04	4.209E−01	8.453E−01
0.600	4.053E−01	4.291E−04	4.057E−01	1.088E−00
0.700	3.956E−01	4.836E−04	3.961E−01	1.337E 00
0.800	3.893E−01	5.407E−04	3.899E−01	1.592E 00
0.900	3.852E−01	6.033E−04	3.858E−01	1.850E 00
1.000	3.826E−01	6.647E−04	3.832E−01	2.110E 00
2.000	3.833E−01	1.360E−03	3.846E−01	4.730E 00
3.000	3.933E−01	2.157E−03	3.955E−0	7.294E 00
4.000	4.029E−01	3.026E−03	4.059E−0	9.789E 00
5.000	4.112E−01	3.954E−03	4.152E−01	1.222E 01
6.000	4.184E−01	4.918E−03	4.234E−01	1.461E 01
7.000	4.248E−01	5.911E−03	4.307E−01	1.695E 01
8.000	4.304E−01	6.931E−03	4.373E−01	1.925E 01
9.000	4.354E−01	7.999E−03	4.434E−01	2.153E 01
10.000	4.400E−01	9.064E−03	4.490E−01	2.377E 01
20.000	4.707E−01	2.042E−02	4.912E−01	4.498E 01
30.000	4.890E−01	3.255E−02	5.216E−01	6.471E 01
40.000	5.013E−01	4.512E−02	5.464E−01	8.343E 01
50.000	5.089E−01	5.797E−02	5.669E−01	1.014E 02
60.000	5.141E−01	7.102E−02	5.852E−01	1.187E 02
80.000	5.211E−01	9.756E−02	6.187E−01	1.520E 02
100.000	5.258E−01	1.245E−01	6.503E−01	1.835E 02

Data selected from M. J. Berger and S. M. Seltzer, "Tables of energy losses and ranges of electrons and positrons," in *Studies in Penetration of Charged Particles in Matter, Publication 1133* (National Academy of Sciences–National Research Council, Washington (1964), p. 205.

Appendix 7 Mass Stopping Powers and Ranges, Electrons—Continued

Carbon, $Z = 6$

	Stopping Power $(MeV \cdot m^2/kg)$			
Energy (MeV)	*Collision*	*Radiation*	*Total*	*Range* (kg/m^2)
0.010	2.015E 00	4.089E–04	2.016E 00	2.819E–03
0.020	1.178E 00	3.952E–04	1.178E 00	9.589E–03
0.030	8.634E–01	3.893E–04	8.638E–01	1.965E–02
0.040	6.958E–01	3.869E–04	6.962E–01	3.264E–02
0.050	5.909E–01	3.885E–04	5.913E–01	4.830E–02
0.060	5.188E–01	3.921E–04	5.192E–01	6.640E–02
0.070	4.661E–01	3.971E–04	4.665E–01	8.676E–02
0.080	4.259E–01	4.029E–04	4.263E–01	1.092E–01
0.090	3.941E–01	4.073E–04	3.945E–01	1.336E–01
0.100	3.685E–01	4.145E–04	3.689E–01	1.599E–01
0.200	2.493E–01	5.042E–04	2.498E–01	5.015E–01
0.300	2.097E–01	6.078E–04	2.103E–01	9.425E–01
0.400	1.907E–01	7.177E–04	1.914E–01	1.443E–00
0.500	1.801E–01	8.295E–04	1.809E–01	1.982E 00
0.600	1.735E–01	9.418E–04	1.745E–01	2.545E 00
0.700	1.693E–01	1.055E–03	1.704E–01	3.126E 00
0.800	1.665E–01	1.169E–03	1.677E–01	3.718E 00
0.900	1.646E–01	1.285E–03	1.659E–01	4.317E 00
1.000	1.634E–01	1.402E–03	1.648E–01	4.922E 00
2.000	1.619E–01	2.617E–03	1.645E–01	1.103E 01
3.000	1.645E–01	3.931E–03	1.684E–01	1.704E 01
4.000	1.670E–01	5.357E–03	1.724E–01	2.291E 01
5.000	1.692E–01	6.878E–03	1.761E–01	2.864E 01
6.000	1.710E–01	8.454E–03	1.795E–01	3.427E 01
7.000	1.726E–01	1.008E–02	1.826E–01	3.979E 01
8.000	1.739E–01	1.174E–02	1.856E–01	4.522E 01
9.000	1.751E–01	1.351E–02	1.886E–01	5.057E 01
10.000	1.761E–01	1.526E–02	1.914E–01	5.583E 01
20.000	1.825E–01	3.388E–02	2.164E–01	1.049E 02
30.000	1.859E–01	5.367E–02	2.396E–01	1.488E 02
40.000	1.882E–01	7.402E–02	2.622E–01	1.887E 02
50.000	1.899E–01	9.475E–02	2.847E–01	2.252E 02
60.000	1.914E–01	1.158E–01	3.071E–01	2.591E 02
80.000	1.936E–01	1.583E–01	3.519E–01	3.198E 02
100.000	1.953E–01	2.013E–01	3.966E–01	3.734E 02

Aluminum, $Z = 13$

Energy (MeV)	*Stopping Power (MeV · m²/kg)* Collision	Radiation	Total	*Range (kg/m²)*
0.010	1.657E 00	8.600E−04	1.658E 00	3.519E−03
0.020	9.885E−01	8.373E−04	9.893E−01	1.165E−02
0.030	7.316E−01	8.276E−04	7.325E−01	2.356E−02
0.040	5.932E−01	8.252E−04	5.940E−01	3.883E−02
0.050	5.059E−01	8.329E−04	5.067E−01	5.714E−02
0.060	4.456E−01	8.446E−04	4.465E−01	7.822E−02
0.070	4.014E−01	8.588E−04	4.023E−01	1.019E−01
0.080	3.676E−01	8.746E−04	3.684E−01	1.279E−01
0.090	3.408E−01	8.928E−04	3.417E−01	1.561E−01
0.100	3.191E−01	9.105E−04	3.200E−01	1.864E−01
0.200	2.188E−01	1.100E−03	2.199E−01	5.772E−01
0.300	1.848E−01	1.317E−03	1.861E−01	1.077E 00
0.400	1.691E−01	1.549E−03	1.706E−01	1.640E 00
0.500	1.603E−01	1.782E−03	1.621E−01	2.243E 00
0.600	1.551E−01	2.011E−03	1.571E−01	2.871E 00
0.700	1.517E−01	2.240E−03	1.540E−01	3.514E 00
0.800	1.496E−01	2.469E−03	1.521E−01	4.168E 00
0.900	1.482E−01	2.704E−03	1.509E−01	4.828E 00
1.000	1.473E−01	2.933E−03	1.502E−01	5.493E−01
2.000	1.476E−01	5.204E−03	1.528E−01	1.212E 01
3.000	1.508E−01	7.612E−03	1.584E−01	1.855E 01
4.000	1.537E−01	1.025E−02	1.639E−01	2.476E 01
5.000	1.561E−01	1.310E−02	1.692E−01	3.076E 01
6.000	1.581E−01	1.604E−02	1.741E−01	3.658E 01
7.000	1.598E−01	1.907E−02	1.789E−01	4.225E 01
8.000	1.613E−01	2.217E−02	1.835E−01	4.777E 01
9.000	1.626E−01	2.545E−02	1.880E−01	5.315E 01
10.000	1.637E−01	2.869E−02	1.924E−01	5.841E 01
20.000	1.709E−01	6.317E−02	2.341E−01	1.054E 02
30.000	1.747E−01	9.973E−02	2.745E−01	1.448E 02
40.000	1.773E−01	1.373E−01	3.146E−01	1.788E 02
50.000	1.792E−01	1.755E−01	3.547E−01	2.087E 02
60.000	1.808E−01	2.141E−01	3.949E−01	2.355E 02
80.000	1.831E−01	2.923E−01	4.754E−01	2.816E 02
100.000	1.849E−04	3.710E−01	5.559E−01	3.204E 02

Appendix 7 Mass Stopping Powers and Ranges, Electrons—Continued

Copper, $Z = 29$

	Stopping Power ($MeV \cdot m^2/kg$)			
Energy (MeV)	*Collision*	*Radiation*	*Total*	*Range (kg/m^2)*
0.010	1.328E 00	1.793E−03	1.329E 00	4.556E−03
0.020	8.120E−01	1.773E−03	8.137E−01	1.453E−02
0.030	6.078E−01	1.818E−03	6.096E−01	2.892E−02
0.040	4.962E−01	1.886E−03	4.981E−01	4.719E−02
0.050	4.252E−01	1.940E−03	4.272E−01	6.896E−02
0.060	3.759E−01	1.988E−03	3.779E−01	9.392E−02
0.070	3.396E−01	2.033E−03	3.417E−01	1.218E−01
0.080	3.118E−01	2.076E−03	3.138E−01	1.524E−01
0.090	2.896E−01	2.105E−03	2.917E−01	1.855E−01
0.100	2.717E−01	2.147E−03	2.738E−01	2.209E−01
0.200	1.882E−01	2.557E−03	1.908E−01	6.739E−02
0.300	1.603E−01	3.022E−03	1.634E−01	1.246E 00
0.400	1.473E−01	3.519E−03	1.508E−01	1.886E 00
0.500	1.396E−01	4.010E−03	1.436E−01	2.567E 00
0.600	1.353E−01	4.489E−03	1.398E−01	3.274E 00
0.700	1.327E−01	4.961E−03	1.376E−01	3.995E−00
0.800	1.310E−01	5.428E−03	1.364E−01	4.725E−00
0.900	1.299E−01	5.837E−03	1.358E−01	5.461E−00
1.000	1.293E−01	6.304E−03	1.356E−01	6.198E 00
2.000	1.305E−01	1.115E−02	1.417E−01	1.344E 01
3.000	1.338E−01	1.635E−02	1.502E−01	2.030E 01
4.000	1.367E−01	2.173E−02	1.584E−01	2.678E 01
5.000	1.391E−01	2.724E−02	1.663E−01	3.294E 01
6.000	1.411E−01	3.292E−02	1.740E−01	3.881E 01
7.000	1.428E−01	3.875E−02	1.815E−01	4.444E 01
8.000	1.442E−01	4.470E−02	1.889E−01	4.984E 01
9.000	1.455E−01	5.104E−02	1.965E−01	5.503E 01
10.000	1.466E−01	5.722E−02	2.038E−01	6.003E 01
20.000	1.535E−01	1.224E−01	2.759E−01	1.020E 02
30.000	1.573E−01	1.917E−01	3.490E−01	1.342E 02
40.000	1.597E−01	2.637E−01	4.234E−01	1.602E 02
50.000	1.616E−01	3.365E−01	4.981E−01	1.819E 02
60.000	1.631E−01	4.101E−01	5.732E−01	2.006E 02
80.000	1.653E−01	5.588E−01	7.242E−01	2.316E 02
100.000	1.670E−01	7.086E−01	8.756E−01	2.567E 02

Lead, $Z = 82$

	Stopping Power (MeV · m²/kg)			
Energy (MeV)	*Collision*	*Radiation*	*Total*	*Range (kg/m²)*
0.010	8.419E−01	4.513E−03	8.464E−01	8.251E−03
0.020	5.450E−01	4.620E−03	5.496E−01	2.335E−02
0.030	4.179E−01	4.827E−03	4.228E−01	4.434E−02
0.040	3.462E−01	5.074E−03	3.513E−01	7.044E−02
0.050	2.997E−01	5.262E−03	3.049E−01	1.011E−01
0.060	2.669E−01	5.426E−03	2.724E−01	1.359E−01
0.070	2.426E−01	5.576E−03	2.481E−01	1.744E−01
0.080	2.237E−01	5.716E−03	2.294E−01	2.164E−01
0.090	2.087E−01	5.812E−03	2.145E−01	2.615E−01
0.100	1.964E−01	5.944E−03	2.024E−01	3.096E−01
0.200	1.389E−01	7.251E−03	1.461E−01	9.100E−01
0.300	1.196E−01	8.460E−03	1.280E−01	1.648E−00
0.400	1.106E−01	9.623E−03	1.203E−01	2.457E−00
0.500	1.059E−01	1.078E−02	1.167E−01	3.303E−00
0.600	1.033E−01	1.194E−02	1.152E−01	4.166E−00
0.700	1.018E−01	1.310E−02	1.149E−01	5.036E−00
0.800	1.010E−01	1.425E−02	1.153E−01	5.906E−00
0.900	1.003E−01	1.545E−02	1.157E−01	6.771E−00
1.000	1.002E−01	1.661E−02	1.168E−01	7.631E−00
2.000	1.036E−01	2.802E−02	1.316E−01	1.571E 01
3.000	1.076E−01	3.999E−02	1.476E−01	2.288E 01
4.000	1.109E−01	5.212E−02	1.630E−01	2.933E 01
5.000	1.135E−01	6.437E−02	1.779E−01	3.519E 01
6.000	1.157E−01	7.678E−02	1.925E−01	4.060E 01
7.000	1.176E−01	8.933E−02	2.069E−01	4.561E 01
8.000	1.191E−01	1.020E−01	2.211E−01	5.028E 01
9.000	1.205E−01	1.146E−01	2.352E−01	5.466E 01
10.000	1.217E−01	1.275E−01	2.493E−01	5.879E 01
20.000	1.293E−01	2.614E−01	3.907E−01	9.060E 01
30.000	1.334E−01	4.003E−01	5.337E−01	1.124E 02
40.000	1.360E−01	5.422E−01	6.783E−01	1.290E 02
50.000	1.380E−01	6.923E−01	8.303E−01	1.423E 02
60.000	1.396E−01	8.434E−01	9.829E−01	1.534E 02
80.000	1.419E−01	1.148E 00	1.290E 00	1.711E 02
100.000	1.436E−01	1.455E 00	1.599E 00	1.850E 02

Appendix 7 Mass Stopping Powers and Ranges, Electrons—Continued

Water, H_2O

	Stopping Power (MeV · m²/kg)			
Energy (MeV)	*Collision*	*Radiation*	*Total*	*Range (kg/m²)*
0.010	2.320E 00	5.069E−04	2.321E 00	2.436E−03
0.020	1.350E 00	4.904E−04	1.351E 00	8.331E−03
0.030	9.879E−01	4.825E−04	9.884E−01	1.712E−02
0.040	7.951E−01	4.788E−04	7.956E−01	2.848E−02
0.050	6.747E−01	4.812E−04	6.751E−01	4.218E−02
0.060	5.919E−01	4.863E−04	5.924E−01	5.804E−02
0.070	5.315E−01	4.932E−04	5.320E−01	7.589E−02
0.080	4.854E−01	5.011E−04	4.859E−01	9.559E−02
0.090	4.491E−01	5.089E−04	4.496E−01	1.170E−01
0.100	4.197E−01	5.184E−04	4.202E−01	1.400E−01
0.200	2.844E−01	6.286E−04	2.850E−01	4.400E−01
0.300	2.394E−01	7.561E−04	2.401E−01	8.263E−01
0.400	2.181E−01	8.921E−04	2.190E−01	1.264E 00
0.500	2.061E−01	1.030E−03	2.071E−01	1.735E 00
0.600	1.989E−01	1.168E−03	2.000E−01	2.227E 00
0.700	1.942E−01	1.306E−03	1.955E−01	2.733E 00
0.800	1.911E−01	1.445E−03	1.926E−01	3.248E 00
0.900	1.890E−01	1.586E−03	1.906E−01	3.771E 00
1.000	1.876E−01	1.727E−03	1.893E−01	4.297E 00
2.000	1.858E−01	3.187E−03	1.889E−01	9.613E 00
3.000	1.884E−01	4.757E−03	1.931E−01	1.485E 01
4.000	1.909E−01	6.458E−03	1.974E−01	1.997E 01
5.000	1.931E−01	8.270E−03	2.014E−01	2.499E 01
6.000	1.949E−01	1.015E−02	2.051E−01	2.991E 01
7.000	1.964E−01	1.209E−02	2.085E−01	3.474E 01
8.000	1.978E−01	1.408E−02	2.119E−01	3.950E 01
9.000	1.989E−01	1.621E−02	2.152E−01	4.418E 01
10.000	2.000E−01	1.829E−02	2.183E−01	4.880E 01
20.000	2.064E−01	4.055E−02	2.470E−01	9.180E 01
30.000	2.100E−01	6.419E−02	2.742E−01	1.302E 02
40.000	2.125E−01	8.849E−02	3.010E−01	1.650E 02
50.000	2.144E−01	1.132E−01	3.276E−01	1.968E 02
60.000	2.160E−01	1.383E−01	3.543E−01	2.262E 02
80.000	2.185E−01	1.890E−01	4.075E−01	2.788E 02
100.000	2.204E−01	2.403E−01	4.607E−01	3.249E 02

Appendix 8 Electron Beam Percent Depth Dose Data

Percent Depth Dose, 10-MeV Electrons*

	Field Size (cm)						
Depth (*cm*)	*4 × 4*	*6 × 6*	*8 × 8*	*10 × 10*	*12 × 12*	*15 × 15*	*20 × 20*
1.0	90.7	92.2	93.2	94.1	94.8	95.6	95.9
1.5	93.7	94.7	95.5	96.4	97.4	98.1	98.1
2.0	100.0	100.0	100.0	100.0	100.0	100.0	100.0
2.5	97.0	98.3	99.1	99.6	99.8	100.0	100.0
3.0	81.0	88.0	91.1	91.7	90.7	89.8	89.7
3.5	59.1	62.3	64.7	66.2	65.7	65.1	64.8
4.0	35.0	37.6	39.6	40.8	40.6	40.4	40.3
4.5	13.7	16.0	17.8	18.8	18.0	17.1	16.8
5.0	3.2	3.8	4.1	4.5	4.5	4.4	4.5
5.5	1.0	1.2	1.3	1.4	1.4	1.5	1.7
6.0	0.9	1.1	1.2	1.3	1.3	1.4	1.6

Percent Depth Dose, 25-MeV Electrons*

	Field Size (cm)						
Depth (*cm*)	*4 × 4*	*6 × 6*	*8 × 8*	*10 × 10*	*12 × 12*	*15 × 15*	*20 × 20*
1	94.5	95.5	95.6	95.6	95.4	94.6	94.4
2	100.0	99.0	98.6	98.4	98.4	98.3	98.2
3	95.6	100.0	100.0	100.0	100.0	100.0	100.0
4	91.2	98.0	100.0	100.0	100.0	100.0	100.0
5	85.4	94.6	98.9	99.7	99.5	99.5	99.5
6	78.3	90.0	96.9	98.0	98.0	98.0	98.0
7	69.6	82.0	90.2	93.7	94.6	94.6	94.7
8	58.7	73.3	82.5	86.5	88.0	88.0	88.0
9	48.0	61.0	70.6	76.0	78.4	78.4	78.5
10	35.6	47.1	55.1	60.3	63.0	64.1	64.4
11	22.6	30.5	36.3	40.0	42.3	43.2	43.4
12	11.6	13.3	15.2	16.8	18.3	19.3	19.7
13	4.9	5.7	6.6	7.5	8.0	8.3	8.5
14	3.0	3.3	3.6	4.0	4.5	5.0	5.0
15	2.8	3.0	3.2	3.5	3.8	3.9	3.9

* 1.00 m = F (distance from origin of beam to water surface).

Appendix 9 Photon Cross Sections and Coefficients

Hydrogen, $Z = 1$ (μ/ρ in m^2/kg = 0.05975 barn/atom)

E(MeV)	*is*	*cs*	*pp(n)*	*pp(e)*	*pe*	*total*	*en*
0.001	0.0852	0.580			11.4	12.1	11.4
0.0015	0.165	0.498			2.93	3.59	2.93
0.002	0.248	0.414			1.11	1.77	1.11
0.003	0.378	0.277			0.281	0.936	0.283
0.004	0.463	0.188			0.106	0.757	0.110
0.005	0.515	0.134			0.0494	0.698	0.0543
0.006	0.547	0.100			0.0265	0.674	0.0328
0.008	0.582	0.0614			0.00981	0.653	0.0186
0.01	0.598	0.0413			0.00455	0.644	0.0157
0.015	0.609	0.0195			0.00116	0.630	0.0180
0.02	0.606	0.0112				0.617	0.0219
0.03	0.592	0.00509				0.597	0.0308
0.04	0.576	0.00288				0.579	0.0384
0.05	0.559	0.00185				0.561	0.0451
0.06	0.544	0.00129				0.545	0.0510
0.08	0.516					0.516	0.0604
0.1	0.492					0.492	0.0679
0.15	0.443					0.443	0.0804
0.2	0.406					0.406	0.0878
0.3	0.353					0.353	0.0952
0.4	0.317					0.317	0.0983
0.5	0.289					0.289	0.0991
0.6	0.267					0.267	0.0983
0.8	0.235					0.235	0.0956

1.	0.211					0.211	0.0928
1.5	0.172		0.000044			0.172	0.0851
2.	0.146		0.000177			0.146	0.0775
3.	0.115		0.000511	0.000040		0.116	0.0668
4.	0.0962		0.000830	0.000164		0.0972	0.0589
5.	0.0831		0.00111	0.000324		0.0845	0.0533
6.	0.0734		0.00136	0.000495		0.0752	0.0487
8.	0.0601		0.00179	0.000848		0.0627	0.0423
10.	0.0511		0.00214	0.00117		0.0544	0.0380
15.	0.0379		0.00279	0.00182		0.0425	0.0311
20.	0.0304		0.00328	0.00233		0.0360	0.0272
30.	0.0221		0.00397	0.00311		0.0292	0.0229
40.	0.0176		0.00446	0.00367		0.0257	0.0206
50.	0.0147		0.00485	0.00413		0.0237	0.0192
60.	0.0126		0.00517	0.00448		0.0222	0.0183
80.	0.00997		0.00566	0.00508		0.0207	0.0172
100.	0.00828		0.00604	0.00557		0.0199	0.0166

Mass coefficients in square meters per kilogram can be obtained by multiplying the tabulated values by the conversion factor given for each absorbing material. The following abbreviations are used:

is, value for incoherent scattering including binding effects
cs, value for coherent scattering
pp(n), value for pair production near a nucleus
pp(e), value for pair production near an atomic electron
pe, value for the atomic photoelectric effect
total, total value for incoherent, coherent, pair, and photoelectric processes
en, total value for energy absorption from the incoherent, pair, and photoelectric processes.

Data selected from E. Storm and H. I. Israel, *Photon Cross Sections from 0.001 to 100 MeV for Elements 1 through 100, LA-3753* (Los Alamos Scientific Laboratory, Univ. of California, 1967), p. 19; and E. Storm and H. I. Israel, "Photon cross sections from 1 keV to 100 MeV for elements $Z = 1$ to $Z = 100$," Nucl. Data Tables **7**, 573 (1970).

Appendix 9 Photon Cross Sections and Coefficients—Continued

Carbon, $Z = 6$ (μ/ρ in $m^2/kg = 0.005014$ barn/atom)

E(MeV)	*is*	*cs*	*pp(n)*	*pp(e)*	*pe*	*total*	*en*
0.001	0.300	21.5			43500.	43500.	43500.
0.0015	0.590	19.1			13800.	13800.	13800.
0.002	0.899	16.4			5980.	6000.	5980.
0.003	1.43	12.2			1780.	1790.	1780.
0.004	1.82	9.20			740.	751.	740.
0.005	2.09	7.19			371.	380.	371.
0.006	2.29	5.83			208.	216.	208.
0.008	2.55	4.19			79.9	86.6	79.9
0.01	2.74	3.24			39.3	45.3	39.4
0.015	3.03	1.96			10.6	15.6	10.7
0.02	3.19	1.29			4.14	8.62	4.25
0.03	3.30	0.672			1.10	5.07	1.27
0.04	3.30	0.409			0.423	4.13	0.643
0.05	3.25	0.274			0.201	3.72	0.463
0.06	3.19	0.196			0.111	3.50	0.410
0.08	3.05	0.114			0.0424	3.21	0.399
0.1	2.92	0.0745			0.0199	3.01	0.423
0.15	2.65	0.0339			0.00516	2.69	0.486
0.2	2.43	0.0193			0.00207	2.45	0.527
0.3	2.12	0.00853				2.13	0.572
0.4	1.90	0.00481				1.90	0.589
0.5	1.73	0.00308				1.73	0.593
0.6	1.60	0.00215				1.60	0.587
0.8	1.41	0.00121				1.41	0.574

1.	1.27					1.27	0.556
1.5	1.03		0.00160			1.03	0.508
2.	0.878		0.00640			0.884	0.467
3.	0.690		0.0184	0.000241		0.709	0.406
4.	0.577		0.298	0.000986		0.608	0.368
5.	0.498		0.0400	0.00194		0.540	0.342
6.	0.441		0.0490	0.00295		0.493	0.322
8.	0.360		0.0642	0.00506		0.429	0.295
10.	0.307		0.0766	0.00702		0.391	0.279
15.	0.227		0.100	0.0108		0.338	0.255
20.	0.182		0.117	0.0140		0.313	0.244
30.	0.133		0.141	0.0187		0.293	0.234
40.	0.105		0.158	0.0222		0.285	0.230
50.	0.0881		0.171	0.0248		0.284	0.227
60.	0.0758		0.181	0.0270		0.284	0.224
80.	0.0598		0.196	0.0303		0.286	0.221
100.	0.0497		0.207	0.0331		0.290	0.218

Mass coefficients in square meters per kilogram can be obtained by multiplying the tabulated values by the conversion factor given for each absorbing material. The following abbreviations are used:

is, value for incoherent scattering including binding effects
cs, value for coherent scattering
pp(n), value for pair production near a nucleus
pp(e), value for pair production near an atomic electron
pe, value for the atomic photoelectric effect
total, total value for incoherent, coherent, pair, and photoelectric processes
en, total value for energy absorption from the incoherent, pair, and photoelectric processes.

Data selected from E. Storm and H. I. Israel, *Photon Cross Sections from 0.001 to 100 MeV for Elements 1 through 100, LA-3753* (Los Alamos Scientific Laboratory, Univ. of California, 1967), p. 19; and E. Storm and H. I. Israel, "Photon cross sections from 1 keV to 100 MeV for elements $Z = 1$ to $Z = 100$," Nucl. Data Tables **7**, 573 (1970).

Appendix 9 Photon Cross Sections and Coefficients—Continued

Aluminum, $Z = 13$ (μ/ρ in m^2/kg = 0.002232 barn/atom)

E(MeV)	*is*	*cs*	*pp(n)*	*pp(e)*	*pe*	*total*	*en*
0.001	0.642	101.			52500.	52600.	52500.
0.0015	1.10	91.3			17800.	17900.	17800.
0.00156	1.15	90.2			16100.	16200.	16100.
					192000.	192000.	186000.
0.002	1.50	82.4			102000.	102000.	99400.
0.003	2.12	68.3			35400.	35500.	34800.
0.004	2.61	58.1			16000.	16100.	15800.
0.005	3.05	50.0			8460.	8510.	8370.
0.006	3.46	43.2			5020.	5070.	4980.
0.008	4.18	32.4			2170.	2210.	2160.
0.01	4.76	24.7			1150.	1180.	1140.
0.015	5.67	14.1			329.	349.	328.
0.02	6.15	9.18			137.	152.	137.
0.03	6.57	4.92			38.2	49.7	38.4
0.04	6.70	3.08			15.4	25.2	15.8
0.05	6.71	2.10			7.58	16.4	8.11
0.06	6.65	1.52			4.23	12.4	4.85
0.08	6.45	0.898			1.68	9.03	2.43
0.1	6.22	0.591			0.808	7.62	1.67
0.15	5.68	0.274			0.215	6.17	1.24
0.2	5.23	0.157			0.0881	5.48	1.22
0.3	4.57	0.0701			0.0254	4.67	1.26
0.4	4.10	0.0396			0.0108	4.15	1.28
0.5	3.75	0.0255			0.00585	3.78	1.30

0.6	3.47	0.0177			0.00365	3.49	1.27
0.8	3.05	0.00999			0.00189	3.06	1.24
1.	2.74	0.00642			0.00117	2.75	1.20
1.5	2.23	0.00288	0.00768			2.24	1.09
2.	1.90	0.00163	0.0303			1.93	1.01
3.	1.50		0.0865	0.000523		1.59	0.907
4.	1.25		0.140	0.00216		1.39	0.844
5.	1.08		0.187	0.00423		1.27	0.807
6.	0.954		0.229	0.00635		1.19	0.784
8.	0.781		0.299	0.0109		1.09	0.761
10.	0.665		0.357	0.0152		1.04	0.753
15.	0.492		0.465	0.236		0.981	0.751
20.	0.395		0.543	0.0303		0.968	0.750
30.	0.288		0.652	0.0404		0.980	0.753
40.	0.229		0.729	0.0481		1.01	0.755
50.	0.191		0.786	0.0533		1.03	0.747
60.	0.164		0.831	0.0580		1.05	0.738
80.	0.130		0.898	0.0650		1.09	0.728
100.	0.108		0.946	0.0700		1.13	0.710

Mass coefficients in square meters per kilogram can be obtained by multiplying the tabulated values by the conversion factor given for each absorbing material. The following abbreviations are used:

is, value for incoherent scattering including binding effects
cs, value for coherent scattering
pp(n), value for pair production near a nucleus
pp(e), value for pair production near an atomic electron
pe, value for the atomic photoelectric effect
total, total value for incoherent, coherent, pair, and photoelectric processes
en, total value for energy absorption from the incoherent, pair, and photoelectric processes.

Data selected from E. Storm and H. I. Israel, *Photon Cross Sections from 0.001 to 100 MeV for Elements 1 through 100, LA-3753* (Los Alamos Scientific Laboratory, Univ. of California, 1967), p. 19; and E. Storm and H. I. Israel, "Photon cross sections from 1 keV to 100 MeV for elements $Z = 1$ to $Z = 100$," Nucl. Data Tables **7**, 573 (1970).

Appendix 9 Photon Cross Sections and Coefficients—Continued

Copper, $Z = 29$ (μ/ρ in m^2/kg = 0.0009478 barn/atom)

E(MeV)	*is*	*cs*	*pp(n)*	*pp(e)*	*pe*	*total*	*en*
0.001	0.632	533.			1100000.	1100000.	1100000.
0.001096	0.730	528.			862000.	863000.	862000.
					985000.	986000.	985000.
0.0015	1.15	507.			460000.	461000.	460000.
0.002	1.69	477.			223000.	223000.	223000.
0.003	2.74	416.			78800.	79200.	78800.
0.004	3.73	359.			36100.	36500.	36100.
0.005	4.65	307.			19500.	19800.	19500.
0.006	5.49	264.			11700.	12000.	11700.
0.008	6.97	198.			5270.	5470.	5270.
0.008981	7.60	174.			3850.	4030.	3850.
					29900.	30100.	20300.
0.01	8.20	153.			22600.	22800.	16100.
0.015	10.3	92.9			7700.	7800.	6220.
0.02	11.6	64.0			3480.	3560.	2980.
0.03	13.0	35.6			1090.	1140.	986.
0.04	13.6	22.4			473.	509.	440.
0.05	13.9	15.5			244.	273.	231.
0.06	13.9	11.4			142.	167.	136.
0.08	13.7	6.86			59.2	79.8	58.7
0.1	13.3	4.52			30.0	47.8	30.9
0.15	12.4	2.10			8.72	23.2	10.8
0.2	11.5	1.21			3.64	16.3	6.08
0.3	10.1	0.544			1.10	11.7	3.81
0.4	9.10	0.308			0.487	9.89	3.29
0.5	8.33	0.198			0.268	8.80	3.12

0.6	7.72	0.138			0.169	8.03	3.00
0.8	6.79	0.0776			0.0862	6.95	2.84
1.	6.11	0.0499			0.0538	6.21	2.70
1.5	4.97	0.0222	0.0423		0.0250	5.06	2.46
2.	4.24	0.0126	0.158		0.0155	4.43	2.29
3.	3.34	0.00576	0.438	0.00117	0.00844	3.79	2.14
4.	2.79	0.00336	0.698	0.00480	0.00573	3.50	2.10
5.	2.41	0.00220	0.928	0.00938	0.00432	3.35	2.12
6.	2.13	0.00152	1.13	0.0146	0.00341	3.27	2.16
8.	1.74		1.46	0.0247	0.00242	3.22	2.24
10.	1.48		1.74	0.0340	0.00187	3.25	2.32
15.	1.10		2.25	0.0534	0.00115	3.40	2.49
20.	0.881		2.63	0.0677		3.58	2.56
30.	0.641		3.15	0.0899		3.88	2.64
40.	0.510		3.51	0.106		4.13	2.64
50.	0.426		3.78	0.117		4.33	2.59
60.	0.367		3.99	0.126		4.49	2.55
80.	0.289		4.30	0.140		4.73	2.48
100.	0.240		4.52	0.151		4.91	2.41

Mass coefficients in square meters per kilogram can be obtained by multiplying the tabulated values by the conversion factor given for each absorbing material. The following abbreviations are used:

is, value for incoherent scattering including binding effects
cs, value for coherent scattering
pp(n), value for pair production near a nucleus
pp(e), value for pair production near an atomic electron
pe, value for the atomic photoelectric effect
total, total value for incoherent, coherent, pair, and photoelectric processes
en, total value for energy absorption from the incoherent, pair, and photoelectric processes.

Data selected from E. Storm and H. I. Israel, *Photon Cross Sections from 0.001 to 100 MeV for Elements 1 through 100, LA-3753* (Los Alamos Scientific Laboratory, Univ. of California, 1967), p. 19; and E. Storm and H. I. Israel, "Photon cross sections from 1 keV to 100 MeV for elements $Z = 1$ to $Z = 100$," Nucl. Data Tables **7**, 573 (1970).

Appendix 9 Photon Cross Sections and Coefficients—Continued

Lead, $Z = 82$ (μ/ρ in m^2/kg = 0.0002907 barn/atom)

E(MeV)	*is*	*cs*	*pp(n)*	*pp(e)*	*pe*	*total*	*en*
0.001	1.24	4320.			1760000.	1760000.	1760000.
0.0015	2.28	4140.			801000.	805000.	801000.
0.002	3.32	3940.			433000.	437000.	433000.
0.002484	4.28	3740.			269000.	273000.	269000.
					720000.	724000.	720000.
0.002586	4.48	3700.			653000.	657000.	653000.
					950000.	954000.	950000.
0.003	5.25	3540.			671000.	675000.	671000.
0.003066	5.37	3510.			630000.	634000.	630000.
					730000.	734000.	730000.
0.003554	6.25	3320.			509000.	512000.	509000.
					540000.	543000.	540000.
0.003851	6.77	3210.			445000.	448000.	445000.
					465000.	468000.	465000.
0.004	7.03	3160.			424000.	427000.	424000.
0.005	8.69	2830.			245000.	248000.	245000.
0.006	10.3	2530.			156000.	159000.	156000.
0.008	13.2	2070.			75000.	77100.	75000.
0.01	15.7	1710.			42200.	43900.	42200.
0.013035	18.9	1330.			21400.	22700.	21400.
					54700.	56000.	43700.
0.015	20.5	1140.			36800.	38000.	30400.
0.0152	20.7	1120.			35900.	37000.	29700.
					50100.	51200.	39300.
0.015861	21.2	1070.			45100.	46200.	35800.
					52200.	53300.	40800.
0.02	23.9	805.			28500.	29300.	23500.
0.03	28.5	474.			9790.	10300.	8650.
0.04	31.3	314.			4530.	4880.	4140.
0.05	32.9	221.			2460.	2710.	2290.
0.06	33.8	165.			1500.	1700.	1410.
0.08	34.4	101.			680.	815.	654.

0.088004	34.4	84.5			526.	645.	509.
					2480.	2600.	857.
0.1	34.3	67.0			1800.	1900.	766.
0.15	33.2	31.6			627.	692.	392.
0.2	31.4	18.3			292.	342.	215.
0.3	28.0	8.27			101.	137.	89.1
0.4	25.4	4.69			48.7	78.8	49.5
0.5	23.3	3.02			28.4	54.7	32.9
0.6	21.7	2.10			18.6	42.4	24.6
0.8	19.1	1.18			9.86	30.1	16.7
1.	17.2	0.760			6.21	24.2	13.0
1.5	14.0	0.337	0.566		2.85	17.8	9.48
2.	12.0	0.191	1.70		1.71	15.6	8.40
3.	9.42	0.0857	3.94	0.00330	0.888	14.3	8.11
4.	7.89	0.0495	5.77	0.0136	0.588	14.3	8.54
5.	6.81	0.0322	7.30	0.0266	0.435	14.6	9.07
6.	6.02	0.0223	8.54	0.0411	0.336	15.0	9.51
8.	4.92	0.0126	10.5	0.0692	0.232	15.8	10.3
10.	4.19	0.00814	12.2	0.0959	0.179	16.7	10.9
15.	3.10	0.00357	15.5	0.148	0.109	18.8	11.9
20.	2.49	0.00170	18.1	0.186	0.0790	20.9	12.4
30.	1.81		21.8	0.242	0.0505	23.9	12.8
40.	1.44		24.3	0.281	0.0370	26.1	12.3
50.	1.20		26.1	0.311	0.0294	27.6	12.1
60.	1.04		27.5	0.334	0.0247	28.9	11.8
80.	0.818		29.6	0.370	0.0179	30.8	11.3
100.	0.679		31.0	0.398	0.0142	32.1	10.6

Mass coefficients in square meters per kilogram can be obtained by multiplying the tabulated values by the conversion factor given for each absorbing material. The following abbreviations are used:

is, value for incoherent scattering including binding effects
cs, value for coherent scattering
pp(n), value for pair production near a nucleus
pp(e), value for pair production near an atomic electron
pe, value for the atomic photoelectric effect
total, total value for incoherent, coherent, pair, and photoelectric processes
en, total value for energy absorption from the incoherent, pair, and photoelectric processes.

Data selected from E. Storm and H. I. Israel, *Photon Cross Sections from 0.001 to 100 MeV for Elements 1 through 100, LA-3753* (Los Alamos Scientific Laboratory, Univ. of California, 1967), p. 19; and E. Storm and H. I. Israel, "Photon cross sections from 1 keV to 100 MeV for elements $Z = 1$ to $Z = 100$,"

Appendix 9 Photon Cross Sections and Coefficients—Continued

Water, H_2O (μ/ρ in m^2/kg)

E(MeV)	*is*	*cs*	*pp(n)*	*pp(e)*	*pe*	*total*	*en*
0.001	0.00135	0.137			402.	402.1	402.
0.0015	0.00271	0.127			128.	128.1	128.
0.002	0.00422	0.116			56.6	56.7	56.6
0.003	0.00705	0.0925			17.9	18.0	17.9
0.004	0.00942	0.0732			7.70	7.78	7.70
0.005	0.0112	0.0581			3.95	4.02	3.95
0.006	0.0126	0.0468			2.28	2.34	2.28
0.008	0.0144	0.0326			0.954	1.00	0.954
0.01	0.0155	0.0242			0.472	0.512	0.472
0.015	0.0170	0.0140			0.128	0.159	0.129
0.02	0.0178	0.00924			0.0499	0.0769	0.0505
0.03	0.0183	0.00489			0.0131	0.0363	0.0140
0.04	0.0183	0.00300			0.00499	0.0263	0.00620
0.05	0.0180	0.00202			0.00238	0.0224	0.00383
0.06	0.0177	0.00145			0.00131	0.0204	0.00296
0.08	0.0170	0.00085			0.00050	0.0183	0.00249
0.1	0.0163	0.00056			0.00024	0.0171	0.00248
0.15	0.0147	0.00025			0.00006	0.0151	0.00274
0.2	0.0135	0.00015			0.00003	0.0137	0.00295
0.3	0.0118	0.00007			0.00001	0.0119	0.00318
0.4	0.0106	0.00004				0.0106	0.00328
0.5	0.00966	0.00003				0.00969	0.00331
0.6	0.00895	0.00002				0.00896	0.00328
0.8	0.00786	0.00001				0.00787	0.00319

1.	0.00707	0.00001				0.00707	0.00310
1.5	0.00573		0.00001			0.00574	0.00283
2.	0.00489		0.00004			0.00493	0.00260
3.	0.00385		0.00011			0.00396	0.00227
4.	0.00322		0.00018	0.00001		0.00341	0.00207
5.	0.00278		0.00024	0.00001		0.00303	0.00192
6.	0.00246		0.00030	0.00002		0.00277	0.00181
8.	0.00201		0.00039	0.00003		0.00242	0.00167
10.	0.00171		0.00046	0.00004		0.00221	0.00158
15.	0.00127		0.00060	0.00006		0.00193	0.00146
20.	0.00102		0.00071	0.00008		0.00181	0.00141
30.	0.00074		0.00085	0.00010		0.00170	0.00135
40.	0.00059		0.00096	0.00012		0.00167	0.00132
50.	0.00049		0.00104	0.00014		0.00166	0.00130
60.	0.00042		0.00109	0.00015		0.00166	0.00128
80.	0.00033		0.00118	0.00017		0.00168	0.00125
100.	0.00028		0.00125	0.00018		0.00171	0.00123

Mass coefficients in square meters per kilogram can be obtained by multiplying the tabulated values by the conversion factor given for each absorbing material. The following abbreviations are used:

is, value for incoherent scattering including binding effects
cs, value for coherent scattering
pp(n), value for pair production near a nucleus
pp(e), value for pair production near an atomic electron
pe, value for the atomic photoelectric effect
total, total value for incoherent, coherent, pair, and photoelectric processes
en, total value for energy absorption from the incoherent, pair, and photoelectric processes.

Data selected from E. Storm and H. I. Israel, *Photon Cross Sections from 0.001 to 100 MeV for Elements 1 through 100, LA-3753* (Los Alamos Scientific Laboratory, Univ. of California, 1967), p. 19; and E. Storm and H. I. Israel, "Photon cross sections from 1 keV to 100 MeV for elements $Z = 1$ to $Z = 100$," Nucl. Data Tables **7**, 573 (1970).

Appendix 9 Photon Cross Sections and Coefficients—Continued

Air (76% N, 23%, O, 1% Ar; μ/ρ in m^2/kg)

E(MeV)	*is*	*cs*	*pp(n)*	*pp(e)*	*pe*	*total*	*en*
0.001	0.0199	0.135			346.	346.1	346.
0.0015	0.0199	0.125			109.	109.1	109.
0.002	0.0198	0.112			48.2	48.3	48.2
0.003	0.0198	0.0881			15.2	15.3	15.2
0.004	0.0197	0.0688			7.49	7.57	7.42
0.005	0.0196	0.0545			3.85	3.91	3.82
0.006	0.0195	0.0440			2.22	2.27	2.20
0.008	0.0194	0.0309			0.921	0.964	0.916
0.01	0.0192	0.0233			0.459	0.495	0.457
0.015	0.0189	0.0137			0.125	0.154	0.125
0.02	0.0186	0.00913			0.0497	0.0744	0.0501
0.03	0.0180	0.00483			0.0131	0.0342	0.0139
0.04	0.0174	0.00295			0.00508	0.0243	0.00616
0.05	0.0169	0.00199			0.00245	0.0205	0.00374
0.06	0.0164	0.00143			0.00135	0.0186	0.00283
0.08	0.0156	0.00084			0.00053	0.0166	0.00231
0.1	0.0148	0.00055			0.00025	0.0154	0.00227
0.15	0.0134	0.00025			0.00007	0.0135	0.00247
0.2	0.0122	0.00014			0.00003	0.0123	0.00265
0.3	0.0106	0.00007			0.00001	0.0107	0.00287
0.4	0.00953	0.00004				0.00953	0.00294
0.5	0.00868	0.00002				0.00870	0.00298
0.6	0.00803	0.00002				0.00805	0.00295
0.8	0.00705	0.00001				0.00706	0.00287

1.	0.00636	0.00001				0.00636	0.00278
1.5	0.00515		0.00001			0.00516	0.00254
2.	0.00438		0.00004			0.00442	0.00234
3.	0.00346		0.00011			0.00357	0.00205
4.	0.00289		0.00018			0.00308	0.00187
5.	0.00250		0.00024	0.00001		0.00275	0.00174
6.	0.00221		0.00030	0.00001		0.00252	0.00165
8.	0.00180		0.00039	0.00003		0.00222	0.00153
10.	0.00154		0.00046	0.00004		0.00203	0.00146
15.	0.00114		0.00060	0.00005		0.00179	0.00136
20.	0.00091		0.00071	0.00007		0.00169	0.00133
30.	0.00067		0.00086	0.00009		0.00162	0.00129
40.	0.00053		0.00096	0.00011		0.00160	0.00127
50.	0.00044		0.00104	0.00012		0.00160	0.00126
60.	0.00038		0.00110	0.00014		0.00161	0.00124
80.	0.00030		0.00119	0.00015		0.00164	0.00122
100.	0.00025		0.00126	0.00016		0.00167	0.00121

Mass coefficients in square meters per kilogram can be obtained by multiplying the tabulated values by the conversion factor given for each absorbing material. The following abbreviations are used:

is, value for incoherent scattering including binding effects
cs, value for coherent scattering
pp(n), value for pair production near a nucleus
pp(e), value for pair production near an atomic electron
pe, value for the atomic photoelectric effect
total, total value for incoherent, coherent, pair, and photoelectric processes
en, total value for energy absorption from the incoherent, pair, and photoelectric processes.

Data selected from E. Storm and H. I. Israel, *Photon Cross Sections from 0.001 to 100 MeV for Elements 1 through 100, LA-3753* (Los Alamos Scientific Laboratory, Univ. of California, 1967), p. 19; and E. Storm and H. I. Israel, "Photon cross sections from 1 keV to 100 MeV for elements $Z = 1$ to $Z = 100$," Nucl. Data Tables **7**, 573 (1970).

Appendix 10 Kerma per Fluence for Neutrons

	Kerma per Fluence ($Gy \cdot m^2$)					
Energy (MeV)	*Hydrogen* (Z = *1*)	*Carbon* (Z = *6*)	*Nitrogen* (Z = *7*)	*Oxygen* (Z = *8*)	*Aluminum* (Z = *13*)	*Water* (H_2O)
2.53E−08	4.20E−10	2.41E−13	7.85E−08	3.56E−15	4.99E−12	
1.00E−07	2.06E−10	1.25E−13	3.83E−08	4.16E−15	2.44E−12	
1.00E−06	7.53E−11	9.66E−14	1.21E−08	2.80E−14	7.74E−13	
1.00E−05	1.27E−10	6.03E−13	3.83E−09	2.75E−13	2.81E−13	1.36E−11
1.00E−04	1.07E−09	5.92E−12	1.22E−09	2.75E−12	4.43E−13	1.23E−10
1.00E−03	1.06E−08	5.91E−11	4.66E−10	2.75E−11	3.69E−12	1.21E−09
1.00E−02	9.89E−08	3.40E−10	8.20E−10	2.75E−10	3.46E−11	1.30E−08
1.00E−01	6.25E−07	2.14E−09	3.99E−09	2.65E−09	1.03E−09	7.23E−08
1.00E 00	2.08E−06	2.78E−08	1.86E−08	4.33E−09	5.14E−09	2.71E−07
2.00E 00	2.80E−06	3.99E−08	3.42E−08	2.14E−08	1.01E−08	3.32E−07
3.00E 00	3.30E−06	6.94E−08	9.39E−08	2.34E−08	1.14E−08	3.91E−07
4.00E 00	3.67E−06	8.45E−08	1.60E−07	4.31E−08	1.35E−08	4.49E−07
5.00E 00	3.94E−06	5.20E−08	1.13E−07	5.84E−08	1.73E−08	4.92E−07
6.00E 00	4.15E−06	6.70E−08	1.06E−07	4.91E−08	2.27E−08	5.06E−07
7.00E 00	4.28E−06	4.50E−08	9.83E−08	6.60E−08	2.92E−08	5.38E−07
8.00E 00	4.40E−06	1.09E−07	1.17E−07	6.72E−08	3.62E−08	5.52E−07
9.00E 00	4.48E−06	1.24E−07	1.18E−07	8.24E−08	4.37E−08	5.75E−07
10.00E 00	4.58E−06	1.27E−07	1.52E−07	1.12E−07	5.90E−08	6.11E−07
15.00E 00	4.71E−06	2.61E−07	2.60E−07	2.00E−07	1.33E−07	7.05E−07
20.00E 00	4.69E−06	3.49E−07	3.35E−07	2.60E−07	1.53E−07	7.55E−07
25.00E 00	4.58E−06	3.80E−07	3.78E−07	2.66E−07	1.93E−07	7.48E−07

Data selected from *Neutron Dosimetry for Biology and Medicine, ICRU Report 26* (International Commission on Radiation Units and Measurements, Washington, 1977) pp 78–91.

Appendix 11 Photon Beam Percent Depth Dose

Percent Depth Dose, ^{60}Co Photons

	Field Size (cm)							
Depth (cm)	*0*	*4 × 4*	*6 × 6*	*8 × 8*	*10 × 10*	*12 × 12*	*15 × 15*	*20 × 20*
Backscatter Factor =	1.00	1.01_4	1.02_1	1.02_9	1.03_6	1.04_3	1.05_2	1.06_1
0.5	100.0	100.0	100.0	100.0	100.0	100.0	100.0	100.0
1	95.4	96.8	97.4	97.8	98.2	98.3	98.4	98.4
2	87.1	90.6	91.9	92.7	93.3	93.6	93.9	94.0
3	79.5	84.7	86.5	87.6	88.3	88.8	89.3	89.6
4	72.7	79.0	81.1	82.5	83.4	84.0	84.7	85.2
5	66.5	73.5	75.9	77.4	78.5	79.3	80.1	80.8
6	60.8	68.1	70.7	72.4	73.6	74.4	75.4	76.4
7	55.6	62.9	65.7	67.5	68.8	69.8	70.8	72.1
8	50.9	58.0	60.8	62.7	64.1	65.3	66.5	68.0
9	46.6	53.5	56.2	58.2	59.7	60.8	62.3	64.0
10	42.7	49.3	52.0	54.0	55.6	56.9	58.4	60.2
11	39.2	45.5	48.1	50.1	51.7	53.0	54.7	56.6
12	35.9	41.9	44.5	46.5	48.1	49.5	51.2	53.2
13	32.9	38.6	41.1	43.2	44.8	46.1	47.9	50.0
14	30.2	35.6	38.0	40.1	41.8	43.2	44.9	47.0
15	27.7	32.9	35.2	37.2	38.9	40.3	42.0	44.2
16	25.4	30.4	32.6	34.5	36.2	37.6	39.3	41.5
17	23.3	28.1	30.2	32.1	33.7	35.1	36.8	39.0
18	21.4	26.0	28.0	29.8	31.4	32.8	34.5	36.7
19	19.6	24.0	26.0	27.7	29.2	30.6	32.3	34.6
20	18.0	22.1	24.0	25.7	27.2	28.5	30.3	32.6
22	15.3	18.9	20.6	22.1	23.7	24.9	26.5	28.8
24	12.9	16.1	17.7	19.1	20.5	21.8	23.2	25.4
26	10.8	13.7	15.1	16.5	17.8	18.9	20.4	22.5
28	9.1	11.7	12.9	14.2	15.5	16.5	17.9	19.9
30	7.7	10.0	11.1	12.3	13.5	14.4	15.7	17.5

0.80 m = *F* (distance from source to water surface).

Tissue-Air Ratio, ^{60}Co Photons

	Field Size (cm)							
Depth (cm)	*0*	*4 × 4*	*6 × 6*	*8 × 8*	*10 × 10*	*12 × 12*	*15 × 15*	*20 × 20*
0.5	1.000	1.014	1.021	1.029	1.036	1.043	1.052	1.061
1	0.965	0.992	1.005	1.017	1.028	1.037	1.045	1.057
2	0.905	0.953	0.972	0.989	1.002	1.014	1.023	1.034
3	0.845	0.913	0.937	0.956	0.971	0.983	0.996	1.009
4	0.792	0.871	0.898	0.921	0.938	0.952	0.966	0.982
5	0.742	0.826	0.858	0.884	0.902	0.918	0.935	0.952
6	0.694	0.783	0.817	0.845	0.864	0.881	0.899	0.921
7	0.650	0.738	0.776	0.804	0.825	0.844	0.862	0.887
8	0.608	0.696	0.732	0.762	0.785	0.804	0.826	0.853
9	0.570	0.654	0.691	0.722	0.746	0.766	0.789	0.819
10	0.534	0.617	0.651	0.681	0.707	0.728	0.755	0.787
11	0.501	0.580	0.614	0.645	0.671	0.693	0.719	0.754
12	0.469	0.544	0.580	0.610	0.635	0.658	0.686	0.722
13	0.440	0.509	0.546	0.577	0.602	0.624	0.652	0.690
14	0.412	0.479	0.515	0.544	0.570	0.595	0.624	0.659
15	0.386	0.452	0.483	0.514	0.539	0.564	0.593	0.629
16	0.361	0.426	0.456	0.485	0.511	0.535	0.564	0.601
17	0.338	0.401	0.432	0.459	0.484	0.507	0.537	0.576
18	0.317	0.376	0.407	0.434	0.459	0.480	0.510	0.550
19	0.297	0.353	0.382	0.409	0.433	0.455	0.484	0.525
20	0.278	0.331	0.360	0.386	0.409	0.431	0.460	0.501
22	0.246	0.292	0.318	0.344	0.366	0.387	0.417	0.455
24	0.215	0.258	0.280	0.303	0.324	0.345	0.374	0.413
26	0.187	0.228	0.247	0.268	0.289	0.309	0.336	0.371
28	0.164	0.201	0.219	0.238	0.256	0.276	0.302	0.337
30	0.144	0.176	0.193	0.210	0.228	0.246	0.271	0.305

Data excerpted from M. Cohen, D. E. A. Jones, and D. Greene, "Central Axis Depth Dose Data for Use in Radiotherapy," Br. J. Radiol. Suppl. **11,** 59 (1972).

Percent Depth Dose, 6-MV Photons

Depth	Field Size (cm)								
(cm)	4 × 4	6 × 6	8 × 8	10 × 10	12 × 12	15 × 15	20 × 20	25 × 25	30 × 30
1	97.8	98.0	98.1	98.2	98.3	98.5	98.7	99.0	99.2
1.5	100.0	100.0	100.0	100.0	100.0	100.0	100.0	100.0	100.0
2	98.1	98.5	98.5	98.5	98.5	98.5	98.5	98.5	98.5
3	93.5	94.0	94.4	94.7	94.8	94.9	95.0	95.0	95.0
4	88.5	89.4	90.0	90.4	90.8	91.1	91.5	91.6	91.6
5	83.5	85.0	86.1	86.8	87.2	87.4	87.7	87.9	88.2
6	79.3	80.8	81.9	82.6	83.2	83.6	84.0	84.2	84.6
7	74.4	76.4	77.8	78.6	79.2	79.7	80.2	80.6	81.1
8	70.1	72.0	73.6	74.6	75.4	76.0	76.8	77.3	77.7
9	65.3	67.6	69.4	70.6	71.4	72.3	73.3	73.8	74.3
10	61.9	64.1	65.6	66.8	67.8	68.8	70.0	70.6	71.0
11	57.6	60.2	62.0	63.4	64.4	65.5	66.7	67.2	67.7
12	53.9	56.2	58.2	59.6	60.6	61.7	63.0	64.0	64.5
13	50.6	52.8	55.0	56.6	57.8	58.9	60.1	60.9	61.6
14	47.5	49.8	51.8	53.4	54.6	55.8	57.2	58.0	58.8
15	44.7	46.7	48.6	50.2	51.6	53.0	54.4	55.4	56.1
16	41.7	43.7	45.6	47.4	48.8	50.2	51.9	53.1	53.8
17	39.2	41.2	43.0	44.5	45.8	47.4	49.2	50.4	51.2
18	36.8	38.6	40.4	41.9	43.2	44.9	46.8	48.0	49.0
19	34.6	36.4	38.4	39.9	41.2	42.6	44.4	45.6	46.5
20	32.6	34.5	36.2	37.7	38.9	40.3	42.0	43.3	44.1
22	28.8	30.6	32.2	33.7	35.0	36.4	38.1	39.2	40.1
24	25.3	27.0	28.6	30.0	31.2	32.6	34.2	35.5	36.4
26	22.4	23.9	25.2	26.4	27.6	28.9	30.8	32.1	33.0
28	19.9	21.3	22.6	23.7	24.8	26.0	27.9	29.1	29.8
30	17.5	18.8	20.0	21.1	22.2	23.4	25.0	26.2	27.0

Data selected from C. W. Coffey II, J. L. Beach, D. J. Thompson, and M. Mendiondo, "X-ray beam characteristics of Varian Clinac 6–100 linear accelerator," Med. Phys. 7, 716 (1980). 1.00 m = F (distance from target to water surface)

Tissue-Maximum Ratio, 6-MV Photons

Depth	*Field Size (cm)*									
(cm)	*0 × 0*	*4 × 4*	*6 × 6*	*8 × 8*	*10 × 10*	*12 × 12*	*15 × 15*	*20 × 20*	*25 × 25*	*30 × 30*
1	0.950	0.966	0.968	0.969	0.970	0.982	0.973	0.977	0.980	0.983
1.5	0.926	1.000	1.000	1.000	1.000	1.000	1.000	1.000	1.000	1.000
2	0.903	0.991	0.992	0.992	0.992	0.993	0.994	0.995	0.996	0.996
3	0.858	0.957	0.964	0.969	0.972	0.974	0.975	0.977	0.979	0.981
4	0.815	0.924	0.934	0.944	0.949	0.953	0.957	0.959	0.960	0.961
5	0.775	0.891	0.904	0.914	0.921	0.926	0.931	0.936	0.940	0.942
6	0.736	0.858	0.874	0.886	0.894	0.901	0.908	0.915	0.920	0.924
7	0.699	0.822	0.841	0.856	0.866	0.874	0.882	0.892	0.897	0.901
8	0.665	0.787	0.806	0.822	0.834	0.842	0.854	0.868	0.875	0.880
9	0.632	0.749	0.769	0.786	0.802	0.814	0.828	0.842	0.851	0.856
10	0.600	0.720	0.740	0.758	0.774	0.788	0.804	0.821	0.829	0.834
11	0.570	0.688	0.710	0.728	0.746	0.760	0.775	0.792	0.802	0.808
12	0.542	0.649	0.672	0.694	0.712	0.728	0.743	0.762	0.773	0.781
13	0.515	0.620	0.644	0.666	0.684	0.701	0.719	0.738	0.749	0.758
14	0.489	0.590	0.614	0.636	0.656	0.672	0.692	0.714	0.726	0.736
15	0.465	0.560	0.584	0.608	0.628	0.646	0.668	0.692	0.703	0.714
16	0.442	0.535	0.559	0.581	0.601	0.620	0.641	0.667	0.682	0.693
17	0.420	0.505	0.530	0.552	0.574	0.592	0.615	0.643	0.660	0.670
18	0.399	0.479	0.504	0.526	0.547	0.566	0.589	0.617	0.634	0.647
19	0.379	0.460	0.484	0.508	0.528	0.547	0.568	0.595	0.614	0.627
20	0.360	0.439	0.464	0.486	0.507	0.525	0.547	0.574	0.590	0.603
22	0.325	0.399	0.422	0.444	0.464	0.482	0.504	0.531	0.549	0.564
24	0.294	0.359	0.384	0.406	0.426	0.444	0.465	0.492	0.514	0.530
26	0.265	0.328	0.350	0.371	0.390	0.407	0.427	0.454	0.473	0.488
28	0.240	0.297	0.319	0.338	0.356	0.373	0.393	0.420	0.440	0.454
30	0.216	0.270	0.290	0.308	0.326	0.342	0.359	0.386	0.406	0.423

Data selected from C. W. Coffey II, J. L. Beach, D. J. Thompson, and M. Mendiondo, "X-ray beam characteristics of Varian Clinac 6–100 linear accelerator," Med. Phys. 7, 716 (1980).

Appendix 12 Absorbed Dose Constants

Technetium-99m

Radiation Type (i)	*Mean Number* w_i	*Mean Energy (MeV)* $\langle E_i \rangle$	Δ_i $\left(\frac{g \cdot rad}{\mu Ci \cdot h}\right)$	*% of dis-integrations*	*Transition Energy (MeV)*
Gamma-1	0.00	0.0021	0.0000	98.6	0.0022
M int. con. electron, 1	0.986	0.0017	0.0036		
Gamma-2	0.883	0.1405	0.2643	98.6	0.1405
K int. con. electron, 2	0.0883	0.1195	0.0225		
L int. con. electron, 2	0.0109	0.1377	0.0032		
M int. con. electron, 2	0.0036	0.1401	0.0011		
Gamma-3	0.0003	0.1427	0.0001	1.4	0.1427
K int. con. electron, 3	0.0096	0.1217	0.0025		
L int. con. electron, 3	0.0030	0.1399	0.0009		
M int. con. electron, 3	0.0010	0.1423	0.0003		
K α-1 x rays	0.0431	0.0184	0.0017		
K α-2 x rays	0.0216	0.0183	0.0008		
K β-1 x rays	0.0103	0.0206	0.0005		
K β-2 x rays	0.0018	0.0210	0.0001		
L x rays	0.0081	0.0024	0.0000		
KLL Auger electron	0.0149	0.0155	0.0005		
KLX Auger electron	0.0055	0.0178	0.0002		
KXY Auger electron	0.0007	0.0202	0.0000		
LMM Auger electron	0.106	0.0019	0.0004		
MXY Auger electron	1.23	0.0004	0.0010		

Appendix 12 Absorbed Dose Constants—Continued

Iodine-131

Radiation Type (i)	*Mean Number* w_i	*Mean Energy (MeV)* $\langle E_i \rangle$	Δ_i $\left(\frac{g \cdot rad}{\mu Ci \cdot h}\right)$	*% of dis-integrations*	*Transition Energy (MeV)*
Beta-1	0.016	0.0701	0.0024	1.6	0.25
Beta-2	0.069	0.0955	0.0140	6.9	0.33
Beta-3	0.005	0.1428	0.0015	0.5	0.47
Beta-4	0.904	0.1917	0.3691	90.4	0.606
Beta-5	0.006	0.2856	0.0037	0.6	0.81
Gamma-1	0.0173	0.0802	0.0030	5.06	0.0802
K int. con. electron, 1	0.0294	0.0456	0.0029		
L int. con. electron, 1	0.0029	0.0751	0.0005		
M int. con. electron, 1	0.0010	0.0792	0.0002		
Gamma-2	0.0001	0.1640	0.0000	0.6	0.1640
K int. con. electron, 2	0.0037	0.1294	0.0010		
L int. con. electron, 2	0.0016	0.1589	0.0005		
M int. con. electron, 2	0.0005	0.1630	0.0002		
Gamma-3	0.0014	0.1772	0.0005	0.18	0.1772
K int. con. electron, 3	0.0003	0.1427	0.0001		
Gamma-4	0.0475	0.2843	0.0288	5.06	0.2843
K int. con. electron, 4	0.0025	0.2497	0.0013		
L int. con. electron, 4	0.0005	0.2793	0.0003		
M int. con. electron, 4	0.0002	0.2834	0.0001		
Gamma-5	0.0017	0.3258	0.0012	0.18	0.3258
Gamma-6	0.833	0.3645	0.6465	85.3	0.3645
K int. con. electron, 6	0.0167	0.3299	0.0117		
L int. con. electron, 6	0.0028	0.3594	0.0021		
M int. con. electron, 6	0.0009	0.3635	0.0006		
Gamma-7	0.0032	0.5030	0.0034	0.32	0.5030
Gamma-8	0.0687	0.6370	0.0932	6.9	0.6370
K int. con. electron, 8	0.0003	0.6024	0.0004		
Gamma-9	0.0159	0.7229	0.0245	1.6	0.7229
K α-1 x rays	0.0252	0.0298	0.0016		
K α-2 x rays	0.0130	0.0295	0.0008		
K β-1 x rays	0.0070	0.0336	0.0005		
K β-2 x rays	0.0015	0.0346	0.0001		
L x rays	0.0078	0.0041	0.0001		
KLL Auger electron	0.0042	0.0245	0.0002		
KLX Auger electron	0.0018	0.0286	0.0001		
KXY Auger electron	0.0003	0.0327	0.0000		
LMM Auger electron	0.0486	0.0032	0.0003		
MXY Auger electron	0.117	0.0009	0.0002		

Data selected from L. T. Dillman, "Radionuclide decay schemes and nuclear parameters for use in radiation-dose estimation," J. Nucl. Med. Suppl. **4,** 22 (1969).

Index

D

E

J

K

N

O

P

Q

R

S

T

X-ray machines

X

Z